HANDBOOK OF BIOMECHATRONICS

HANDBOOK OF BIOMECHATRONICS

JACOB SEGIL

ELSEVIER

ACADEMIC PRESS
An imprint of Elsevier

Academic Press is an imprint of Elsevier
125 London Wall, London EC2Y 5AS, United Kingdom
525 B Street, Suite 1650, San Diego, CA 92101, United States
50 Hampshire Street, 5th Floor, Cambridge, MA 02139, United States
The Boulevard, Langford Lane, Kidlington, Oxford OX5 1GB, United Kingdom

Library of Congress Cataloging-in-Publication Data
A catalog record for this book is available from the Library of Congress

British Library Cataloguing-in-Publication Data
A catalogue record for this book is available from the British Library

ISBN 978-0-12-812539-7

Publisher: Mara Conner
Acquisition Editor: Chris Katsaropoulos
Editorial Project Manager: Charlotte Rowley
Production Project Manager: Kamesh Ramajogi
Cover Designer: Christian J. Bilbow

Typeset by SPi Global, India

Contents

Contributors

Ahmed R. Arshi
Biomedical Engineering, Amirkabir University of Technology, Tehran, Iran

Lilach Bareket
Faculty of Engineering and Information Technologies, University of Sydney, Sydney, NSW, Australia

Alejandro Barriga-Rivera
Faculty of Engineering and Information Technologies, University of Sydney, Sydney, NSW, Australia; Division of Neuroscience, University Pablo de Olavide, Seville, Spain

Georgios A. Bertos
National Technical University of Athens, Athens, Greece; Northwestern University Prosthetics-Orthotics Center, Physical Medicine & Rehabilitation, Feinberg School of Medicine; Bionic Healthcare, Inc, Chicago, IL, United States

Graham Brooker
Australian Centre for Field Robotics, University of Sydney, Sydney, NSW, Australia

Jeff Christenson
Research and Development, Motion Control, Salt Lake City, UT, United States

Adson Ferreira da Rocha
Biomedical Engineering Program, University of Brasilia, Brasilia, Brazil

Borna Ghannadi
Systems Design Engineering, University of Waterloo, Waterloo, ON, Canada

Reva E. Johnson
Mechanical Engineering and Bioengineering, Valparaiso University, Valparaiso, IN, United States

Alberto Lopez-Delis
Medical Biophysics Center, University of Oriente, Santiago de Cuba, Cuba

Nigel H. Lovell
Graduate School of Biomedical Engineering, University of New South Wales, Sydney, NSW, Australia

John McPhee
Systems Design Engineering, University of Waterloo, Waterloo, ON, Canada

Naser Mehrabi
University of Washington, Seattle, WA, United States

Domen Novak
Department of Electrical & Computer Engineering, University of Wyoming, Laramie, WY, United States

Evangelos G. Papadopoulos
National Technical University of Athens, Athens, Greece

Jeffrey V. Rosenfeld
Monash Institute of Medical Engineering and Department of Surgery, Monash University, Clayton; Department of Neurosurgery, Alfred Hospital, Melbourne, VIC, Australia; Department of Surgery, F. Edward Hébert School of Medicine, Uniformed Services University, Bethesda, MD, United States

Andres F. Ruiz-Olaya
Faculty of Electronics and Biomedical Engineering, Antonio Nariño University, Bogotá, Colombia

Jonathon W. Sensinger
Institute of Biomedical Engineering, Department of Electrical & Computer Engineering, University of New Brunswick, Fredericton, NB, Canada

Reza Sharif Razavian
Systems Design Engineering, University of Waterloo, Waterloo, ON, Canada

Gregg J. Suaning
Faculty of Engineering and Information Technologies, University of Sydney, Sydney, NSW, Australia

Ahmet Fatih Tabak
Max-Planck Institute for Intelligent Systems, Stuttgart, Germany

Preface

In the absence of any other proof, the thumb alone would convince me of God's existence

Sir Isaac Newton

The merging of man and machine has captured our collective imaginations for centuries. Popular entertainment created memorable characters from Mary Shelley's *Frankenstein* (1818) to the *Six Million Dollar Man* (1974) to the android hosts in the modern television series *Westworld* (2016). We are entertained by imagining our innate abilities augmented by technology. Our species has evolved to be bipedal, erect in posture, endowed with complex manual dexterity, and able to perform high-level cognitive functions including language and problem solving. But, we are still subject to innumerable pathologies that limit our abilities and lifespan. Can we develop technologies that measure, actuate, rehabilitate, augment, restore, or even replace our native physiological systems? The answer is yes. The field of biomechatronics is the integration of human physiology with electromechanical systems. This *Handbook of Biomechatronics* presents the foundational principles of this flourishing field and a series of case studies describing specific applications and technologies.

The *Handbook of Biomechatronics* will provide a resource for readers with a wide range of scientific and engineering backgrounds. The handbook will begin with a broad presentation of biomechatronic design and components followed by detailed case studies of specific biomechatronic devices spanning brain-machine interface to artificial hearts. The case studies span most physiological systems in the body, including the:

(1) muscular system (Chapters 3, 6–9, 13, 14)
(2) nervous system (Chapters 5, 6, 10)
(3) skeletal system (Chapters 6–9)
(4) digestive system (Chapter 11)
(5) reproductive system (Chapter 12)
(6) circulatory system (Chapters 13, 14)

Equally, the technology within these case studies spans an array of diverse fields like anatomy, physiology, electrical engineering, mechanical engineering, computer engineering, neuroscience, and more. The inherent interdisciplinary nature of biomechatronics presents challenges to all researchers and requires collaborative efforts to produce impactful results.

When successful, these discoveries promote the health and quality of life for generations to come.

Sir Isaac Newton founded our understanding of the laws of motion and gravitation, but saw the thumb to be divine. Our work in biomechatronics is founded with this same respect for our beautiful abilities. We simply have the hubris to believe that we can extend our abilities further. The *Handbook of Biomechatronics* will provide a glimpse into this field and hopefully motivate future inventors to attempt to make the divine even better.

Jacob Segil
Lead Editor

PART ONE

Biomechatronic Design and Components

CHAPTER ONE

Introduction

Ahmed R. Arshi
Biomedical Engineering, Amirkabir University of Technology, Tehran, Iran

Contents

Mechatronics is a fascinating field of study. It challenges the mind to think in multiple disciplines. No other engineering concept is so adept in encouraging instantaneous jumps from one field of engineering to another. An experienced mechatronic designer is in reality composing a piece of music for an orchestra of engineers. As individual musicians have a fluent command over their instrument, the mechatronic specialist is the script writer and produces the blue print for the route. Learning to play in, work with, or even lead a team of engineers is therefore an inseparable part of being a mechatronics specialist.

Handbook of Biomechatronics
https://doi.org/10.1016/B978-0-12-812539-7.00001-5

Mechatronics as the name implies brings mechanical concepts, electronic solutions, control strategies, and software technologies together under the same roof. A growing volume of literature provides ample supply of details addressing issues on each one of these fields. The subject aims at providing a wide range of technical competencies necessary to face multidisciplinary projects. Mechatronics mode of thought requires a systematic approach and the key role is perhaps played by experience in integration of diverse subsystems. A mechatronic specialist considers integration as an important part of design stage. Interaction with other systems is where the design or modeling teams define the outskirts of integration. In industrial or domestic environments, mechatronic systems assist interactions through action and response using actuator and control systems by processing information gained from sensory constructs. Such systems rely on feedback in closed circuits and prediction in open control strategies. That is why the nature and the characteristics of the environment with which the system is interacting play a key role.

Biological systems on the other hand, are inherently multiscale and multidisciplinary. Biologically inspired mechatronic or biomimetic systems are always eye-catching items on show at science and engineering exhibitions. The most fascinating technologies are however, those that interact with human body. Human body as a biological system is exceptionally sophisticated and when efforts are made to decipher its functional principles it turns out to be an awe-inspiring engineering system. One that imitating or surpassing its intricate potentials is exceedingly difficult. Today's technological advances are yet to grow to the level of sophistication exhibited by biological and in particular, physiological systems.

Human body as a physiological system is susceptible to deviations from physiological or normal states. Deviations in function better known as pathological states could be observed in individual organs or could even adversely affect the entire system. Changes in physiological states commonly encountered in human body are accompanied by an unending and ever-increasing necessity for identification, categorization, diagnosis, or intervention by engineering and in particular, mechatronic solutions. This amazing multidisciplinary physiological environment is in fact quite suitable for the implementation of mechatronic systems.

The simple but highly effective electrocardiogram or ECG test for example, which is routinely performed in cardiological assessments provides a portrayal of the electro-mechanochemical interactions taking place in the heart. A complete ECG test is a window to electrophysiological performance of all cellular groups in that organ. The device output, in the shape of an ECG signal, could be considered as an indicator of electrophysiological interactions at the cellular level. It is also an indication of the manner by which electrical signals are propagated throughout various cell families leading to contractions throughout the muscular structure and resulting in blood flow output. The traditional engineering approach when facing a uniquely challenging environment of this complexity, requires fundamental metamorphoses by subscribing to a new mode or school of thought.

Biomechatronics is the discipline that aims to integrate mechatronic and biological and in particular the human physiological systems. The potentials offered by human body are so diverse that traditional approach to engineering solutions is routinely challenged. Cellular characteristics leading to the functioning of different organs create situations where established engineering principles are easily overstretched. The traditional mechatronic educational programs may thus require an overhaul and other contributions to make the new generations of mechatronic specialists fully versed with characteristics of human body and biological systems in general. The new generation of multidisciplinary specialists will have to be prepared to help biorobotic, biotechnology, and biomechatronic startups as well traditional robotic and automation forums.

Hand Book of Biomechatronics aims at establishing the infrastructure for this school of thought.

1 ENGINEERING APPROACH

The design of multiscale and multidisciplinary systems evolves around an efficient integration of both biomechatronic and human body systems. A successful integration requires an appreciation of how engineering principles could be adopted to provide a mathematical description of function and performance of anatomical and physiological systems. The human body should in effect be viewed as a sophisticated engineering system. There are numerous instances to support this argument.

Nonspecific low-back pain, which is experienced by many at some point throughout their lives, with no tangible medical solution, could be viewed as a structural problem with a biomechanical solution in the form of design of exercise programs. Phenomena such as heat and mass transfer, fluid flow, translational and rotational movements are areas where Newtonian and non-Newtonian mechanics govern the functions of organs. Biomechanics provides the constitutive equations describing physical characteristics of both the soft and hard tissues. The constitutive equation for a soft tissue for example, could describe the organ characteristics using a psuedoviscoelastic approach.

Biomechatronic specialists, on the other hand, might be required to mimic biological systems. Biomechanics as a pillar of biomimetics is not simply the application of mechanical principles to biological systems. The concept has far reaching implications as the nature of biological systems dictate a more complex version of basic fundamental principles. Nonhomogenous anisotropic composite tissues with elastic properties modulated by age, sex, and pathological or environmental factors create exceptional challenges. The traditional engineering principles, in isolation, might therefore fail to provide convincing designs for the interface between biomechatronic and physiological systems.

This is where the biomechatronic specialist makes an inspiring contribution to engineering sciences; and this contribution can be best manifested at the design stage.

2 FUSION OF BIO AND MECHATRONICS

An energetically optimized solution to fusion of physiological and mechatronic systems relies heavily on the design of interfaces. Design of an interface will have to embrace biocompatible combinations of mechanical, electromagnetic, electronic, optical, and audio systems. The interface of such systems with the intended physiological system is growing in sophistication. The newly developed robotic systems imitate horse movements used in hippotherapy or therapeutic riding, taking advantage of the dynamic input by the horse, to the human neuromuscular system. This is achieved through simulation of three-dimensional mechanical inputs exerted to human upper extremity during horse gait. In other instances, continuously developing retina tracking systems used in transportation or military systems represent a prime example of an effective interface. Here, the contributions made by subjects such as man-machine interface and optomechatronics have made biomechatronics even richer in content. Fusion of bio and mechatronics should further address biocompatibility guidelines to ensure complete functionality and reliability.

Fusion of biomechatronic systems with human body has roots in four areas of manipulation, locomotion, sensory interactions, and finally processing and control.

2.1 Manipulation

The ability to manipulate objects in daily tasks is often hindered by injuries or neuromuscular disorders. Robotics is the recognized domain responsible for the development of manipulators in industrial environments. The multidisciplinary approach embedded in robotics is the most widely followed forum for mechatronic research. The fusion of mechatronic and physiological systems is perhaps best manifested in the field of bio-robotics, which is growing in two avenues of bio-mimetics and rehabilitation robotics with many overlapping areas. The former aims at providing services to human issues by imitating a suitable biological system such as an animal, whereas the latter focuses on interventional potentials for robotic devices.

In robotic surgery, the accuracy and precision exhibited in the manipulation of an array of instruments during surgical procedures poses some of the most exciting challenges in decades to come. Interventional radiology as a specialized medical field is also a ripe environment for the implementation of telechiric robotic systems when navigation, interaction, and tactile recognition are corner stones of autonomous robotic surgery.

2.2 Locomotion

Biomechatronic specialists have been fascinated with animal and human locomotion for many years. Human motion studies, from sit-to-stand tasks, to heavy load manipulation and agile skilled athletic performances are still at the forefront of opportunities and promise new horizons. A major contribution is also found in walking or running gait by biomechatronic designs. In the most advanced biomechatronic laboratories the focus is placed on human locomotion from walking gait to remarkable solutions to above-knee amputee requirements. Biomimetics is also used to imitate biodynamic characteristics of physiological systems.

Walking or running gait however, present enormous engineering challenges. In human gait, a large number of muscles are recruited in coordination so that the lower extremity can exhibit an almost symmetric dynamic behavior. This highly influential aspect of human mobility is governed by uniquely adaptable neuromuscular control strategies which rely on variability in foot placement and neural plasticity to entertain learning and skill enhancement. Here, balance and dynamic stability present the core of any optimizations of cost functions in the design of biomechatronic systems.

2.3 Sensory Interactions

Human body in both physiological and pathological states can be assumed as a closed system with an array of input/output ports through which energetic interactions occur with the surrounding. Information regarding the nature of interaction is translated from a variety of energy domains by neuromechanical sensory systems. The design of suitable biomechatronic interfaces with neuromechanical sensory systems require an in-depth understanding of the neuroanatomy. Sensors in body transform external or internal stimuli from multitudes of energy domains to an information carrying signal. Involuntary actuation signals are also transformed into other energy domains to control the operation of cellular structures through biochemical interactions like metabolism, whereas complex movements such as skilled performance encountered in athletic agility drills, require a different array of actuation signals. Body sensors rely on identification and quantification of internal or external stimuli like pressure, heat, texture, vibration, and tensile or compressive deformations. Highly dedicated mechanoreceptors for example, take advantage of biomechanical deformations to produce time-dependent neuromechanical signals. Such systems are interesting for those involved in biomimetics and biosensor design as well as those involved in rehabilitation robotics or smart skin technologies.

2.4 Processing and Control

Body sensors are considered as a highly advanced data acquisition and information gathering system. The biophysical/biochemical mechanisms governing processing of gathered data result in involuntary mechanical movements like heart rate control or voluntary artistic movements such as in painting. Design of interactive interfaces which rely on this data will attract more attention in biomechatronic circles in the years to come. Current efforts rely on noninvasive physiological techniques like those used in electroencephalogram (EEG), electromyogram (EMG) or through nerve conduction studies. The information obtained using these devices require advanced real-time signal processing and matching control algorithms. The data gathered provides a complex array of real-time signals which could be utilized in real-time operational biomechatronic systems. A gap between the undecipherable large data and often ingenious solutions to control problems requires an alternative approach. This alternative mode of thought needs new biosensor technologies to access the neuromechanical systems with much better defined data gathering algorithms. Here, combinations of implantable myoelectric sensors and predictive controller approach using learning strategies can contribute toward real-time user intent recognition. Advances in neuroscience are the key component of a sound and solid biomechatronic future. Neuroscience is the holy grail of biomechatronics. The propositions made by perceptual control theories are an example of possibilities in developing control strategy.

Neuromechanical biomechatronic systems are and will be in a good position to offer true personalized solutions to many human concerns. The signal processing and control problems in personalized biomechatronic systems need to address cognitive and perception issues through emphasis on integration with motor control and motor learning concepts. Although the core of current research funding is directed at such systems as all terrain autonomous vehicles and exoskeletons, the subject will be gradually moving toward a new generation of integration with the human neuromusculoskeletal system. This is where proprioception and enhancement of peripheral information acquisition systems could provide remarkable design opportunities for biomechatronics.

3 MODELING

The multidisciplinary nature of mechatronic systems when combined with an exceptionally unique and diverse set of not totally understood neurophysiological systems dictate the necessity for a suitable multilingual modeling technology. The multiscale, multidimensional, and pseudodeterministic nonlinear dynamic characteristics of such systems pose immense challenges to established intradisciplinary modeling methodologies. Electrophysiological energetic interactions taking place at the cellular level are governed by multi-domain energetic paths encompassing biochemical, ionic, heat and mass transfer across cellular membranes, and broadly, initiation and propagation of action potentials throughout the cellular structures. Any inter- or intradisciplinary modeling apparatus should be well equipped with the potentials to include nonlinearities in a model which is based on a linear analysis scaffold. To include all different modeling languages in a biomechatronic design project is rather challenging, if not difficult. An ensuing outcome of this multilingual approach to modeling is restrictions on communications among disciplinary project managers. Mathematical models capable of embracing aspects such as electrophysiology which govern neuromechanical functions require mastery, fluency, and command over complex interacting biochemical, biomagnetic, bioelectrical, heat and mass transfer, biofluid dynamics, and movement biomechanics. Tissue biomechanics in conjunction with neuromusculoskeletal descriptions are required at times to allow full investigations of the manipulation and locomotion while a large set of data is being processed to implement any control strategies by the central nervous system.

There are two basic approaches to modeling in biomedical engineering. The first utilizes classical disciplinary mathematical modeling where a description of a combination of function and structure are produced to simulate the system. The second approach is in favor of looking at the physiological systems as a black box and various algorithms such as neural networks are adopted to learn the dynamics of the system. These two, often conflicting modes of thought, should in biomechatronics be considered as two sides of the same coin. The importance of constructional modeling cannot be over emphasized as the current applications of such intelligent algorithms or soft computing in design of biomechatronic systems is in need of further development. The black box approach, however, can be used effectively in design of the control strategies. The fundamental problem with the current knowledge of human physiology is that although a vast array of knowledge is constantly being produced by biological, physiological, or electrophysiological laboratories, there still is a wealth of knowledge to be gained so that the existing gaps are covered. Furthermore, the current mathematical tools used in modeling also require further developments. The continuous advancements of microprocessors are reaching the state where principles of predictive controller could be revisited so that real-time simulation results could predict immediate necessary responses of the biomechatronic system in daily interaction of human subject with his/her environment. Here, the mentality of a generalized mathematical model could shift toward tailored solutions. Tailored biomechatronic systems require individualized and personalized models of the system which could in turn play an important role in control strategy.

Furthermore, problems such as intent are increasingly recognized as high-level cost functions against which standard neurophysiologically obtained parameters do not necessarily lead to suitable models. Intent recognition could require real-time integration and processing of a multitude of sensory inputs. Modeling of such complex systems require an alternative but reliable technology. Bond graph technology could provide a measurable solution to modeling and design problems.

4 VARIABILITY

Stand with an arm stretched out facing and just touching a white board with a marker pen. Close your eyes for a minute or two until the end of the exercise. With every exhalation, place a point on the white board with stretched arm and then hang your arm down and relax (be careful not to leave a mark on your garments). Repeat the exercise until some 40 points are placed on the board. You can then open your eyes and look at your masterpiece. You are now facing a cluster of spreading points. You might even be pleasantly surprised by how wide spread the points are. The spreading marks on the white board are a reflection of how your neuromuscular system is capable or rather incapable of repeating a simple task with any degree of accuracy and precision in the absence of visual feedback. This is variability. Variability is the culmination of functional characteristics of a highly nonlinear physiological system. The complexities and nonlinearities associated with electrochemical/neuromechanical aspects of physiological systems are not the only challenge facing the fusion of mechatronics and human body. The mathematical constructs which form the back bone of engineering concepts are also not fully equipped to handle the variabilities inherent in physiological systems. As an example, to fully describe the dynamic characteristics of human locomotion, parametric modeling is required to describe the functions using nonexact individual coefficients with a range of values to cater for a wide spectrum of possibilities from genetic disorders to Olympic standard athletes. Recent studies on variability attribute this dynamic behavior to neural plasticity and thus a necessary trait in learning new skills. How is variability tackled in biomechatronics?

5 INTEGRATION

For a newly setup biomechatronics laboratory or design center, it is paramount to take advantage of valuable experiences gained in different engineering industries. In handling projects large and small, engineers adopt a systematic methodology known as project management. The approach provides a guideline for the new laboratory to exhibit an efficient dynamic behavior and to perform and deliver products as planned and reach intended goals. The guidelines could be used in formation of a specific organizational dynamic behavior to address sponsor's and stakeholder's requirements.

ISO 21500 could be considered as an industrially acceptable guide on how to manage the multidisciplinary projects. Technical integration in such laboratories or research centers require biomechatronic management. Any new concept has to go through a diversification stage until an optimal solution is identified. In biomechatronic centers, this search, research, and finally development consumes time, technical resources, and funds, which are the basic building blocks of a "project". Integration has roots in the initiation stage and solidify during the planning stage of a project where modeling acts as the essence of design. Once again, a multilingual approach to mechatronic design could hamper integration by adversely affecting the communications between the members of the project team and hence a unifying technological approach is crucial.

6 ANATOMY OF DESIGN

In solving human problems, the engineer began a practical manipulation of scientific values resulting in new ideas and tools. The inventiveness and creativity accompanying this practical manipulation are considered as the foundation stones of what is called design. Although design represents a profound intellectual achievement, it has not until recently been approached as a distinct discipline or a science on its own right. The barrier to such an approach has always been mounted on two pillars, one of which is deeply embedded in subjectivity, and the other in specialization. The former is nourished by what is against structuring of inventiveness and adoption of a set of unique criteria, and the latter would force the design concept to be cloaked by intradisciplinary established routines.

Intuition and creativity form a part of design hierarchy known as synthesis. The causal structure of mental process behind spontaneity in synthesis is not tangible and defies any structuring attempts. Spontaneity in design could be a personal skill and an organizational asset. The challenge in promoting design as a discipline or science is how to approach design and in particular the synthesis, systematically.

Biomechatronic design, in the current context, is primarily concerned with functionality and reliability. The approach adopted by biomechatronic school of thought embarks on associating all attributes of design to the engineering aspects. For this association to materialize, a common ground in the shape of a general design methodology is required. The lack of an effective general methodology for design in biomechatronic systems is an insufficient emphasis upon general methodologies in engineering. This has never been more pronounced than in fusion of mechatronics and biological systems; resulting in an evident challenge faced by existing intradisciplinary design tools and methodologies. An appreciation of design anatomy could therefore assist in distinguishing a qualified location for the design methodology within a biomechatronic project environment.

The thought process involved in a design takes an iterative shape resulting in a three-phase pattern of divergence, systematization, and convergence. Each of these phases has proposed techniques, methods, and procedures within individual engineering disciplines. The total sum of these phases are termed as the design process. In this process, analysis of the search space and generation of solution variants form the general content of the diverging phase. This is followed by the structuring activities which are primarily a combination of synthesis, intradisciplinary methodologies, and designer skills and experience. The converging phase, on the other hand, predominantly consists of the selection process with two steps of evaluation and decision. Here an important factor is the nature of the criteria which is used in determining the value scales and the basis of comparison for assessing the range of solution variants. Evaluation emerges as the central element in the design process where the design tools begin to play their role.

7 DEVELOPMENTS IN DESIGNS

The progressive advancements of biomechatronic systems are occasionally marked by groundbreaking contributions of unique designs. A closer scrutiny, however, reveals that in practice a step by step and incremental development of already proven technologies is the norm. It might therefore prove substantially more tangible to place the emphasis of a design methodology on integration of devices and exiting elemental constituents in obtaining a new system. Systematic synthesis as the core of design would therefore be affected by what comes "before" and "after" it. Formulation and appropriate packaging of design requirements is what comes before the synthesis stage and evaluation of a proposed idea is what comes after. A methodical and systematic approach to these two parts can provide an insurance and a safety net for the multidisciplinary designer. A systematic approach to physically realizable solutions requires a solid platform upon which all else is built. Modeling based on mathematical isomorphism is the natural platform for multidisciplinary specialists to examine the solution space. This mode of thought brings the argument back to the invaluable potentials of modeling in design. A suitable platform for design through modeling provides a forum for the evaluation of the possible solutions to a particular problem. Here, it is important to rely and take advantage of an already proven approach to modeling of multidisciplinary systems. A suitable methodology would have to be based on a set of fundamental principles upon which all energetic engineering systems evolve.

8 ENERGETIC INTERACTIONS

All engineering systems rely on interchanges of energy. This fundamental concept is the key to a unified language necessary for modeling and design of biomechatronic systems. From electrophysiological exchanges to injury prevention in man-machine systems, it is the flow, storage, connectivity, and changes in energetic structures that govern all activities. Mathematical descriptions of flow of energy and power within system elements is perhaps the most vivid and tangible portray of how the system is performing.

Bond graph technology is an approach to multidisciplinary modeling. The term technology is used to indicate considerable strength in adaptive capacities. It is also used to indicate tangible, repeatable, and reliable methodologies when dealing with well-established systems as well as ill-defined problems in all energetic engineering spheres.

When a biomechatronic designer is looking at the neuromuscular systems of an animal or human being, he or she is facing a fascinating engineering system; fascinating but expansively complex, an engineering system which is uniquely adaptive while extremely sensitive to perturbations. A long list of chemical, physical, and other factors interact to build this ultimate engineering temple. The complexities, some known and many still unknown, encountered in human body encourages an engineer to adopt a simplifying approach while being fully aware of the inability of current scientific forums in providing efficient descriptions for many biological or physiological events. This simplifying mode of thought is precisely what should be addressed when design is promoted in biomechatronic educational platforms. Simplification of a complex system requires a well-coordinated pattern and a well-tested approach. A solid foundation for simplification could be mathematical descriptions of energetic interactions among the elemental constituents.

Bond graph technology could provide a mathematical description of exchanges of energy throughout the entire system and between the system and its environment. Biomechatronic designer can construct a suitable and scale-based model of subsystems and interfaces while feeling confident that all possible nonlinearities could be added, as and when required. A simplified and linearized starting point is what the dreams are made of, in modeling and design. From mechanoreceptors to right-hand punch kinematics, and from concert pianist to degenerative diseases, it is the flow of energy and power that is the common denominator. It is the energetic descriptions that provide a tangible insight into the often clogged mechanisms governing performance of this sophisticated engineering system known as human body. The mathematical descriptions, however, are based on fundamental rules of interaction set by causal laws. Causality provides the skeleton for simplicity while allowing nonlinearities of all shape, size, and form to be included at a later stage. The concept of causality is based on basic bidirectional relationship between the two systems where the first system is exerting an "effort" or "flow," and the second system responds by exerting flow or effort onto the first.

9 DESIGN PHILOSOPHY

The relationship between any event (consequence) and its cause (antecedent) is primarily dependent on the observer field (discipline) and his sphere of realm. It is he/she, based on intradisciplinary criteria, who establishes the connections and performs the selection for an individual cause. Any discipline, on the other hand, inherently shapes the boundaries for the generic causal relations. Here, mathematical isomorphic relations could be adopted to define and describe the characteristics and properties of systems and subsystems. This makes the elemental causal characteristics independent of the observer and the disciplinary criteria. Such formalization would have a direct bearing on any synthesizing technique which may adopt these elements as the building blocks. An element, with a set causal structure, can therefore indicate a particular antecedent within but independent of the nature of the system in which the event has taken place. The two problems of causal connection and causal selection may therefore be solved through:

1. Question on the existence of causal relations (causal connections) is by-passed; by the virtue of existence of an element, the existence of causal relations has already been established.
2. The relative importance of antecedents with direct bearing on an event may be established by appropriate backtracking of the set causal relations

(e.g., the cause for element A to behave in a particular manner, is the effect of element B whose cause may be the algebraic summation of the effects of a number of elements). The intradisciplinary subjectivity in the causal selection process may thus be eliminated. The system elements defined and described by logico-mathematical causal relations may also be adopted in the synthesis of new systems, as well as in the analysis of already existing ones. This approach to synthesis may embrace a number of characteristics such as the following:

a. Introduction (existence) of an element independent of the observer (discipline) would indicate a predictable effect (consequence). Therefore, the causal relationships presented in the synthesis of a conceptual design would represent a structure which may not change when the observer is altered. This remains true unless the nature of functional connectedness is altered.

b. Contribution of individual elements to the system output can be established and critically analyzed, independent or within the structure of a discipline.

c. Existence of causal conflicts is an indication of missing or unaccounted relations, and leads to model expansion or reconfiguration.

Combination of the two concepts of causality and systems isomorphism would qualify an alternative approach to bond graph description techniques with emphasis on synthesis as opposed to analysis.

10 COHESION IN DESCRIPTIONS

A purely mathematical approach to systems analysis is all too often an inadequate means of providing full appreciation of interactions present in a system. In engineering, however, a view that a picture is worth a thousand words has generally prevailed and the starting point of analysis of any dynamic system is commonly a systematic diagram or other graphical or pictorial representations. Excellent graphical representations and corresponding analytical techniques already exist in different domains. When multidisciplinary systems are under investigations and biophysical domains are coupled, the coherency of graphical representation techniques evaporates and the situation is no longer routine.

The graphical or pictorial descriptions of such complex systems are commonly extremely generalized mixture of disciplinary notations. Here simple linguistic phrases are often used in crucial coupling junctions. The multitudes of connections in biomechatronic systems could thus result in a nonuniform combination of schematic diagrams, equations, words, and semipictorial representations.

11 MECHANISM OF INTERCONNECTIONS

The bond graph technology used for studying dynamic systems consists of subsystems linked together by "lines" representing power bonds. When major subsystems are being modeled by "words," the subsequent system description would be called a "word graph," an example of which is shown in Fig. 1. This type of description would be very important at the elementary stages of synthesis in establishing structures in the way they bonded effort and flow variables at the subsystem ports, sign conventions, and power interchanges. In bond graph notation, a bond with half arrow (⇀) indicates the direction of positive flow of power and a full arrow (⇢) indicates an active bond or a signal flow (low-power information bonds). A word bond graph is very useful for sorting true power interactions from the one-way influences of active bonds. To distinguish which of the excitation and response variables at a power port are actually input to the multiport, a further piece of information must be supplied which is the causal stroke, denoted by a small vertical line at the end of the bond.

A study of excitation-response causalities is the unique feature of bond graphs. Comparison of the two connections, shown in Fig. 2, presents the way causal strokes are implemented. The position of the causal stroke at either end of a bond indicates direction of effort. Flow would consequently be in the opposite direction.

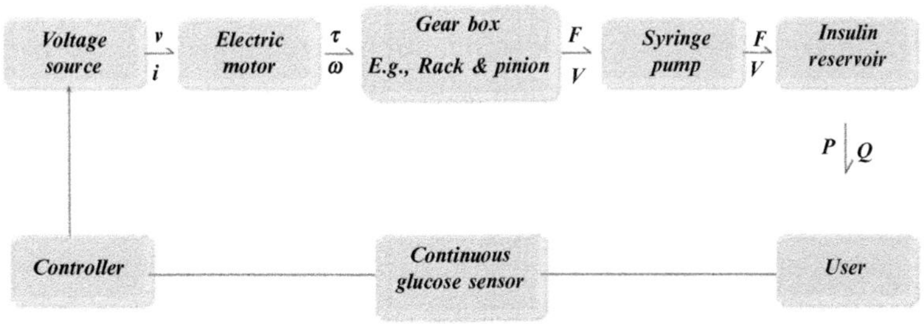

Fig. 1 A word graph representation of an automatic insulin injection device.

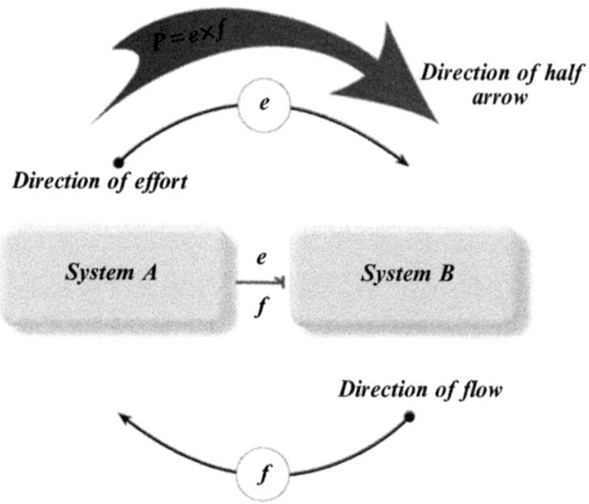

Fig. 2 Schematic description of the relationship between Power direction and causality between two systems A and B.

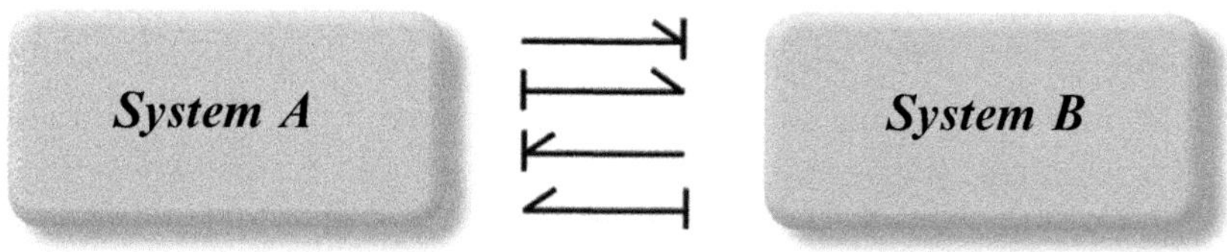

Fig. 3 Alternative relationships between causality and power between two systems A and B.

In general, whether the effort is entering or leaving a system determines position of causal strokes on a bond. A distinction should at this point be made between a half arrow placed at the end of a bond and a causal stroke. The four possible causal combinations are shown in Fig. 3.

Here each bond implies the existence of both excitation and response signals. This is important since power interactions require a pair of bilaterally oriented signals. Bond graphs are a more efficient means of describing models in comparison to other conventional techniques on the basis of quantity and quality of information which is being conveyed. System visualization through bond graph notation is far more effective than that permitted by state equations or other multidisciplinary graphical representations. The subsystems considered from the point of view of power exchanges and external port variables could be categorized by a limited number of fundamental multiports. These functioning components of a model are idealized mathematical versions of real components of material and physical models such as resistive, capacitive, inertial, transducing, and transmission elements. Although it may not be possible to provide full descriptions for every probable system through this reductionist approach, a vast majority can be comfortably synthesized, analyzed, and handled. The important issue at hand is that in all such elements, manipulation of the Poynting vector provides an established mathematical expression defining an energetic characteristic which is used to describe the causal structure.

The approach provides a solid mode of thought toward modeling which could also be adopted as the basis for synthesis. If the rules for interconnections that are based on causality are observed, it is possible to conceptualize many novel biomechatronic systems which are physically realizable and causally valid but independent of any disciplinary constraints.

12 GENERAL DESIGN METHODOLOGY

12.1 Modification of Systems Approach

Among many contributions to systematic design, the systems approach takes a superior position due to its inherent harmony with the concept of systematic design. The systems approach aims at producing the optimum design for complex systems and it reflects the general appreciation that complex problems are best tackled in a series of defined steps. These being problem definition, goal setting, solution development, solution analysis, solution evaluation, optimum decision, and finally preparation for physical realization.

A brief study of the proposed steps makes it quite clear that the aim of the approach is a broad and generalized outline or a frame of action. The apparent overgeneralization is that particular attribute which renders the approach open to criticism due to an inherent inability to address specific design issues. Although conceptually acceptable, the generality has left the most important steps of goal identification and synthesis to the designer's discretion and his understanding of disciplinary design techniques. To obtain a general methodology for biomechatronic design, the overgeneralization associated with the systems approach, or any other systematic approach, has to be overcome.

To begin with, each of the steps set out in the systematic approach take up on themselves a unique configuration and meaning when applied to a particular discipline. Although the pattern of the process might have remained conceptually similar for various designs, the actual process would be quite different within the structure of various disciplines. To formalize a structure applicable to many if not all disciplines, it is essential to concentrate upon those areas in the systems approach or stages of the design process that are most likely to be affected when applied to different fields of science. The objective is to identify and isolate all such areas and more importantly to abstract and structuralize their common characteristics.

The series of activities commonly performed by practicing designers are affected by the nature of design environment and the designer's intuition and creativity. The ultimate objective of a designer has always been to obtain some optimum solution in the face of imposed constraints. The search area, on the other hand, may already be isolated by the existence of such constraints and these can be primarily dictated by the disciplines involved. Thus by keeping the constraints away from the most elementary stages of the design process, it is possible to synthesize the system independent of any discipline or any energy domain. Adoption of this approach in synthesis is acceptable in a general conceptual term and in an optimum form, an ideal conceptual design (ICD). It follows that the designer can be encouraged to produce an ICD independent of any discipline. This abstract model of a system, however, must represent true and intended functionality (Figs. 4 and 5).

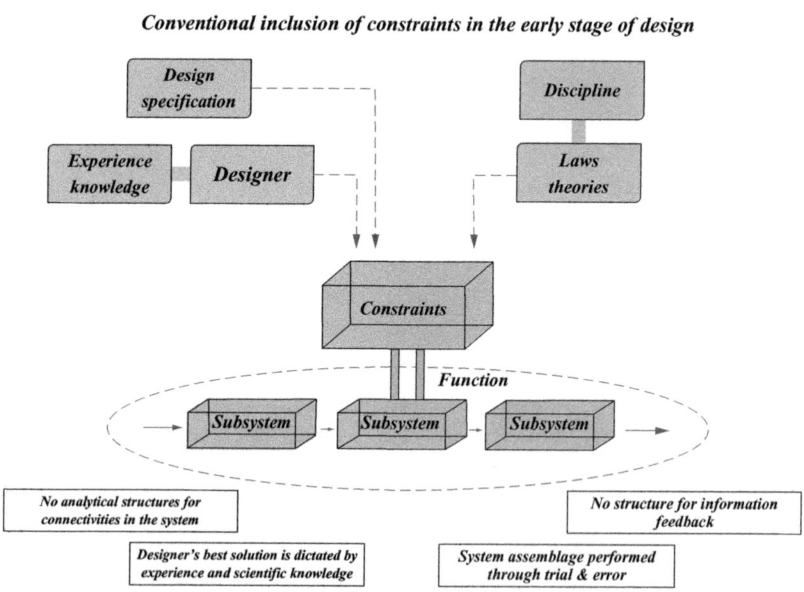

Fig. 4 Conventional inclusion of constraints in the early stages of design.

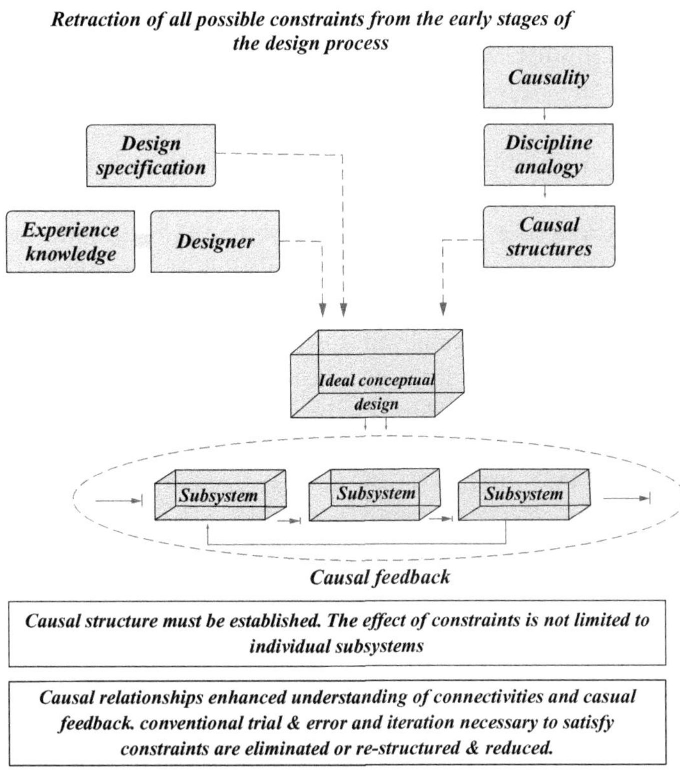

Fig. 5 Retraction of all possible constraints from the early stages of the design process.

12.2 Intuition and Creativity in ICD

During system evaluation stage of a design, solution variants may be compared against some form of a criterion function manifested in the shape of mathematical functions or a series of statements and figures or a list of objectives. Adoption of ICD as the criterion function, on the other hand, could provide a solid platform to make quantitative comparisons among solution variants. The proposition rests on the distinction made by the systems approach between the stages of solution variant identification and solution evaluation. Although such distinctions are commonly quite valid, there is no analytical structure to the content of evaluation stage. The efficient approach to any design problem is to design an ideal conceptual system irrespective of any energy domain that may be involved. The proposed design can then be analyzed to establish the fundamental characteristics of the model.

The relationship between the constituent elements and in particular their effects on one another, within the causal structure of the system as a whole, could be critically analyzed and well understood. As a result, the designer gains a clear understanding of the system and is able to benefit from the information feedback to improve on the design or even reshape the original structure of the problem. The design can therefore, to a large extent, be completed and the root characteristics established prior to introduction to any discipline or energy domain. The formation of the solution variants which may be the consequence of introducing the ICD into alternative energy domains can be achieved through appropriate substitution of corresponding disciplinary elements. The particular advantages or disadvantages of individual disciplines quickly becomes apparent. Here, the extent and range of solution variants has already been decided upon through the complexities of the criterion function. Any necessary extensions of the model to cater for any particular requirement associated with any one discipline could also affect the choice of disciplinary elements. Handling of overriding design specifications and the general decision-making process are all based on an analytical structure which is derived from a causally valid and mathematically described model and hence reduces the reliance on lists or linguistic constructs.

The designer can therefore formulate or design an optimum system to begin with in an ideal form and could even optimize the design at this elementary stage. All possible ideas could be implemented, tested, analyzed, and simulated using the ICD.

12.3 Bond Graph Technology in Synthesis

A word graph in its conventional form cannot convey sufficient detailed information about a system. It simply set out to describe the essence of the system. Perhaps not unlike preliminary sketches drawn by architects. It is, however, the first step in a line of progression to a detailed design with its origins in an idea. The word graph could then be augmented to establish underlying causal relationships among constituent elements. Augmentation at this point means an introduction of the basic causal structure, thereby indicating input-output relations governing the interchanges of energy and flow of power in the system. Inclusion of causal relations in the structure of word graphs is not intended to give rise to formulation of mathematical relations but to encourage feedback during abstract and conceptual stages of synthesis. In such a sphere of conceptuality, adoption of conceptual causality within the skeleton of a conceptual model should not be considered as anything more than an idealistic approach to structure the thoughts. Having taken the disciplinary constraints away from the designer, the methodology must substitute some form of a safety net. The concept of causality as it was described through the definition of mathematical isomorphism, presents an ideal safety net for the system synthesizer to ensure inclusion of fundamental physical laws. The ideal and conceptual system model is therefore unable to contravene fundamental disciplinary laws.

The substitution of bond graph multiports for elements of word graph is the next stage in advancing toward a detailed final design. A simple substitution may, however, prove insufficient in the formulation of a valid bond graph model, since individual elements of a word graph can quite often represent rather complex systems. A reasonable multiport may thus greatly expand the initial model structure. Expansion within the sphere of conceptuality must be limited to the introduction of absolutely essential details ensuring that the bond graph will undergo a minimum reticulation process through multiport substitution. Inclusion of necessary multiports with their associated bonds will change the system description from simple word graph to a more detailed symbolic structure. For the development of the new system to be coherent, ideal junction structures that are fundamental to assertion of causality are introduced. The properties of bond graphs, particularly, the necessity for correct causal structure, quite often dictate alterations to perceived ideal structure. Such dictated changes are valuable in establishing the functionality and reliability at earliest stage of a product life cycle. Here, the possible existence of causal conflicts could direct attention toward unaccounted factors. Additional elements to cater for insufficiencies can contribute to further expansions. This process of presenting an idea through causal word graphs is iterative in the progression of an ICD. Fig. 6 is a diagrammatical presentation of such a recursive reticulation process.

12.4 Design Criterion

For the reticulation process to sustain an effective progress, the ICD must satisfy an objective beyond what is imposed by specific disciplinary objective functions. The functional connectedness which might have been arrived at through conventional design procedures may not hold any longer. The biomechatronic problem in its initial proposed format could have more than a single solution. As a result, some form of criterion that could be applicable to a great majority of engineering systems is needed to assess alternative

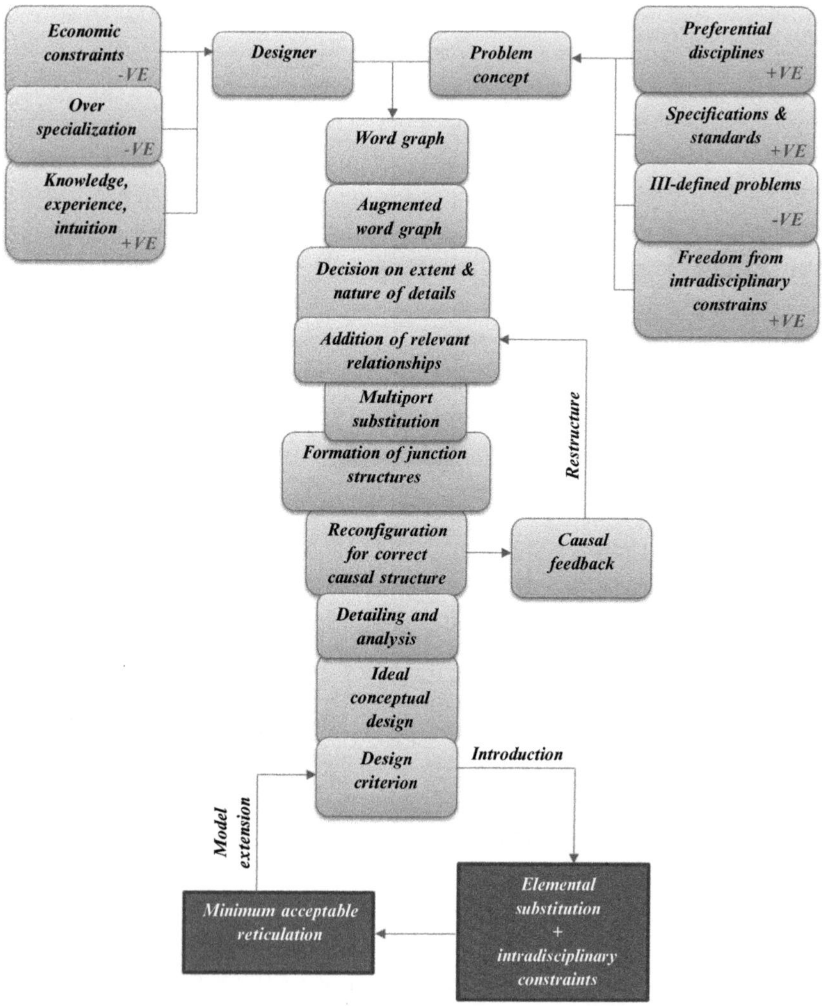

Fig. 6 Recursive reticulation in causal synthesis.

models of solutions. Therefore, a criterion for the absolute minimum objective function should be adopted. An ICD could quite adeptly address the energy balance and the energetic efficiency in a design. The characteristics of any deviation from optimum energetic efficiency, on the other hand, is primarily dictated by system impedance. System impedance could act as a measure of optimality for a proposed design and any attempt toward maximizing the efficiency would be direct at system impedance. Furthermore, introduction of such concepts as system controller would in effect aim at optimization and manipulation of the total impedance inherent in the system which is a consequence of devised functional connectedness. Optimization of the system impedance is therefore the objective function from which a set of generalized constraints can be abstracted. Optimization of impedance could mean minimization of structures associated with energy dissipation and inertial optimization could be addressed through minimum connectedness. The connectedness optimization, on the other hand, is directly pointing at minimum number of elements and optimum configuration.

Following a minimalistic approach, the first design criterion could thus be stated as: the energy consumption of a proposed system for performing a given task must be optimized. Within the structure of such criterion, concepts like energy density in a system, the effectiveness of the power sources, or transforming modules may be investigated. In thermodynamic terms, entropy generation must be minimized or unnecessary irreversibilities must be eliminated. The attention could thus be directed at advanced modes of decision-making using switching mechanism to avoid classical energy consumption problems. The next stage is the identification of the modulating elements through which the performance of the system is manipulated, regulated, and controlled.

Conventionally, the controllers are considered only after the nature of functional connectivity has to a great degree been established. In a multidisciplinary approach to design, however, the controlling system is all part of the complete functional connectivity and is developed simultaneously. The recognition of modulatory constituents at the outset of design will contribute toward a minimalistic criterion function. Here the root characteristics of the controlling mechanism is readily provided by the bond graph model of the synthesized system. Engaging in controller design during the synthesis of biomechatronic systems induces a harmony with dissipation minimization approach. The preliminary criterion function can thus be based on flow of energy, materials, and signals. The constraints imposed on the ICD could therefore be stated as minimum energy consumption, minimum functional connectivity, and minimum restriction to energy flow.

The designer, however, must leave the sphere of conceptuality and advance toward physical realization. The results of the synthesis and the natural expansion of the initial model will lead to physically realizable systems as bond graph technology is inherently adept in catering for physical realization. Once disciplinary elements are substituted for ICD constituents and possible expansion of the model has taken place, other issues such as economics and industrial constraints could be addressed.

The steps suggested here for a "General Design Methodology" are outlined diagrammatically in Figs. 7 and 8.

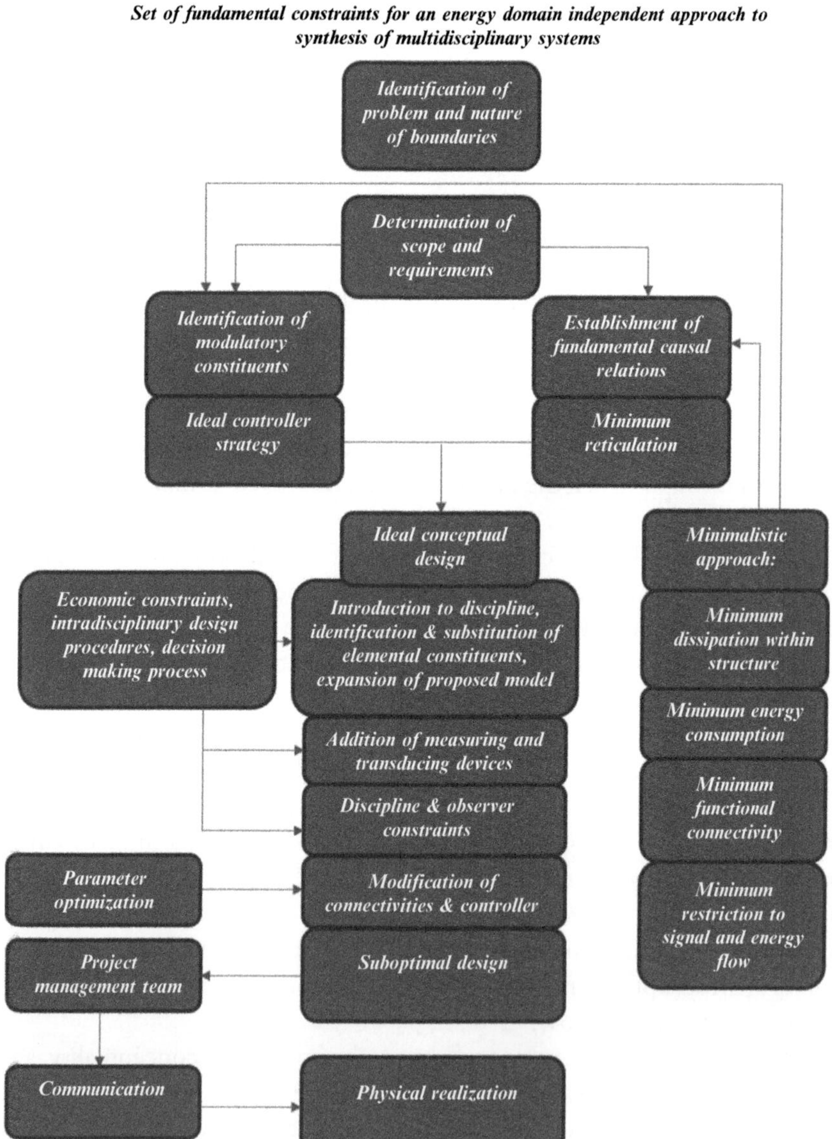

Fig. 7 Fundamental set of constraints leading to optimum design of biomechatronic systems.

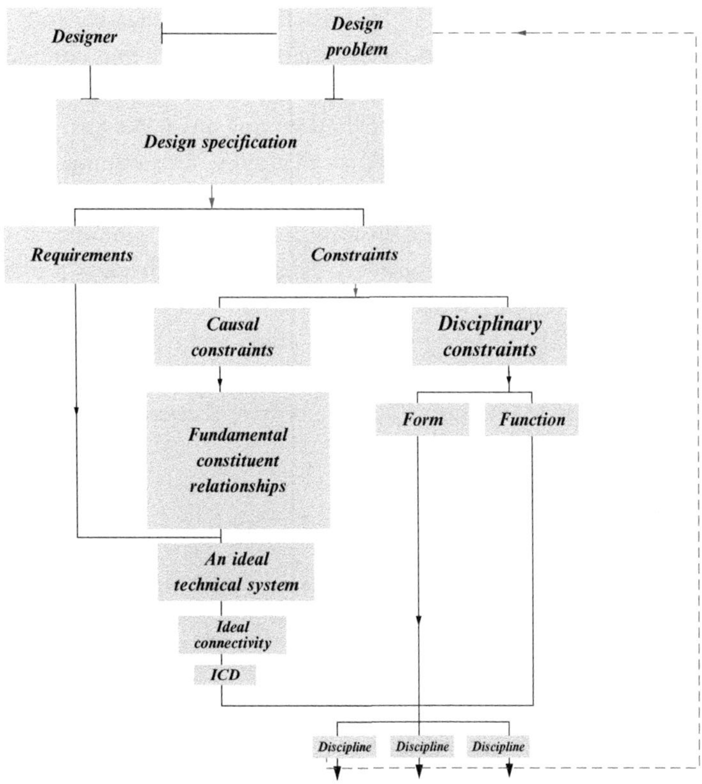

Fig. 8 Relationship between constraints and requirements.

13 SUMMARY

Design of biomechatronic systems is a complex process and as such, it would achieve a greater degree of economic, technical, and aesthetic excellence when cloaked by logic and rationality. The influential and complimentary concepts of systematic design and systems approach to design reflect general appreciation that complex problems are best tackled in a series of defined steps. Such structuring is governed by the nature of design environment and directed at obtaining an optimum solution in the face of imposed limitations. The boundaries of a design problem are therefore dictated by the disciplines involved and the associated constraints. In other words, limitations are formed by two sets: (a) intradisciplinary constraints and, (b) specific problem constraints. The combined limitations of these two sets would adversely affect a biomechatronic designer and even refrain her/his creativity and intuition. This is particularly disadvantageous when a variety of alternative combinations of disciplines and system configurations are capable of satisfying the objective function. To encourage creativity and intuitiveness, in producing efficient, better, and novel designs, the core of the problem is abstracted using causal word graphs. The ensuing transformation to bond graphs provide a solid analytical platform for further manipulations including possible expansions, inclusion of nonlinearities, and extraction of variables and parameters. Result of synthesis is presented as an ICD which is uniquely suitable to be adopted as the criterion for further evaluations. The ICD derived using bond graph technology has an adaptive capacity in producing an energy domain-independent solution which is optimized in terms of functional connectivity and energetic management.

FURTHER READING

Bashir, H.A., Thomson, V., 1999. Metrics for design projects: a review. Des. Stud. 20 (3), 263–277.

Chang, A.S.T., 2002. Reasons for cost and schedule increase for engineering design projects. J. Manag. Eng. 18 (1), 29–36.

Cross, N., Roy, R., 1989. Engineering Design Methods. vol. 4. Wiley, New York.

Dieter, G.E., 1991. Engineering Design: A Materials and Processing Approach. McGraw-Hill, Boston.

Dutson, A.J., Todd, R.H., Magleby, S.P., Sorensen, C.D., 1997. A review of literature on teaching engineering design through project-oriented capstone courses. J. Eng. Educ. 86 (1), 17–28.

Dym, C.L., Agogino, A.M., Eris, O., Frey, D.D., Leifer, L.J., 2005. Engineering design thinking, teaching, and learning. J. Eng. Educ. 94 (1), 103–120.

Dym, C.L., Little, P., Orwin, E.J., Spjut, E., 2009. Engineering Design: A Project-Based Introduction. John Wiley and Sons, New York.

Finger, S., Dixon, J.R., 1989. A review of research in mechanical engineering design. Part I: descriptive, prescriptive, and computer-based models of design processes. Res. Eng. Des. 1 (1), 51–67.

Haik, Y., Sivaloganathan, S., Shahin, T.M., 2015. Engineering Design Process. Nelson Education, Boston.

Hirtz, J., Stone, R.B., McAdams, D.A., Szykman, S., Wood, K.L., 2002. A functional basis for engineering design: reconciling and evolving previous efforts. Res. Eng. Des. 13 (2), 65–82.

Kalpakjian, S., Schmid, S.R., 2014. Manufacturing Engineering and Technology. Pearson, Upper Saddle River, NJ, p. 913.

Karnopp, D., Rosenberg, R.C., 1968. Analysis and Simulation of Multiport Systems: The Bond Graph Approach to Physical System Dynamics. MIT Press, Cambridge, MA.

Karnopp, D.C., Margolis, D.L., Rosenberg, R.C., 2012. Basic bond graph elements. In: System Dynamics: Modeling, Simulation, and Control of Mechatronic Systems, fifth ed, John Wiley & Sons, Inc., Hoboken, NJ, pp. 37–76

Lewis, W., Samuel, A., Cleland, R.D., Maffin, D., 2002. Engineering Design Methods: Strategies for Product Design. John Wiley & Sons Ltd, Chichester.

Pahl, G., Beitz, W., 2013. Engineering Design: A Systematic Approach. Springer Science & Business Media, London.

Samuel, A.E., 2006. Make and Test Projects in Engineering Design: Creativity, Engagement and Learning. Springer Science & Business Media, Massachusetts.
Sydenham, P.H., 2004. Systems Approach to Engineering Design. Artech House, Boston.
Tayal, S.P., 2013. Engineering design process. Int. J. Comput. Sci. Commun. Eng. 1–5.
Thompson, G., Lordan, M., 1999. A review of creativity principles applied to engineering design. Proc. Inst. Mech. Eng. E 213 (1), 17–31.

CHAPTER TWO

Actuator Technologies

Reva E. Johnson*, Jonathon W. Sensinger†
*Mechanical Engineering and Bioengineering, Valparaiso University, Valparaiso, IN, United States
†Institute of Biomedical Engineering, Department of Electrical & Computer Engineering, University of New Brunswick, Fredericton, NB, Canada

Contents

1 INTRODUCTION

In this chapter, we focus on actuators that generate movement for biomechatronic systems. Actuators are subsystems that transform various types of energy into mechanical movement or force. In typical control systems (shown in Fig. 1), the actuator receives a signal from the controller and responds by acting on the plant or process in some desired way. In biomechatronic systems, the actuator usually converts supplied energy into mechanical movement or force.

We begin this chapter by discussing broad goals and specific metrics for designing biomechatronic actuators. We then introduce and categorize types of biomechatronic actuators. We end by describing common purposes of biomechatronic systems with examples of typical actuators.

Handbook of Biomechatronics
https://doi.org/10.1016/B978-0-12-812539-7.00002-7

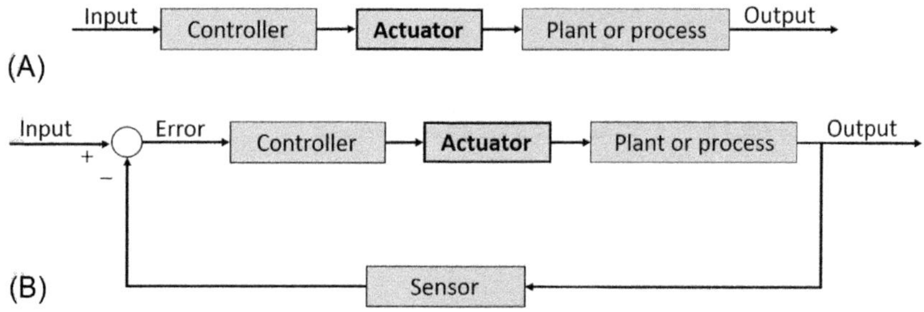

Fig. 1 Block diagram of typical open-loop (A) and closed-loop (B) control system. The actuator receives a control signal that dictates how the supplied energy should be converted into a mechanical movement or force that acts on the plant or process.

2 DESIGN GOALS OF ACTUATORS

Below we discuss three broad design goals that apply to every biomechatronic actuator: safety, performance, and ease of use. Within each broad design goal are specific metrics, whose desired values depend on the purpose of the overall system. These metrics help quantify the trade-offs of design choices. For example, there is often a trade-off between safety and performance. One strategy to improve actuator safety is to decrease the stiffness, so that interaction with humans is more flexible and injury-causing impacts are minimized. However, a decrease in stiffness can also worsen performance by reducing bandwidth. When faced with this common trade-off, how can we minimize injury while still designing a useful actuator?

Quantitative metrics enable us to optimize the system for several design goals. One example of a design optimization for a PUMA 560 robot is shown in Fig. 2. The PUMA 560 is an articulated robot, originally designed for industrial assembly lines and now widely used for research and education. The PUMA often operates alongside or directly interacts with humans; so, minimizing injury risk is an important design goal. Fig. 2B is an example of how a plot can be used to show how design parameters (in this case, actuator stiffness and effective inertia) influence design metrics (in this case, head injury risk). The designer then selects a combination of parameters to achieve the desired outcome metric. Similar multivariable optimizations can be used to choose the type and characteristics of other biomechatronic actuators (another example is provided in Fig. 3B).

2.1 Safety

How do I design an actuator to interact with humans safely?

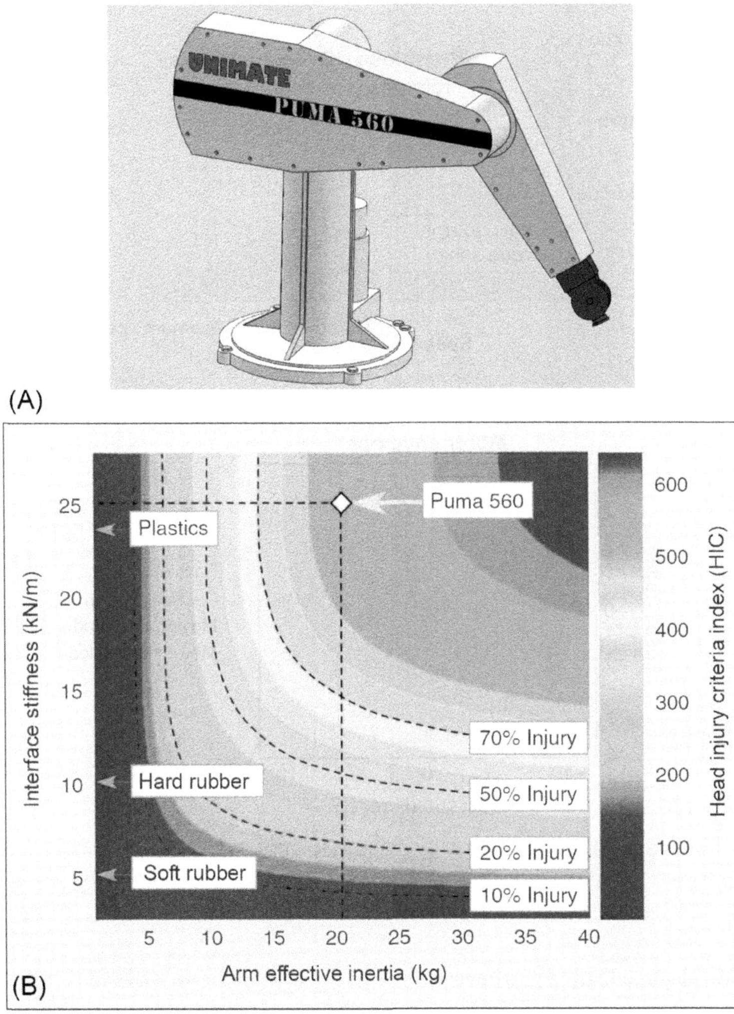

Fig. 2 (A) PUMA 560 robot, often used alongside humans in industrial assembly lines or research applications. (B) Multivariable optimization of the stiffness and effective inertia of the PUMA 560 robot. *((A) Courtesy of Gonzalo Loredo Neri; (B) From Zinn, M., Khatib, O., Roth, B., Salisbury, J.K., 2004. Playing it safe, IEEE Robot. Autom. Mag. 11 (2), 12–21, with permission.)*

For biomechatronic systems that interact with humans, safety is a vital design goal. If the injury risk of a device outweighs the functional benefit, the device will not be widely used, commercially viable, or covered by insurance. Even the perception of injury risk can alter the human operator's behavior. If a human perceives a biomechatronic system as being dangerous, they will not fully utilize the device. For example, if a person using a

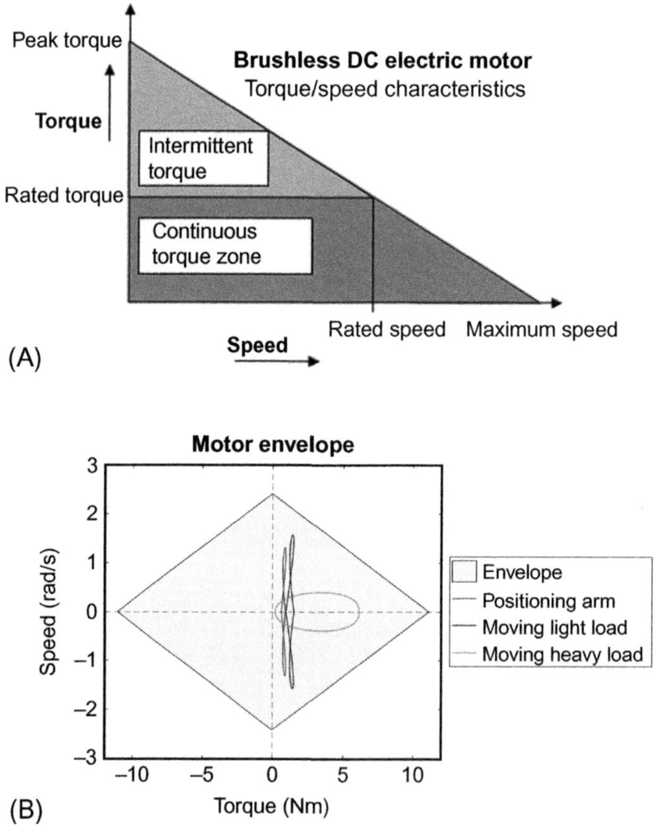

Fig. 3 (A) Torque-speed curve of brushless DC motor and (B) example of motor envelope plot. *((A) Reproduced with permission from Wikimedia Commons.)*

powered lower limb prosthesis does not trust that the device will fully support them, they will avoid placing their full weight on the prosthesis. On the other hand, there is also some danger in perceiving an inappropriately low injury risk. When people see an anthropomorphic device, they often overestimate the human-like capabilities of the device. This perception can lead to a false feeling of safety, and should be considered in the design of anthropomorphic devices (Murashov et al., 2016).

The field of robotics has traditionally assumed the safest design strategy is to separate robots and people; however, since the late 1990s there has been increased interest in human-robot interaction. New strategies for designing safe robotic systems include removing pinch points, reducing size and weight, limiting the operating speeds and forces, and implementing control strategies that minimize high-speed collisions (Zinn et al., 2004). There are several sets of recently developed safety regulations for robots that interact with humans; for example, the International Standards Organization (ISO) has developed requirements for robots that perform surgical, rehabilitation, personal care, and industrial tasks (ISO, 2011, 2014, 2017a,b). Below we introduce several quantitative measurements of safety that can be helpful in designing biomechatronic actuators.

2.1.1 Impedance and Compliance

Mechanical impedance is the frequency-dependent relationship between forces and motions. When you impose motion on an object, the amount of force generated in response is determined by the object's impedance. Low impedance is typically desirable for actuators that interact with humans.

Compliance is the ratio of the force generated by an elastic element in response to deformation. Compliance is the reciprocal of stiffness. One of the most common methods of designing a low-impedance actuator is by including compliant elements in the transmission.

Intuitively, humans are safest when the objects around them give way or yield upon contact. When objects are very rigid or heavy, a high-speed impact with humans can cause serious injury. Typical industrial robots are both rigid and heavy, and are programmed to follow precise positions with no consideration for obstacles (or humans) that may impede motion. If humans are located in the path of such a robot moving at high speeds, they will be subjected to dangerously large forces. One way to avoid injury is to limit the speed and torque of the robot. Another way is to lower the impedance.

2.1.2 Head Injury Criterion

One of the most severe risks of a biomechatronic actuator is that of a head injury to the human operator or bystander. The potential for head injury can be quantified using the Head Injury Criterion (HIC), which is calculated as a function of the magnitude and time duration of head acceleration (Gao and Wampler, 2009). The HIC metric was first developed for automotive applications, later applied to athletics, and is now often used for human-robot interaction. The values of HIC have been correlated to the abbreviated injury scale (AIS), which encodes the severity of injuries to all body regions.

2.1.3 Voltage, Current, and Heat

Caution must be taken to ensure that the human operators are shielded from electrical circuits of the actuator. The potential injury from electric shock depends on the magnitude and time duration of current. The current that flows through the body is a ratio of applied voltage to the body's electrical impedance. Thus, safe levels of voltage depend on how the actuator contacts the body—for example, dry skin has very high impedance, whereas internal body cavities have much lower impedances (Fish and Geddes, 2009). Acceptable levels of current and voltage have been developed by regulatory bodies such as the International Electrotechnical Commission (B. S. Publication, 2016). These guidelines are especially important for actuators such as electroactive polymers (EPAs) that often require high voltages for operation. Some actuators, such as electric motors, generate significant amounts of heat during operation. Heat dissipation should be considered in the design of the actuator system.

2.2 Performance

How do I make this actuator perform as well or better than the equivalent human system?

Whether humans or actuators move in free space or interact with the environment, there is an interplay between forces and motions. Performance can be thought of in terms of these forces and motions: can my fingers produce 100 N of force? Can I move my arm over my head? Can I rotate my elbow 180 degrees/s? Can I give a burst of acceleration fast enough to stand up before I fall over? Any mechanical performance metric can be thought of as a function of these four generalized parameters (force/torque, position, velocity, acceleration).

Many activities for which we wish to evaluate performance require combinations of force and motion parameters. For example, can I rotate my elbow 180 degrees/s while holding a 10 N weight? Can I do that same activity when my elbow starts from rest fully extended, and stop before my hand collides with my upper arm at 145 degrees of flexion? The capabilities of actuators also often depend on combinations of those four parameters. For example, the maximum torque that an electric motor can produce is a function of its speed: it cannot produce high torques at high speeds, and it cannot produce high speeds at high torques (Sensinger et al., 2011). Other actuators, such as shape-memory alloys (SMAs) and human muscles, have maximum forces that are dependent on their position or percent contraction. It accordingly makes sense to think of both the activities we wish to perform, and the capabilities of actuators, in terms of force as a function of motion (position, velocity, and acceleration).

The capabilities of many actuators can be visualized as a subset of this interplay between force and motion. For example, ideal electric motors (e.g., without friction or inertia) have a relationship between torque and velocity that is independent of position or acceleration. This relationship can be visualized as an envelope of the maximum capabilities of the actuator in terms of torque and speed (Sensinger, 2010a) (e.g., Fig. 3A). The demands of the task and nonidealized portions of the actuator (e.g., friction and inertia) can be calculated over time, from which the net torque, position, velocity, and acceleration can be calculated. Net torques associated with this motion often include the inertial torque of the motor, gear, and load caused by the acceleration, the viscous force of the transmission, and the gravitational force of load. The task can then be overlaid on the actuator envelope as a parametric function of torque vs speed (since both total task torque and speed were calculated as functions of time). If the profile of the task falls within the envelope of the motor, then the motor is capable of performing the task (e.g., see Fig. 3B). This visualization between forces and motions represents the most accurate understanding of the ability of an actuator to perform a task, but it is a fairly involved calculation, and is task specific.

Often, designers wish to use a proxy for performance that conveys a general sense of whether or not an actuator will be capable of performing a given task. These proxies often fail to convey important information relevant to biomedical tasks. For example, most conventional actuators run at constant speed, whereas most biomechatronic actuators start and stop at rest, with substantial acceleration/deceleration in between. Designers often look to proxies that are either particular to their specific applications, or that best generalize across the many desired attributes. These proxies are a good way to quickly compare different actuators, but an envelope technique should often be used in the final stages of verification that takes into account the dynamical properties of the task. Proxies can either be given as a final value, or as a normalized value (e.g., density), depending on the type of comparison being made. Several useful metrics for describing actuators will be discussed below.

2.2.1 Stall Torque and No-Load Speed Density

The maximum torque, and the maximum speed, that a motor can produce are both often-considered metrics. For many actuators, maximum torque occurs when the actuator is not moving, and the maximum speed occurs when there is no applied torque or acceleration. Because of this, designers often look at stall torque (the torque when no motion is occurring), or no-load speed. Two common normalizations are often used. The first normalization is with respect to mass (e.g., stall torque per kg of actuator, or no-load speed per kg of actuator), as larger, heavier actuators can typically produce better performance at the expense of larger, heavier designs. It is important when reviewing these specifications to clarify whether the mass includes all of the components (e.g., compressors for hydraulic actuators, power sources, etc.) to ensure a fair comparison.

2.2.2 Torque and Speed Constant

The second normalization is with respect to electrical input. Electrical motors produce more torque and spin faster if they are provided with more voltage (which, for a given electrical resistance, in turn permits more current). The torque constant (K_t) is a measure of how much torque per amp a motor can produce. If SI units are used, it is nearly equivalent to the reciprocal of the speed constant K_v, which is a measure of how much speed per volt the motor can produce.

2.2.3 Mechanical Power

These metrics of maximum speed and torque convey useful information regarding the performance of an actuator. However, most actuators must produce a range of torques across a range of speeds. Many designers accordingly turn to the maximum mechanical power the motor can produce (typically in Watts). Maximum power for electric motors occurs at half of the no-load speed and half of the no-load torque (Alciatore and Histand, 2003). Mechanical power can also be normalized by mass (e.g., W/kg). If applications are in this vicinity of torque and speed it is a useful metric, but if actuators operate away from that region, it can misrepresent the comparative performance of different actuators. For example, many biomechatronic actuators must either produce high torque (e.g., sit to stand), or high speed (e.g., walking), but rarely both at the same time. For these activities, the metric of maximum mechanical power is a poor proxy for task performance (Sensinger, 2010a).

2.2.4 Envelope Visualizations

Envelope visualizations capture all four relevant parameters (torque, position, velocity, and acceleration), but they do not provide a compact number and cannot be easily normalized. The other metrics we have discussed (e.g., stall torque and mechanical power) provide compact, normalizable numbers, but they are only accurate for specific portions of the applicable space (e.g., only when not moving, or at one-half maximum speed/torque). There is, however, an easily accessible metric that incorporates the ability of an actuator to achieve any combination of torque, position, speed, and acceleration, in a compact normalizable metric. That metric is the speed ratio–the reciprocal of the mechanical time constant (a metric that is often reported in actuator specification sheets and is equal to the amount of time for an unloaded motor to rise to 63.2% of its final velocity after application of a command voltage). The speed ratio can be expressed in various forms, as shown in the equation below. Although not well understood and rarely used, the speed ratio incorporates each of those four parameters (Sensinger, 2010a), and can be used to streamline the design of biomechatronic actuators.

$$SR = \frac{K_t^2}{J_m R} = \frac{K_m^2}{J_m} = \frac{1}{\tau_m}$$

where K_t is the torque constant, K_m is the motor constant, J_m is the inertia of the motor, R is the resistance of the motor windings, and τ_m is the mechanical time constant.

2.2.5 Efficiency

Efficiency is another useful metric. Efficiency is defined as the amount of output power (typically, mechanical) divided by the amount of input power (typically, electrical). Peak efficiency for electrical motors does not occur in the same region as peak mechanical power–it occurs at higher speeds (Alciatore and Histand, 2003). Although efficiency is a useful metric, its use as a biomechatronic design metric is often eclipsed by total weight.

2.2.6 Total Weight

The total weight of biomechatronic actuators is often an afterthought, but it is actually a powerful metric, if used properly (Sensinger, 2010a). Imagine that you are trying to compare a series of actuators for a given application, and that you have the ability to generate envelope visualizations or access to torque and speed densities. The mechanical properties of the task can be used to calculate the weight of the actuator needed to perform the task (e.g., if the task requires 10 Nm of torque, and a particular actuator design has a torque density of 10 Nm/kg, then your actuator weight is 1 kg). The electrical requirements of the task can also be calculated. This energy draw can be multiplied by the energy density of the supply (e.g., a battery), and added

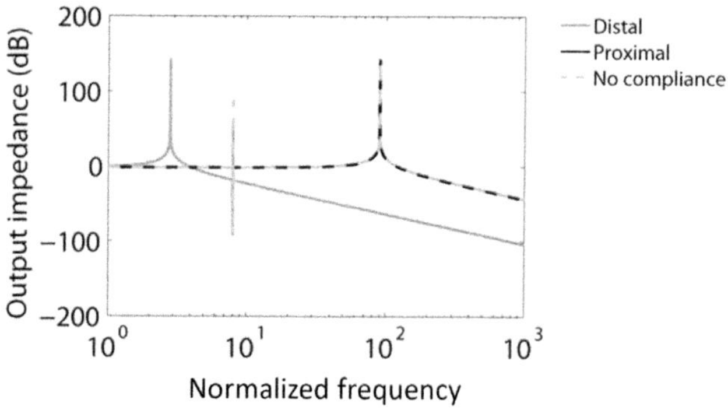

Fig. 4 Bode plot used to represent output impedance of a series elastic actuator across a range of frequencies, as well as compliance locations (distal, proximal, none). See the section on variable and low-impedance actuators and Fig. 12 for more details on series elastic actuators and compliance location. *(Modified from Sensinger, J.W., Burkart, L.E., Pratt, G.A., Weir, R.F. ff, 2013. Effect of compliance location in series elastic actuators. Robotica 31 (8), 1313–1318.)*

to the weight of the actuator. In this way, both the efficiency of the mechanism, along with its ability to produce the torques and speeds in the appropriate region, are taken into account. The actuator technology with the lowest total weight is then selected as having the best performance. This is a powerful design tool that has enabled a new era of biomechatronic actuator design (e.g., Lenzi et al., 2016; Johannes et al., 2011), with performance much closer to the human counterpart it seeks to replace.

We have presented the interplay between force and motion above as a parametric function across time in which position, speed, and acceleration all play a role. There is another way that this interplay may be expressed, however, and that is force as a function of motion frequency. Shown on a Bode plot (e.g., Fig. 4), it is easy to visualize the ability of an actuator to render a variety of impedances, and this is a useful performance metric for biomechatronic actuators, particularly in the field of haptics. Several compact metrics have been developed based on this foundation, including *Z*-width, which is a measure of the frequency range over which the actuator can produce stable renderings (Weir and Colgate, 2008).

2.2.7 Summary

There are many available metrics to assess performance. Many designers are familiar with compact notations, such as stall torque or maximum mechanical power, along with their normalized counterparts (such as torque density). For biomechatronic applications in which acceleration occurs, the speed ratio is a more relevant compact notation, and motor envelope visualizations are even better, at the expense of being less compact a comparison. Weight calculation (as a function of torque densities, and including electrical energy density necessary to accomplish the task), although more involved, is a powerful metric that has enabled significant improvements in the field.

2.3 Ease of Use

How do I make this actuator easy for humans to use?

Potentially the greatest challenge of biomechatronics is designing a system that is intuitive and comfortable. Actuators play an important role in achieving this challenge: they should generate movement in a way that people can learn to predict. Actuators with hard nonlinearities (e.g., stiction and backlash) feel unnatural to the human user. Actuators with soft nonlinearities (e.g., quadratic viscous drag) are fine as long as the human user is able to control and predict them.

One strategy to test ease of use is to study the closed-loop performance of the human and the mechatronic device together. Testing with the human operator is of course the gold standard; however, the challenge is that a device needs to be designed and functional before testing is possible. Oftentimes, modifying actuator parameters may require complete redesigns of hardware systems. To overcome these challenges and decrease the time required for design iterations, a flexible testing platform can be very helpful. For example, one group developed a prosthesis emulator that allows testing of a wide range of prosthesis parameters without the need to design multiple devices (Caputo and Collins, 2014).

Another strategy is to study human perception and control, and use that knowledge to choose system characteristics. Psychophysics methods can be used to quantify how precisely human users are able to predict movements with the biomechatronic system (Johnson et al., 2017). System identification methods can be used to quantify typical human body parameters such as mechanical joint impedance (Rouse et al., 2014), which can then be mimicked by biomechatronic actuators.

3 TYPES OF BIOMECHATRONIC ACTUATORS

There have been several helpful categorizations of actuators that drive biomechatronic systems. Hannaford and Winters categorized actuators by their effort-flow relationships into self-induction machines, slip-driven machines, linear effort-controlled machines, linear flow-controlled machines, and concave effort-flow machines (muscle-like machines) (Hannaford and Winters, 1990). Hollerbach et al. categorized actuators into macro-motion (electromagnetic, hydraulic, and pneumatic), micro-motion (piezoelectric and magnetostrictive), and muscle (nature's easily scalable actuator) (Hollerbach et al., 1991). In our discussion of actuators, we separate the technology used for the core actuator technologies (the motors) and that used for the structure of how the actuator interacts with the overall system (the transmissions). An example of an emerging core technology, or motor, is that of EPAs: materials that convert electrical energy to mechanical deformation. An example of novel transmission design is that of variable impedance strategies—many of which use traditional electric motors. In an attempt to minimize confusion and cleanly categorize technologies, we separate the motors and transmissions in the further sections.

3.1 Motors

The motor is the subsystem that converts one type of energy (electrical, fluidic, thermal, and chemical) to mechanical energy. In this section, we categorize motors according to the type of input energy.

3.1.1 Electromagnetic Actuators

Electromagnetic actuators take advantage of Lorentz's force law, which states that when a current-carrying conductor is moved in a magnetic field, a force is produced in a direction perpendicular to the current and magnetic field directions (Alciatore and Histand, 2003). The magnetic field may either be produced by permanent magnets, or by another energized coil.

There are many types of electromagnetic actuators, although some are used more than others in biomechatronic actuators. Solenoids and relays are simple devices with a stationary iron core and coil (Fig. 5A), and a movable armature core attached to the stationary core through a spring. These are rarely used for biomechatronic actuators because they cannot produce large forces or high frequencies. Voice coils are a similar concept that has a stationary iron core and permanent magnet, and a movable coil. Voice coils cannot produce large forces either, but can produce high frequencies, and are thus used for some biomechatronic actuators such as tactors (e.g., Schultz et al., 2009).

Electric motors have a stationary housing, called the stator, and a part that rotates, termed the rotor. In contrast to solenoids and voice coils, electric

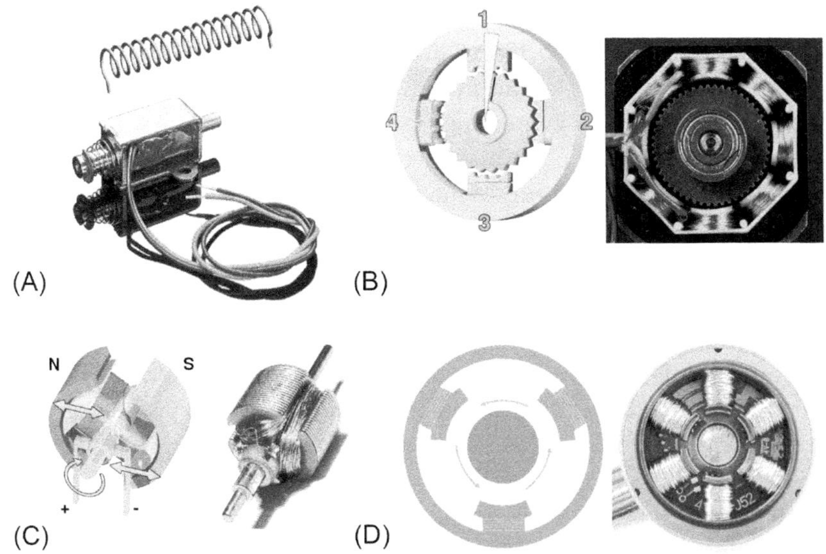

Fig. 5 Electromagnetic actuators: (A) illustration of solenoid coil, and photograph of packaged solenoid; (B) illustration of a stepper motor with four electromagnets, and photograph of stepper motor; (C) illustration of brushed DC motor, and photograph of DC motor rotor; and (D) illustration and photo of a brushless DC motor. *((A) and (B) Reproduced with permission from Adafruit (www.adafruit.com); (C) From Wikipedia; (D) From Wikimedia Commons.)*

motors can spin continuously. Electric motors are ubiquitous in our lives, and there are cascading levels through which they may be grouped. At the highest level, these include direct-current (DC) motors and alternating current (AC) motors. There are a variety of subsets for each (e.g., for AC there are single phase vs polyphaser vs universal, induction vs synchronous; squirrel cage vs wound rotor; for DC, there are permanent magnet, series wound, shunt wound, and compound wound) (Alciatore and Histand, 2003). The majority of biomechatronic actuators use permanent magnet DC motors, and there are three subsets worth exploring.

Stepper motors are a type of permanent magnet DC motor that can rotate in both directions, move in precise angular increments, sustain a holding torque at zero speed, and be controlled with digital circuits (Alciatore and Histand, 2003) (Fig. 5B). However, they typically cannot produce high torques, high speeds, or high frequencies, and are accordingly only used for a subset of biomechatronic actuators.

Brushed permanent-magnet motors use electrical brushes to switch the direction of current in the coils (Fig. 5C). The coils are located on the rotor, and the permanent magnets are located on the stator. Compared with many other motors, these motors have a high torque to weight ratio, because the field strength of permanent magnets is very high. The current that can be delivered to the coils is limited by sparking across the brushes; this in turn limits the torque that the motors can produce. Some brushed motors have hollow rotors (no iron core), termed coreless motors. These motors have reduced inertia (enabling greater acceleration), but can also produce reduced torque.

In contrast, brushless permanent-magnet motors use sensors such as Hall sensors to determine when to reverse the polarity of current across the coil (Fig. 5D). These actuators require more complicated and delicate circuitry that handles commutation of the phases, but can produce greater torque as there are no brushes. They are increasingly being preferred over brushed motors in biomechatronic applications for this reason. There are two variants–internal-rotor motors (which are more common) and exterior-rotor motors (which are common in applications like remote-control quad copters), and are increasingly being used in biomechatronic applications (Lenzi et al., 2016) for their superior torque-generating capabilities (Sensinger et al., 2011).

It is worth noting that the term "mechatronic" was coined (and trademarked) to describe the innovative concept of decoupling sensing from actuation in brushed permanent-magnet motors, resulting in a brushless motor.

3.1.2 Fluidic Actuators

Pneumatic artificial muscles convert pneumatic energy (pressurized gas) to mechanical motion and force. The most common pneumatic artificial muscles are called McKibben muscles, which are composed of tubes surrounded by woven threads (Fig. 6). When inflated with pressurized air, the tubes expand radially and contract axially, generating tensile forces. McKibben muscles were originally designed to mimic natural muscle function for prosthetic devices (Chou and Hannaford, 1994). For a review, see Daerden and Lefeber (2002).

A major advantage of McKibben muscles is that they have inherently variable compliance, depending on the pressure of the gas. They are lightweight and unaffected by magnetic fields, which makes them attractive choices for imaging applications (e.g., inside MRI machines).

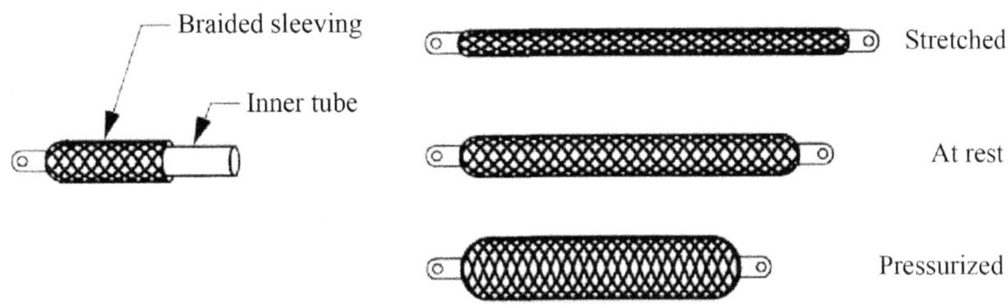

Fig. 6 McKibben muscles. *(Reproduced with permission from Daerden, F., Lefeber, D., 2002. Pneumatic artificial muscles: actuators for robotics and automation. Eur. J. Mech. Environ. Eng. 47 (1), 11–21.)*

3.1.3 Shape Memory Alloys

The term "smart materials" is often used to describe materials with inherent transduction behavior. These materials all change shape in response to applied energy, with different mechanisms governing the properties of each material: SMAs change shape when exposed to temperature or magnetic field changes, piezoelectric materials deform in response to an electric field (and vice versa), and magnetostrictive materials deform in response to magnetization (and vice versa). These transduction behaviors enable smart materials to be used for both actuation and sensing—sometimes simultaneously. Each class of materials has different advantages and disadvantages—and sometimes, combinations of materials provide the best blend of features.

SMAs earned their name from their ability to "remember" an original shape: when in a deformed state, they respond to thermal or magnetic stimuli by returning to their original shape. This shape memory effect is possible because SMAs have two stable solid phases with different crystal structures. The phase transformation is stimulated by temperature changes, which are typically achieved by applying electrical current (certain types of alloys also respond to magnetic fields). The phase transformation occurs even in the presence of heavy loading, which makes SMAs good candidates for actuators. For two excellent reviews of SMAs, see Mohd Jani et al. (2014) and (2017).

The advantages of SMAs include their high power-to-weight ratio, noiseless operation, biocompatibility, and the ability to form nearly any shape. Because of their inherent material properties, SMAs can be formed into three-dimensional actuators with unique shapes such as helical springs (Figs. 7 and 8 show two possible forms). The most common SMA, Nitinol, is often used in medical devices such as stents, catheters, and surgical tools.

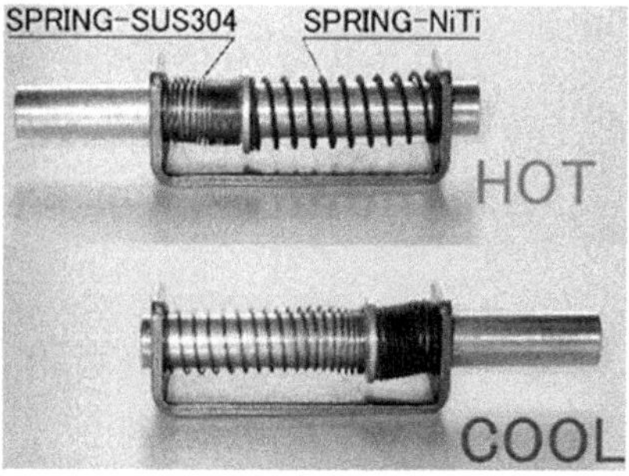

Fig. 7 Shape memory alloy actuators are often coupled with a bias element. In this example, there is a Nitinol (NiTi) spring on the right and stainless steel spring on the left. When Nitinol is heated, it returns to its original shape and deforms the stainless steel spring. *(Reproduced with permission from Nespoli, A., Besseghini, S., Pittaccio, S., Villa, E., Viscuso, S., 2010. The high potential of shape memory alloys in developing miniature mechanical devices: a review on shape memory alloy mini-actuators. Sens. Actuators A: Phys. 158(1), 149–160.)*

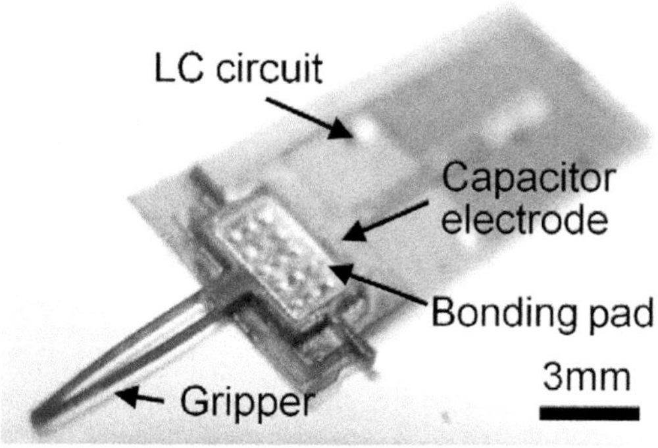

Fig. 8 Microgripper with shape memory alloy (SMA) actuator. *(Reproduced with permission from Mohamed Ali, M.S., Takahata, K., 2010. Frequency-controlled wireless shape-memory-alloy microactuators integrated using an electroplating bonding process. Sens. Actuators A: Phys. 163(1), 363–372.)*

The main disadvantage of SMAs is that bandwidth is limited and operational frequency is low, due to slow cooling processes. The phase transitions require both heating and cooling processes, and because most SMAs have high heat capacities, they heat up rapidly but cool down slowly. Much of the work on SMAs focuses on improving bandwidth, and one successful development is that of magnetic shape memory alloys (MSMAs), which have higher operating frequencies. Other strategies to improve bandwidth include changing the wire shape to improve heat dissipation, dividing wires into individually controllable segments (Selden et al., 2006), and active cooling. Other challenges of SMAs are low-energy efficiencies (10%–15%) and high costs.

3.1.4 *Electroactive Polymers*

EAPs are another category of smart materials that have been called "artificial muscles." Of all current actuator technologies, EAPs are the most functionally similar to natural muscle. EAPs are polymer materials that transduce electrical energy into mechanical energy (and vice versa). Similar to the above smart metal alloys, they have high power-to-weight ratios, but also have the benefits of lower costs, inherent compliance, and much larger strain capabilities. For a great introduction to EAPs for bioinspired applications, see Bar-Cohen (2001).

There are two categories of EAPs: ionic EAPs, which respond to ion flow, and electronic EAPs, which respond to electrostatic forces. Ionic EAPs require wet environments; so, electronic EAPs are generally more appropriate for biomechatronic applications. Of the electronic EAPs, a recently developed but highly promising type of EAP is the dielectric elastomer.

A dielectric elastomer is composed of two compliant electrodes that sandwich an insulative polymer film. When voltage is applied across the electrodes, electrostatic forces squeeze the dielectric film, causing a decrease in thickness and increase in area (Fig. 9). This dielectric behavior enables capabilities that approach that of natural muscle: strains of 10%–100% and stress levels of 0.1–9 MPa (Carpi et al., 2008). The advantages of dielectric elastomers are many: they are compliant, lightweight, inexpensive, quiet, and have high power densities. The major disadvantage is that they must be prestrained to reach full performance, which requires mechanisms that increase weight and packaging size. However, dielectric elastomers are in the early stages of commercialization, and show promise for further improvements.

3.2 Transmissions

A motor outputs mechanical energy to the overall system through some coupling or transmission. The transmission may be as simple as a clamp that connects a motor shaft to the load, or as complex as a system of variable springs and dampers. The transmission is often designed to have some dynamic behavior that improves the overall system safety or performance.

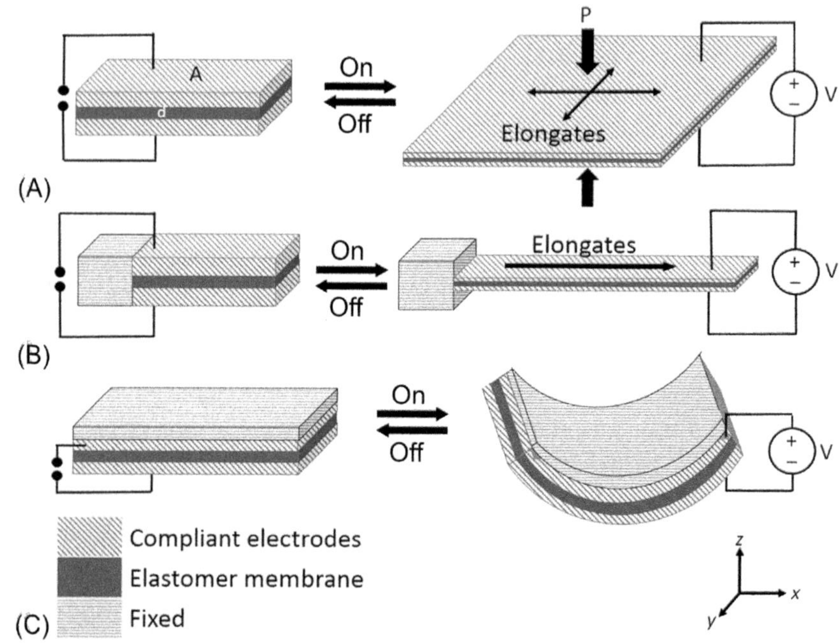

Fig. 9 Illustrations of dielectric elastomer configurations: (A) two-degree-of-freedom planar actuation, (B) one-degree-of-freedom planar actuation, and (C) bend actuation. *(Reproduced with permission from Mohd Ghazali, F.A., Mah, C.K., AbuZaiter, A., Chee, P.S., Mohamed Ali, M.S., 2017. Soft dielectric elastomer actuator micropump. Sensors Actuators A Phys. 263, 276–284.)*

Several terms should be defined when discussing transmissions. A transmission scales forces and velocities, and this scaling ratio is typically defined by a gear ratio stated as N:1. It is common to say that forces are reflected from one side of the transmission to the other, implying that the reflected force is scaled by the transmission ratio. Many transmissions have static friction (commonly called stiction) that affects their efficiency and controllability. Many transmissions also have backlash, in which over a brief range of movement, movement of the input of the transmission does not generate movement of the output of the transmission.

Many transmissions improve one attribute at the expense of others. For example, compared with the range of forces and motion exerted by humans, most motors produce too much speed but not enough force. Thus, the most common reason to include a transmission is to convert the range of forces and speeds the motor can produce to the range of forces and speeds used by humans. However, many of these transmissions decrease efficiency and substantially increase weight (transmissions often weigh twice as much as the motor). In addition, they have a disproportionate effect on reflected inertia: whereas the force is scaled up by N:1 and the speed is scaled down by N:1, the inertia of the motor is scaled up by N^2:1. Thus, a motor with a relatively insignificant inertia can end up having a substantial inertia if coupled to a transmission with a 1000:1 ratio, which is not uncommon. Finally, many transmissions introduce hard[1] nonlinearities that are perceptible to humans and difficult to control, including static friction and backlash. It is important to choose a transmission that achieves the desired goals while introducing a minimum of these undesirable attributes.

As an aside, designers typically focus on the maximum force a transmission can produce. However, many transmissions have a maximum speed as well (typically necessitated by the bearings in the design), and this limit should not be overlooked.

3.2.1 Linear Transmissions

Linear transmissions convert the rotation of the motor to a linear output. This linear output may either be used to produce linear motion, or coupled to a linkage to produce a rotary motion. Linear transmissions typically have lower output inertia than rotary counterparts with comparable specifications, and can often withstand higher loads. However, they often take up a considerable amount of space, and they typically introduce soft nonlinearities into the transmission ratio across the range of motion. Many biomechatronic designs use linear transmissions, particularly in powered prostheses and orthoses.

The simplest form of a linear transmission is a lead screw, which is composed of a threaded rod and a nut (Fig. 10A). The input of the transmission (coupled to the motor) is the threaded rod. The output of the transmission is the nut. The nut is prevented from rotating by guiding rods or other structural components, such that when the threaded rod is rotated, the nut moves along the threaded rod. Compared with a standard screw, a lead screw's tooth profile is engineered to be more efficient and withstand greater loads. Most lead screws have noticeable static friction, and are inherently non-backdrivable, which is often undesirable in biomechatronic applications, but can be useful for things like powered upper limb prostheses. Ball screws

[1] A hard nonlinearity has undefined derivatives. Examples include backlash and static friction. In contrast, soft nonlinearities have defined derivatives. Examples include quadratic viscous drag, or the sinusoidal effects of gravity on linkages.

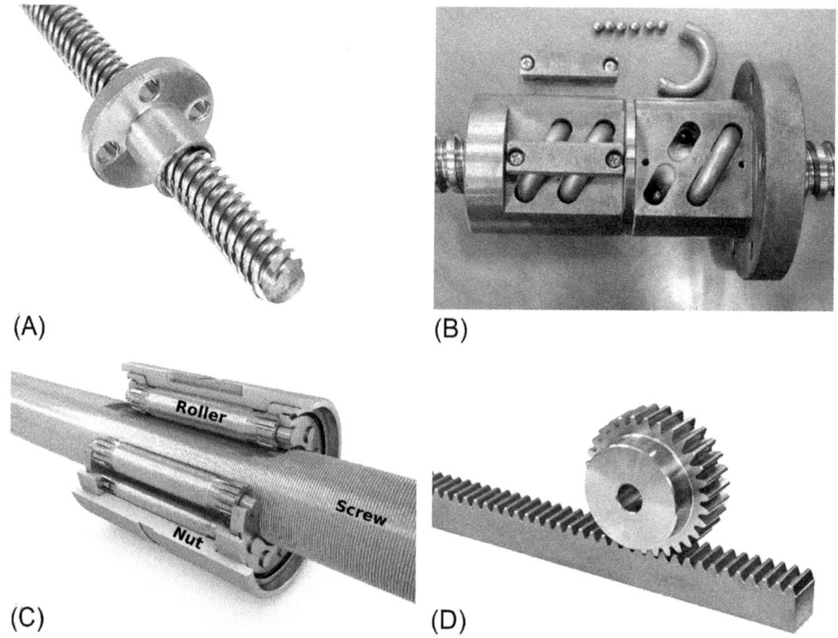

Fig. 10 Linear transmissions: (A) lead screw, (B) ball screw, (C) roller screw, and (D) rack and pinion. *((B) Photograph by Glenn McKechnie; (C) Reproduced with permission from Sandu, S., Biboulet, N., Nelias, D., Abevi, F., 2018. An efficient method for analyzing the roller screw thread geometry. Mech. Mach. Theory 126, 243–264.)*

are conceptually similar to lead screws, but the threaded rod has grooves, and the nut has balls in it that rotate in the channels of the screw (Fig. 10B). Thus, ball screws roll rather than slide, and accordingly have substantially better efficiency than lead screws. This efficiency, however, comes with a more complicated design; greater cost; and often with greater noise and backlash. In addition, the contact between the balls and the grooved rod are point-contacts vs the line contacts of lead screws, which substantially decreases the forces that can be handled.

Differential roller screws combine the best attributes of lead screws and ball screws in that they can withstand high loads and yet have high efficiency (Fig. 10C). However, their manufacture requires high precision and is accordingly expensive. They are used in some prosthetic devices (e.g., Lenzi et al., 2016).

Designs are occasionally seen with a rack and pinion design (Fig. 10D), although other transmissions typically have a broader range of desirable attributes.

3.2.2 Rotary Transmissions

Rotary transmissions are typically concentric with the motor, although in some designs they are not. Rotary transmissions tend to be more compact than linear transmissions, but often have greater reflected inertia.

The simplest form of a rotary transmission is simply a train of spur gears linked to each other (Fig. 11A). Helical gears are similar but do not have

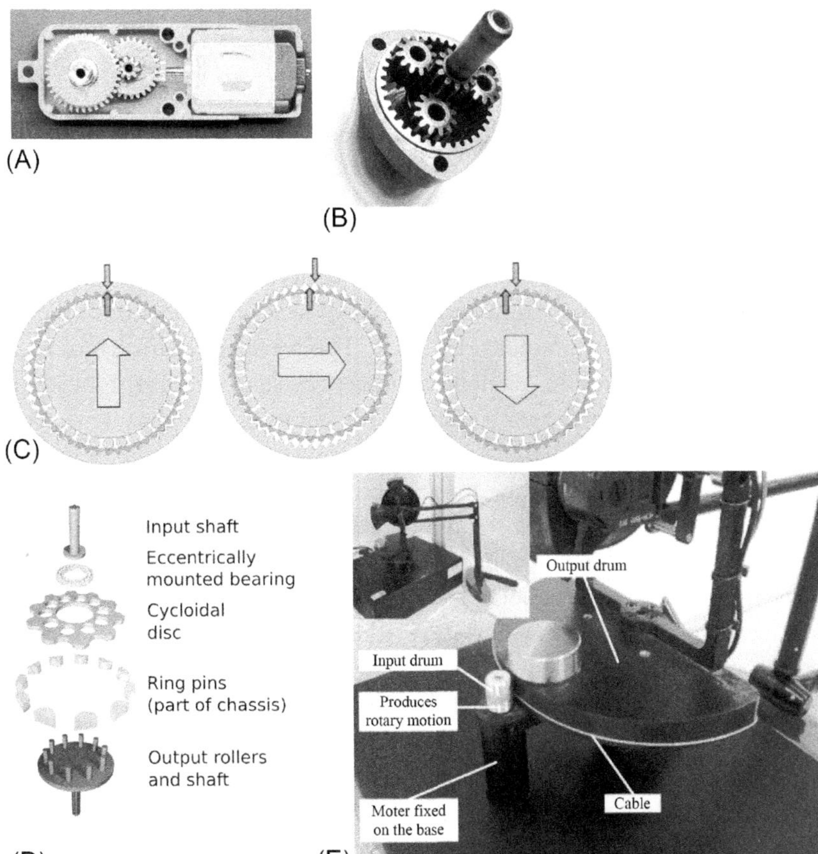

Fig. 11 Rotary transmissions: (A) spur gear train, driven by a DC motor, (B) planetary gear train, (C) diagram showing the basic operation of harmonic drives, (D) cycloid drive, and (E) Capstan drive on a Phantom haptic device from Geomagic, Inc. (formerly SensAble Technologies Corp.). *((A) www.adafruit.com; (C) Reproduced with permission from Tjahjowidodo, T., Al-Bender, F., Van Brussel, H., 2013. Theoretical modelling and experimental identification of nonlinear torsional behaviour in harmonic drives. Mechatronics 23 (5), 497–504; (D) Illustration by Petteri Aimonen; (E) Reproduced with permission from Baser, O., Ilhan Konukseven, E., 2010. Theoretical and experimental determination of capstan drive slip error. Mech. Mach. Theory 45 (6), 815–827.)*

teeth that are parallel to the axis of rotation, enabling them to run more smoothly and quietly, but at the cost of introducing thrust and decreasing efficiency. Bevel and worm gears are sometimes used in biomechatronic applications, and hypoid gears, in which the shaft axes do not intersect, have recently been considered by some groups.

Planetary gears are similar to spur gears, but use a set of three or more planets that revolve around a central sun gear (Fig. 11B). They are also in contact with a fixed annulus. Compared with spur gears, planetary gears have a higher strength-to-weight ratio, and multiple stages can be stacked within the same annulus. Some groups have used friction planetary gears, in which none of the components are toothed. Friction planetary gears have great potential as they do not have static friction or backlash, but they require pretensioning that reduces the lifespan of the components. They have been considered for use in some biomechatronic actuators (e.g., Sup et al., 2008).

Many high-torque biomechatronic actuators use harmonic drives, in which an elliptical input cam deforms a flexible wave generator, which is in contact with a rigid annulus (Fig. 11C). The wave generator has two less teeth than the rigid annulus, such that for every cycle of rotation of the input, the wave generator shifts two teeth with respect to the annulus. Thus, very high gear ratios may be achieved in a compact package. Examples include the LTI Boston Elbow and Sensinger and Weir (2008). Harmonic drives are often favored in robotics because they do not have backlash. However, they have substantial inertia, since they require a large, heavy, elliptical cam on the input side of the gear ratio (which is then reflected by the square of the large gear ratio). They also introduce torque ripple, due to the elliptical nature of the input cam.

Cycloid drives (Sensinger, 2010b) are topologically equivalent to harmonic drives, but use an offset input rather than an elliptical input (Sensinger, 2013) (Fig. 11D). They are less common in biomechatronic applications, although easier to manufacture (Lenzi et al., 2016). Cycloid drives have less reflected inertia than harmonic drives, and better efficiency (since they can operate on rolling, vs sliding contact). However, they have backlash, and their gear ratio fluctuates depending on the position of the input shaft (Sensinger and Lipsey, 2012) (Of historical note, the involute tooth profile used in most modern gear teeth was invented to achieve a constant gear ratio, in contrast to the cycloidal tooth profile previously used).

3.2.3 Other Transmissions

Some biomechatronic actuators use differential gear transmissions, including wolfram transmissions, Ikona transmissions, and differential cycloids.

However, the differential typology typically comes at the cost of substantial reduction in efficiency (Sensinger, 2013).

For biomechatronic actuators that do not need large transmission ratios, but that cannot afford backlash or stiction, capstan transmissions are often the design of choice (e.g., Fig. 11E and Brown et al., 2012). These designs couple the motion of one pulley to that of another through pinned cabling. They can produce high forces even in the absence of pretension between the cables because the cables are pinned. They can only be used for small gear ratios and for limited range of motion.

Some biomechatronic actuators use chains or belts and pulleys for their transmission (e.g., Lawson et al., 2014), although for most designs these are not compact enough and introduce substantial vibration at higher speeds.

Some biomechatronic actuators use linkages themselves–particularly in parallel configurations, as a form of transmissions. Manipulandums, commonly used in assessing human motor control, are one such example.

3.2.4 Variable and Low Impedance

Variable and low-impedance actuators are increasingly important for both industrial and research applications. They enable safer interaction with humans and a more robust interaction with unknown environments.

The mechanical impedance of an object refers to the ratio of force an object exerts relative to the frequency-dependent displacement of the object. Impedance is the generalization of related concepts including stiffness, viscosity, and inertia. Ratios that are constant across frequencies (and thus only depend on displacement) can be expressed solely as stiffness. Ratios that rise at 20 dB/decade can be expressed solely as viscosity. Ratios that rise at 40 dB/decade can be expressed solely as inertias. Objects that have multiple springs and inertias will typically "look" like a spring, or an inertia, in various ranges of the frequency spectrum (Fig. 4).

Motors typically have a large output impedance that looks like an inertia at high frequencies. This large output impedance is often caused by the reflected inertia generated by a transmission, and leads to high forces during impacts (high frequencies), which in turn can cause damage either to the mechanism or the person. However, if a spring or damper is placed at the end of the actuator, it acts as the "weakest link," saturating the impedance seen at the output.

The intentional introduction of compliance within an otherwise-rigid electromechanical actuator has conventionally been avoided, because it reduces high-frequency/high-magnitude force generation and creates the potential for sensor/actuator de-colocation in some control strategies.

However, for many applications these disadvantages are outweighed by the advantages, which include improved force rendering, improved force-sensing fidelity, larger stable feedback gains (Whitney, 1977), improved power densities using a catapult effect (Albu-Schäffer et al., 2011), and saturation of maximum impedance at high frequencies (Pratt and Williamson, 1995). Series elastic actuators reduce environmental impact forces for actuators without substantial endpoint inertia (Zinn et al., 2004) (paragraph excerpt from Sensinger et al., 2013).

One of the most common and oldest types of inherently low-impedance actuators is the series elastic actuator (Pratt and Williamson, 1995) (Fig. 12). Newer strategies include the distributed macro-mini (DM^2) approach, which attempts to solve the trade-off between low impedance and high bandwidth by combining two compliant mechanism strategies (Zinn et al., 2004).

Variable impedance actuators take advantage of the benefits of both stiff and compliant actuators (Walker and Niemeyer, 2010). Impedance can be varied by changing the effective stiffness, damping, and inertia, or by active control methods. For a collaborative review and classification of different types of variable-impedance actuators, see Vanderborght et al. (2013).

Variable-stiffness actuators are the most common subset of variable-impedance actuators. Stiffness may be varied by changing lever lengths, adjusting spring preloads, changing spring lengths, or by using a continuously variable transmission. For a review of variable-stiffness actuators and their design processes, see Van Ham et al. (2009) and Wolf et al. (2016).

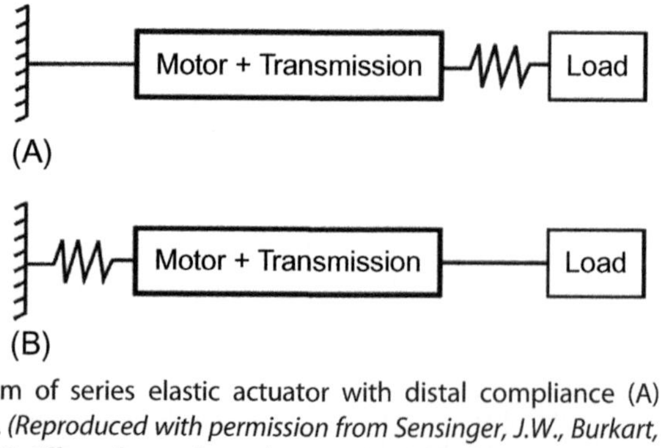

Fig. 12 Diagram of series elastic actuator with distal compliance (A) and proximal compliance (B). *(Reproduced with permission from Sensinger, J.W., Burkart, L.E., Pratt, G.A., ff Weir, R.F., 2013. Effect of compliance location in series elastic actuators. Robotica 31 (8), 1313–1318.)*

4 PURPOSES OF BIOMECHATRONIC ACTUATORS

4.1 Biological Function Replacement

One purpose for biomechatronic systems is to replace a missing function or component of the body. There are high expectations for these systems, because the body's intact actuators—muscles—are strong, lightweight, flexible, and closely integrated with the neural control system. There is not yet any artificial actuator that approaches the capabilities of natural muscle. Furthermore, the body's natural communication systems are missing, and recognition of the user's movement intentions remains a huge challenge.

Historically, the actuators used to drive prostheses were body-powered, meaning the user acts as the motor and inputs power to a transmission. For upper-limb amputees, body-powered prostheses typically feature a Bowden cable transmission (Weir and Sensinger, 2009). For lower-limb amputees, passive spring-based prostheses offer energy storage and return, while linkage-based prostheses such as polycentric knees and multi-axis feet offer stability over uneven terrain. An advantage of body-powered actuators is that because the user acts as the motor and transmits force through a mechanical transmission, they receive direct sensory feedback—a phenomenon called extended physiological proprioception (Doubler and Childress, 1984). Because of this direct sensory feedback, low cost, and high durability, body-powered prostheses remain a commonly used type of prosthesis.

For powered upper-limb prostheses, size and weight are especially critical for the user's comfort (Biddiss et al., 2007). Most commercially available prosthetic hands are actuated by DC motors with geared transmissions such as worm gears and lead screws (Belter et al., 2013). To decrease weight and size further, researchers are designing custom exterior rotor motors, harmonic drives, and cycloid drives (Lenzi et al., 2016). There are also research efforts focusing on compliant and underactuated devices.

The main advantage of powered lower-limb prostheses is that providing net positive power enables more functionality, such as sit-to-stand movements and stair climbing. The actuators of lower-limb prostheses must sustain body-weight loading, and typically feature conventional electric motors with gear train transmissions. For reviews on the actuators of lower-limb prostheses, see Pieringer et al. (2017) and Windrich et al. (2016).

Other applications of biological function replacement are less focused on large-scale movements but still include actuator technologies. For example, biomechatronic devices are being developed to replace organs or tissues, such as heart valves.

4.2 Biological Function Augmentation

Another purpose of biomechatronic systems is to augment the human body in some way. The goal may be to regain function diminished by motor disorders (e.g., stroke or traumatic brain injury), or to amplify typical human function for people in demanding environments.

Orthoses (also called exoskeletons) may be used as either stationary or wearable devices. Stationary devices such as the Lokomat focus on rehabilitation. They are typically very large and stable, with traditional electric motors as actuators. Wearable orthoses are used for a variety of applications ranging from assistance for people with motor disorders, to support for soldiers traveling long distances with heavy loads. The actuators vary widely, especially in research systems. Many use compliant or variable-impedance actuators. For a review of actuators for orthoses, see Veale and Xie (2016).

There are many other applications of biological-function augmentation. For example, surgical tools and medical devices augment the capabilities of physicians, and haptic interfaces enable people to interact with virtual, small-scale, remote, or dangerous environments.

5 CONCLUSION

The capabilities of biomechatronic actuators have been increasing rapidly due to a number of factors. Traditional actuator technologies such as electric motors have been decreasing in size and weight (their power supplies, typically batteries, have also been shrinking). Newer actuator technologies such as SMAs and dielectric elastomers are moving out of research labs and into commercial applications. These improvements enable closer integration with humans across a broad range of applications. However, communicating intentions from the human to the machine remains a significant challenge in many systems.

This chapter introduced the design goals, categories, and applications of biomechatronic actuators. The applications of biomechatronic actuators range widely, from microfluidic implantable devices to industrial robots that interact with people. Because of the wide range of applications, we did not provide specific quantitative guidelines for designing biomechatronic actuators but recommended several important factors to consider. For further information, we suggest reading the several excellent reviews on more specific types of biomechatronic actuators, referenced below.

REFERENCES

Albu-Schäffer, A., Eiberger, O., Fuchs, M., Grebenstein, M., Haddadin, S., Ott, C., Stemmer, A., Wimböck, T., Wolf, S., Borst, C., Hirzinger, G., 2011. Soft robotics: from torque feedback controlled lightweight robots to intrinsically compliant systems. Int. J. Robot. Res. 70, 185–207.

Alciatore, D.G., Histand, M.B., 2003. Introduction to Mechatronics and Measurement Systems, second ed. McGraw-Hill, Boston, MA.

B. S. Publication, 2016. IEC TS 60479-1 Consolidated.

Bar-Cohen, Y., 2001. In: Electroactive polymers as artificial muscles—reality and challenges. Proceedings of the 42nd AIAA Structures, Structural Dynamics, and Materials Conference (SDM).

Belter, J.T., Segil, J.L., Dollar, A.M., Weir, R.F., 2013. Mechanical design and performance specifications of anthropomorphic prosthetic hands: a review. J. Rehabil. Res. Dev. 50 (5), 599.

Biddiss, E.A., Beaton, D., Chau, T., 2007. Consumer design priorities for upper limb prosthetics. Disabil. Rehabil. Assist. Technol. 2 (6), 346–357.

Brown, J.D., Gillespie, R.B., Gardner, D., Gansallo, E.A., 2012. In: Co-location of force and action improves identification of force-displacement features.IEEE Haptics Symposium, pp. 187–193.

Caputo, J.M., Collins, S.H., 2014. A universal ankle–foot prosthesis emulator for human locomotion experiments. J. Biomech. Eng. 136 (3), 35002.

Carpi, F., De Rossi, D., Kornbluh, R., Pelrine, R., Sommer-Larsen, P., 2008. Dielectric Elastomers as Electromechanical Transducers: Fundamentals, Materials, Devices, Models and Applications of an Emerging Electroactive Polymer Technology, Elsevier.

Chou, C., Hannaford, B., 1994. In: Static and dynamic characteristics of McKibben pneumatic artificial muscles.Robot. Autom. 1994. vol. 3, pp. 281–286.

Daerden, F., Lefeber, D., 2002. Pneumatic artificial muscles: actuators for robotics and automation. Eur. J. Mech. Environ. Eng. 47 (1), 11–21.

Doubler, J., Childress, D., 1984. An analysis of extended physiological proprioception as a prosthesis-control technique. J. Rehabil. Res. Dev. 21 (1), 5–18.

Fish, R.M., Geddes, L.A., 2009. Conduction of electrical current to and through the human body: a review. Eplasty 9, e44.

Gao, D., Wampler, C.W., 2009. Head injury criterion (HIC). IEEE Robot. Autom. Mag. 16 (4), 71–74.

Hannaford, B., Winters, J., 1990. Actuator properties and movement control: biological and technological models. In: Multiple Muscle Systems: Biomechanics and Movement Organization. Springer-Verlag, New York.

Hollerbach, J.M., Hunter, I., Ballantyne, J., 1991. A comparative analysis of actuator technologies for robotics. Robotics Rev. 2, 299–342.

ISO, 2011. Robots and robotic devices—Safety requirements for industrial robots—Part 1: Robots. International Standards Organization. ISO 10218-1:2011.

ISO, 2014. Robots and robotic devices—Safety requirements for personal care robots. International Standards Organization. ISO 13482:2014.

ISO, 2017a. Medical electrical equipment—Part 2-77: Particular requirements for the basic safety and essential performance of robotically assisted surgical equipment. International Standards Organization. IEC/DIS 80601-2-77.

ISO, 2017b. Medical electrical equipment—Part 2-78: Particular requirements for the basic safety and essential performance of medical robots for rehabilitation. International Standards Organization. IEC/DIS 80601-2-78.

Johannes, M.S., Bigelow, J.D., Burck, J.M., Harshbarger, S.D., Kozlowski, M.V., 2011. An overview of the developmental process for the modular prosthetic limb. Johns Hopkins APL Tech. Dig. 30 (3), 207–216.

Johnson, R.E., Kording, K.P., Hargrove, L.J., Sensinger, J.W., 2017. EMG versus torque control of human-machine systems: equalizing control signal variability does not equalize error or uncertainty. IEEE Trans. Neural Syst. Rehabil. Eng. 25 (6), 660–667.

Lawson, B.E., Mitchell, J., Truex, D., Shultz, A., Ledoux, E., Goldfarb, M., 2014. A robotic leg prosthesis: design, control, and implementation. IEEE Robot. Autom. Mag. 21 (4), 70–81.

Lenzi, T., Lipsey, J., Sensinger, J.W., 2016. The RIC arm—a small, anthropomorphic transhumeral prosthesis. IEEE/ASME Trans. Mechatron. 21 (6), 2660–2671.

Mohd Jani, J., Leary, M., Subic, A., Gibson, M.A., 2014. A review of shape memory alloy research, applications and opportunities. Mater. Des. 56, 1078–1113.

Mohd Jani, J., Leary, M., Subic, A., 2017. Designing shape memory alloy linear actuators: a review. J. Intell. Mater. Syst. Struct. 28 (13), 1699–1718.

Murashov, V., Hearla, F., Howard, J., 2016. Working safely with robot workers: recommendations for the new workplace. J. Occup. Environ. Hyg. 13(3).

Pieringer, D.S., Grimmer, M., Russold, M.F., Riener, R., 2017. In: Review of the actuators of active knee prostheses and their target design outputs for activities of daily living. 2017 Int. Conf. Rehabil. Robot., no. July, pp. 1246–1253.

Pratt, G.A., Williamson, M.M., 1995. Series elastic actuators. In: Proceedings 1995 IEEE/RSJ International Conference on Intelligent Robots and Systems. Human Robot Interaction and Cooperative Robots, vol. 1, pp. 399–406.

Rouse, E.J., Hargrove, L.J., Perreault, E.J., Kuiken, T.A., 2014. Estimation of human ankle impedance during the stance phase of walking. IEEE Trans. Neural Syst. Rehabil. Eng. 22 (4), 870–878.

Schultz, A.E., Marasco, P.D., Kuiken, T.A., 2009. Vibrotactile detection thresholds for chest skin of amputees following targeted reinnervation surgery. Brain Res. 1251, 121–129.

Selden, B., Cho, K., Asada, H.H., 2006. Segmented shape memory alloy actuators using hysteresis loop control. Smart Mater. Struct. 15 (2), 642–652.

Sensinger, J.W., 2010a. In: Selecting motors for robots using biomimetic trajectories: optimum benchmarks, windings, and other considerations.IEEE Conference on Robotics and Automation. Anchorage, Alaska, pp. 4175–4181.

Sensinger, J.W., 2010b. Unified approach to cycloid drive profile, stress, and efficiency optimization. ASME J. Mech. Des. 132 (2), 1–5.

Sensinger, J.W., 2013. Efficiency of high-sensitivity gear trains, such as cycloid drives. J. Mech. Des. 135 (7), 71006.

Sensinger, J.W., Lipsey, J.H., 2012. In: Cycloid vs. harmonic drives for use in high ratio, single stage robotic transmissions. IEEE Conference on Robotics and Automation, pp. 4130–4135.

Sensinger, J.W., Weir, R.E.F., 2008. User-modulated impedance control of a prosthetic elbow in unconstrained, perturbed motion. IEEE Trans. Biomed. Eng. 55 (3), 1043–1055.

Sensinger, J.W., Clark, S.D., Schorsch, J.F., 2011. In: Exterior vs. interior rotors in robotic brushless motors. IEEE Conference on Robotics and Automation. Shanghai, China, pp. 2764–2770.

Sensinger, J.W., Burkart, L.E., Pratt, G.A., Weir, R.F.f., 2013. Effect of compliance location in series elastic actuators. Robotica 31 (8), 1313–1318.

Sup, F., Bohara, A., Goldfarb, M., 2008. Design and control of a powered transfemoral prosthesis. Int. J. Robot. Res. 27 (2), 263–273.

Van Ham, R., Sugar, T.G., Vanderborght, B., Hollander, K.W., Lefeber, D., 2009. Compliant actuator designs: review of actuators with passive adjustable compliance/controllable stiffness for robotic applications. IEEE Robot. Autom. Mag. 16 (3), 81–94.

Vanderborght, B., Albu-Schaeffer, A., Bicchi, A., Burdet, E., Caldwell, D.G., Carloni, R., Catalano, M., Eiberger, O., Friedl, W., Ganesh, G., Garabini, M., Grebenstein, M.,

Grioli, G., Haddadin, S., Hoppner, H., Jafari, A., Laffranchi, M., Lefeber, D., Petit, F., Stramigioli, S., Tsagarakis, N., Van Damme, M., Van Ham, R., Visser, L.C., Wolf, S., 2013. Variable impedance actuators: a review. Robot. Auton. Syst. 61 (12), 1601–1614.
Veale, A.J., Xie, S.Q., 2016. Towards compliant and wearable robotic orthoses: a review of current and emerging actuator technologies. Med. Eng. Phys. 38 (4), 317–325.
Walker, D.S., Niemeyer, G., 2010. In: Examining the benefits of variable impedance actuation. IEEE/RSJ 2010 Int. Conf. Intell. Robot. Syst. IROS 2010—Conf. Proc, pp. 4855–4861.
Weir, D.W., Colgate, J.E., 2008. Stability of haptic displays. In: Haptic Rendering: Foundations, Algorithms, and Applications. A K Peters/CRC Press, pp. 151–189.
Weir, R.F.f., Sensinger, J.W., 2009. The design of artificial arms and hands for prosthetic applications. In: Kutz, M. (Ed.), second ed. In: Biomedical Engineering and Design Handbook, vol. 2. McGraw-Hill, New York, NY, pp. 537–598.
Whitney, D.E., 1977. Force Feedback Control of Manipulator Fine Motions. J. Dyn. Syst. Meas. Control. 99(2).
Windrich, M., Grimmer, M., Christ, O., Rinderknecht, S., Beckerle, P., 2016. Active lower limb prosthetics: a systematic review of design issues and solutions. Biomed. Eng. Online 15 (S3), 140.
Wolf, S., Grioli, G., Eiberger, O., Friedl, W., Grebenstein, M., Vanderborght, B., Visser, L., Bicchi, A., Albu-Schaffer, A., 2016. Variable stiffness actuators: review on design and components. IEEE/ASME Trans. Mechatron. 21 (5), 2418–2430.
Zinn, M., Khatib, O., Roth, B., Salisbury, J.K., 2004. Playing it safe. IEEE Robot. Autom. Mag. 11 (2), 12–21.

FURTHER READING

Baser, O., Ilhan Konukseven, E., 2010. Theoretical and experimental determination of capstan drive slip error. Mech. Mach. Theory 45 (6), 815–827.
Mohamed Ali, M.S., Takahata, K., 2010. Frequency-controlled wireless shape-memory-alloy microactuators integrated using an electroplating bonding process. Sens. Actuators A: Phys. 163 (1), 363–372.
Mohd Ghazali, F.A., Mah, C.K., AbuZaiter, A., Chee, P.S., Mohamed Ali, M.S., 2017. Soft dielectric elastomer actuator micropump. Sensors Actuators A Phys. 263, 276–284.
Nespoli, A., Besseghini, S., Pittaccio, S., Villa, E., Viscuso, S., 2010. The high potential of shape memory alloys in developing miniature mechanical devices: a review on shape memory alloy mini-actuators. Sens. Actuators A: Phys. 158 (1), 149–160.
Sandu, S., Biboulet, N., Nelias, D., Abevi, F., 2018. An efficient method for analyzing the roller screw thread geometry. Mech. Mach. Theory 126, 243–264.
Sun, L., Huang, W.M., Ding, Z., Zhao, Y., Wang, C.C., Purnawali, H., Tang, C., 2012. Stimulus-responsive shape memory materials: a review. Mater. Des. 33 (1), 577–640.
Tjahjowidodo, T., Al-Bender, F., Van Brussel, H., 2013. Theoretical modelling and experimental identification of nonlinear torsional behaviour in harmonic drives. Mechatronics 23 (5), 497–504.

CHAPTER THREE

Sensors and Transducers

Jeff Christenson
Research and Development, Motion Control, Salt Lake City, UT, United States

Contents

Handbook of Biomechatronics
https://doi.org/10.1016/B978-0-12-812539-7.00003-9

1 INTRODUCTION

An important element of a biomechatronic system is the method in which the device determines what is happening in the surrounding environment. By monitoring the environment, systems can be built which can enable, improve, and enhance the user's experience. To make this determination, sensors and transducers are used. A sensor is any "device which detects or measures a physical property and records, indicates, or otherwise responds to it" (https://en.oxforddictionaries.com/definition/sensor, Accessed 21 August 2017). A transducer is a "device that converts variations in a physical quantity… into an electrical signal, or vice versa" (https://en.oxforddictionaries.com/definition/transducer, Accessed 21 August 2017).

Sensors are critical elements of any biomechatronic device, since they allow systems to be built which respond to biological input. An electric prosthesis with no user input, such as control cables or sensors of muscle signals, becomes an ill-formed tennis racket. A load cell with no sensor to measure the load becomes a paperweight. A brain array with no brain activity sensing capability becomes an expensive surgery with no beneficial outcome. Biomechatronic devices need sensors to be useful devices.

The process of selecting what type of sensor to use is not a trivial matter and requires careful consideration of form, function, and environment. When discussing sensors, there is some general terminology often used to define a sensor's performance. These terms are: accuracy, precision, resolution, range, and hysteresis (Bolton, 2003a).

Accuracy refers to how close a sensor measures a defined standard (Bolton, 2003a). For instance, the accuracy of a ruler can be determined by measuring a block which conforms to a standard dimension and comparing the results of the ruler with the standard.

Precision refers to the density of repeat measurements (Bolton, 2003a). Tolerance gives a numeric value to the type of spread which can be expected from the sensor. Consider the ruler and standard block example. A ruler with high precision will give very close to the same number for each repeat measurement, even if the numbers are not accurate.

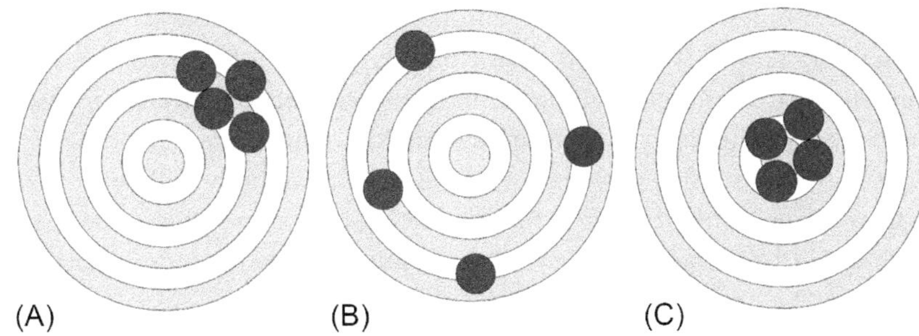

Fig. 1 Dartboard example. (A) shows lows accuracy, high precision. (B) shows high accuracy, low precision. (C) shows high accuracy and high precision.

Accuracy and precision are often explained using a dartboard (Fig. 1). A low accuracy, high precision thrower will have tight spread, but not around the bullseye. A high accuracy, low precision thrower will have a wide spread, but the spread will be centered around the bullseye. A high accuracy, high precision thrower will have a tight spread around the bullseye.

Resolution refers to how small the sensor can measure (Bolton, 2003a). A ruler with gradations every inch has low resolution compared to the one with gradations every 1/16 of an inch. Range refers to the useful spectrum of measurement of the sensor (Bolton, 2003a). A yard stick has a range of 3 ft.

Hysteresis is a characteristic most often observed in electrically powered sensors and is the separation of the signal when testing the full range of the sensor (Bolton, 2003a). The path of the sensor is different going from the top of the range to the bottom than starting at the bottom and going to the top. Fig. 2 illustrates the hysteresis of an electrically powered sensor.

This chapter is separated into four sections: passive sensors, standard sensor elements, common sensors, and biological sensors. In the passive sensors

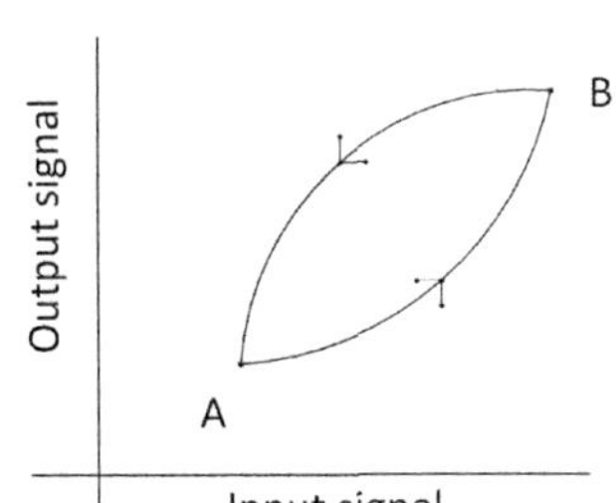

Fig. 2 Example of hysteresis of a sensor. As the sensor progresses from point A to point B, the sensor tracks a different path than when going from point B to point A.

section, examples of nonelectrical sensors are discussed. In the standard sensor elements section, some of the basic building blocks of many of the common sensors are described. In the common sensors section, electrical sensors which are found in many devices are described. In the biological sensors section, systems for sensing various outputs of the human body will be detailed. These sections are not meant to be complete lists of all possible sensors, but are designed to be an introduction to possible sensors for biomechatronic systems.

As each sensor is discussed, examples of biomechatronic devices will be presented which implement said sensors. To facilitate this discussion, let us suppose there is a theoretical person, named Jacob, who had his leg amputated above the knee about 2 years ago. He has been wearing a passive prosthetic knee and foot system about a year, but has experienced some discomfort and hopes you can provide him a better biomechatronic device.

2 PASSIVE SENSORS

Often when designing a biomechanical system, the default inclination is to implement an electrical sensor. It is important to consider all options in the design processes, even those that may appear to be less technical. There are several pure mechanical sensor options which can be successfully implemented to result in a cheaper and often more reliable device. However, these devices are often bulkier and may require more attention by the user.

2.1 Ruler

A ruler (Fig. 3) is used to measure distance (https://en.oxforddictionaries.com/definition/ruler, Accessed 21 August 2017). Through visual inspection of the indicators on the ruler, the distance can be determined. For long distances, tape measures are used. Rulers are readily available in Standard and Metric gradations, or both.

Before designing Jacob a new prosthetic system, you decide to evaluate his current device to determine if it might be adjusted better to fit Jacob's needs.

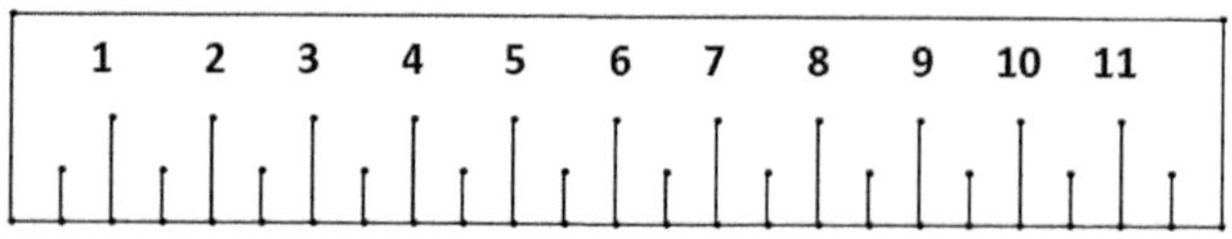

Fig. 3 Example of a ruler.

The first thing you do with Jacob is to measure many of his body's dimensions, specifically the difference between the sound side of his body and side with the amputation. A ruler is a convenient tool for such a task.

2.2 Protractor

A protractor (Fig. 4) is used to measure angles (https://en.oxforddictionaries.com/definition/protractor, Accessed 21 August 2017). By aligning the center point and the 0 axis of the protractor with the center of rotation and the reference plane, an angle can be measured through visual inspection of the angle indicators.

You measure the angle of alignment between Jacob's prosthetic foot and his prosthetic knee using a protractor. You align the central axis of the foot/ankle system with the axis of the protractor and hold the edge of the protractor horizontal to the ground. Then, you read the angle off the protractor.

2.3 Goniometer

A goniometer (Fig. 5) is a device used to measure angles, similar to a protractor, but more specifically designed for measuring body joint angles (https://en.oxforddictionaries.com/definition/goniometer, Accessed 21 August 2017).

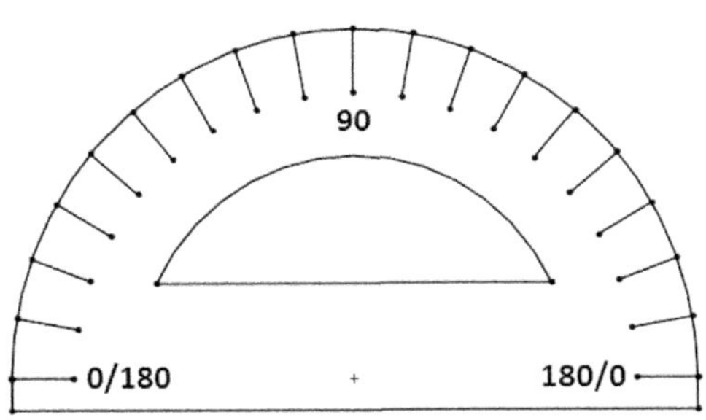

Fig. 4 Example of a protractor.

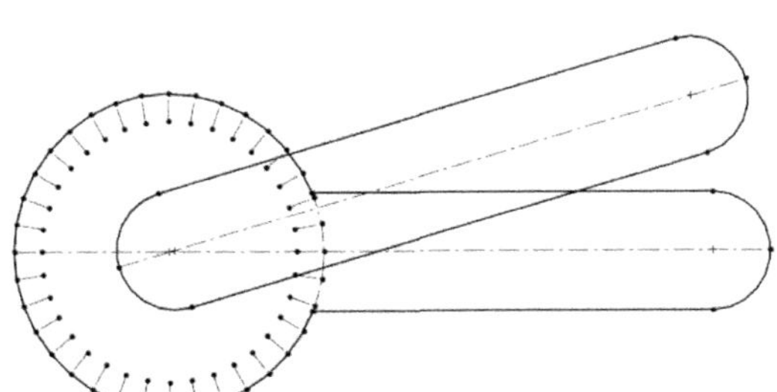

Fig. 5 Example of a goniometer.

By aligning the axis of rotation of the goniometer with the central axis of the joint and each leg along the joint segments in question, visual inspection of the indicators will show the angle.

To understand what range of motion (ROM) Jacob currently achieves with his sound-side knee and with his current prosthetic knee, you use a goniometer. To do so, you align the central axis of the goniometer with the axis of the knee being measured and take measurements at the extremes of the ROM of both knees.

2.4 Lever

A lever (Fig. 6) is a device which consists of a beam with a fulcrum, or pivot point (https://en.oxforddictionaries.com/definition/lever, Accessed 21 August 2017). Levers are used to sense force at one end of the beam and transfer that force to the other end. By varying the fulcrum position along the beam, the force can be amplified or reduced, based on the sum of torques about the fulcrum.

A prosthetic foot is essentially a lever. Through heel strike to toe off, forces are generated in the foot which are transferred to the knee through the fulcrum of the ankle. You analyze Jacob's ankle and find the heel strike and toe-off forces.

2.5 Cable

A cable (Fig. 7) is a device which is able to sense a signal (force, torque, position, etc.) at one end and transfer that signal to the other end (https://www.merriam-webster.com/dictionary/cable, Accessed 21 August 2017). Cables can be routed through devices and can modulate the signal strength through the use of pulleys. Standard cable materials are steel or spectra cable.

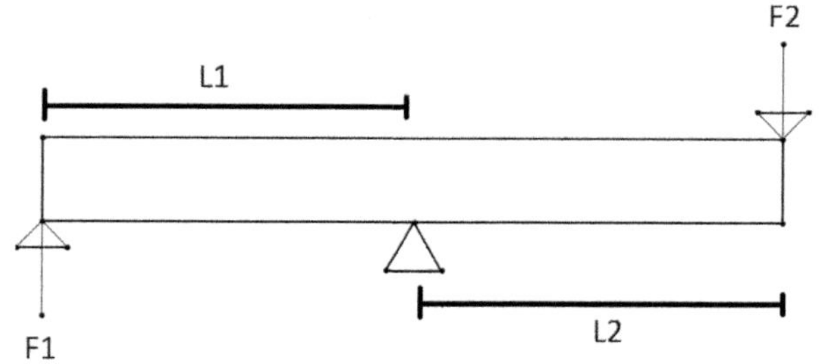

Fig. 6 Schematic representation of a lever.

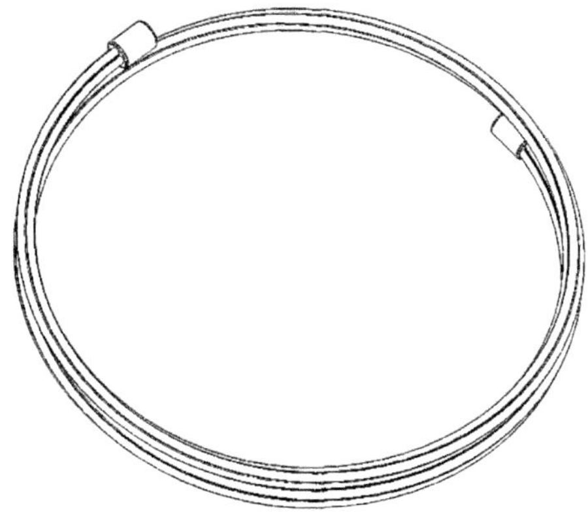

Fig. 7 Example of a cable.

To evaluate the force which Jacob can control, you use a system of cables and pullies to apply various forces to Jacob's residual limb. With feedback from Jacob, you determine an acceptable range of forces.

With your evaluation of Jacob's physical capabilities and the capabilities of his current prosthetic system, you decide to design Jacob an electrically powered prosthetic system.

3 SIMPLE SENSORS

With the invention of the integrated circuit, standard electrical sensors continue to get smaller, more efficient, cheaper, and easier to use. The ability to develop electrical-mechanical systems on the microscopic scale, micro-electrical-mechanical systems (MEMS) have also been a great benefit to sensor technology (Lamers and Pruitt, 2011).

A standard electrical sensor consists of a minimum of three electrical lines: a supply voltage, a ground, and the sense voltage. The supply voltage provides power to the sensor. The ground is the electrical reference, value zero. The sense voltage is the response of the sensor to the environment. As the senor responds to external stimuli, generally a resistance value will change which causes a proportional change in the sense voltage through Ohm's law, the current draw of the sensor being constant.

There are many different types of electrical sensors and many different uses of these sensors in biomedical design applications. What follows is a review of some of the more common electrical sensors.

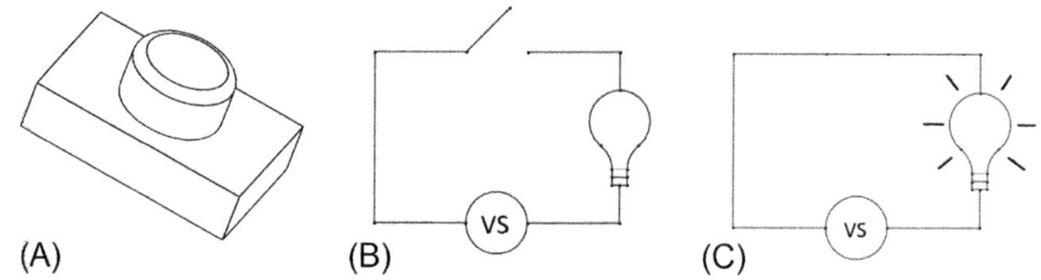

Fig. 8 (A) Example of a button. (B) Circuit diagram showing when a button is open. (C) Circuit diagram showing when a button is closed.

3.1 Mechanical Button

Perhaps, the simplest of electrical sensors, mechanical buttons (Fig. 8) are useful for sensing user intent. There are two main types of buttons: on/off and multistate switches (Digikey—Pushbutton-Switches, n.d.). In each case, a button either closes or opens a circuit to indicate intent of the user.

An on/off button will switch between on and off. Some on/off buttons will remain in a certain state until an input is applied. Then, the button switches to the opposite state and remains in that state until an input is again sensed. Other on/off buttons remain in a certain state and will only switch when the input is sensed and will return to the original state once the input is removed.

Multistate switches are those buttons which have more than two states. There are many different types of latching states which are useful for selecting different inputs or modes in a device.

Mechanical buttons are different than the capacitive buttons which are found on touch screens such as cellphones, iPads, and many tablets. Capacitive buttons will be discussed in Section 3.8.

In Jacob's new prosthetic system, you specify an on/off button to turn on and off the device. You consider where to place the button to minimize accidental on/offs.

3.2 Potentiometer

A potentiometer (Fig. 9) is an electrical device which provides a unique resistance with an associated position. As the potentiometer dial is turned, a wiper moves along a variable resistor such that the resistance of the device changes, which varies the sensor voltage. It is the practical application of a voltage divider (https://www.merriam-webster.com/dictionary/potentiometer, Accessed 21 August 2017).

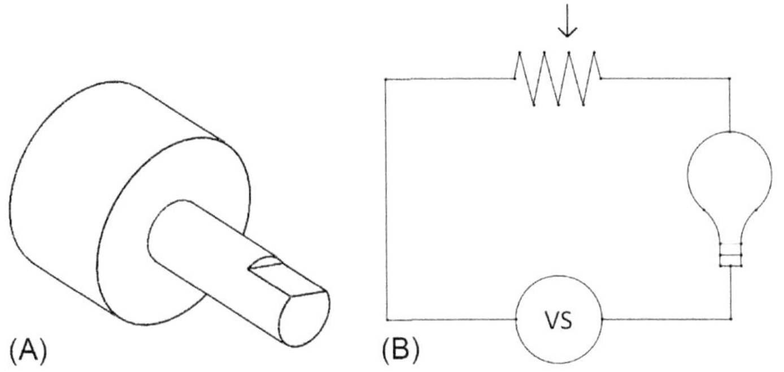

Fig. 9 (A) Example of a potentiometer. (B) Circuit diagram with a potentiometer. As the potentiometer is adjusted, the resistance changes, varying the intensity of the light bulb in the circuit.

There are several types of potentiometers (Newark, n.d.). There are single turn, partial turn, and multiturn potentiometers. There are also linear and logarithmic, referring to the rate of change of the resistance. They come in a multitude of sizes and shapes, from very large for heavy machinery sensing, to tiny integrated circuit chips for setting microprocessor inputs. They are useful for sensing device position or for setting controls on a device.

You select a potentiometer for Jacob's motor-driven knee which allows him to adjust how freely the knee moves. When Jacob turns the potentiometer, the resistance changes which varies the voltage signal sent to the microprocessor. The microprocessor uses that signal to adjust system values which effects the function of the knee.

3.3 Photoresistor

A photoresistor or light-dependent resistor (Fig. 10) is composed of photoconductor material. When light hits this material, the material absorbs the radiation and electrons move from the valance band of the semiconductor to the conduction band. The more electrons in the conduction band of the resistor, the less the resistance of the resistor (http://www.resistorguide.com/photoresistor/, Accessed 22 August 2017).

In Jacob's prosthetic foot, you install a photoresistor on the side. When this resistor is light activated, the system assumes that Jacob is not wearing a shoe in his foot. When the resistor is not light activated, the system assumes Jacob is wearing a shoe. Minor variations in prosthetic system performance are introduced into the system control algorithm based on the light sensor input.

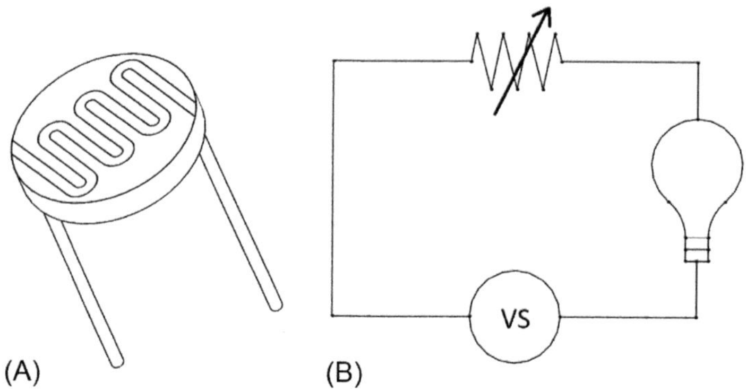

Fig. 10 (A) Example of a light-dependent resistor and (B) circuit diagram with a light-dependent resistor. As the intensity of the light sensed varies, the brightness of the light changes.

3.4 Hall Effect Sensor

A Hall effect sensor is an electrical sensor which responds to change in a magnetic field. Named after Edwin Hall who is credited with first observing a voltage potential across a current carrying conductive plate in the presence of a magnetic field in 1879 (Milano, 2009). In the sensor, there is a thin rectangular piece of semiconductor which constantly is carrying a current when the sensor is turned on. As a magnetic field is introduced, the electrons passing across the plate deviate from the center due to the Lorentz force. The electron deviation creates a voltage difference across the two ends of the plate which is proportional to the electric field (Fig. 11).

There are several types of Hall effect sensors (Allegromicro, n.d.). On/off Hall effects are used for binary signals. If the magnet is close to the sensor, a high voltage is output. If the magnet is far away, the voltage is low.

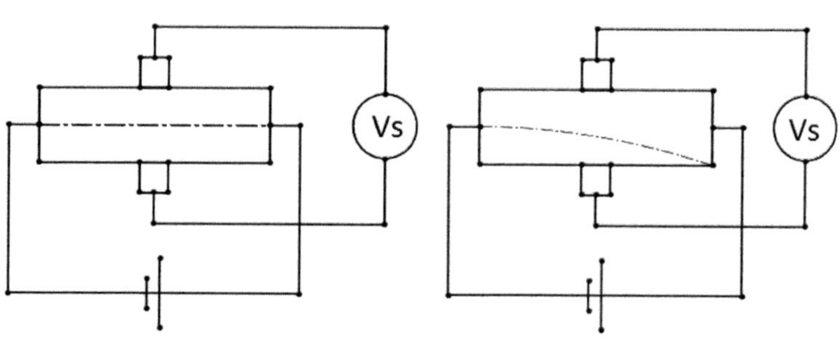

Fig. 11 Hall effect sensor diagram. When no magnetic force is influencing the sensor, the current, represented by the dashed line, remains in the center *(left)*. When a magnetic field is introduced, the electrons deviate from the center *(right)*.

Proportional Hall effect sensors return a voltage proportional to the position of the magnet within a certain range, generally at a fixed orientation. Rotary Hall effect sensors are designed to sense the change in orientation of the magnet, generally at a fixed position. For both proportional and rotary Hall effect sensors, the orientation or position, respectively, do not have to be fixed, but doing so often helps for calibration of the signal, minimizing the number of setup variables.

Hall effect sensors do tend to drift and can be affected by the environment, such as when other magnets or magnetic materials are close to the sensor. The orientation and position of the magnet being used with the Hall effect sensor is critical to the function of the sensor.

You decide to use an on/off Hall effect sensor to determine when the knee is at the limits of its range. To do so, you install two magnets in the rotating joint of the knee and mount the Hall effect sensor on the stationary side of the joint. You position the magnets such that when at the limits, the Hall effect sensor senses the magnets, but not before.

3.5 Strain Gauge

A strain gauge (Fig. 12) is a sensor which responds to the expansion or contraction of a material, or the strain. A strain gauge consists of a long thin piece of metal which folds back on itself, or zig zags across the sensor. As the material expands or contracts, the long thin piece of metal gets longer or shorter with the material, changing the resistance of the metal. The voltage output of the sensor corresponds to the change in resistance (Omega—Straingages, n.d.). Strain gauges work well for most metals but are seldom successful in use with plastic.

A single strain gauge is highly susceptible to environment temperature changes and placement on the material. To normalize environmental

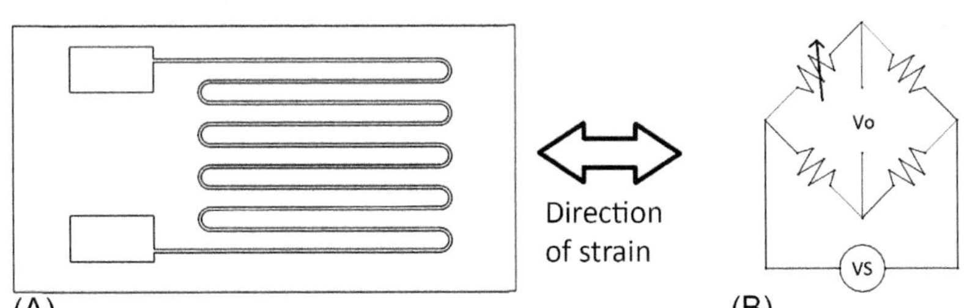

Fig. 12 (A) Schematic representation of a strain gauge and (B) circuit diagram of a strain gauge in a Wheatstone bridge.

conditions, multiple strain gauges are used, with the gauges placed at various orientations with respect to the direction of desired strain sensing. Temperature effects can be measured and removed from the signal by placing a strain gauge orthogonal to the direction of desired measured strain. The orthogonal strain gauge will expand and contract with temperature, but only minimally respond to the orthogonal strain.

Strain gauges are often wired in a Wheatstone bridge configuration (Fig. 12B). The Wheatstone bridge provides temperature compensation since all the resistors in the bridge experience the same environmental temperatures. Each leg of the bridge can have either a sensor or a dummy resistor of similar resistance (Omega—Straingages, n.d.).

The application of strain gauges is difficult. Special care must be taken for material surface preparation, orientation of the gauges, gauge to material attachment, and postattachment sensor handling.

On the pylon, or tibial section of Jacob's prosthetic system, you mount a strain gauge to sense the load being born by the prosthesis. To do so, you carefully prepare the surface of the pylon, select a strain gauge which is rated for the expected strain range, determine the appropriate orientation to mount the gauge, and decide to mount several gauges at various angles to be able to sense various directions of strain. Once glued in place, you test your gauges to verify adhesion and performance.

3.6 Thermistor

A thermistor, or temperature-dependent resistor, is used to sense a change in temperature. A thermistor consists of a material which is highly responsive to temperature. The resistance of the material is proportional to the expansion and contraction of the material, which can be calibrated to the temperature (Fig. 13) (Omega—Straingages, n.d.).

In Jacob's knee, you select a temperature-dependent resistor for the processor board which controls the knee motor. You will use this sensor to monitor the temperature of the motor. If the resistor gets too hot, the processor can limit motor power to keep the system within specified temperature parameters.

3.7 Current Sensor

Often in electrical systems, it is useful to know the current being used by the system. There are two ways to measure current, either directly or indirectly (Fig. 14) (Omega—Thermistor, n.d.).

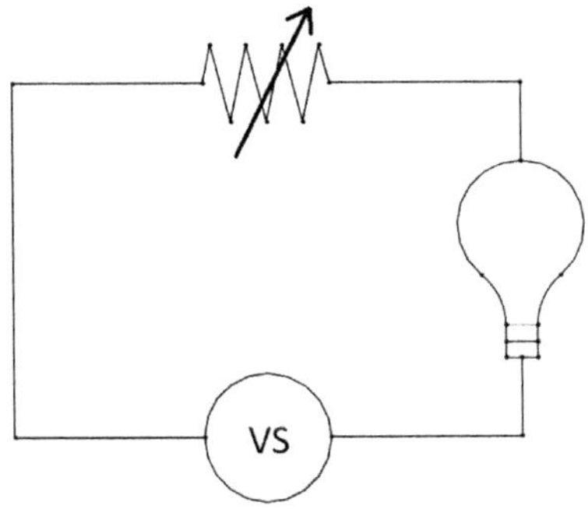

Fig. 13 Circuit diagram with a thermistor. As the temperature sensed varies, the brightness of the lightbulb changes.

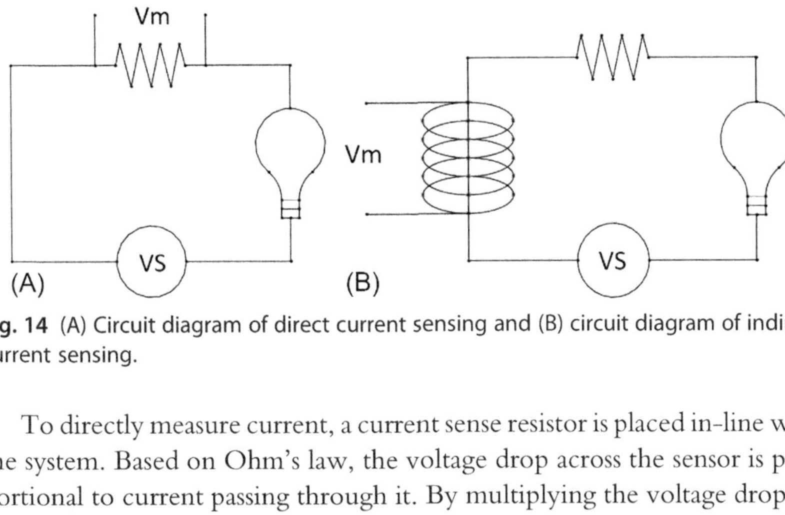

Fig. 14 (A) Circuit diagram of direct current sensing and (B) circuit diagram of indirect current sensing.

To directly measure current, a current sense resistor is placed in-line with the system. Based on Ohm's law, the voltage drop across the sensor is proportional to current passing through it. By multiplying the voltage drop by the value of the sense resistor, the current can be calculated. Direct current is easily implemented, but effects the current itself since the sensor is a part of the system.

To indirectly measure the current, a coil is wrapped around a current carrying wire. Based on Ampere's and Faraday's laws, an inductive voltage will be generated in the coil which is proportional to the current. Indirect current sensors tend to be more accurate, but are harder to implement on printed circuit boards.

Along with the temperature-dependent resistor, you select a current sensor resistor to monitor motor current of Jacob's knee. Before the current gets too high, the processor can limit the current drawn by the motor to protect the system and the person with amputation.

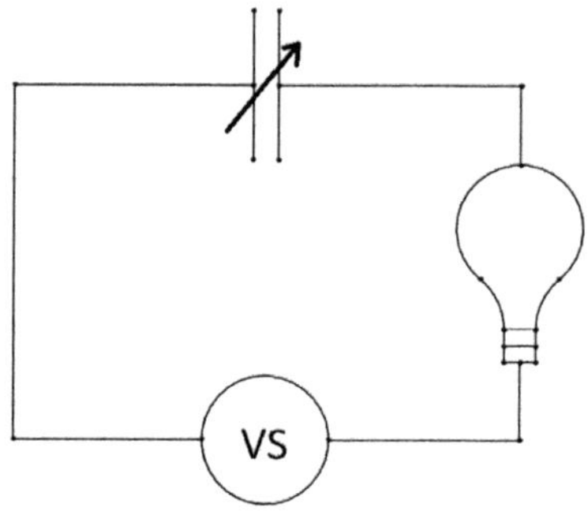

Fig. 15 Circuit diagram with a capacitance sensor. As the conductance between the conductors is varied, the brightness of the lightbulb changes.

3.8 Capacitance Sensor

Capacitance defines how the strength of a magnetic field is affected by the gap between two conductors. There are three factors which influence capacitance: the size of the conductors, the size of the gap between them, and the material between them (the dielectric). The bigger the conductors, the bigger the capacitance. The smaller the gap, the bigger the capacitance. The dielectric is chosen based on the range of capacitance being sensed (Digikey—Current Sensors, n.d.).

When a voltage is applied to the conductors, positive and negative charges accumulate on each conductor. By alternating the voltage on the conductors, the charge also alternates, generating a current that is proportional to capacitance. By allowing modulation of the distance between the conductors, the current of the sensor will go up or down (Fig. 15). The distance changes are applied to what is being sensed (Bolton, 2003b).

The capacitive touch screens found on devices such as cellphones, iPads, and tablets use capacitive technology. By measuring the charge on each corner of the screen, the location of capacitive disturbances can be determined.

After further review of Jacob's knee design, you decide to switch the on/off button with a capacitive button. The capacitive button gives the knee a more sleek and modern feel and also allows you to design a more water-resistant device, which is more suited for Jacob's lifestyle.

4 COMMON SENSORS

For the purpose of this discussion, common sensors refer to standard applications of simple sensors. Generally, a common sensor is composed of

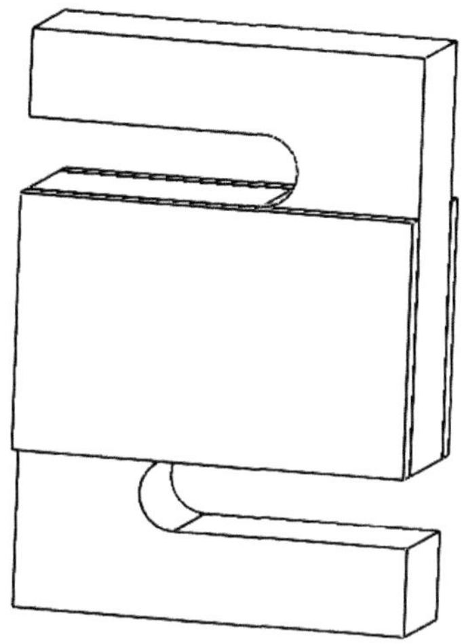

Fig. 16 Load cell.

one or more simple sensors, supporting hardware for signal amplification, filtering, and power management, and mechanical structure support.

4.1 Load Cell/Force Plates

A load cell is used for measuring force (Fig. 16). A load cell consists of a spring with a known spring constant and a way to sense the displacement of the spring. Given the spring constant the displacement, Hook's law can be applied to find the force (Wang, 2014). A force plate is a load cell that is configured for measuring ground reaction forces.

The spring used is often either a coil-type spring or a cantilever spring. Many different simple sensors can be used to sense the displacement. For instance, some load cells have strain gauges mounted on the deflection arm of the cantilever spring. Others have a magnet and Hall effect configuration. Some load cells have a potentiometer which measures spring displacement. Another option is to measure the change in capacitance between to conductors, where one conductor is at the end of the spring.

Recall the strain gauge mounted on the pylon of Jacob's prosthetic system. To enhance the usefulness of the strain gauge signal, you add some support electronics to reduce the noise, amplify the signal, and deal with the drift of the signal overtime. Also, you apply various loads to the end of the pylon and record the filtered and amplified strain gauge output to calibrate the sensor. The work accomplished results in a reliable load cell.

4.2 Pressure Sensors

Pressure sensors are used for measuring pressure. There are two types of pressure that is of interest, either pressure from a gas or pressure from a touch (Bolton, 2003c).

Gas pressure sensors work by measuring the displacement or strain of a diaphragm over a given area using simple sensor components. Pressure is calculated by dividing the force by the displacement.

A common way of measuring tactile pressure is to use force-sensitive resistors (FSR) (Fig. 17). A FSR consists of a grid of small thin wires. When the grid is touched and pressure applied, the wires become longer which changes the output voltage (Bolton, 2003d).

In order to provide a sleeker design, you switch the knee resistance adjustment method from potentiometers to two FSRs. One FSR is used for increasing the resistance, the other is for decreasing the resistance. As Jacob taps the FSR, the resistance increments by a set amount defined in the microprocessor code.

4.3 Accelerometer

Accelerometers measure the acceleration of an object. A MEMS accelerometer consists of a mass suspended in an elastic support system (Fig. 18). When the sensor is moved, the elastic support responds and the force is measured

Fig. 17 Example of a force-sensitive resistor.

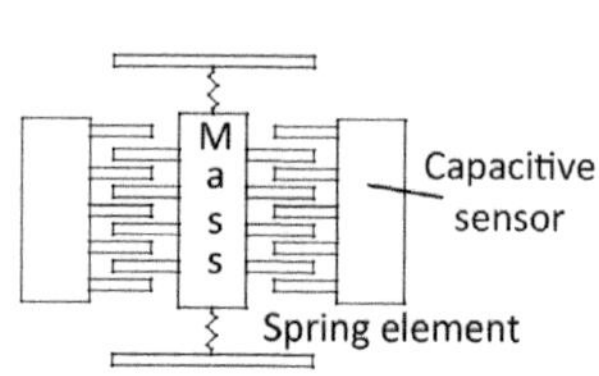

Fig. 18 Schematic representation of MEMS accelerometer. As the mass moves, the capacitive sensor measures the deflection of the spring supporting the mass. Based on that deflection, the force can be calculated.

through displacement using a capacitor sensor element. The displacement measured through capacitance can be used to derive the acceleration (ElProCus, n.d.).

4.4 Inclinometer

Inclinometers measure the angle of an object, or inclination with respect to gravity. MEMS inclinometers consist of a MEMS accelerometer which is small enough and sensitive enough to measure the change in the pull of gravity of a mass based on the orientation of the sensor (Omega—Accelerometers, n.d.).

4.5 Gyroscope

A MEMS gyroscope measures the angular velocity from the effect of the Coriolis force applied to a vibrating element. The gyroscope has two vibrating legs and two sensing legs (Fig. 19). When the gyroscope is rotated, the resultant Coriolis force causes an unbalanced force in the vibrating legs which gets transmitted to the sensing legs. The motion of the sensing legs is measured through a capacitance element (https://www.posital.com/en/products/inclinometers/mems/MEMS-Technology.php, Accessed 22 August 2017).

You select an accelerometer, inclinometer, and gyroscope and design them into the foot of Jacob's prosthesis. From the accelerometer, you are able to determine when the foot is moving and when it is stationary. From the inclinometer, you can determine what type of terrain Jacob is traversing, flat, downhill, or uphill. From the gyroscope, you can determine the foot stability.

Based on the information from these three sensors, you develop an algorithm which determines if Jacob is in the swing (foot not on the ground) or

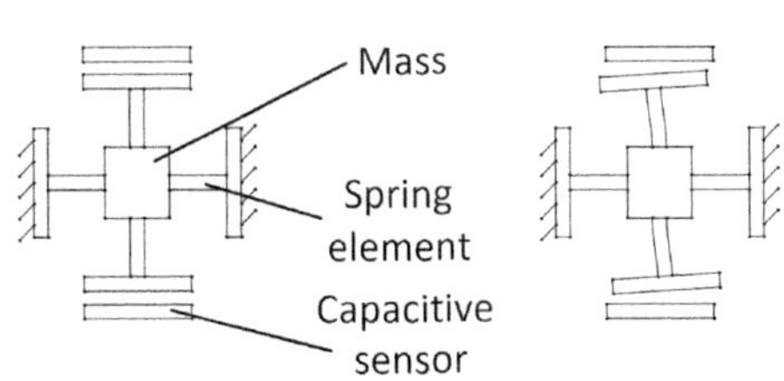

Fig. 19 Schematic representation of a MEMS gyroscope. When not rotating, there is constant capacitance measured across the top and bottom legs (left). When rotating, the Coriolis effect causes the top and bottom legs to vibrate, varying the capacitance distance on the top and bottom legs (right).

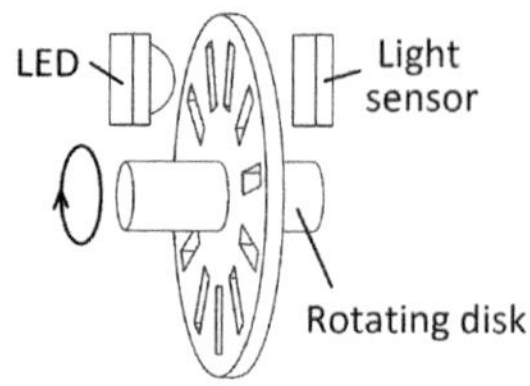

Fig. 20 Encoder schematic. As the slit disk rotates, the light sensor intermittently senses light which are counted and used to determine rotary position and numbers of rotation.

stance phase (foot on the ground) and some useful information about the terrain he is traversing, what he is doing in the terrain, and what he is trying to do. The algorithm you develop adjusts certain parameters of the foot and knee system which support him in his desired motion.

4.6 Encoder

An encoder is a sensor which measures rotary position (Fig. 20). It is a sensor which can again use a variety of different simple sensors. Commonly, either a Hall effect or light resistive sensor is used to count impulses from either a magnet or light source, respectively. The number of impulses is related to rotary position (http://www5.epsondevice.com/en/information/technical_info/gyro/, Accessed 22 August 2017). Noncontact sensor elements tend to have a longer life, but tend to use more power.

You select an encoder for the motor in the knee which adjusts the knee resistance. By sensing motor position, you can know the resistance of the knee, and calculate how much and which direction to move the motor in order to achieve the desired resistance.

5 BIOLOGICAL SENSORS

The capturing and processing of biological signals are some of the most critical elements of the design of biomechatronic devices and an understanding of how the nervous system works is helpful in understanding biological signals. Section 5.1 is a brief overview of neuromuscular anatomy. There is much to be learned on this topic and the information presented here is a shallow skim. The rest of the sections in Chapter 5 follow the capturing and processing methods of motor signals starting at the surface of the skin and proceeding up to the brain.

5.1 Neuromuscular Anatomy

The neuromuscular system consists of three main elements: the central nervous system, nerves, and muscles. The central nervous system is comprised of the brain and spinal cord. Muscles include not only standard muscles such as the biceps and calves, but also muscles in the heart, lungs, eyes, etc. Efferent biological signals, signals which are generated in the central nervous system and travel to periphery systems, originate in the brain or spinal cord and travel through thousands of nerves until it reaches its final destination and causes the body to respond. Along the way, that signal passes through the spinal cord, down nerve cords, to nerve centers and into the target organ or muscle. Afferent biological signals, those generated in the periphery systems and sent to the central nervous system, travel a similar but opposite path to the brain (Bolton, 2003e).

Nerves (Fig. 21), the main unit of the nervous system, are cells which create an electrochemical communication system. Each nerve has at least one axon, nucleus, and dendrite. When a signal is generated in the brain, the signal travels across a nerve electrically, starting at the axon and ending at the dendrite. The signal is called an action potential (Fig. 22). Between the dendrite of one nerve and the axon of the next, there is a small gap called the synaptic cleft. To cross the gap, a chemical process is used where ionic receptors are released from the axon and accepted by the dendrite. This electrochemical process continues on until the action potential reaches the desired muscle or organ and causes a response (Bolton, 2003e).

Decades of research have been dedicated to accessing these afferent and efferent biological signals. Much progress has been made, but there is still more to be learned. This discussion will begin at the peripheries of the body system and work up the signal pathway.

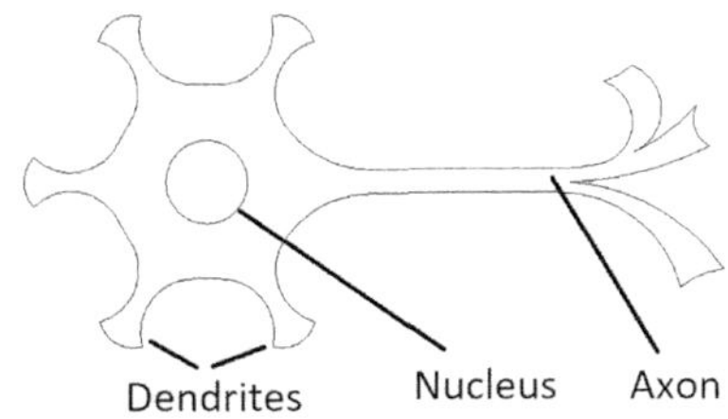

Fig. 21 Nerve cell.

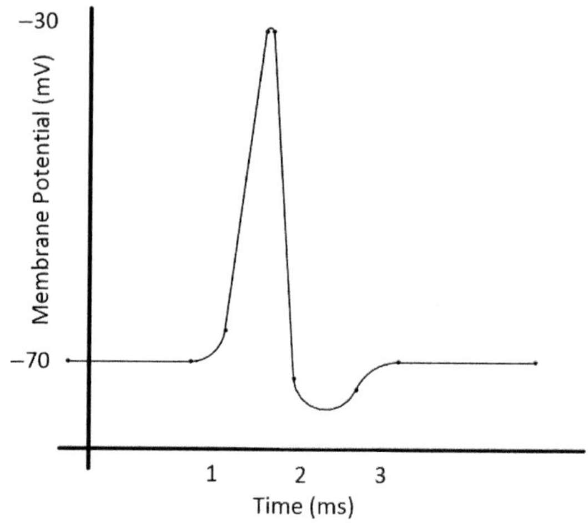

Fig. 22 Action potential.

5.2 Surface Electromyographic Sensors

The electromyographic (EMG) signal is a small voltage signal generated by a skeletal muscle when electrically or neurologically activated. Francisco Redi is credited with being the first to study EMG when he discovered and studied a highly specialized muscle in the electric eel which generates electricity when contracted (Purves et al., 2008a). Since that time in the mid-1600s, EMG has been studied with constantly improving technology and methodology.

EMG is a byproduct of muscle contraction. Recall how the nervous system communicates through a chain of nerves. When the action potential reaches a muscle, the depolarization threshold of the motor nerve is reached and the muscle fibers contract. Depolarization produces an electromagnetic field and the action potential is measurable as a voltage. The voltage generated is a summation of all the muscle fibers enervated by the motor neuron. The greater number of cells enervated, the greater the electrical signal (Basmajian and de Luca, 1985).

Therefore, as muscles contract, they are constantly releasing small voltages into the surrounding environment. The EMG signal is proportional to motor activity, meaning that the higher the voltage, the more the muscles are contracting. EMG sensors have been developed to measure these voltages (Fig. 23). There are two common types of surface EMG sensor: button electrode and preamp.

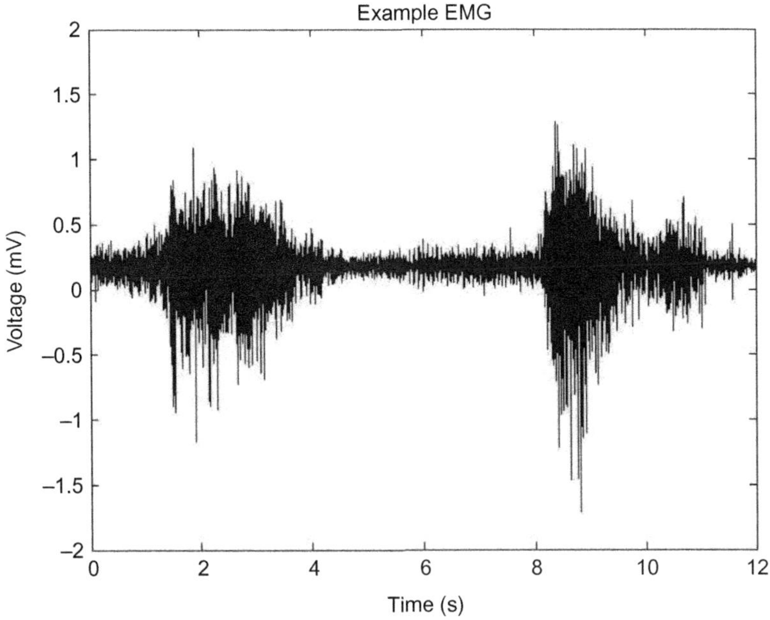

Fig. 23 Example electromyographic (EMG).

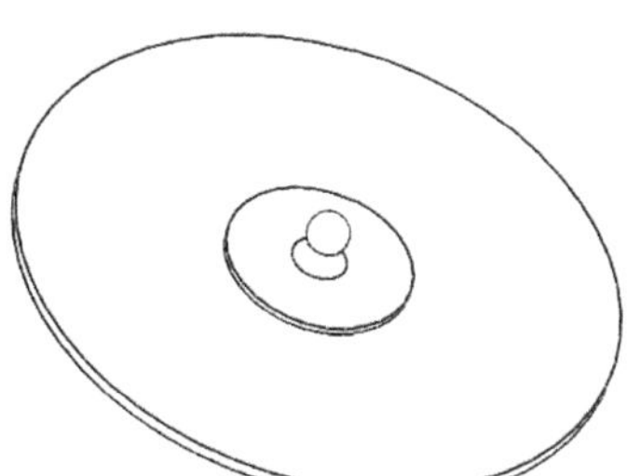

Fig. 24 Button electrode.

A button electrode (Fig. 24) is an electrode which is placed anywhere on the skin. The sensing end is attached to the skin with tape or has a surrounding stick membrane. The button connects to an EMG amplification and processing system. When using such a system, two electrodes are paired to create a voltage differential. One electrode is required to be a reference to ground the signals. These systems have more flexibility since the electrodes can be placed anywhere and electrode sets can be self-selected. However, they require more effort to setup and are often less portable.

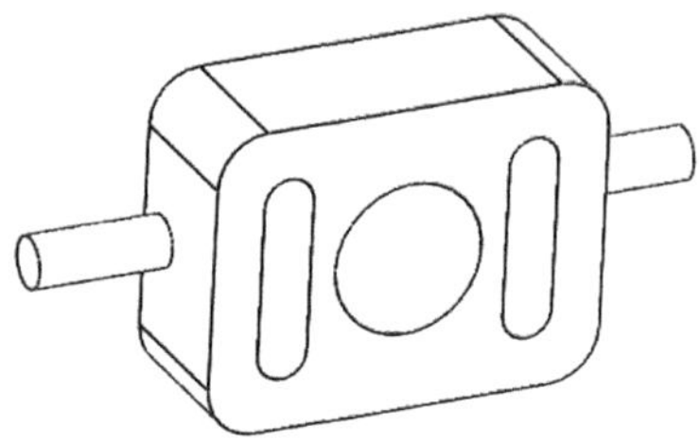

Fig. 25 Preamplifier electrode.

A preamp electrode is short for preamplifier electrode (Fig. 25). These devices contain two electrodes and a ground. The preamp electrode collects the signal and also amplifies it, although further amplification may occur downstream. Many preamp electrodes have an input for adjusting the amplification of the signal. The metal electrodes of the preamp electrode also must contact skin. The advantage of preamp electrodes is the ease of setup and portability.

Once sensed and amplified, the EMG signal is still very noisy, and must be filtered and rectified. EMG is often filtered through a band-pass filter within the power spectrum of the signal, with a low passband of around 10 Hz and a high band pass of 500 Hz (Raez and Hussain, 2006; Soares et al., 2003). The EMG is then rectified (the absolute value is taken) and filtered through a third-order low pass filer of around 5 Hz to envelope the signal (Fig. 26). These processing techniques return a proportional signal which can be used in biomechatronic devices.

Surface EMG has been used successfully in many biomechatronic devices. However, even after processing, surface EMG is a noisy signal and does not necessarily correlate to one muscle. The surface EMG sensor will pick up all muscle activity in surrounding area, meaning a combination of different muscle neurons and their fibers.

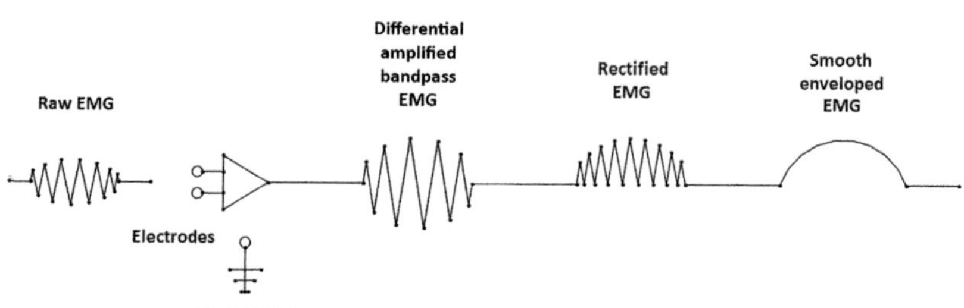

Fig. 26 EMG processing.

For Jacob's prosthetic system, you decide again to change how he varies the knee resistance. You decide to use two surface EMG preamplifiers mounted in the socket for the residual limb of his leg. You configure the system such that when he flexes his residual limb muscles, the resistance of the knee goes down. When he extends his residual limb muscles, the knee resistance goes up.

5.3 Intramuscular EMG

Instead of collecting noisy EMG signals from the surface, EMG signals can be collected by placing EMG electrodes directly into the muscle. There are two methods of collecting this intramuscular EMG. The first is to insert an intramuscular needle electrode into the muscle site of interest. Depending on the type of needle electrode used, a ground needle electrode is often required.

By collecting the signals directly from the muscles, the signal tends to be cleaner. Although useful for short-term experiments, overtime the body tends to reject the wire electrodes. Also, in areas such as the forearm, where multiple muscles are closely located, specific muscle targeting can be difficult.

An alternative method of collecting signals directly from the muscles is by using an intramuscular myoelectric sensor (IMES) (Fig. 27). An IMES consists of a self-contained surface EMG sensor which has been miniaturized and placed in a biologically inert package. Electronic resonance is used for communicating with the device as well as for charging the battery. An IMES is about the size of a piece of long-grain rice (Farrell and Weir, 2007).

Since IMES are biologically inert, the body does not reject the sensor. However, there are still the issues of device placement, with the added

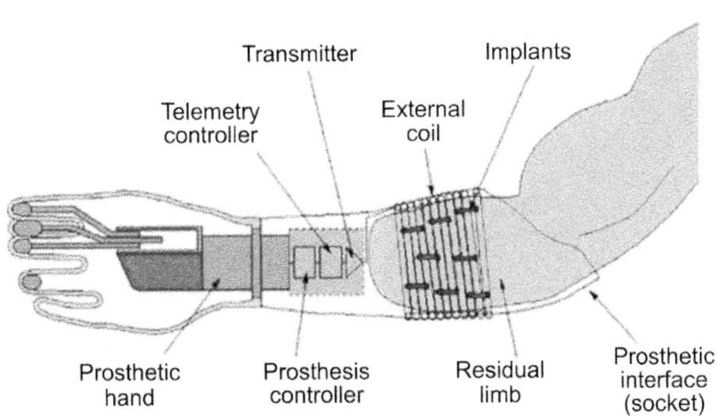

Fig. 27 Intramuscular myoelectric sensor (IMES) illustration. *From Weir, R., Troyk, P.R., DeMichele, G.A., Schorsch, J.F., Maas, H., 2009. Implantable myoelectric sensors (IMES) for intramuscular electromyogram recording. IEEE Trans. Biomed. Eng. 56 (1), 159–171.*

complications of communication and charging. Research still progresses in the further development of IMES.

Due to sweat, a poor fitting socket, and changes in the muscle structure, the surface EMG system you installed for Jacob was unreliable and caused more problems than they solved. You consult with Jacob and he agrees to undergo a minor, but experimental, surgery to have IMES implanted in his residual limb. You implement the same algorithm as with the surface EMG, and find the IMES system to be more reliable.

5.4 Nerve Cuff

The next step up in the nervous system is to access signals from the nerve bundle before it enters the muscles. To do this, a sensor called the nerve cuff has been developed (Fig. 28) (Weir et al., 2009). A nerve cuff consists of biologically inert wrap which has small electrodes embedded into the wrap. The nerve cuff is surgically inserted into the body.

During surgery, first the desired nerve branch is discovered. Then, the nerve cuff is wrapped around the nerve bundle and a suture is used to sow the cuff together, being careful not to pierce the nerve cuff membrane (to prevent scaring). The fine wires attached to the electrodes are brought out and a junction point is created on the skin. After the patient has healed, the nerve cuff is ready to be used (Weir et al., 2009).

Since a nerve bundle is just that, a bundle of nerves, the nerve cuff must be calibrated. By either stimulating the electrodes individually and recording what part of the body the patient feels is like being touched, or by asking the patient to think about moving a certain limb or joint and recording what electrodes are activated, a map can be developed of which electrode corresponds to which nerve or nerves. This system is currently being used to attempt to restore sensation and is still experimental (Tyler and Durand, 2002).

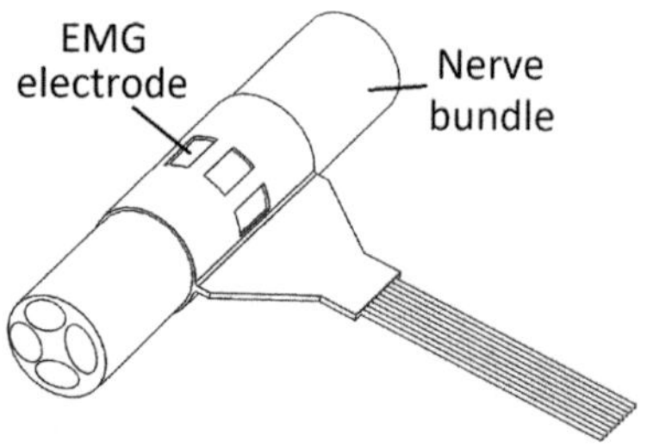

Fig. 28 Nerve cuff electrode.

After a month or two of the IMES, Jacob's control is improving, but now he is having trouble with sensing what is happening at the prosthetic foot. An avid rock climber preaccident, he is trying to get back into the sport, but lacks the foot feedback he desires. After consultation with his medical team and with you, he decides to undergo an invasive surgery to install a nerve cuff electrode as part of a Food and Drug Administration (FDA) field trial.

After the surgery and necessary recovery time, you determine which nerve areas are stimulated by which embedded nerve cuff electrode. You install two FSR sensors on his prosthetic foot, one on the toe and one on the heel. You connect and program the system such that when one of the FSR sensors is touched, the processor sends a signal to an appropriate, or close to appropriate, nerve area so that Jacob now can "feel" his toe and heel.

5.5 Brain Array

Perhaps, the most invasive method of collecting muscle and sensory signals is to use a brain implant. This device generally consists of 4 mm × 4 mm plate with an array of 100 needle electrodes (Fig. 29). The array is surgically implanted into the brain, generally in the motor cortex, needles interacting with the gray matter. The electrode array is then calibrated by having the patient think about performing a variety of motor tasks. Using pattern

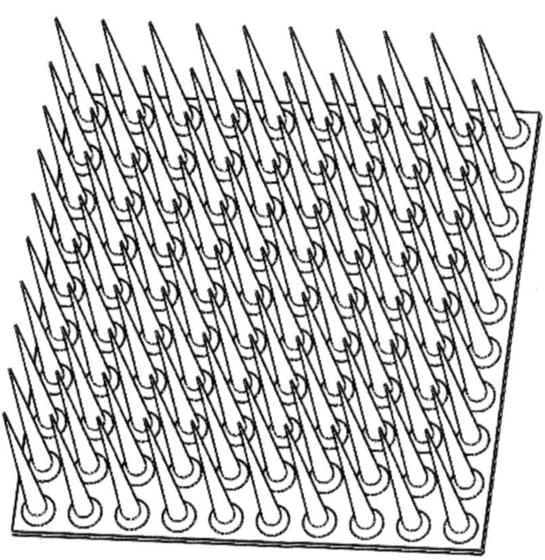

Fig. 29 Brain electrode array.

recognition techniques, the brain activity observed on each of the electrodes can be correlated with intent (Tyler, 2017).

Once trained, brain arrays can be very efficient at deciphering intent and returns much information. However, the surgery is quite invasive and the risk of brain infection can be high. Portability of the system is generally low due to the intensive computer processing required for the quantity amounts of data obtained.

After about a year of Jacob's IMES and nerve cuff system, Jacob begins desiring even greater control and sensory feedback. After another lengthy discussion with his medical team and with you, Jacob decides to participate in a FDA study to use the brain array to control and receive feedback from his prosthetic system. The surgery and recovery times are even longer, but successful. Jacob then goes to a highly specialized lab at a top-notch university and undergoes many days of nerve mapping and training. As the neural map is determined, you integrate more sensors on the prostheses, develop the control algorithms, and determine how to use the nerve outputs and inputs available through the brain array, all in coordination with the university lab. Finally, Jacob and the device are ready to begin moving together.

6 OTHER BIOLOGICAL SIGNAL TRANSDUCERS

Signals traveling along the motor neuron pathway are not the only body signals for which transducers have been developed. The body is constantly generating various types of signals. The following is an overview of some of the systems developed.

Our theoretical friend Jacob first came to you to get a better device than his passive prosthetic knee and ankle system. In this section, we will use the sensing technology described to illustrate how to use these systems to determine if a better system for Jacob has been developed.

6.1 Electroencephalography

Electroencephalography (EEG) was first observed in humans by Hans Berger, a psychiatrist at the University of Jena (Fernandez et al., 2014). It is a method of sensing brain activity, specifically the voltage variations resulting from ionic current within the brain neurons. It is performed by placing EEG electrodes on various standard positions around the scalp. All the electrode signals are feed into a computer for signal analysis and recording. Overtime, various patterns become apparent, often referred to as waves. Four basic waves have been defined and correlated to various states of attention. The alpha wave indicates the person is awake with his or her eyes closed. The beta wave signifies mental activity and attention. The theta and delta waves are correlated with drowsiness, sleep, or a pathological condition (Fernandez et al., 2014).

Although easy to setup, since the EEG electrodes are placed externally, and relatively inexpensive, EEG is often criticized for its low resolution. EEG is useful in recording brain signals, but it has been difficult interpret the individual electrode signals.

To determine if you have built Jacob a better system, you decide to have Jacob do two EEG recordings, one where he pedals a stationary bike while using his passive leg, and another where he pedals using his new leg. After the test, you compare the EEG recordings, focusing specifically on beta waves to see which device required greater mental focus.

6.2 Electrocardiogram

Since the heart is a muscle, as the four chambers contact and relax, the muscle responses release voltage signals, similar to standard muscle contraction. The heart also has special Purkinje fibers, or "pacemaker" cells, which initiate electrical signals regulate systematic chamber contraction. An electrocardiogram, EKG or ECG, is the transducer system which is used to measure the voltage output of the heart (Purves et al., 2008b).

By placing an electrode on the skin over the heart and recording the voltage sensed overtime, an electrocardiogram is produced. During one heartbeat, there are six standard elements of the wave form, labeled P, Q, R, S, T, and U (Fig. 30). The P wave represents atrial depolarization.

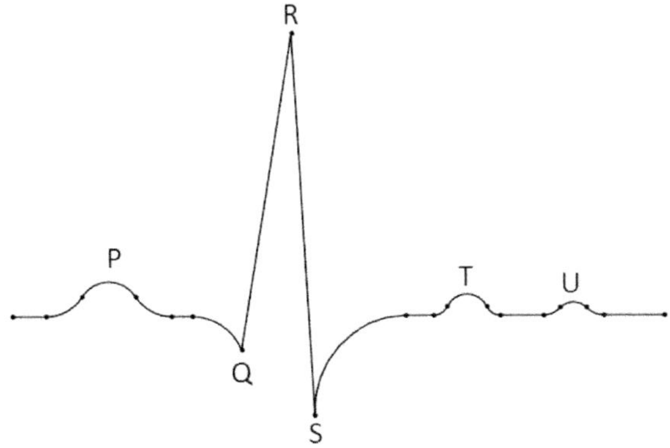

Fig. 30 Standard electrocardiogram wave.

The QRS complex indicates ventricle depolarization and contraction. The T wave shows ventricular repolarization. The U wave indicates recovery of the Purkinje conduction fibers (Purves et al., 2008b).

Heart rate is often correlated with physical exertion, so you also request Jacob to have two EKG recordings performed, one with his old system while walking at various rates on a treadmill, and one with his new at the same speeds and durations. You hope to see that while using the new system, Jacob has a slower heart rate.

6.3 O_2 Light Sensors

A less refined method of heart rate sensing is through use of an optical sensor. By shining a near infrared (IR) light into a finger or earlobe and measuring the intensity of the light on the opposite side, the pulse rate of the person can be calculated, as well as how oxygenated is the blood. As more blood is in the reflection zone, the less light will be shine through, facilitating pulse rate calculation. Also, the more oxygen which the hemoglobin has absorbed, the less the light intensity, allowing for oxygenation calculation (Silverthorn et al., 2007).

While doing the EKG recordings, you decide to also use an O_2 light sensor to corroborate the data. You place an O_2 light sensor on Jacob's thumb and record his heart rate while he uses the two prosthetic devices.

6.4 Oxygen Consumption Sensor

Respiration is constantly output as a signal to the environment when breathing. It is useful to analyze the rate and oxygen content of inhalation and exhalation. Such metrics are often related to energy consumption (Chan et al., 2013).

Oxygen sensors are a fuel cell with a gas permeable membrane at one end. The cell contains an electrolyte, anode, and cathode. As oxygen passes through the membrane, a chemical reaction occurs between the elements, creating a voltage which can be monitored and recorded by a computer (Nieman et al., 2003). The person whose breath is being monitored must wear a mask which controls the flow of inhaled and exhaled air.

With Jacob on the treadmill, you fit him with an oxygen mask to measure the oxygen content of his inhalations and exhalations. You hope to see that he consumes less oxygen with the new system, indicating that the new device is easier for him to use.

6.5 Eye Movement

Where a person is looking is another potentially useful signal. Eye tracking can be done through three main ways: tracking the movement of something attached to an eye, generally a special contact lens, visual observation, or electrooculogram (EOG) (eye muscle electrodes) (PASCO, n.d.).

A special contact lens can be fit with electrical coils. As the coils move, they disturb the electrical field. These disturbances can be measured and converted to movement profiles.

Video cameras with tracking algorithms can be setup to record and track the position of the eye. Postprocessing of the video data is performed for eye movement analysis and interpretation.

EOG electrodes perform a similar function as EMG. The muscles contract which release a voltage. Through precise placement of the EOG electrodes, the orientation of the eye can be sensed (Fig. 31).

Another measure of prosthesis performance is mental load. You decide to measure the mental load Jacob experiences while using his old prosthesis compared with his new prosthesis. To do so, you setup a screen in front of the test treadmill. As Jacob is using the treadmill at various speeds, you show an image on the screen and measure the time it takes for Jacob to notice the image. Using EOG electrodes, you are able to measure a precise time between the appearance of the image and Jacob noticing it. You hope that the new prosthesis will require less mental load, or a smaller time difference between displaying the image and Jacob's eyes moving toward the image.

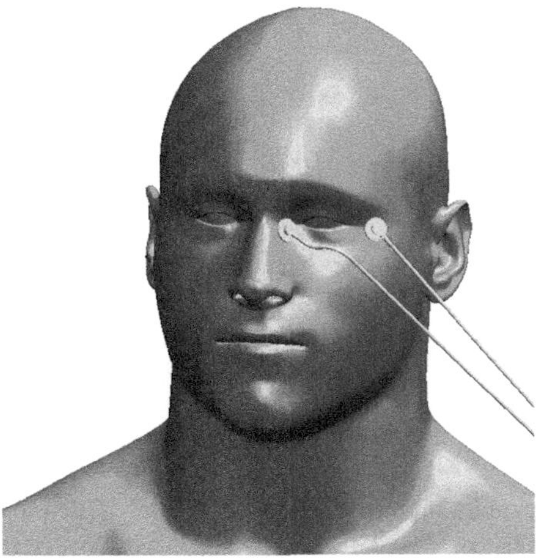

Fig. 31 Placement of electrooculogram (EOG) electrodes.

6.6 IR Body Markers and Camera Tracking Three-Dimensional Motion Capture

When discussing EMG and muscle movements, the focus was on capturing one muscle group. However, the body is composed of many muscle groups functioning in coordination. In order to capture entire body movements, three-dimensional (3D) motion capture is used.

A motion capture system consists of a minimum of four IR cameras (generally six) mounted such that they view the laboratory observation area from multiple angles (Fig. 32). The subject is instrumented with IR reflection balls, generally about 25 in. in diameter. These balls are placed on the subject over critical joints or limbs such as the base of the neck, shoulder blades, elbow, wrist, hand, hip, knee, and ankle (Lupu and Ungureanu, 2013).

When the subject is instrumented, a calibration routine is performed to map the joint space and create a model in the software. After calibration, the combined data from the IR cameras give precise positioning the body. Using joint-space algorithms, Newtonian physical effects can be calculated, such as joint accelerations and torques (https://www.vicon.com/what-is-motion-capture, Accessed 23 August 2017).

Motion capture systems are often combined with force plates to measure ground reaction forces or treadmills to observe motions within the 3D motion capture space.

An alternative to the IR cameras and indicators is to use a single IR camera with no joint markers, such as is used with the Microsoft Kinect motion capture system. This method is not as precise since the joint positions are

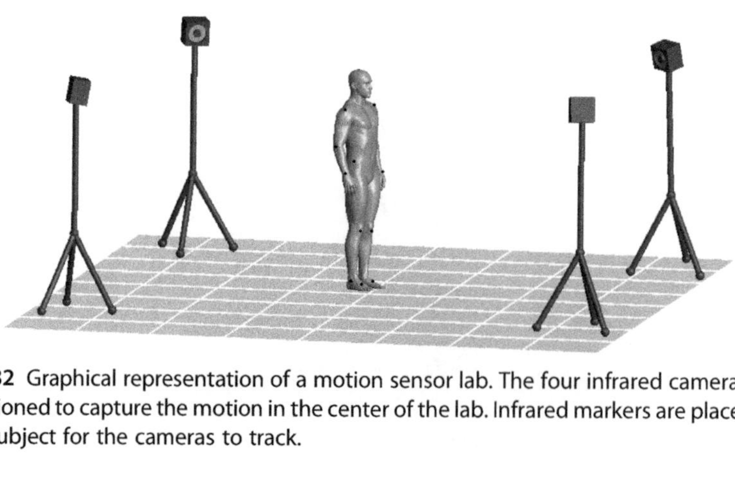

Fig. 32 Graphical representation of a motion sensor lab. The four infrared cameras are positioned to capture the motion in the center of the lab. Infrared markers are placed on the subject for the cameras to track.

calculated, not marked but requires less setup time, is more portable, and allows for greater movement freedom for the subject (Sigal et al., 2010).

Despite all the other tests Jacob has performed, you decide that his gait data will be the most telling for determining how his new prosthesis compares with his old one. You coordinate with the local university and arrange to get access and support for evaluating Jacob's gait with the two prostheses.

On the specified day, you and Jacob arrive at the lab and the lab technicians help Jacob get himself and his two prostheses marked with the IR reflection balls. Once the balls are in place and the system is calibrated to Jacob, the lab supervisor and technicians ask Jacob to walk back and forth along a specific line. Using a self-selected gait, they also ask Jacob to try to time his steps such that his prosthetic leg lands squarely on the load cell on the floor of the lab. Jacob complies and is able to get some good gait data for both prosthetic devices.

With the help of the lab supervisor, you are able to analyze the gait data and are able to determine many things about Jacob's gait with his two devices. You can determine the rate of his self-selected gait, the joint loads and moments, velocities, and accelerations. With the gait data, you can tell Jacob how his gait changes between each device and how those gait patterns compare with the gait of a person without limb loss.

7 CONCLUSION

In this chapter, we have discussed what a sensor is and some key characteristics of all sensors. We have also discussed some passive sensors, simple sensors, common sensors, and a variety of biological sensors. As was mentioned previously, this chapter does not cover all sensor technologies, but is an introduction to those sensors and systems which many be encountered in biomechanical design. Use this information to begin your own explorations into sensing technologies.

REFERENCES

Allegromicro, n.d. List of available Hall effect sensors from Allegro Microsystems. Available from: https://www.allegromicro.com/en/Products/Magnetic-Digital-Position-Sensor-ICs.aspx (Accessed 22 August 2017).

Basmajian, J.V., de Luca, C.J., 1985. Muscles Alive – The Functions Revealed by Electromyography. The Williams & Wilkins Company, Baltimore.

Bolton, W., 2003a. Mechatronics - Electrical Control Systems in Mechanical and Electrical Engineering, third ed. Pearson Education Limited, Harlow, England (Section 2.2).

Bolton, W., 2003b. Mechatronics - Electrical Control Systems in Mechanical and Electrical Engineering, third ed. Pearson Education Limited, Harlow, England (Section 2.3.3).

Bolton, W., 2003c. Mechatronics - Electrical Control Systems in Mechanical and Electrical Engineering, third ed. Pearson Education Limited, Harlow, England (Section 2.5.1).

Bolton, W., 2003d. Mechatronics - Electrical Control Systems in Mechanical and Electrical Engineering, third ed. Pearson Education Limited, Harlow, England (Section 2.6).

Bolton, W., 2003e. Mechatronics - Electrical Control Systems in Mechanical and Electrical Engineering, third ed. Pearson Education Limited, Harlow, England (Section 2.3.7).

Chan, E., Chan, M., Chan, M., 2013. Pulse oximetry: understanding it basic principles facilitates appreciation of its limitations. Respir. Med. 107, 780–799.

Digikey—Current Sensors, n.d. Digikey current sensor guide. Available from: https://www.digikey.com/en/articles/techzone/2012/sep/the-basics-of-current-sensors (Accessed 22 August 2017).

Digikey—Pushbutton-Switches, n.d. List of types available push buttons at Digikey. Available from: https://www.digikey.com/products/en/switches/pushbutton-switches/199?k=button (Accessed 21 August 2017).

ElProCus, n.d. Force sensitive resistors information Available from: https://www.elprocus.com/force-sensing-resistor-technology/ (Accessed 22 August 2017).

Farrell, T., Weir, R., 2007. The optimal controller delay for multifunctional prostheses. 2007 IEEE Trans. Neural. Syst. Rehabil. Eng. 15 (1), 111–118.

Fernandez, E., Greger, B., House, P., Aranda, I., Botella, C., Albisua, J., Soto-Sanchez, C., Alfaro, A., Normann, R., 2014. Acute human brain responses to intracortical microelectrode arrays: challenges and future prospects. Front. Neuroeng. 7, 24.

Lamers, T., Pruitt, B., 2011. The MEMS design process. In: MEMS Materials and Processes Handbook. Springer US, New York, pp. 1–36.

Lupu, R., Ungureanu, F., 2013 A survey of eye tracking methods and applications. Bul. Inst. Polit. Iaşi, t. LIX (LXIII), f. 3.

Milano, S., 2009. Allegro Hall-effect sensor ICs. Product information.

Newark, n.d. List of available types of rotary potentiometers at Newark Available from: http://www.newark.com/c/passive-components/potentiometers-trimmers-accessories/rotary-potentiometers (Accessed 21 August 2017).

Nieman, D., Trone, G., Austin, M., 2003. A new handheld device for measuring resting metabolic rate and oxygen consumption. J Am Diet Assoc 103 (5), 588–593.

Omega—Accelerometers, n.d. Omega accelerometer guide. Available from:: http://www.omega.com/prodinfo/accelerometers.html (Accessed 22 August 2017).

Omega—Straingages, n.d. Omega strain gauge guide Available from: http://www.omega.com/prodinfo/straingages.html (Accessed 22 August 2017).

Omega—Thermistor, n.d. Omega thermistor guide Available from: http://www.omega.com/prodinfo/thermistor.html (Accessed 22 August 2017).

PASCO, n.d. Pasport oxygen gas sensor instruction sheet, 012-11736C. Available from: https://www.pasco.com/file_downloads/Downloads_Manuals/PASPORT-Oxygen-Gas-Sensor-Manual-PS-2126A.pdf (Accessed 24 August 2017).

Purves, D., Augustine, G., Fitzpatrick, D., Hall, W., LaMantia, A., McNamara, J., White, L., 2008a. Neuroscience, fourth ed. Sinauer Associates, Sunderland, MA (Chapter 1).

Purves, D., Augustine, G., Fitzpatrick, D., Hall, W., LaMantia, A., McNamara, J., White, L., 2008b. Neuroscience, fourth ed. Sinauer Associates, Sunderland, MA (Chapter 28).

Raez, M., Hussain, M., Mohd-Yasin, F., 2006. Techniques of EMG signal analysis: detection, processing, classification and applications. Biol. Proced. Online 8, 11–35.

Sigal, L., Balan, A., Black, M., 2010. HumanEva: synchronized video and motion capture dataset and baseline algorithm for evaluation articulated human motion. Int. J. Comput. Vis. 87, 4.

Silverthorn, D., Ober, W., Garrison, C., Silverthorn, A., Johnson, B., 2007. Human Physiology An Integrated Approach, fourth ed. Pearson Benjamin Cummings, San Francisco, USA (Chapter 14).

Soares, A., Andrade, A., Lamounier, E., Carrijo, R., 2003. Development of a virtual myoelectric prosthesis controlled by an emg pattern recognition system based on neural networks. J. Intell. Inf. Syst. 21 (2), 127–141.

Tyler, D., 2017. In: iSens - progress and prospect of a fully implanted system for sensorimotor integration.Myoelectric Controls Symposium.

Tyler, D., Durand, D., 2002. Functionally selective peripheral nerve stimulation with a flat interface nerve electrode. IEEE Trans. Neural Syst. Rehabil. Eng. 22 (4), 294–303.

Wang, D., 2014. FDC1004: Basics of Capacitive Sensing and Applications. Texas Instruments, Dallas.

Weir, R.F., Troyk, P.R., DeMichele, G.A., Kerns, D.A., Schorsch, J.F., Maas, H., 2009. Implantable myoelectric sensors (IMES) for intramuscular electromyogram recording. IEEE Trans. Biomed. Eng. 56 (1), 159–171.

FURTHER READING

Duong, S., Choi, M., 2013. Interactive full-body motion capture using infrared sensor network. Int. J. Comput. Graph. Anim. 3 (4), 41–56.

CHAPTER FOUR

Model-Based Control of Biomechatronic Systems

Naser Mehrabi*, John McPhee†
*University of Washington, Seattle, WA, United States
†Systems Design Engineering, University of Waterloo, Waterloo, ON, Canada

Contents

1 BIOMECHATRONIC SYSTEM MODELS

Biomechatronics is an applied multidisciplinary science that integrates biology, mechanics, and electronics to develop devices that support and assist humans. Based on this broad definition, biomechatronic devices include a wide range of applications, from human prostheses and exoskeletons to driver-assist systems in vehicles. These devices usually consist of a mechanical system actuated with electrical actuators, wherein a controller coordinates the mechatronic system response based on the user's intention and predefined logic. In this chapter, we focus on the model-based design of these controllers.

https://doi.org/10.1016/B978-0-12-812539-7.00004-0

1.1 Mechatronic System Modeling

The first step in the design of a biomechatronic device is to understand how the device will interact with its user and the environment. A successful device considers physiology to reasonably enhance human body movements or compensate for lack of movement. A dynamic model of a biomechatronic device can provide in-depth insight into its dynamic behavior and can be used to design and evaluate model-based control systems. The system model can also be used in model-in-the-loop (MIL) simulations to improve systems design. MIL simulations accelerate the design process by saving time in developing and revising the design on the computer rather than physically creating new prototypes. MIL simulations offer many other advantages such as flexibility (i.e., allow various scenarios) and repeatability (i.e., perform the same experiments repeatedly). Various methods can be used to derive dynamic equations of motion of an multidisciplinary device such as energy-based methods, linear graph theory (McPhee, 1996), and bond graph theory (Karnopp et al., 2012).

1.2 Biomechanical Modeling

To design a device for assisting human movements, it is crucial to understand how the human body works. By only contracting the skeletal muscles, our body can produce very complex and meaningful movements such as walking and reaching. All these actions are initiated by thoughts in the brain and then conveyed through the nervous system to the muscles attached to our skeleton. Some brain activities (i.e., readiness potential) can be produced up to 1 s before the actual volitional movement, and can be captured using electroencephalography (EEG) (Brinkman and Porter, 1979; Deecke and Kornhuber, 1978). EEG is a method that captures the brain's electrical activities by placing noninvasive electrodes along the scalp. These movement initiations are transmitted through the central nervous system (CNS) to the motor neurons that innervate muscle fibers. Then, after a sequence of chemical reactions, the muscle fiber contracts and produces a change in potential in the muscle membrane. This electrical activity produced during muscle contraction can be picked up through electromyography (EMG) using an electrical sensor placed on or under the skin above the muscle of interest. EEG and EMG are windows to our brain because they record signals originating from the brain and thus can be used to capture user intention. There are several assistive devices available in the market [e.g., prostheses and brain-computer interfaces (BCIs)] that take advantage of these signals to understand user intention and control a device.

The biomechanics of human movement can be simulated in computers through inverse and forward dynamics simulations. The natural flow of human motion starts from the motor-neuron spikes in the CNS (i.e., including the brain and spinal cord) leading to the production of muscle twitches and a force pulling the bones to reach the desired position. A forward dynamic simulation can properly capture these neuromuscular dynamics since it follows the same natural flow. Equations of motion are integrated forward in time to obtain motion trajectories from neuromuscular inputs. In contrast, an inverse dynamics approach processes information in the opposite direction: the measured joint trajectories and limb motion and external loads from a motion capture system and force sensors are the simulation inputs, and the muscle twitches are the simulation outputs. While an inverse dynamics approach is useful for clinical decision making, it cannot explain the underlying cause-and-effect relationships between motor neuron-spikes and system kinematics. The forward dynamic simulation can also be used to simulate what-if scenarios such as what happens if the stiffness of a foot-ankle orthoses increases? The biomechanical model parameters can be adjusted to represent different individuals with various physical abilities and disorders.

1.2.1 Inverse Dynamic Simulation

To study the biomechanics of a task, one can measure the kinematics (motion) and perhaps a portion of kinetics (e.g., external loads) of that particular task in the laboratory. The kinematics can be measured using optical movement-monitoring systems with active or passive markers (e.g., Optotrak and Vicon motion capture systems, respectively) or with a markerless system (e.g., Microsoft Kinect), or using other movement assessment tools such as electro-goniometers and inertial measurement units (e.g., MVN suit). Force sensors can measure external loads applied to the body (e.g., foot-ground reaction forces during walking). Knowing the kinematics and external forces acting on the system, one can compute the required generalized forces (e.g., net joint torques and forces) to perform the given task by means of an inverse dynamic simulation. Before an inverse dynamic simulation can be performed, the equations of motion representing the task should be extracted using a dynamic modeling method:

$$\dot{x}(t) = f(x(t), T(t), F(t)) \tag{1a}$$

$$g(x(t), T(t), F(t)) = 0 \tag{1b}$$

where x represents the model coordinates (e.g., positions and velocities) and T and F represent net joint torques and the external loads acting on the system. Eq. (1b) represents the kinematic constraints that restrict the movements of two rigid bodies relative to each other (e.g., joints). Since x and F have been measured beforehand in the laboratory, the joint torques (T) can be computed by simply substituting the measured kinematics and external loads into Eq. (1) and evaluating T at each time step. The torque and force requirement of a task are important to know when designing a system controller and its actuator capacity. For example, the maximum ankle torque during normal walking can be used to select the stiffness or the actuator power of an ankle-foot orthosis (AFO), which is a wearable assistive device that supports and corrects ankle motion.

If muscle-level information is required, a static optimization can be performed to resolve the muscle indeterminacy problem and compute the share of each muscle contributing to the resultant joint torque. The muscle indeterminacy problem results from the number of muscles crossing a joint exceeding the degrees of freedom of that joint; it is difficult to identify individual muscle forces because different combinations of forces can produce the same net joint torque. To resolve this problem, the static optimization is subjected to the torque equilibrium equation:

$$T_j = \sum_{i=1}^{n} r_{i,j}^{m} F_i \tag{2}$$

where $r_{i,j}^{m}$ represents the moment arm of the muscle force F_i about the joint j, and index i refers to the individual muscles crossing the joint of interest. A unique muscle activation pattern similar to that of humans can be achieved by minimizing a physiological cost function during static optimization, such as

$$J = \sum_{i=1}^{n} a_i^p \tag{3}$$

where a is the muscle activation level at the current time step, n is the number of muscles crossing the joint, and p is an exponent (usually, $p=2$). The inverse dynamics can only provide insight into a task whose kinematics and kinetics have already been measured in the laboratory, and it cannot predict the dynamics of a new task based on previously measured data.

1.2.2 Predictive Simulation

A predictive simulation is a forward dynamic simulation that can predict the kinematics and kinetics of a task of interest based on the underlying physiological phenomena governing its dynamics. In these simulations, a mathematical controller representing the human CNS coordinates the movements of the biomechanical model for that task. However, to develop such a controller, we should first understand how our CNS controls our body. As first formulated by Bernstein (1967), the CNS simultaneously coordinates the kinematics and kinetics of body motions, despite uncertain (future) trajectories and the redundancy in muscle actuators. As an example, during reaching and pointing tasks, where only the final position of the hand is specified, an infinite number of hand trajectories (and muscle activation patterns) can be expected to reach the target. However, despite the possible variations, individuals usually choose a similar trajectory. The early observations of reaching and pointing tasks led to the well-known "Minimum-X" models (e.g., minimum-jerk model (Flash and Hogan, 1985; Wada et al., 2001), minimum-torque-change model (Uno et al., 1989), minimum-variance model (Harris and Wolpert, 1998), and minimum-work model (Soechting et al., 1995)) to predict the hand trajectory. These models hypothesize that the CNS coordinates the body movement such that an exertion (X) is minimized. Later, this hypothesis was extended to consider physiologically motivated exertions such as muscle activation effort (Crowninshield and Brand, 1981; Ackermann and van den Bogert, 2010; Happee and Van der Helm, 1995), metabolic energy expenditure (Anderson and Pandy, 2001; Peasgood et al., 2006), and muscle fatigue (Sharif Razavian et al., 2015).

In computer simulations, the Minimum-X model has been successfully implemented using dynamic optimization (DO) to predict the normative human motion for a given task. A common DO approach parameterizes the muscle activation profiles for the period of motion and searches the feasible space to find the profiles that minimize X (Anderson and Pandy, 2001; Davy and Audu, 1987; Yamaguchi and Zajac, 1990; Neptune and Hull, 1998; Kaplan and Heegaard, 2001; Sha and Thomas, 2013). This approach provides an open-loop (feedforward) command of muscle activations to control the given task. This command can represent the descending command of a well-repeated/well-learned task (e.g., platform diving (Koschorreck and Mombaur, 2011)). In this approach, the CNS only recalls the learned information, and does not intelligently adjust the commands in real time.

However, during conscious voluntary movements, the CNS has to continuously update the motor commands to correct for errors (Todorov, 2004). For example, previous studies on pointing and reaching (Sarlegna and Pratik, 2015) have shown that the CNS constantly updates the hand trajectory based on sensory (feedback) information. This sensory information can be received from vision, proprioception, audition, the vestibular system, and internal models that can predict the motion (Desmurget and Grafton, 2000). A few studies have used feedback controllers to coordinate the movements of a musculoskeletal model. The linear quadratic regulator (LQR) and linear quadratic Gaussian (LQG) optimal feedback control methods have been applied to a linear arm model to describe the hand trajectory (Harris and Wolpert, 1998; Todorov and Jordan, 2002; Liu and Todorov, 2007). Later, to control the nonlinear dynamics of the neuromuscular system, an iterative LQG (iLQG) controller has been developed, in which the nonlinear model is iteratively linearized (Todorov and Li, 2005). Recently, a nonlinear model predictive control (NMPC) has been used to mimic the CNS during reaching tasks (Mehrabi et al., 2017). This near-optimal controller uses a nonlinear model to predict the reaching dynamics over a finite horizon ahead of the current time, and uses the sensory information as feedback to correct the prediction errors. Depending on the application, the CNS can be modeled as either a feedforward or feedback controller, or as a combination of both. A control system with both feedforward and feedback components is preferred because it performs better and is more robust to external disturbances.

1.3 Integrated Biomechatronic Models

Having a clear understanding of the dynamical system is crucial in designing a controller, since not only does it strengthen our knowledge about the system but also it reduces development time and cost. A predictive simulation of an integrated model of the biomechatronic device and its user for the task under study allows replicating the user-device interaction in silico (Ghannadi et al., 2017; Mehrabi et al., 2015a). This platform can be used to improve the device and controller design without going through the conventional and cumbersome trial and error design methods. Now that we introduced different approaches to develop and simulate biomechanical models, we will describe the benefits and deficiencies of different model-based control techniques that can be used to operate various biomechatronic devices.

2 MODEL-BASED CONTROL DESIGN

Control systems can be categorized as either open loop or closed loop, depending on their structure. A closed-loop control system regulates control actions based on the information received through a feedback loop. There is no feedback loop in an open-loop system; thus, no further control action adjustments can be made. Both of these control systems, depending on their design methodology, can be categorized into model-based or error-based controllers. Model-based controllers exploit a physical or nonphysical model to estimate system dynamics and predict the system's response to a control action. This category includes optimal, robust, and nonlinear control methods. Error-based controllers use only an error signal (the difference between the desired and actual trajectories) to control the system. Classic proportional-integral-derivative (PID), sliding mode, and fuzzy controllers are the most well-known controllers in this category. Since model-based control methods can consider physiological constraints and often outperform their counterparts, we will focus on model-based control methods in this section.

2.1 Model-Based Open-Loop Control

An open-loop control system is a control system in which the system output does not influence the control actions [shown in Fig. 1A]. In an open-loop control system, a sequence of control actions is precomputed and stored in a feedforward controller, then executed when a trigger is activated. Once the control action is initiated, it cannot be adjusted based on the system response or external loads acting on the system. Feedforward controllers are usually used when there is no feedback available or the interaction with the

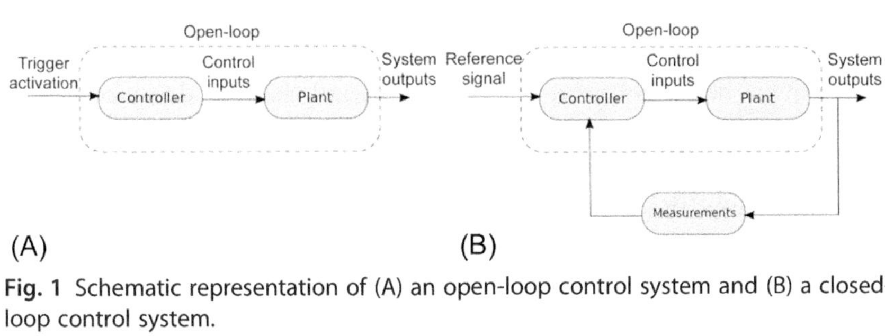

Fig. 1 Schematic representation of (A) an open-loop control system and (B) a closed-loop control system.

environment is reasonably predictable. For example, in functional electrical stimulation (FES) of foot drop, a predefined sequence of electrical impulses stimulates the appropriate muscles to raise the forefoot at the appropriate time during a gait cycle (when a trigger is activated) (Stein et al., 2006). Foot drop is a pathological gait disorder in which the forefoot drags on the ground during walking. Foot drop usually occurs because of muscle weakness or neuromuscular disorders.

The control sequence can be achieved through trial and error experiments or by using a DO method. For example, for FES of gait, an optimal sequence of muscle activations can be achieved through DO of a biomechanical model of gait. The major advantage of this method over the trial and error approach is that in DO, a criterion such as applied electrical stimulation can be minimized so that the onset of muscle fatigue occurs later in therapy.

A DO can be solved through direct and indirect optimal control methods. An indirect method finds an optimal solution by reformulating the original control problem such that the necessary conditions of the optimality are satisfied. In the indirect methods (optimize and then discretize), the optimal control problem is converted to a two-point boundary value problem (2PBVP) by applying Pontryagin's minimum principle. The solution of the 2PBVP provides an optimal solution for the original problem. In a typical direct solution (discretize and then optimize), the dynamic equations are discretized using a numerical integrator; combined with the cost function, the result is a relatively large nonlinear programming (optimization) problem, or NLP. These NLPs can be solved using specially designed optimizers (e.g., IPOPT (Wachter and Biegler, 2006) and SNOPT (Gill et al., 2005)) that exploit the sparsity pattern that exists in such problems. Although this is one of the most common techniques for formulating a direct optimal control problem, there are many other methods (e.g., multiple-shooting and direct collocation) that exist in the literature. Overall, indirect methods may be very sensitive to the initial values and to the changes of the unspecified boundary conditions in the 2PBVP. In contrast, direct methods usually have better convergence properties, and the user doesn't need to worry about the costate variables that appear in indirect methods. However, in the presence of many local extrema, direct methods may converge to a local extremum (Betts, 1998). Although these approaches employ different philosophical approaches, the techniques may ultimately merge. Interested readers are referred to Rao (2009) for more information about indirect and direct optimal control techniques.

2.2 Model-Based Closed-Loop Control

As shown in Fig. 1B, in a closed-loop control system with a feedback controller, the system outputs feedback to the controller to regulate the control action. Feedback controllers based on system dynamics can be categorized into linear and nonlinear feedback controllers.

2.2.1 Linear Control Theory

A linear system is a system whose dynamics obey the superposition principle and whose equations of motion are composed of linear differential equations. Optimal and robust control theories of linear systems with quadratic cost functions have been well developed over decades and have been used in many practical applications (Kirk, 2013; Doyle et al., 2013). In this section, the linear quadratic (LQ) optimal control theory is presented. Consider the linear time-varying system with a state differential equation:

$$\begin{aligned} \dot{x}(t) &= A(t)x(t) + B(t)u(t) \\ z(t) &= C_1(t)x(t) + D_1(t)u(t) \end{aligned} \tag{4}$$

where x, z, and u are system state variables, controlled variables, and control inputs; A, B, C_1, and D_1 are the time-varying matrix functions of time; and x_0 is the state initial condition.

Linear-Quadratic Control

The LQ control law is optimal concerning a quadratic integral performance criterion, as shown below:

$$J = x^T(t_1)P_1x(t_1) + \int_{t_0}^{t_1} \left[z^T(t)R_3(t)z(t) + u^T(t)R_2(t)u(t)\right]dt \tag{5}$$

Here, $R_3(t)$ is a nonnegative-definite symmetric matrix that determines the weighting of each element of the controlled variable z. The quantity $z^T(t)R_3(t)z(t)$ shows the error of the controlled variable z with respect to zero at time t. $R_2(t)$ is a positive-definite symmetric weighting matrix that is used to reduce the control effort. If needed, a terminal state condition can be added to the objective function with a nonnegative-definite symmetric matrix P_1 [see the first term in Eq. 5] such that the state $x(t)$ at the final time t_1 is as close as possible to zero. The optimal feedback controller with respect to the performance criterion shown in Eq. (5) is in the form of a linear full-state feedback controller (Kirk, 2013) as shown in Fig. 2, and the optimal control law is

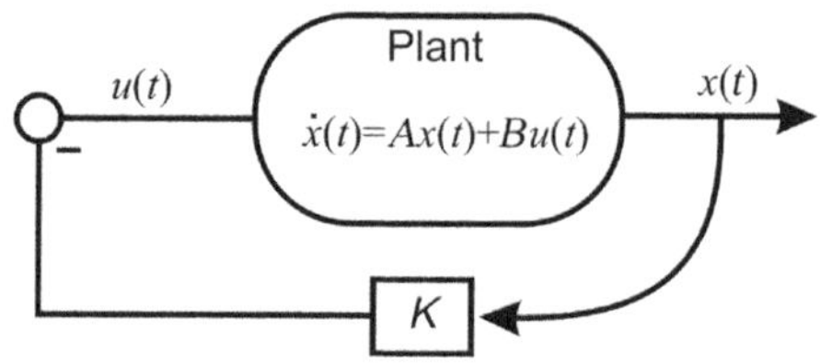

Fig. 2 Schematic representation of a full-state feedback controller.

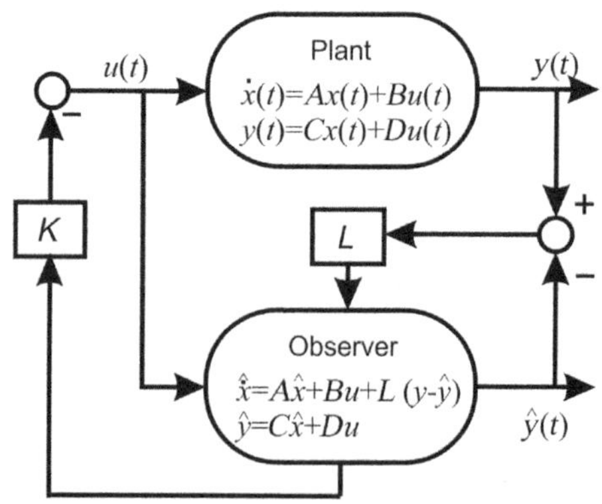

Fig. 3 Schematic representation of a full-state feedback controller with a state observer.

$$u(t) = -K(t)x(t) \tag{6}$$

where

$$K(t) = R_2^{-1}(t)B^T(t)P(t) \tag{7}$$

and $P(t)$ is computed from the solution of the following matrix Riccati equation:

$$-\dot{P}(t) = R_1(t) - P(t)B(t)R_2^{-1}(t)B^T(t)P(t) + A^T(t)P(t) + P(t)A(t) \tag{8}$$

where R_1 is equal to $D^T(t)R_3(t)D(t)$, and the Riccati equation should be solved backward in time with the final condition of $P(t_1) = P_1$.

It is not easy and sometimes even infeasible to measure all the individual state variables required for a full-state feedback controller. In many cases, the measurements are restricted or are a function of a few different state variables, and they may also include measurement noise. One solution is to construct unavailable states from the available measurements (y) and controls (u) using a dynamic system called an *observer* (Fig. 3).

Linear State Estimation

In this section, we introduce an optimal state observer called the Kalman filter (KF). A KF is a data processing algorithm that estimates the current value of the state variables of interest using the available information. A KF incorporates all the available measurements to estimate the current state variables by considering the system and measurement device dynamics, the statistical significance of the measurement and system noise, and the available information about the system's initial condition. Here, consistent with continuous LQ control, a continuous KF is introduced. Consider a linear time-varying continuous-time system:

$$\begin{aligned} \dot{x}(t) &= A(t)x(t) + B(t)u(t) + w(t) \\ y(t) &= C(t)x(t) + v(t) \end{aligned} \tag{9}$$

Here, $y(t)$ is the measurement variable, and C is a continuous time-varying matrix; $w(t)$ and $v(t)$ are Gaussian white noise with zero mean value and Q and R are covariance matrices that represent process noise and sensor noise, respectively. The process noise represents the uncertainty in the system model, and sensor noise is usually used to show uncertainty in the measurements. $Q(t)$ and $R(t)$ are symmetric and nonnegative definite matrices in which each element represents the covariance of the corresponding measurement or system noise. The initial state $x(t_0)$ is also assumed to be Gaussian random variable with a mean value of x_0 and a covariance P_{e0}. A KF is an optimal state observer in which the state estimation $\hat{x}(t)$ is computed in a way that the expected value of the estimation error squared is minimized (i.e., $E\left[(x(t)-\hat{x}(t))(x(t)-\hat{x}(t))^T\right]$). The continuous-time KF observer is in the following form:

$$\begin{cases} \dot{\hat{x}}(t) = A\hat{x}(t) + Bu(t) + L(t)(y - C\hat{x}(t)) \\ \hat{x}(0) = E\{x(t_0)\} \end{cases} \tag{10}$$

where $L(t)$ is often called the Kalman gain from:

$$\begin{aligned} L(t) &= P_e(t)C^T R^{-1} \\ \dot{P}_e &= AP_e + P_e A^T + B_w Q B_w^T - P_e C^T R^{-1} C P_e \end{aligned} \tag{11}$$

which solves forward in time with the boundary condition $P_e(t_0) = P_{e0}$.

Based on the separation principle (Kirk, 2013), the optimal control input can be determined by feeding the estimated states instead of the measurements into Eq. (6). Then, the optimal feedback control becomes:

$$u(t) = -K(t)\hat{x}(t) \tag{12}$$

Here, $K(t)$ is the same gain array obtained from optimal control feedback in Eq. (7). With the substitution of the control law with the observer Eq. (10), the controller equations take the following form:

$$\dot{\hat{x}}(t) = [A(t) - B(t)K(t) - L(t)C(t)]\hat{x}(t) + L(t)y(t) \quad u(t) = -K(t)\hat{x}(t) \tag{13}$$

KFs are used to estimate the system state variables from indirect and noisy measurements that are common in mechatronic systems (e.g., force sensors). LQRs in conjunction with KF can be used to implement a biomechatronic system device control logic while minimizing a cost function (e.g., electrical energy consumption). As an example, this method can increase the battery life of untethered biomechatronic devices or just simply decrease the device energy consumption.

2.2.2 Nonlinear Control Theory

Nonlinear control theory covers a larger class of systems and can be used for a wider range of real-life problems. Nonlinear systems do not obey the superposition principle, and the equations of motion are governed by nonlinear differential-algebraic equations (DAEs). A nonlinear system can be linearized (approximated with a linear system) by use of Taylor series expansion or perturbation methods around an operating point, and then a linear control theory can be applied to design a controller for the nonlinear system. However, the linear model is only valid if the model varies in the sufficiently small range about the operating point, while nonlinear controllers can incorporate nonlinear models to guarantee performance under nonlinear phenomena (e.g., limit cycles, multiple equilibria). In this section, we focus on the NMPC method that has attracted attention both in industry and academia in recent years. NMPC has been widely used in the chemical industry, where a lower sampling rate is required, but recently it has been applied in other industries such as automotive and assistive devices.

A NMPC can be considered as the general form of the LQ control method in which the controller uses a nonlinear model and can account for constraints on inputs and states. Moreover, the NMPC is not required to have a quadratic performance criterion. The NMPC includes both feedforward and feedback control schemes. The NMPC uses a control-oriented model (COM) representing the physical system to predict the optimal dynamics in a finite time interval ahead of current time called the *prediction horizon*, and feedback information to correct the prediction errors.

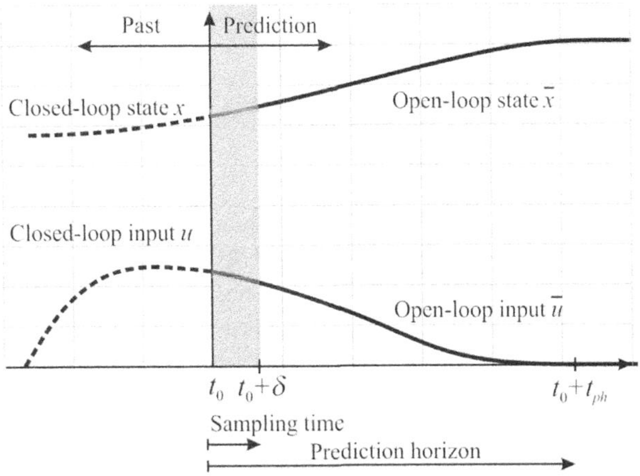

Fig. 4 Schematic representation of the nonlinear model predictive control (NMPC).

The NMPC predicts the optimal dynamics of the system $(\overline{x}, \overline{u})$ over a prediction horizon as shown in Fig. 4 by minimizing the following cost function subjected to the nonlinear dynamic equations of motion:

$$J = \Psi\left(t_0 + t_{ph}\right) + \int_{t_0}^{t_0 + t_{ph}} \psi\left(x(t), u(t)\right) dt \tag{14}$$

where Ψ is the cost evaluated at the end of the prediction horizon, ψ is the cost evaluated during the prediction horizon, and t_{ph} is the length of prediction horizon. As shown in Fig. 4, the state variables at the current time (t_0) are obtained from the current measurements or estimated with the aid of an observer. The input $(\overline{u})$ is an optimal open-loop solution over the prediction horizon. If there are no external disturbances and no model uncertainty in the system, with infinitely long prediction horizon, the open-loop solution can be applied to the system for all time $t > t_0$. However, for the finite horizon case and in the presence of noise and uncertainty, the open-loop solution should only be applied until the next sampling time $(t_0 + \delta)$. At the new time step, the optimal solution is re-evaluated with the new initial conditions for the receding horizon and iteratively applied to the system. By incorporating the feedback information, the NMPC is converted from a completely open-loop controller to an optimal closed-loop controller. The NMPC can handle constraints on both the states and the inputs.

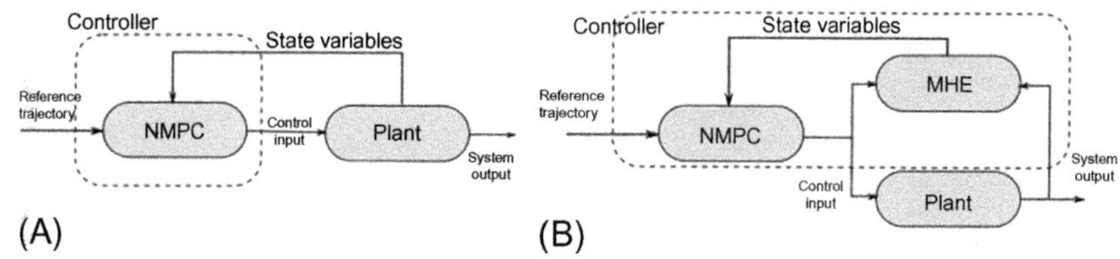

Fig. 5 Schematic representation of a NMPC in conjunction (A) with moving horizon estimator (MHE) and (B) without moving horizon estimator.

The optimal dynamics over the prediction horizon can be calculated using any optimal control method. Several software packages can automatically formulate and execute an NMPC controller (e.g., YANE (Grune and Pannek, 2011), MUSCOD-II (Schafer et al., 2007), ACADO (Diehl et al., 2002), MPsee (Tajeddin and Azad, 2017), SCDE (Walker et al., 2016)).

In the presence of incomplete measurements and for a constrained nonlinear system, an optimization method can be used to estimate the state variables. If all the measurements from the initial to the current time are used to estimate the state at the current time, the observer is called a full-information estimator. However, this technique is not suitable for real-time implementation, since the computational burden grows exponentially with time. By only considering the information in a window moving behind the current time, and approximating older information by a simple function, the computation time can be significantly reduced. This so-called "moving horizon estimator" (MHE) has been shown to work for real-time vehicle dynamics applications and rehabilitation robots with current computational resources (Fig. 5). The required online solution of the optimization problem can be computationally demanding, but can provide significant benefits in estimator accuracy and rate of convergence (Soechting et al., 1995). The optimal estimations at each given horizon (window) can be computed using indirect or direct optimal control methods (Crowninshield and Brand, 1981).

3 CASE STUDY: DESIGN OF POPULATION-BASED ELECTRIC POWER STEERING SYSTEMS

In this section, we examine a case study in which a systematic model-based method to design individualized electric power steering (EPS) systems for different driver populations is introduced. An EPS system is a biomechatronic driver-assist device because it is a mechatronic system that interacts with a human driver, and supports the driver to have a better driving experience. The driver-assist systems receive sensory feedback from the vehicle, and commands such as acceleration, brake, and steering from the driver. Here, four neuromuscular driver models representing drivers with different physical strength, age, and gender were developed. These models were used to design a new model-based EPS controller that adjusts the steering assistance based on the driver's physical strength. In the proposed controller, the EPS characteristic curves (determining the steering assistance) were precomputed for the predefined driver populations and stored in the controller. The characteristic curves were optimized such that the drivers within different populations performing the same steering maneuver have a similar targeted "steering feel." The steering feel was defined by a combination of drivers' muscular effort and road feel. Finally, the new EPS controller was evaluated in MIL simulations using a high-fidelity integrated driver-vehicle model. The results showed that the tuned EPS controller could equally assist drivers with different physical strengths and abilities.

3.1 Introduction

Emerging research has resulted in new models of the interaction dynamics between the vehicle and its driver, the results of which have given rise to new driver-assistance technologies—haptic gas pedals, lane keeping, artificial steering wheel torque feedback (Abbink, 2006), and EPS systems (Mehrabi and McPhee, 2014a; Farrelly et al., 2007). Steering feel and vehicle stability are two commonly used criteria in the design of EPS controllers. Vehicle stability measures are well documented in the vehicle dynamics literature (Karnopp, 2003), while there is only a limited literature available on quantifiable steering feel measures. Previous research has found correlations between steering feel and vehicle handling characteristics; however, these investigations were limited to a specific driver population (i.e., truck drivers) (Rothhämel et al., 2011, 2014). Vehicle manufacturers typically employ professional drivers to tune steering systems to provide "good" steering feel. However, this approach has numerous drawbacks. Such experiments can be expensive, time consuming, and are subject to human error. In addition, the preferred steering feel is different for vehicles with different handling characteristics (e.g., sport vs luxury cars) (Bertollini and Hogan, 1999), and simultaneously the optimum steering feel may vary between driver populations (i.e., drivers with different physical abilities). For example, young drivers generally have stronger muscles, and thus greater ability to overcome resistive torques at the wheel, than elderly drivers. Therefore, it is unlikely to find a unique steering setting that provides optimum steering feel for the general population. This work does not directly deal with setting of the preferred steering feel for a specific vehicle type. However, when a preferred steering feel is set, the proposed EPS system can provide equal steering feel across different predefined driver populations.

Realistic driver models can play a major role in accelerating the development of driver-assistance technologies by reducing the cost and time associated with physical experiments. The driver models are usually developed to assess the vehicle performance and not the driver preference (e.g., path-following driver model). Few studies have developed driver-centered models that consider the driver's physiology (i.e., neuromusculoskeletal system) (Mehrabi et al., 2015a; Cole, 2012). These models can be used to give insight about how our body interacts with the steering system. Understanding and quantifying these interactions facilitates the development of the next generation of driver-assistance technologies. A forward dynamic simulation can simulate the interaction between driver and vehicle, and also provide a platform to ask "what if" questions such as "what if a stronger driver steers the same vehicle." These predictive simulations can support the design of individualized EPS controllers for different driver populations. Accordingly, the following work presents a systematic approach to standardize EPS systems (e.g., steering feel) for various driver populations by considering the human physiology.

3.2 Dynamic Model of Biomechatronic System

In this section, we present the models and methods used to develop and verify an individualized EPS system. We have two integrated models of driver and vehicle that we will refer to as (1) high-fidelity and (2) simplified models. The simplified model was used to design the EPS system, and the high-fidelity model was used in MIL simulations to verify the performance of the EPS controller. Finally, the characteristic curves and the EPS controllers used in these models are presented.

3.2.1 High-Fidelity Driver-Vehicle Model

The high-fidelity integrated driver-vehicle model described in Mehrabi et al. (2015a) and shown in Fig. 6A was used to simulate real-world driving conditions. This model consists of a multibody dynamic model of a vehicle and a three-dimensional (3D) neuromusculoskeletal model of a driver. The muscle activities predicted by the neuromusculoskeletal driver model were verified against the electromyographic activities of a driver's arm muscles

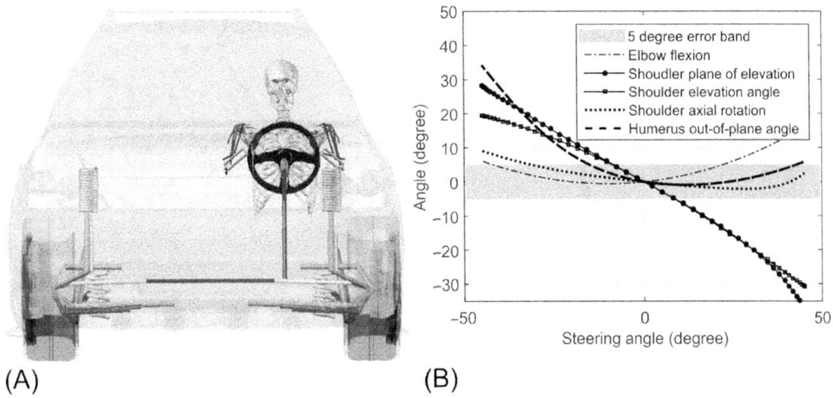

Fig. 6 (A) High-fidelity integrated driver-vehicle model and (B) variation of the shoulder and elbow angles and the rotation of the humerus about the vertical axis for a sinusoidal steering wheel angle. The presented angles are consistent with the definitions recommended by the International Society of Biomechanics (ISB) (Wu et al., 2005).

during steering experiments (Mehrabi and McPhee, 2014b). In the first experiment, the driver was instructed to hold the steering wheel stationary against external torques (indicative of on-center steering); in the second experiment, a sinusoidal steering maneuver was performed to simulate a slalom maneuver (Hayama et al., 2012). Since real-life steering usually is a combination of these two tasks, this driver model can realistically predict muscle activities during everyday steering maneuvers.

The DAEs used to describe the high-fidelity integrated driver-vehicle model are very complex and computationally expensive, and thus not suitable to be used within a real-time optimal control. Therefore, a simplified version of this model that conveys the important dynamics of the system has been developed.

3.2.2 Simplified Driver-Vehicle Model

The simplified integrated driver-vehicle model consists of a linear vehicle model with a column-assist EPS system and a two-dimensional (2D) neuromuscular driver model. To develop the simplified driver model, we first studied the kinematics of the high-fidelity 3D driver model performing a sinusoidal steering maneuver. The modeled driver is holding the steering wheel at the 3 and 9 o'clock positions as suggested by Hayama et al. (2012), and the steering axis is parallel to the line connecting the shoulder to the steering wheel as shown in Fig. 7A. The suggested driver's posture can be changed without substantially affecting the method and simulation

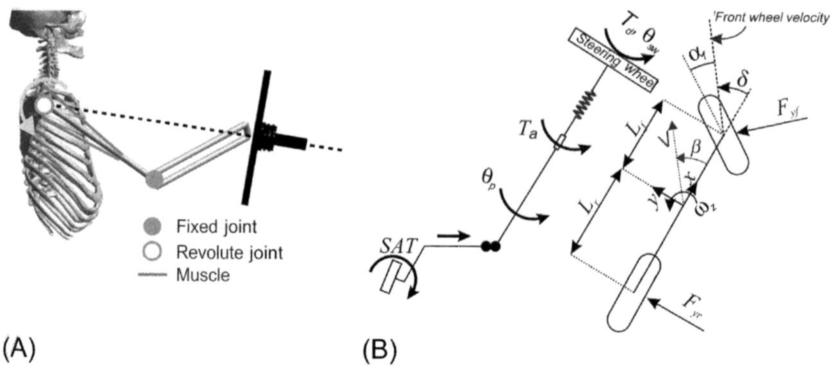

Fig. 7 The simplified driver-vehicle model: (A) the two-dimensional (2D) musculoskeletal driver model and (B) the simplified vehicle model with a column-assist electric power steering (EPS) system.

results. The effect of changing the grip position on the moments of muscle forces on the shoulder and elbow was briefly discussed in Mehrabi et al. (2014).

Fig. 6B shows the variation of elbow and shoulder angles when the 3D driver model performs a sinusoidal steering wheel angle with an amplitude of 45 degrees. In this research, we have used the Euler XYX convention to represent the shoulder's plane of elevation angle (PEA), elevation angle (EA), and axial rotation, respectively, where X is along the humerus and Y is normal to X and toward the humerus lateral direction. The change of shoulder's PEA and EA is significantly larger than the elbow flexion and extension angle. The standard deviation of the elbow angle from its mean value during this simulation is about 5 degrees while it is 22 and 18 degrees for the shoulder's PEA and EA. As expected, the standard deviation of the shoulder's axial rotation is small, around 3 degrees. The humerus rotation about the vertical axis (i.e., parallel to torso) is less than 5 degrees when steering wheel angle varies 14 degrees, depicting a mostly planar motion of the arm for small steering angles. Therefore, the shoulder in the simplified model was reduced from a spherical joint to a revolute joint, and the elbow joint has been assumed to be fixed. Based on these assumptions, a simplified 2D driver model as shown in Fig. 7A was developed, in which the arm segments move only in the sagittal plane of the driver's body, pivoting at the shoulder.

As shown by Jonsson and Jonsson (1975), the shoulder muscles are the prime movers in steering tasks and can be classified into two groups: the muscles providing clockwise torque and muscles providing counterclockwise torque on the steering wheel (Sharif Razavian et al., 2015).

Accordingly, in the simplified driver model, two representative muscles, one flexor and one extensor, were used to actuate each arm segment to represent the muscles producing clockwise and counterclockwise torques. A model inspired from the popular Hill muscle model (Mehrabi et al., 2014; Thelen, 2003) was used to simulate the muscle contraction dynamics. The Hill muscle model consists of a contractile element (CE) and a parallel elastic (PE) element in series with a series elastic (SE) element. In this study, the SE dynamics representing the tendon were neglected because the steering motion is relatively slow and the amount of energy transfer to tendons is small. Therefore, the muscle model was reduced to the CE element, and the muscle force (F_{TM}) was computed as follows:

$$F_{TM} = F_0^{\max}\left(F_{PE}(t, L_M) + F_{CE}(t, a, L_M, V_M)\cos\left(\alpha_p\right)\right) \tag{15}$$

where F_{CE} represents the active force of the muscle and L_M, V_M, α_p, and $F_0^{\max}$ are the muscle length, contraction velocity, pennation angle, and maximum isometric muscle force, respectively. The muscle activation level (a) represents the number of active motor units in the muscle (between 0% and 100%), and since the SE element was removed, the pennation angle for all muscles was assumed to be zero. The force generated by F_{CE} can be separated into force-length and force-velocity relations scaled by the muscle activation command (a):

$$F_{CE} = a(t)F_{CE}^L(t, L_M)F_{CE}^V(t, a, L_M, V_M) \tag{16}$$

where the force-length (F_{CE}^L) and force-velocity (F_{CE}^V) relations are:

$$F_{CE}^L = e^{-\left(\frac{L_M}{L_M^{\text{opt}}}-1\right)^2/\gamma} \tag{17}$$

$$F_{CE}^V = \begin{cases} \dfrac{\dfrac{V_M}{V_M^{\max}L_M^{\text{opt}}} + AV_M^{\max}}{-\dfrac{V_M}{V_M^{\max}L_M^{\text{opt}}A_f} + AV_M^{\max}} & V_M < 0 \\[2ex] \dfrac{\dfrac{V_M B\overline{F}_{\max}^{\text{len}}}{V_M^{\max}L_M^{\text{opt}}} + ACV_M^{\max}}{-\dfrac{V_M B}{V_M^{\max}L_M^{\text{opt}}A_f} + ACV_M^{\max}} & V_M > 0 \end{cases} \tag{18}$$

where γ, A, B, and C are shape factors, $V_M^{\max}$ is the maximum fiber velocity, L_M^{opt} is the optimal length of fiber at which F_{CE} is a maximum, and $\overline{F}_{\max}^{\text{len}}$ is the maximum normalized muscle force during lengthening. The numerical values of the muscle parameters used in this research are reported in Mehrabi and McPhee (2014a). The PE force of muscle (F_{PE}) is represented by an exponential function:

$$F_{PE} = \frac{e^{k_{pe}\left(\frac{L_M}{L_M^{\text{opt}}}-1\right)/\epsilon_0^m} - 1}{e^{k_{pe}} - 1} \tag{19}$$

where k_{pe} (=0.5) is a shape factor and ϵ_0^m is passive muscle strain at maximum isometric force.

For steering with two hands, the total torque T_d generated at the steering wheel is as follows:

$$T_d = 2\,G_{SHS}\bar{r}\begin{cases} F_f(a, L_M, V_M) & T_d \geq 0 \\ -F_e(|a|, \theta, \cdot\theta) & T_d < 0 \end{cases} \tag{20}$$

where F_f and F_e are flexor and extensor muscle forces that, respectively, produce a clockwise and counterclockwise torque at the steering wheel, and θ and $\bar{r}$ are the shoulder angle and the average moment arm of flexor and extensor muscles, respectively; G_{SHS} is a fixed ratio that projects the moment of muscles produced at shoulder to the steering wheel. For simplicity, the muscle length and velocity, and moment arms, are rearranged and simplified to be only a function of shoulder angle and angular velocity (i.e., $L_M = L_0 \pm \bar{r}\theta$ and $V_M = \pm\bar{r}\dot{\theta}$). Here, we assume that there is no muscle co-contraction between flexor and extensor muscles, and the positive and negative values of T_d are produced by the flexor and extensor muscles, respectively.

A simplified single-track model with a column-assist EPS steering system as shown in Fig. 7B was developed to speed up the optimization procedures. The driver torque T_d transfers through a torsion bar to the steering pinion and rotates the tires. The torque sensor measures the torsion bar twist and sends it to the EPS system that regulates the assist torque (T_a). The following equation describes the steering wheel, and the torque sensor dynamics:

$$J_{sw}\ddot{\theta}_{sw} = -b_{sw}\dot{\theta}_{sw} + T_{tb} + T_d \tag{21}$$

$$T_{tb} = K_{tb}\left(\theta_{sw} - \theta_p\right) \tag{22}$$

where T_d and T_{tb} are the driver and the torsion bar torques, θ_p is the pinion angle, and θ_{sw}, J_{sw}, and b_{sw} are the angle of rotation, the moment of inertia, and the viscous damping coefficient of the steering column, respectively.

The rack and its connection to the wheel spindle, as well as the intermediate steering shaft, are combined and represented as a single inertia at the pinion. The dynamics of the steering pinion are described by

$$J_p\ddot{\theta}_p = -K_p\theta_p - b_p\dot{\theta}_p + T_{tb} + T_a + T_{SAT} \tag{23}$$

where θ_p, J_p, and b_p are, respectively, angular displacement, inertia, and damping of the pinion, and K_p is the stiffness induced by the inclined kingpin axis on the rack displacement. T_{SAT} and T_a represent the self-aligning torque (SAT) and the assist torque provided by the EPS system, respectively.

In the single track, the vehicle's center of mass velocity (V) makes an angle β with the longitudinal direction of the vehicle. Considering the side-slip angle (β) and yaw rate (ω_z) of the vehicle as the state variables of the single track model, the equations of motion are expressed as follows:

$$mv_x\left(\dot{\beta} + \omega_z\right) = F_{yf}\cos(\delta) + F_{yr} \tag{24}$$

$$I_{zz}\dot{\omega}_z = L_f F_{yf}\cos(\delta) - L_r F_{yr} \tag{25}$$

where F_{yf} and F_{yr} are front and rear lateral force of the wheels and are approximated by a linear tire model (in contrast to a nonlinear tire model used in the high-fidelity model):

$$F_{yf} = C_{\alpha f}\alpha_f \tag{26}$$

$$F_{yr} = C_{\alpha r}\alpha_r \tag{27}$$

Assuming small steer angles, the front and rear slip angles can be approximated as follows:

$$\alpha_f = \frac{v_y + L_f\omega_z}{v_x} - \delta \tag{28}$$

$$\alpha_r = \frac{v_y - L_r\omega_z}{v_x} \tag{29}$$

where v_x and v_y, respectively, are the longitudinal $[v_x = V\cos(\beta)]$ and lateral $[v_y = V\sin(\beta)]$ components of the vehicle mass center velocity, and v_x is assumed to be constant during the simulations. The steering angle of the front wheel is represented by $\delta = \theta_p / G_{\text{steering}}$, and G_{steering} is the ratio of the rotation of steering wheel angle to the average value of left and right wheel steer angles. The SAT, which is created by the interaction between the tire and the road, is a linear function of slip angle (α_f) for small slip angles ($T_{SAT} = C_{T\alpha}\alpha_f$), where $C_{T\alpha}$ is a SAT coefficient.

3.3 Electric Power Steering (EPS) Control Design

The main responsibility of an EPS system is to reduce the driver physical effort. As a result, almost all power steering systems have a component in their logic to magnify driver torque by generating an assist torque proportional to the driver torque. The relation that assigns an EPS assist torque to each driver steering torque is presented in so-called characteristic curves. Typically, the steering characteristic curves are multilinear functions of the driver steering torque at different vehicle speeds. In this research, we used a bilinear characteristic curve at each given speed as shown in Fig. 8. This characteristic curve consists of an unassisted zone to avoid the off-center feeling, a steering assistance zone, and a maximum assist value that is restricted by maximum motor torque. The bilinear characteristic curves can be expressed as follows:

$$T_a = \begin{cases} 0 & 0 < T_d < T_{d0} \\ K_a(T_d - T_{d0}) & T_{d0} < T_d < T_d^{\max} \\ T_m^{\max} & T_d^{\max} < T_d \end{cases} \tag{30}$$

where T_a, $T_m^{\max}$, and K_a, respectively, represent the assist torque, the maximum torque of the motor, and the assist gain. T_d, T_{d0}, and $T_d^{\max}$ represent the driver's steering torque, the driver's steering torque when the motor begins to assist, and the driver's steering torque when the motor assist reaches the maximum assistance ($T_d^{\max} = \frac{T_m^{\max}}{K_a} + T_{d_0}$), respectively. The coefficient K_a is an adjustable shape factor that represents the rate of assist. Note that K_a reduces as vehicle speed increases. In the high-fidelity integrated

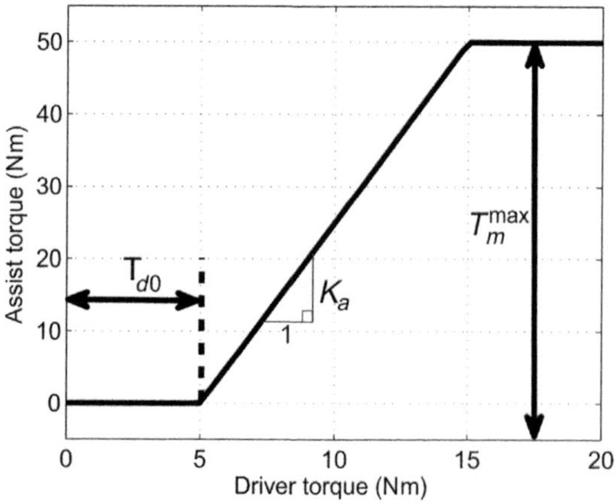

Fig. 8 Bilinear EPS characteristic curve.

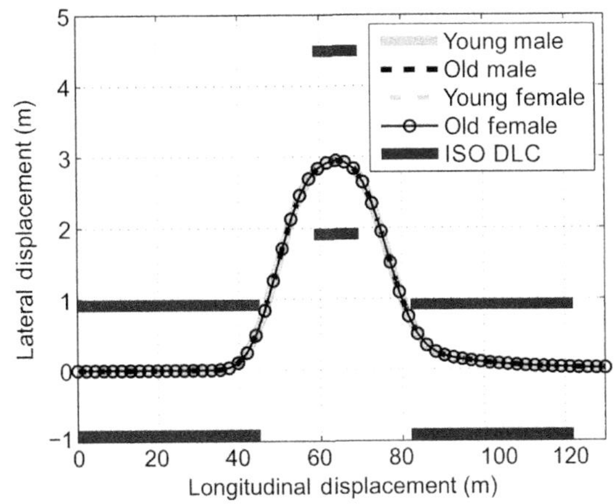

Fig. 12 The vehicle trajectory of the four driver types performing an ISO DLC maneuver. Vehicle trajectories of all four driver types are shown.

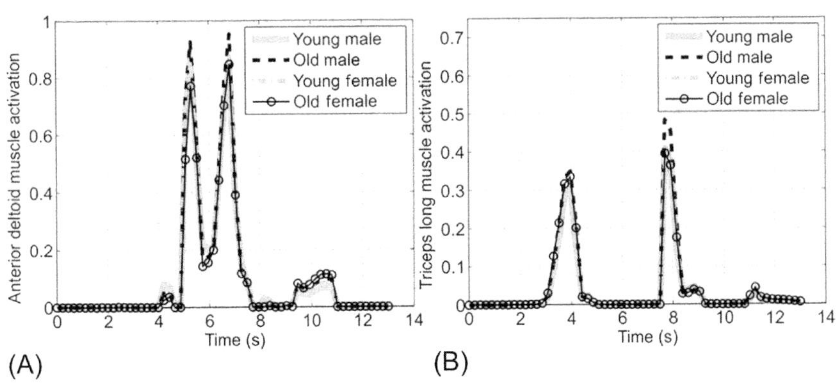

Fig. 13 Right arm's muscle activities during a double lane-change maneuver for the four driver types (A) anterior portion of deltoid and (B) long head of triceps. Muscle activities of all four groups are shown.

the high-fidelity vehicle model equipped with an EPS controller tuned for that specific group at the speed of 10 m/s.

As shown in Fig. 12, the vehicle lateral displacements of all groups are similar to each other and to the desired trajectory, and they are all within the ISO double-lane change maneuver constraints. Therefore, the steering loads in all of the simulations are the same, since the driving conditions in all of the simulations are the same.

Fig. 13 shows the predicted muscle activities of the anterior portion of deltoid and the long head of triceps of the driver's right arm for the four

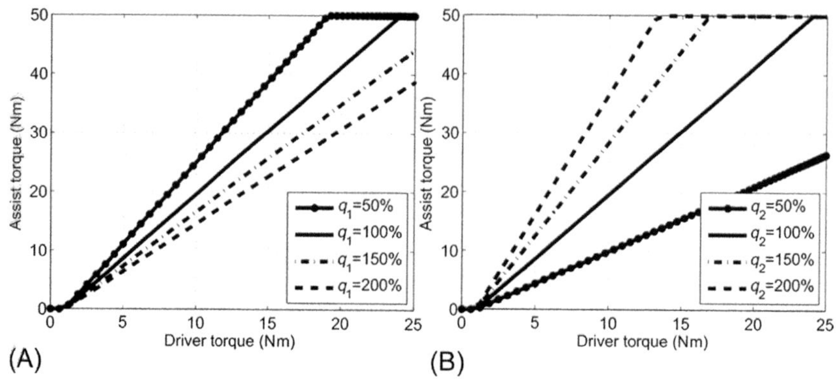

Fig. 14 Sensitivity of the optimal characteristic curve to the variation of optimization weights, (A) variation of q_1 and (B) variation of q_2.

predefined driver types. Although other muscle activations are not presented here, a similar behavior can be seen in other muscles. As shown in this figure, the magnitude and trend of these patterns are very similar. Although young male drivers have higher physical strength than old female drivers, the portion of motor units that have been recruited by the CNS are the same as for other drivers. In conclusion, the drivers' muscular efforts are equal, thereby satisfying the controller objective to provide the same targeted steering feel to all drivers.

Fig. 14 shows the sensitivity of the characteristic curve to the variation of cost function weighting factors. The cost function weights are modified proportional to their nominal values. The results demonstrate that the variation of muscle fatigue weight (q_2) has a greater effect on the characteristic curve's assist gain than the variation of road feel weight (q_1), because the cost function is a linear function of the road feel but a quadratic function of muscle activations. These cost function weights can be used to adjust the target steering feel. For example, for a sports car, the driver expects to have stiffer steering than in a comfortable car. Then, to have a sportier feel, the road feel weighting factor should be increased as shown in Fig. 14A, which results in less assistance and a steering system, that is, therefore more sensitive to road forces.

4 CONCLUSIONS

In this chapter, we introduced various tools for the model-based design of biomechatronic systems. Included in these tools are integrated biomechatronic system models, model-based controllers, and inverse and predictive simulations. The biomechatronic model is an integrated model of the user's biomechanics and a dynamic model of the assistive device, which can be used to simulate the human-machine interactions. The biomechatronic model parameters can be adjusted to represent a specific individual or groups of individuals. This biomechatronic model facilitates the design of individualized model-based controllers, and can be used to improve the device and controller design through MIL inverse or predictive simulations.

In the case study, a systematic method to consider the driver's physical characteristics in the design of a driver-specific EPS controller is proposed. To design such an EPS controller, first, the high-fidelity driver-vehicle model is simplified to reduce the computational burden associated with the multibody and biomechanical systems. The muscle parameters in the high-fidelity and simplified integrated driver-vehicle models have been adjusted to represent drivers with different physical abilities (young male, old male, young female, and old female). A steering feel optimization procedure is used to tune the EPS controller for each group. Simulation results using the high-fidelity biomechatronic driver-vehicle model showed that it is possible to develop a model-based EPS controller that considers the physical characteristics of a driver and delivers a targeted steering feel to a predefined driver population. Evaluation of the tuned EPS controller also showed that, although the EPS controller has been tuned based on the simplified model, the controller shows the same expected behavior in high-fidelity simulations.

REFERENCES

Abbink, D., 2006. Neuromuscular Analysis of Haptic Gas Pedal Feedback During Car Following. Delft University of Technology, Delft, The Netherlands.

Ackermann, M., van den Bogert, A.J., 2010. Optimality principles for model-based prediction of human gait. J. Biomech. 43 (6), 1055–1060.

Anderson, F.C., Pandy, M.G., 2001. Static and dynamic optimization solutions for gait are practically equivalent. J. Biomech. 34 (2), 153–161.

Bernstein, N.A., 1967. The Co-ordination and Regulation of Movements. Pergamon Press, Oxford, UK.

Bertollini, G., Hogan, R., 1999. Applying driving simulation to quantify steering effort preference as a function of vehicle speed. SAE technical paper 1999-01-0394. https://doi.org/10.4271/1999-01-0394.

Betts, J.T., 1998. Survey of numerical methods for trajectory optimization. J. Guid. Control. Dyn. 21 (2), 193–207.

Brinkman, C., Porter, R., 1979. Supplementary motor area in the monkey: activity of neurons during performance of a learned motor task. J. Neurophysiol. 42 (3), 681–709.

Cole, D., 2012. A path-following driver–vehicle model with neuromuscular dynamics, including measured and simulated responses to a step in steering angle overlay. Veh. Syst. Dyn. 50 (4), 573–596.

Crowninshield, R.D., Brand, R.A., 1981. A physiologically based criterion of muscle force prediction in locomotion. J. Biomech. 14 (11), 793–801.

Davy, D., Audu, M., 1987. A dynamic optimization technique for predicting muscle forces in the swing phase of gait. J. Biomech. 20 (2), 187–201.

Deecke, L., Kornhuber, H.H., 1978. An electrical sign of participation of the mesial 'supplementary' motor cortex in human voluntary finger movement. Brain Res. 159 (2), 473–476.

Desmurget, M., Grafton, S., 2000. Forward modeling allows feedback control for fast reaching movements. Trends Cogn. Sci. 4 (11), 423–431.

Diehl, M., Bock, H.G., Schloder, J.P., Findeisen, R., Nagy, Z., Allgower, F., 2002. Real-time optimization and nonlinear model predictive control of processes governed by differential-algebraic equations. J. Process Control 12 (4), 577–585.

Doyle, J., Francis, B., Tannenbaum, A.R., 2013. Feedback Control Theory. Courier Corporation, Macmillan Publishing Co., New York

Farrelly, J.O.P., Stevens, S.D., Barton, A.D., 2007. Haptic controller for road vehicles. Patent US patent 7,234,564.

Flash, T., Hogan, N., 1985. The coordination of arm movements: an experimentally confirmed mathematical model. J. Neurosci. 5 (7), 1688–1703.

Forkenbrock, G., Elsasser, D., 2005. An assessment of human driver steering capability.

Forster, E., Simon, U., Augat, P., Claes, L., 2004. Extension of a state-of-the-art optimization criterion to predict co-contraction. J. Biomech. 37 (4), 577–581.

Ghannadi, B., Mehrabi, N., Sharif Razavian, R., McPhee, J., 2017. In: Nonlinear model predictive control of an upper extremity rehabilitation robot using a two dimensional human-robot interaction model.IEEE/RSJ International Conference on Intelligent Robots and Systems, Vancouver.

Gill, P.E., Murray, W., Saunders, M.A., 2005. SNOPT: an SQP algorithm for large-scale constrained optimization. SIAM Rev. 47 (1), 99–131.

Grune, L., Pannek, J., 2011. Nonlinear Model Predictive Control. Springer, London.

Happee, R., Van der Helm, F.C.T., 1995. The control of shoulder muscles during ing goal directed movements, an inverse dynamic analysis. J. Biomech. 28 (10), 1179–1191.

Harris, C., Wolpert, D., 1998. Signal-dependent noise determines motor planning. Nature 394, 780–784.

Hayama, R., Liu, Y., Ji, X., Mizuno, T., 2012. In: Preliminary research on muscle activity in driver's steering maneuver for driver's assistance system evaluation.Proceedings of the FISITA 2012 World Automotive Congress.

Jonsson, S., Jonsson, B., 1975. Function of the muscles of the upper limb in car driving. Ergonomics 18 (4), 375–388.

Kaplan, M.L., Heegaard, J., 2001. Predictive algorithms for neuromuscular control of human locomotion. J. Biomech. 34, 1077–1083.

Karnopp, D., 2003. Vehicle Stability. CRC Press, Boca Raton, FL.

Karnopp, D.C., Margolis, D.L., Rosenberg, R.C., 2012. System Dynamics: Modeling, Simulation, and Control of Mechatronic Systems. John Wiley & Sons, Hoboken, NJ.

Kirk, D.E., 2013. Optimal Control Theory: An Introduction. Courier Corporation, Englewood Cliffs, NJ.

Koschorreck, J., Mombaur, K., 2011. Modeling and optimal control of human platform diving with somersaults and twists. Optim. Eng. 13 (1), 29–56.

Liu, D., Todorov, E., 2007. Evidence for the flexible sensorimotor strategies predicted by optimal feedback control. J. Neurosci. 27 (35), 9354–9368.

McPhee, J., 1996. On the use of linear graph theory in multibody system dynamics. Nonlinear Dyn. 9 (1), 73–90.

Mehrabi, N., McPhee, J., 2014a. In: Steering feel improvement for different driver types using model-based control.Proceedings of the ASME 2014 International Design Engineering Technical Conferences, August 17–20, Buffalo, USA.
Mehrabi, N., McPhee, J., 2014b. In: Evaluation of a musculoskeletal arm model for automobile drivers using electromyography.7th World Congress of Biomechanics, July 6–11, Boston, MA.
Mehrabi, N., Sharif Razavian, R., McPhee, J., 2014. A physics-based musculoskeletal driver model to study steering tasks. J. Comput. Nonlinear Dyn. 10 (12), 021012.
Mehrabi, N., Sharif Razavian, R., McPhee, J., 2015a. Steering disturbance rejection using a physics-based neuromusculoskeletal driver model. Veh. Syst. Dyn. 53 (10), 1393–1415.
Mehrabi, N., McPhee, J., Azad, N.L., 2015b. Observer-based disturbance rejection control of electric power steering systems. Proc. IMechE D: J. Auto. Eng. 230 (7), 867–884.
Mehrabi, N., Sharif Razavian, R., Ghannadi, B., McPhee, J., 2017. Predictive simulation of reaching moving targets using nonlinear model predictive control. Front. Comput. Neurosci. 10, 143.
Neptune, R.R., Hull, M.L., 1998. Evaluation of performance criteria for simulation of submaximal steady-state cycling using a forward dynamic model. J. Biomech. Eng. 120, 334–341.
Peasgood, M., Kubica, E., McPhee, J., 2006. Stabilization and energy optimization of a dynamic walking gait simulation. ASME J. Comput. Nonlinear Dyn. 2 (1), 149–159.
Rao, A., 2009. A survey of numerical methods for optimal control. Adv. Astronaut. Sci. 135 (1), 497–528.
Rothhämel, M., IJkema, J., Drugge, L., 2011. A method to find correlations between steering feel and vehicle handling properties using a moving base driving simulator. Veh. Syst. Dyn. 12 (49), 1837–1854.
Rothhämel, M., IJkema, J., Drugge, L., 2014. Influencing driver chosen cornering speed by means of modified steering feel. Veh. Syst. Dyn. 52(4).
Sarlegna, F.R., Pratik, K.M., 2015. The influence of visual target information on the online control of movements. Vis. Res. 110, 144–154.
Schafer, A., Kuhl, P., Diehl, M., Schloder, J., Bock, H.G., 2007. Fast reduced multiple shooting methods for nonlinear model predictive control. Chem. Eng. Process. 46 (11), 1200–1214.
Sha, D., Thomas, J., 2013. An optimisation-based model for full-body upright reaching movements. Comput. Methods Biomech. Biomed. Eng. 18 (8), 847–860.
Sharif Razavian, R., Mehrabi, N., McPhee, J., 2015. A model-based approach to predict muscle synergies using optimization: application to feedback control. Front. Comput. Neurosci. 9.
Soechting, J., Buneo, C., Herrmann, U., Flanders, M., 1995. Moving effortlessly in three dimensions: does donders law apply to arm movement? J. Neurosci. 1, 27–32.
Stein, R., Chong, S., Everaert, D., Rolf, R., Thompson, A., Whittaker, M., Robertson, J., Fung, J., Preuss, R., Momose, K., Ihashi, K., 2006. A multicenter trial of a footdrop stimulator controlled by a tilt sensor. Neurorehabil. Neural Repair 20 (3), 371–379.
Tajeddin, S., Azad, N., 2017. In: Ecological cruise control of a plug-in hybrid electric vehicle: a comparison of different GMRES-based nonlinear model predictive controls. American Control Conference (ACC), Seattle, USA.
Thelen, D., 2003. Adjustment of muscle mechanics model parameters to simulate dynamic contractions in older adults. J. Biomech. Eng. 125 (1), 70–77.
Todorov, E., 2004. Optimality principles in sensorimotor control. Nat. Neurosci. 907–915.
Todorov, E., Jordan, M.I., 2002. Optimal feedback control as a theory of motor coordination. Nat. Neurosci. 5 (11), 1226–1235.
Todorov, E., Li, W., 2005. In: A generalized iterative LQG method for locally-optimal feedback control of constrained nonlinear stochastic systems.American Control Conference, Portland, Oregon.

Uno, Y., Kawato, M., Suzuki, R., 1989. Formation and control of optimal trajectory in human multijoint arm movement. Biol. Cybern. 61, 89–101.

Wachter, A., Biegler, L.T., 2006. On the implementation of a primal-dual interior point filter line search algorithm for large-scale nonlinear programming. Math. Program. 106 (1), 25–57.

Wada, Y., Kaneko, Y., Nakano, E., Osu, R., Kawato, M., 2001. Quantitative examinations for multi joint arm trajectory planning—using a robust calculation algorithm of the minimumcommanded torque change trajectory. Neural Netw. 14, 381–393.

Walker, K., Samadi, B., Huang, M., Gerhard, J., Butts, K., Kolmanovsky, I., 2016. Design environment for nonlinear model predictive control. SAE Technical Paper, Paper# 2016-01-0627, vol. 04. .

Wu, G., van der Helm, F.C., DirkJan Veeger, H., Makhsous, M., Van Roy, P., Anglin, C., Nagels, J., Karduna, A.R., McQuade, K., Wang, X., Werner, F.W., Buchholz, B., 2005. ISB recommendation on definitions of joint coordinate systems of various joints for the reporting of human joint motion - Part II: shoulder, elbow, wrist and hand. J. Biomech. 38 (5), 981–992.

Yamaguchi, G.T., Zajac, F.E., 1990. Restoring unassisted natural gait to paraplegics via functional neuromuscular stimulation: a computer simulation study. IEEE Trans. Biomed. Eng. 37, 886–902.

Zaremba, A., Davis, R., 1995. In: Dynamic analysis and stability of a power assist steering system.American Control Conference, Seattle, WA.

FURTHER READING

Mizuno, T., Hayama, R., Kawahara, S., Lou, L., Liu, Y., Ji, X., 2013. Research on relationship between steering maneuver and muscle activities. eb-cat.ds-navi.co.jp, 13–18.

Pick, A., Cole, D., 2007. Driver steering and muscle activity during a lane-change manoeuvre. Veh. Syst. Dyn. 45 (9), 781–805.

PART TWO

Biomechatronic Devices

CHAPTER FIVE

Biomechatronic Applications of Brain-Computer Interfaces

Domen Novak
Department of Electrical & Computer Engineering, University of Wyoming, Laramie, WY, United States

Contents

Brain-computer interfaces (BCIs), which measure a human's brain activity and use it to control machines, have nearly limitless potential in biomechatronics. Indeed, such biomechatronic applications of BCIs have been a staple of science fiction for decades: BCIs were used to connect to the Matrix in the 1999 movie of the same name, they were used by a paralyzed Captain Pike to control his wheelchair in a 1966 episode of Star Trek,

Handbook of Biomechatronics
https://doi.org/10.1016/B978-0-12-812539-7.00008-8

and they were used by Robocop to control his prosthetic limbs in the 1987 movie. While these applications may have seemed far-fetched at the time, scientists have now developed actual functioning prototypes of BCI-controlled wheelchairs, prostheses, and other biomechatronic devices. However, real-life BCIs are also prone to errors and lack intuitiveness, and thus have not yet achieved widespread use. In this chapter, we briefly review the functional principles of BCIs, their advantages and disadvantages, and existing prototypes in a number of biomechatronic applications.

1 BCI MODALITIES AND SIGNALS

Most state-of-the-art BCIs are based on electroencephalography (EEG), a noninvasive measurement of the brain's electrical activity obtained from the scalp (Section 1.1). However, BCIs can also utilize invasive electrical measurements (Section 1.2) or hemodynamic measurements (Section 1.3), and multiple sensing modalities can be combined for better performance (Section 1.4).

1.1 Electroencephalography

EEG is the use of electrodes placed on the scalp to measure the electrical activity of the brain (Jackson and Bolger, 2014). This electrical activity arises from synchronized synaptic activity in populations of cortical neurons

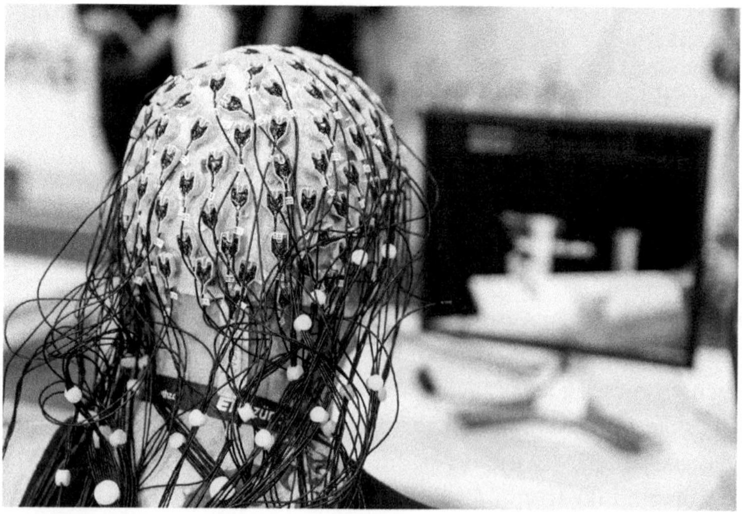

Fig. 1 A person uses an electroencephalography system to play a computer game. *(Courtesy Cybathlon, ETH Zurich. Photographer: Alessandro Della Bella.)*

(Da Silva, 2010) and can be detected using electrodes placed on the scalp (Fig. 1). However, since the brain contains many different neurons and is separated from the electrodes by layers of tissue (dura, skull, and skin), any scalp electrode essentially measures the summed activity of thousands of individual neurons. Furthermore, the signal obtained from the electrode does not necessarily only reflect the activity of the neurons directly beneath the electrode, but may also contain components originating from other regions of the brain (Jackson and Bolger, 2014). Finally, the tissues between the brain and electrode essentially act as a low-pass filter, attenuating high-frequency components of brain activity. Thus, high-quality hardware and signal-processing approaches are required to obtain useful data from EEG.

EEG can be recorded from many locations on the scalp, depending on the brain region of interest. To standardize EEG electrode placement, researchers have developed the International 10–20 system to describe different electrode locations. A standard 10–20 layout is shown in Fig. 2, and labels electrode sites according to their region and distance from the central line of the head. For example, F sites are located in the frontal region (close to the forehead) while C sites are located in the central region. Cz (C-zero) is located in the center of the scalp while C1 is located slightly to the left of Cz and C3 is located farther to the left of Cz; conversely, C2 is located slightly to the right of Cz and C4 is located farther to the right.

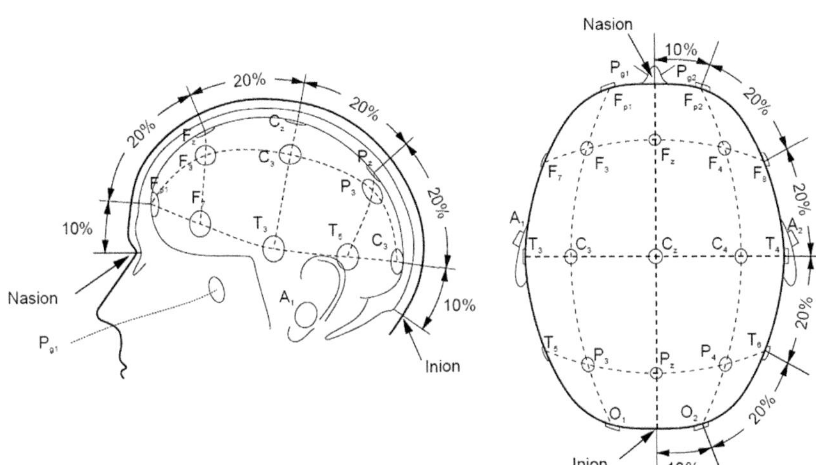

Fig. 2 Electroencephalogram electrode placement on the scalp according to the International 10–20 system. *(From Nicolas-Alonso, L.F., Gomez-Gil, J., 2012. Brain computer interfaces, a review. Sensors 12, 1211–1279, reused under the Creative Commons Attribution License.)*

1.1.1 EEG Paradigms

Before focusing on the technical aspects of EEG measurements, let us first look at the waveforms of interest in the EEG signal as well as ways of eliciting them. The most important waveforms for biomechatronics are steady-state visually evoked potentials (SSVEPs), the P300, and motor/mental imagery, all of which are used to actively send commands through a BCI (Novak and Riener, 2015). However, BCIs can also measure a user's mental workload or error-related brain potentials without the user's active participation or even awareness, as we shall see in the following sections.

Steady-State Visually Evoked Potentials

SSVEPs are the brain's natural responses to visual stimulation at different frequencies (Nicolas-Alonso and Gomez-Gil, 2012). In brief, if a person looks at a light that is flashing with a particular frequency, their visual cortex responds with EEG activity at the same frequency. This principle is used in BCIs as a gaze-tracking method: multiple symbols are shown to the user on a screen, with each symbol flashing at a different frequency. By measuring the SSVEP frequency using electrodes close to the visual cortex, the machine can identify which symbol the user is looking at. Depending on the number and complexity of possible commands, this can be done either in a single stage (the final command is directly selected from all possible ones) or in multiple stages (a subset of commands is first selected from all possible ones, and the final specific command is then selected from the subset).

SSVEPs are commonly used in biomechatronics to send commands to a device. The user is presented with multiple commands on a screen (e.g., move robot forward, stop) and selects one by looking at it. The user can also choose not to send a command by simply not focusing on the screen. The approach is noninvasive and easy to use with little or no training, and the number of possible commands can be quite high–the main limitations are keeping the symbols on the screen far enough apart so that the user is not looking at two flashing lights at once as well as keeping the different symbols flashing at sufficiently different frequencies that they can be separated in the EEG. The main disadvantage of the SSVEP approach is that a screen must be added to the device, which may not be optimal for all situations (e.g., portable devices). Furthermore, it is prone to false positives since users still see the screen at the edge of their vision even if they do not wish to control the device (Ortner et al., 2011).

The P300

The P300 is an electrical potential that appears about 300 ms after the user has observed a rare relevant stimulus (Nicolas-Alonso and Gomez-Gil, 2012). For example, if a person is told to listen for animal types and is then read the words "house," "apartment," "shark," and "building," a P300 response can be expected about 300 ms after the word "shark." This method of eliciting P300 responses by mixing a relevant stimulus with several other irrelevant stimuli is known as the oddball paradigm.

Similarly to SSVEPs, the P300 is used to select among multiple possible commands. Possible commands flash on the screen, and the command that the user desires will evoke a P300 response since it is the relevant "oddball" command. The timing of the P300 response can then be analyzed to determine what command likely triggered the response. When many possible commands are available (e.g., the user is selecting the next possible letter for an e-mail), the selection is generally done in a two-stage process. First, all possible commands are displayed in a two-dimensional grid, and the columns of the grid flash one after the other. The user's brain generates a P300 in response to the column that contains the command of interest. Once the correct column has been identified, the rows of the grid begin to flash one after the other, and the user's brain generates a P300 in response to the row of interest, allowing the correct command to be identified as the intersection of the correct row and column. This process is illustrated in Fig. 3. If the system is unsure what command should be selected (e.g., two columns

Fig. 3 The principle of a P300-controlled spelling device. The user is thinking of the letter "P." The different columns of the grid flash one after the other, and the column containing the relevant letter evokes a P300 response (A). The different rows then flash one after the other, and the row containing the relevant letter evokes a P300 response (B). The relevant letter can then be identified as the intersection of the column and row that evoked the P300 (C).

generated a P300), the procedure can be repeated until the system is sufficiently certain of the correct command.

The P300 requires no training to utilize, but has a lower information transfer rate than SSVEPs in state-of-the-art BCIs, ~20–25 bits/min compared to 60–100 bits/min with SSVEPs (Nicolas-Alonso and Gomez-Gil, 2012). Again, false positives are problematic, as P300 responses also occur naturally in the absence of visual stimuli. Furthermore, the P300 suffers from the same disadvantage as the SSVEP: a screen must be used to present the stimuli.

Motor Imagery

Unlike SSVEPs and the P300, motor imagery has the advantage that no devices or other external stimuli are required for it. Its principle is simple: the user thinks of making a motion, and the activity of the motor cortex changes as a result of the imagined motion even if no movement is actually performed. This activity can be measured and used to control biomechatronic devices. For example, imagined left-arm movement could be used to move the left arm of a full-body exoskeleton. However, effective use of motor imagery requires special user training, and only a small number of motor images can be distinguished using EEG (Nicolas-Alonso and Gomez-Gil, 2012). For example, the user may be able to select whether to move the left or right arm of an exoskeleton, but would not be able to choose the specific movement that should be performed with that arm.

Mental Imagery

Mental imagery is similar to motor imagery, but instead of imagining motions, the user performs different types of cognitive activities: mental subtraction, auditory imagery, spatial navigation, etc. (Friedrich et al., 2012) As the frequency distribution of the EEG changes depending on the user's mental workload (Herrmann et al., 2004; Antonenko et al., 2010), BCIs can use this information to determine whether or not the user is performing a certain cognitive activity. Furthermore, since different cognitive activities are connected with different regions of the brain (e.g., frontal regions for mental subtraction), it is possible to differentiate between them using EEG recorded from different regions. By programming the BCI to perform specific commands in response to specific mental imagery (e.g., start moving a wheelchair if mental subtraction is detected), we can thus allow users to control biomechatronic devices through different cognitive activities.

Workload Indicators

The spectral distribution of EEG activity broadly reflects the alertness of the user. For example, activity in the alpha band (7.5–12.5 Hz) tends to indicate a relaxed mental state while activity in the beta (12.5–30 Hz) and gamma (30–70 Hz) bands tend to indicate focused attention and mental workload (Herrmann et al., 2004; Antonenko et al., 2010). Furthermore, some specific waveforms change their amplitude as a function of workload: for example, the P300 amplitude is lower in cases of high workload (Brouwer et al., 2012). This brain activity is generated subconsciously without any action from the user and can thus provide an unobtrusive measure of mental workload while the user is performing a task. Such measurements can then be used to, for example, adapt the level of automation in complex tasks such as uninhabited air vehicle control (Wilson and Russell, 2007) where monitoring the level of user workload is critical but should be done unobtrusively, without interrupting the user.

BCIs that react to mental workload are often referred to as passive BCIs, as they can perform actions even if the user remains completely unaware of them (Zander and Kothe, 2011). This is in contrast to active BCIs based on the previous four paradigms, where the user must either consciously observe visual stimuli (SSVEP and P300), consciously imagine different motions, or consciously perform different mental tasks.

Error-Related Potentials

Humans generate error-related potentials (ERPs) in the EEG when they realize that they have performed an erroneous action (Chavarriaga et al., 2014). ERPs typically appear as large negative deflections in EEG recorded from frontal and central regions of the brain, and are proportional to the awareness of the error and its importance: for example, when users are told to prioritize task accuracy over speed, their ERPs typically have higher amplitudes than when they are told to prioritize speed (Gentsch et al., 2009). Furthermore, they are produced by both self-generated errors (i.e., user has made a mistake) and externally generated errors (i.e., a device has produced the incorrect response to a correct user command) (Gentsch et al., 2009).

By detecting these ERPs and their amplitudes, biomechatronic devices could determine whether an error has been during human-machine interaction, and could take corrective actions. For example, if a user has accidentally input an erroneous command (either via the BCI or via another input), the device could detect the associated ERP and prevent the command from being fully executed or revert its outcome (Chavarriaga et al., 2014). Alternatively, if the error was made by the device, the device could take steps to reduce the probability of the error occurring in the future. For example, if the user performed motor imagery of the right arm and the BCI interpreted it as imagery of the left arm, moving the left arm of a full-body exoskeleton would evoke an ERP. The detected ERP could then be used to trigger an adjustment of the BCI pattern-recognition rules so that similar future imagery would be correctly classified as imagery of the right arm.

1.1.2 EEG Amplifiers and Electrodes

As EEG signals have an amplitude in the microvolt range and are vulnerable to different artifacts, it is critical to capture them with amplifiers and multiple electrodes with a high signal-to-noise ratio (SNR). Classic EEG systems generally use reusable electrodes made of silver-silver chloride (Ag/AgCl) (Sinclair et al., 2007), with a desired electrode-scalp contact impedance of 1–10 kΩ (Usakli, 2010). Furthermore, the electrodes are generally active: they include a preamplifier immediately next to the electrode that amplifies the low-amplitude EEG signal, making it less vulnerable to cable motion artifacts. To reduce impedance, classic EEG systems make use of electrode gel; however, this greatly increases the setup time and is often uncomfortable for users since they must wash their hair afterwards. Newer BCIs have thus begun using water-based (Volosyak et al., 2010) and ungelled (dry) (Chi et al., 2010; Guger et al., 2012) electrodes. These have been shown to provide comparable performance to traditional gelled electrodes, but still remain relatively uncommon, for example, at the Cybathlon 2016 BCI competition, all the competing teams used gelled electrodes (Novak et al., 2018).

Laboratory-grade EEG systems generally include 4–64 electrodes (Nicolas-Alonso and Gomez-Gil, 2012), with newer high-resolution systems allowing as many as 256 or 512 electrodes (Petrov et al., 2014). This allows better localization of brain activity as well as the use of signal-processing approaches such as spatial filtering, but does result in a long setup time–15–60 min, depending on number of electrodes (Novak et al., 2018). Consumer-grade EEG systems such as those from Neurosky (United States) and Emotiv Systems (Australia), on the other hand, may capture only one or two EEG channels, sacrificing accuracy for ease of use. However, the practical usefulness of such consumer-grade devices for biomechatronics is hotly contested–some studies have found them to be significantly worse than laboratory-grade devices (Duvinage et al., 2013) while others have found them to be sufficiently accurate for use in real-world conditions (Lin et al., 2014).

The placement of electrodes depends on the EEG paradigm used and has a huge effect on BCI performance. While some researchers prefer to place electrodes at evenly spaced location across the scalp (thus obtaining both relevant and irrelevant information, which is useful for, e.g., filtering), electrodes can also be placed only at locations relevant to the EEG paradigm of interest. For example, electrodes for SSVEP detection are commonly placed near the visual cortex (Nicolas-Alonso and Gomez-Gil, 2012), electrodes for motor imagery are commonly placed near the motor cortex, and electrodes for workload recognition are commonly placed near the frontal lobe (Novak et al., 2014).

1.1.3 Signal Processing and Pattern Recognition

EEG signal processing generally begins with a bandpass filter that removes very low-frequency artifacts as well as high-frequency noise. However, many artifacts cannot be removed using simple bandpass filtering. For example, eye artifacts such as blinks appear in EEG measured from the frontal lobe since the eyes are located near the front of the brain, but these artifacts overlap with the frequency bands of the EEG (Vaughan et al., 1996). Similarly, head movement causes artifacts in EEG measured from electrodes near the back of the head due to activation of the neck muscles. These artifacts can be reduced using secondary sensors. For example, eye artifacts can be removed from the EEG by using the electrooculogram (EOG) as a reference for noise-removal algorithms (Croft and Barry, 2000); similarly, head movement can be detected using accelerometers or neck electromyography (EMG) and used as a reference input to adaptive filtering algorithms. If secondary sensors are not available, we can instead use spatial-filtering methods such as Laplacian filtering, which enhance localized activity while suppressing components that are present in many signal channels (such as blink artifacts, which are present in all signals measured from frontal areas).

Once the SNR has been improved, patterns corresponding to different desired commands or mental states must be identified from the EEG. This can be done in one of two different operating modes: synchronous or asynchronous. In synchronous mode, commands are only accepted by the BCI at specific times that are clearly communicated to the user (e.g., via visual signal). At each of these specific times, a window of the EEG is analyzed by the BCI. In asynchronous mode, commands are accepted by the BCI at any time, and a sliding window of the EEG signal (with lengths ranging from 250 to 1000 ms for SSVEPs, P300, and motor imagery (Novak and Riener, 2015) and 1–5 min for workload indicators (Novak et al., 2014)) is constantly analyzed for the presence of the EEG waveform of interest (e.g., motor imagery). Asynchronous operation is thus significantly more complex, as it must account for the fact that the system is likely in a "no command" state the majority of the time. This is acknowledged to be a significant challenge in BCIs, and was the subject of a BCI signal-processing competition in 2008 (Tangermann et al., 2012). At the same time, the asynchronous mode is more realistic and commonly used in, for example, assistive devices: the user may require assistance at any point in time, but will likely spend long periods of time not needing it (Ortner et al., 2011; Pfurtscheller et al., 2005).

In both synchronous and asynchronous modes, the pattern-recognition method depends on the paradigm being used:

- For SSVEPs, the goal is to measure the dominant frequency in the EEG, which can be done using any established power spectral density (PSD) calculation method (Rangayyan, 2015). The dominant frequency in the EEG can then be matched to the closest frequency shown on the screen: for example, if symbol A flashes with 6 Hz and B flashes with 12 Hz, a measured dominant frequency of 6.5 Hz is interpreted as the user choosing symbol A.
- For the P300 wave and ERPs, the goal is to detect a specific waveform, which can be done with any standard event detection and classification method (Rangayyan, 2015). Once the event has been detected and identified as a P300 or ERP, its cause can be determined. For example, to find the cause of the P300, we look for a stimulus that was presented to the person ~300 ms prior to the P300.
- Motor or mental imagery causes EEG power to decrease in some frequency bands and at some electrode locations while increasing in other bands and at other electrode sites. Thus, to recognize imagery, several features are extracted from PSD estimates and input into classification algorithms such as linear discriminant analysis (Horki et al., 2011) or support vector machines (Xu et al., 2011). Among such "classic" algorithms, particularly support vector machines have been recommended for the synchronous mode of operation (Lotte et al., 2007). However, recent years have seen extensive development of new types of classification algorithms for motor and mental imagery, including adaptive classifiers, matrix and tensor classifiers, transfer learning, and deep learning (Lotte et al., 2018). Among these, particularly adaptive classifiers have been shown to outperform most other algorithms (Lotte et al., 2018).

- For workload indicators, it is common to record EEG for 1–5 min, calculate the PSD over that time period, extract features such as mean frequency from the PSD, and use classification algorithms to translate those features into different levels of workload (Novak et al., 2014). This workload level is then assumed to apply to the entire 1–5-min time period. Similarly to motor/mental imagery, popular classification algorithms include, for example, linear discriminant analysis, support vector machines, and artificial neural networks (Novak et al., 2014). However, compared to motor/mental imagery, there has been little development of advanced algorithms and little comparison of different algorithms to each other. Thus, workload classification is still largely based on factors such as ease of implementation and developers' personal preferences.

The different paradigms can also be combined to some degree in order to improve BCI performance. One classic example is to use SSVEPs to control the elbow function of an artificial limb and motor imagery to control the grasp function of the same limb (Horki et al., 2011). Similarly, a wheelchair can be controlled by using motor imagery of the left and right hands to trigger left/right turns and by using the P300 to control the acceleration (Long et al., 2012). A different example is to use SSVEPs and the P300 response simultaneously using a screen that shows P300 visual stimuli on one part of the screen and SSVEP stimuli on another part of the screen (Bi et al., 2014).

1.2 Electrocorticography and Intracortical Electrodes

The electrocorticogram (ECoG) is similar to the EEG, but is recorded invasively with electrodes placed on the surface of the brain using a surgical procedure. This results in a significantly higher SNR than in EEG; however, due to invasiveness, the biomechatronic applications of ECoG are largely limited to severely impaired users (e.g., tetraplegics). Similarly, intracortical electrodes are placed inside the brain itself, resulting in an even higher SNR than ECoG and allowing measurement of the electrical activity of small, very specific regions of the brain. However, they are again very invasive and are frequently rejected by the cortical tissue surrounding them, gradually resulting in loss of the signal (Groothuis et al., 2014).

Signal processing for the ECoG and intracortical electrodes can be similar to that seen in the EEG, but is characterized by less noise and higher pattern-recognition accuracy. For example, while EEG is commonly bandpass-filtered between 5 and 30 Hz, the lower cutoff frequency for ECoG can be as low as 0.1 Hz (Novak and Riener, 2015). Most of the EEG paradigms can then also be applied to ECoG. However, due to its higher SNR, it is possible to use additional signal analysis paradigms that achieve much more accurate estimation of the user's desired motions. While EEG-based motor imagery can only identify broad classes such as "move left arm" vs "move right arm," ECoG and intracortical electrodes allow "movement decoding": reconstruction of the detailed movement trajectory (actual or desired) from the brain signal. Similarly to motor imagery analysis, this process usually begins by extracting frequency features from a PSD estimated over a sliding window. These features are transformed into an estimate of the desired motion trajectory by means of linear regression (Chao et al., 2010) or more advanced methods such as Kalman filters (Hochberg et al., 2012) and then used as direct inputs to a biomechatronic device, for example, as the trajectory of a BCI-controlled robotic arm.

1.3 Functional Near-Infrared Spectroscopy

Functional near-infrared spectroscopy (fNIRS) differs from EEG and ECoG in that it measures the hemodynamic activity rather than electrical brain activity, that is, it is a measure of blood flow. Specifically, it measures the degree of tissue oxygen saturation and changes in hemoglobin volume using near-infrared light (Ferrari et al., 2004). Near-infrared light (700–1000 nm) penetrates the skin, subcutaneous fat, skull, and underlying muscle/brain, and is either absorbed or scattered within the tissue, with the degree of absorption and scatter dependent on, among other things, the ratio of oxyhemoglobin to total hemoglobin within the tissue (Ferrari et al., 2004). Since this ratio changes as a result of increased oxygen consumption due to, for example, higher mental workload, fNIRS can be used to measure the degree of activation of different brain regions.

A typical fNIRS sensor consists of a light source and a light detector, with the two commonly placed on the scalp 3–5 cm apart (Ferrari et al., 2004; Naseer and Hong, 2015). The source emits a known amount of infrared light through the scalp and skull toward the brain, and the detector measures the amount of scattered light. Tissue oxygen saturation and brain blood flow are then estimated from these optical density measurements via the modified Beer-Lambert law (Naseer and Hong, 2015). While the response is slower than EEG (often appearing a few seconds after a stimulus), it has the advantage that it is less susceptible to data corruption by artifacts (e.g., blinks, muscle activity) and offers better spatial resolution, allowing localization of brain responses to specific cortical regions (Naseer and Hong, 2015; Lloyd-Fox et al., 2010). When measured properly, the fNIRS signal closely correlates with the blood oxygen level dependent (BOLD) signal from functional magnetic resonance imaging (Huppert et al., 2006), but can be measured with relatively simple, portable hardware.

1.3.1 fNIRS Paradigms

The most common fNIRS paradigm is to measure mental workload using methods similar to EEG: fNIRS of the prefrontal cortex is recorded over 1–5 min, different features are extracted from it, and classification algorithms are used to translate the features into different levels of workload (Naseer and Hong, 2015; Girouard et al., 2013). Less commonly, it is also possible to use fNIRS to measure motor imagery—using multiple fNIRS channels over the human motor cortex allows observation of distinctly different hemodynamic responses to, for example, imagery of the left hand and the right hand (Naseer and Hong, 2015; Sitaram et al., 2007).

1.3.2 Signal Processing and Pattern Recognition

Regardless of the paradigm, fNIRS signals still contain various types of noise that are not related to brain activity. These are commonly reduced by preprocessing the optical density signals before converting them into oxygen saturation signals, and can be roughly divided into instrumental noise (e.g., instrumental degradation), experimental error (e.g., sudden head motions), and physiological noise (e.g., effects of heartbeat and respiration on blood pressure fluctuations) (Naseer and Hong, 2015). Some of these (e.g., high-frequency instrumental noise) can be removed using simple bandpass filters while others require more advanced methods such as principal/independent component analysis or adaptive filtering (Naseer and Hong, 2015).

After noise removal, it is common to convert the optical density signals into oxygen saturation signals via the modified Beer-Lambert law, then extract different features from the oxygen saturation signals as a basis for pattern recognition (Naseer and Hong, 2015). The most frequently used features are those related to the signal shape (signal mean, signal slope, signal variance, skewness, kurtosis, zero crossing rate, etc.) though more advanced feature extraction methods such as wavelet transforms have been used with some success (Naseer and Hong, 2015). These features are then input into standard classification algorithms such as linear discriminant analysis, support vector machines, and artificial neural networks (Naseer and Hong, 2015).

1.4 Combining Multiple Sensor Types

The different BCI signal modalities (EEG, ECoG, and fNIRS) can also be combined with each other or with other signals (not originating in the brain) in order to improve BCI performance. Such approaches are called hybrid BCIs, and have been reviewed in detail in a recent paper by Hong and Khan (2017); a few representative examples are provided in the following sections.

1.4.1 EEG and fNIRS

EEG offers a rapid response to stimuli but poor spatial resolution; conversely, fNIRS offers poor temporal resolution but good spatial resolution. Thus, combining them has the potential to harness the advantages of each modality and increase overall BCI performance. One of the first studies on this topic indeed showed that simultaneously recording both EEG and fNIRS during motor imagery allows better classification of different motor images (left vs right arm) than using either modality alone (Fazli et al., 2012). As such classification of motor imagery requires both EEG and fNIRS sensors to be placed over roughly the same area of the brain (motor cortex), it necessitates the use of specialized devices designed to measure both modalities simultaneously.

As an alternative to measuring both EEG and fNIRS from the same part of the brain (e.g., the motor cortex), it is possible to use different paradigms for each modality and thus measure each signal from a different region. For example, a user can send one type of command by performing mental arithmetic (which is monitored at the prefrontal lobe using fNIRS) and send another by imagining left or right-hand movements (which are monitored at the motor cortex using EEG) (Khan et al., 2014). While this does not necessarily increase the speed with which the user must send commands (since it can be difficult to simultaneously perform mental arithmetic and imagine hand movements), it can increase overall BCI accuracy by making it easier to differentiate between different types of commands.

1.4.2 EEG and EOG

The EOG measures the electrical activity generated by the eyes using electrodes placed to the left/right as well as above/below the eyes. This results in two different EOG channels, of which one is proportional to the vertical angle while the other is proportional to the horizontal angle of the eyes.

Thus, the EOG can be considered a form of eye tracker. Furthermore, blinks are easily identifiable as very large, brief changes in the signal value.

Perhaps the most common use of EOG is to remove blink artifacts from EEG data using methods such as regression and independent component analysis (Hong and Khan, 2017). However, many other interesting EEG-EOG fusion approaches have been developed. For example, since eye measures such as blink frequency are correlated with workload and fatigue, they can be used together with EEG-based workload indicators to obtain a more accurate estimate of a person's workload or fatigue (Khushaba et al., 2013; Novak et al., 2015). Alternatively, EEG and EOG can be used as two independent control channels: one command (e.g., raise/lower robotic arm) is performed using EOG while the other (e.g., open/close robotic hand) is performed using EEG paradigms such as motor imagery (Hortal et al., 2015; Ma et al., 2014).

EEG and EOG can even be combined without the use of dedicated EOG electrodes: since eye artifacts appear in the EEG, it is possible to estimate EOG "traces" from EEG electrodes. For example, Ramli et al. (2015) developed a wheelchair controller where EOG traces in EEG are used to estimate whether the eyes are open or closed. If the eyes are closed, no wheelchair movement is allowed; if the eyes are open, the wheelchair is controlled based on the EEG. However, while this approach reduces the number of required electrodes, it is currently unclear whether the increase in convenience is large enough to outweigh any decreases in BCI accuracy caused by not having access to a "true" EOG signal.

1.4.3 EEG and Electromyography

EMG is the measurement of electrical signals generated by individual muscles. Such electrical muscle activity frequently acts as a source of noise in EEG: for example, EEG electrodes placed near the back of the head are frequently contaminated by neck muscle EMG while EEG electrodes placed near the front of the head are contaminated by jaw EMG. As with EOG, the most common use of EMG in BCIs is thus to remove muscle artifacts from the EEG. However, other sensor fusion methods exist and are similar to those used to combine EEG and EOG. For example, one input channel of a device can be controlled using EEG while the other can be controlled using intentionally generated jaw EMG (Foldes and Taylor, 2010).

1.4.4 EEG/fNIRS and Autonomic Nervous System Responses for Workload Analysis

As previously mentioned, both EEG and fNIRS can be used as indicators of mental workload. Since the mental workload estimate is obtained by extracting several features from multiple EEG or fNIRS channels and inputting those features into a classification algorithm, it would be possible to increase the classification accuracy using additional signals whose features would provide complementary information about mental workload. One popular type of signal are autonomic nervous system responses such as heart rate, respiration, and peripheral skin conductance, all of which are correlated with both physical and mental workload. Features from these signals can be combined with features from the EEG and/or fNIRS using standard classification algorithms such as linear discriminant analysis or neural networks, as reviewed in a survey paper by the author of this book chapter (Novak et al., 2012).

2 BIOMECHATRONIC APPLICATIONS

Regardless of the exact sensor(s), BCI paradigm, and signal-processing methods, the outputs of a BCI are essentially the commands that the user wants to send to a biomechatronic device (for most BCIs) or an estimate of the user's mental state (for passive BCIs such as those mentioned in "Mental Imagery" section). Currently, BCIs are primarily used in assistive applications by people with disabilities who are unable to use other control methods. For example, people with tetraplegia are paralyzed from the neck down and thus cannot use devices such as keyboards, but can still control biomechatronic devices using BCIs since this requires no movement below the neck. However, nonassistive applications of BCIs also exist, and we present a few examples of each application in the following sections.

2.1 Control of Powered Wheelchairs

Millions of people worldwide suffer from mobility impairments, and many of them rely on powered wheelchairs to perform everyday activities. Such powered wheelchairs are equipped with strong motors that allow them to drive around quickly and climb ramps or even stairs. However, many patients who could benefit from powered wheelchairs are not able to use them since severe impairments (e.g., tetraplegia) prevent them from using conventional wheelchair interfaces such as joysticks. Instead, such patients could use a BCI to control the wheelchair only with their mind, thus moving around with any assistance from a caretaker.

Depending on how much authority is left to the users, several wheelchair BCI architectures can be considered. For example, one P300-based BCI wheelchair includes a screen (mounted on the front or side of the wheelchair) that displays a 3×3 grid of possible destinations in the user's house (e.g., the bathroom) (Rebsamen et al., 2010). The rows and columns are sequentially highlighted, and the desired destination triggers a P300 response. Once the BCI has identified the desired destination, the wheelchair autonomously moves to that room along a predefined route, though the user can send a mental "emergency stop" command to terminate the movement. This greatly simplifies the BCI functioning, but limits the user to a few predefined locations that they can access.

Wheelchairs with more autonomy allow the user to perform individual commands such as "move forward," "turn left," etc. This can be done with several different BCI paradigms. For example, a common strategy for wheelchair control is via SSVEPs induced by a screen mounted on the front or side of the wheelchair. Several buttons labeled "move forward," "turn left," etc. are presented to the user on the screen, with each button flashing at a different frequency. The user selects the desired command by gazing at the corresponding button, causing an SSVEP of the same frequency in the occipital lobe, which is detected by the BCI or sent to the wheelchair. The wheelchair then has different options regarding how to respond:

- it can carry out one discrete command (e.g., move 3 feet forward), stop, and wait for the next one,
- or it can keep executing the command until the user either stops looking at the screen (resulting in no SSVEP observed the BCI) or looks at a different button on the screen.

Both approaches have their own advantages and disadvantages. If the wheelchair executes discrete commands, it tends to be stationary much of the time while waiting for the next command. Conversely, if the wheelchair keeps executing the command until the user changes their gaze point, there is higher potential for accidents, for example, the user may keep looking at the screen and not realize that the wheelchair is about to hit an obstacle.

An advanced approach that utilizes motor imagery and aims to reduce the user's mental workload was presented by Carlson and Millán (2013). In brief, the wheelchair responds to two different types of motor imagery that correspond to turning the wheelchair left or right. However, if neither type of imagery is detected by the BCI, the wheelchair continues moving forward on its own, thus requiring the user to only input actions if they want to change the wheelchair's behavior. Obstacle avoidance is achieved by means of cameras and sonar sensors attached to the wheelchair; these sensors constantly scan the area around the wheelchair, creating an "occupancy grid" of nearby obstacles. If an obstacle is detected partially in the wheelchair's path, it is treated as a repeller in the occupancy grid, causing the wheelchair to automatically swerve to avoid it and then continue on its original path. However, if an obstacle is directly in front of the wheelchair, the wheelchair will slow down and smoothly stop in front of it, then remain stationary until the user executes a turn command via the BCI. This allows the user to "dock" with an object of interest (e.g., a table or sink) by aiming the wheelchair directly for it. Such a shared control paradigm successfully combines the intelligence and desires of the user with the precision of the machine, allowing experienced unimpaired users to complete tasks using the BCI approximately as fast as using a two-button manual input. We believe that such shared control, where users give high-level commands through a BCI and the machine takes care of low-level details, represents the future of practical BCI control and will be adopted by a broad range of applications.

2.2 Control of Mobile Robots and Virtual Avatars

The same principles described in the previous section can be used to control not only wheelchairs, but also all other types of mobile robots and even avatars in virtual environments. For example, in a classic study by Millán et al. (2004), two participants were taught to steer a mobile robot through multiple rooms using motor and mental imagery. Specifically, three images (relax, move left arm, move right for one participant; relax, move left arm, mental cube rotation) were translated into different robot commands by the BCI, with the exact interpretation of the mental state depending on the location of the robot. For example, if the robot was located in an open area, the "move left arm" motor image caused the robot to turn left; however, if there was a wall to the robot's left, "move left arm" caused the robot to follow the wall. In all situations, the "relax" image caused the robot to move forward and automatically stop when an obstacle was detected in front of it. Finally, three lights on top of the robot were always visible to the participants and indicated which of the three motor or mental images was currently being detected by the BCI. Using this control approach, the two participants were able to complete steering and navigational tasks nearly as well as using manual control. A later study by the same research group (Leeb et al., 2015) asked nine participants with motor disabilities (tetraplegia, myopathy, etc.) to control a telepresence robot using a shared control strategy similar to the one used by Carlson and Millán (2013) for powered wheelchairs. The participants were able to successfully complete navigational tasks in an unfamiliar environment, demonstrating that people with disabilities could use such technology to interact with friends, relatives, and health-care professionals in other buildings and perhaps even cities.

In a related example, Riechmann et al. (2016) trained participants to move an avatar through a three-dimensional virtual kitchen environment using codebook visually evoked potentials (cVEP), a method similar to SSVEPs. The virtual kitchen was presented on a screen from the avatar's perspective (similarly to a first-person computer game), and 8–12 different cVEP stimuli were overlaid on top of the kitchen. The cVEP stimuli consisted of four movement buttons (move forward/backward/right/left), four buttons for looking around (up/down/left/right), and up to four action buttons (oven, cup, coffee machine, sink). Each button flashed at a different frequency and could be selected by looking at it, as in the standard SSVEP control paradigm. When the avatar moved, the view of the kitchen scene changed, but the cVEP stimuli remained in the same place. Furthermore, the movement and looking buttons were shown at all times while the action buttons were only shown if the corresponding kitchen item was within the view of the participant's avatar. Participants were asked to use the cVEP interface to move around the kitchen and prepare cups of coffee using a sequence of five actions (get cup, put cup into machine, get water from sink, put water into coffee machine, turn coffee machine on). Individual desired commands (among the 8–12 buttons) were correctly classified with accuracies of around 80%, and well-trained participants were able to complete the task with the BCI in approximately twice the time they needed when using a keyboard. While this may not seem like an impressive result, it is encouraging for participants with severe impairments, who would not be able to use manual commands to perform such tasks.

A final interesting example of this application was recently presented at the Cybathlon 2016, a competition for participants with disabilities who compete against each other using assistive technologies. In the BCI discipline, 11 participants with tetraplegia competed against each other in a virtual environment where their avatars raced along a virtual obstacle course (Novak et al., 2018) (Fig. 4). The course had multiple repetitions of three different types of obstacles, and participants thus had to send one of three

Fig. 4 The brain-computer-interface-controlled racing game for four people that was used at the Cybathlon 2016. Competitors use the brain-computer interface to send commands that avoid obstacles on the racecourse. *(From Novak, D., Sigrist, R., Gerig, N.J., Wyss, D., Bauer, R., Götz, U., Riener, R., 2018. Benchmarking brain-computer interfaces outside the laboratory: the Cybathlon 2016. Front. Neurosci. 11, 756, reused under the Creative Commons Attribution License.)*

different commands (jump, slide, spin) at the correct times to avoid being slowed down by obstacles. However, there were also stretches of the course without any obstacles, and participants had to avoid accidentally sending any command during those times in order to avoid penalties. Since external visual stimuli were not allowed at the Cybathlon, participants could not make use of SSVEPs and P300, and instead relied on motor and/or mental imagery to control their avatars (Novak et al., 2018). As expected, the results varied strongly between the 11 participants, with the best participant completing the race in 90 s and the worst completing it in 196 s (Novak et al., 2018). However, though the participating teams used different hardware and different pattern recognition for mental and motor imagery, there was no clear advantage to any hardware/software approach. While this was undoubtedly due to the small sample size, it suggests that other factors besides hardware and software have major effects on BCI performance. Nonetheless, some conclusions can still be drawn. For example, every team used gelled electrodes, indicating that they did not consider dry or water-based electrodes reliable enough for use in uncontrolled environments. Similarly, every team used laboratory-grade EEG amplifiers, suggesting that no team trusted consumer-grade devices to provide sufficiently good performance. Furthermore, the competition emphasized the importance of effective BCI training for the user–the teams all had very different participant-training strategies, and the winning team stated that their effective BCI training regimen (which included mock audiences and loud noises) likely had a major effect on their success (Perdikis et al., 2017).

2.3 Control of Artificial Limbs

Artificial limbs that can be controlled using only brain signals are a staple of science fiction and would be extremely useful for amputees. State-of-the-art powered limb prostheses are generally controlled by the EMG of residual muscles, but often include unintuitive and complicated control schemes that require significant user training, which limits user acceptance (Farina et al., 2014). BCI-controlled prostheses could be significantly more unintuitive, as they could directly interpret desired commands from the motor cortex, making the user feel as if they are controlling their own limb. A step in this direction, but without BCIs, was taken by the surgical technique of targeted muscle reinnervation: motor nerves that previously led from the brain to the missing limb are surgically reattached to a different muscle, controlling that muscle's behavior, and the EMG of that muscle is then used to control the prosthesis (Cheesborough et al., 2015). However, BCIs could streamline the process further by directly connecting the brain to the prosthetic limb.

Unfortunately, noninvasive BCI methods are too inaccurate, unintuitive, and/or nonportable for control of artificial limbs. SSVEPs and P300 responses, which rely on an additional screen to provide visual stimuli, cannot be used with a prosthetic limb due to mobility issues, though they could be used with a fixed artificial limb such as a robotic arm that is attached to a dinner table and assists with self-feeding. For example, Ortner et al. (2011) developed an assistive orthosis that moved a paralyzed user's arm via SSVEP control. The system was tested with participants with tetraplegia and achieved reasonable performance rates, though participants complained about the flickering lights required to evoke SSVEP responses.

Both motor and mental imagery could, in principle, be used with prosthetic arms and have actually been used to control the behavior of a stationary robotic arm (Hortal et al., 2015). BCI users were successfully able to pick up boxes and move them to a different location using the arm, but the classification accuracy was relatively low–significantly worse than state-of-the-art EMG-based prosthesis control. In a related study, motor imagery was combined with SSVEPs for robotic arm control: imagery was used to open and close the hand while the SSVEP was used to move the arm to different locations (Horki et al., 2011). Again, however, the system was not suitable for use with prosthetic arms due to its lack of mobility and inaccurate response.

If the use of a BCI with truly mobile prosthetic limbs is desired, we should instead turn to the ECoG and intracortical electrodes, which provide sufficient signal quality for continuous control of a prosthetic arm via movement decoding (rather than simple classification). This was demonstrated in multiple studies where intracortical electrodes were surgically implanted into people with tetraplegia and used to control an advanced robotic arm with multiple degrees of freedom (Hochberg et al., 2012; Collinger et al., 2013). The studies found that, after training, people with tetraplegia could use the intracortical BCI to effectively perform reach-and-grasp motions. While the arm in these studies was stationary, future studies could attach it to the body of an amputee and use it as a prosthesis since the BCI did not depend on any external stimuli. However, the need for intracortical electrodes may limit the adoption of this technology, as many amputees may prefer to use simpler prostheses rather than undergo brain surgery.

2.4 Restoration of Limb Function After Spinal Cord Injury

While the previous section demonstrated the use of BCIs for control of artificial limbs, a similar principle could be used by people with spinal cord injury, who still have all their limbs but have lost the nerves connecting the brain and the limb. In the past, restoration of limb function in people with spinal cord injury was frequently done with functional electrical stimulation, where the remaining muscles were artificially stimulated in a coordinated pattern (generated by, e.g., a finite state machine) in order to move the limbs (Ho et al., 2014). However, such electrical stimulation frequently results in unnatural and/or unstable motion patterns (e.g., "robotic" gait). A more natural alternative would be to use a BCI to guide functional electrical stimulation of the limb, thus achieving more intuitive and stable control than could be achieved with an artificial control system. The same approach could also be used with other assistive devices such as exoskeletons.

As with artificial limbs, such BCI-guided restoration of limb function mainly relies on invasive systems to achieve the necessary signal quality. A proof-of-concept BCI system that used intracortical electrodes to control an implanted functional electric stimulator was recently presented by Ajiboye et al. (2017) for reaching and grasping motions in tetraplegia. 463 days after device implantation, the single study participant was able to drink a mug of coffee; 717 days after implantation, he was able to feed himself. While the participant still needed a mobile arm support (which was also BCI-controlled) to help move his weakened arm, such technology represents a promising step toward restoring independence of people with severe disabilities. A simpler noninvasive BCI-stimulation combination was recently presented by Gant et al. (2018), who used a motor-imagery-based BCI to control only the opening and closing of the hand through electrical stimulation with a classification accuracy (open vs close) of 75%. Furthermore, a similar noninvasive system by Soekadar et al. (2016) combines a motor-imagery-based BCI to control a hand exoskeleton (rather than electrical stimulator) that opens and closes the hand of individuals with tetraplegia. While not as effective as implanted BCI systems, such imagery-based BCIs may still become popular among users who wish to restore their limb function but are unwilling to undergo brain surgery.

An approach similar to the one of Ajiboye et al. (2017) was recently also presented for the lower limbs by Capogrosso et al. (2016), who implanted intracortical electrodes and an epidural spinal cord stimulation system into a monkey with a corticospinal tract lesion at the thoracic level. Six days after the spinal cord injury, the monkey was able to walk again without any training, both on a treadmill and over normal ground. Similar results have also been achieved in rats (Knudsen and Moxon, 2017); while no successful tests have been performed with humans, first experiments are expected in the near future, and the technology has great potential to further increase the functional independence of people with tetraplegia.

2.5 Communication Devices

BCIs can also be used for communication by people with severe disabilities that prevent them from both moving their limbs and speaking. As long as users can still move their eyes and read, they can make use of BCI spellers–devices that allow them to spell out letters and words via SSVEPs, P300 responses, and motor imagery (Rezeika et al., 2018). While the speed of such communication is not very fast compared to typing on a keyboard by able-bodied people (with information transfer rates of BCI spellers ranging from 5 to 25 bits/min in users with disabilities (Rezeika et al., 2018)), it nonetheless serves as a valuable tool for users with, for example, locked-in syndrome, who cannot communicate in any other way.

One of the earliest BCI spellers was a matrix-based P300 speller developed by Farwell and Donchin (1988). Users are given a screen that shows a matrix of letters, and the individual columns of the matrix light up one after another. The user focuses on the letter that they wish to select, and this triggers a P300 response when the column containing that letter lights up. Once the correct column has been identified, the screen next lights up each individual row of the matrix one after another; again, a P300 response is triggered when the row containing the letter of interest lights up. The selected letter is then added to the message and the process repeats with the next letter that the user wishes to select. This matrix-based speller achieved a mean letter selection accuracy of 95% and a mean information transfer rate of 12 bits/min. This principle is shown in Fig. 3.

Significant work on P300 spellers has been performed since their introduction in the 1980s and has included innovations in both EEG processing (e.g., improved P300 recognition) as well as user interface design. For example, researchers have experimented with different letter layouts in both two and three dimensions, have added "autocomplete" functions similar to those on mobile phones, and have developed letter matrices for different languages (Rezeika et al., 2018). In a particularly interesting variation, Kaufmann et al. (2011) superimposed faces of different famous people such as Albert Einstein over individual letters, allowing participants to focus on both faces and letters for stronger P300 elicitation. Such improved P300 spellers now achieve information transfer rates of up to 50–60 bits/min in able-bodied users (Rezeika et al., 2018), though it is often necessary to perform multiple identification trials per letter if the signal quality is low.

Aside from P300 spellers, spellers based on SSVEPs and motor imagery have also been gaining in popularity. One of the best-known SSVEP spellers is the Bremen BCI speller (Volosyak et al., 2010), which presents a virtual keyboard on the screen next to five buttons flashing at different frequencies: up, left, down, right, and select. Participants can use these buttons (via the SSVEP BCI) to control a cursor on the keyboard and thus select the desired letter. The letters are arranged according to their usage frequency in the English language, and each selected letter is spoken out loud by the system as a form of confirmation. As with P300 spellers, the interface can be expanded with word prediction algorithms that automatically complete the word and/or suggest the next word in the sentence. Furthermore, newer versions of the Bremen speller have added visual feedback about the strength of the SSVEP signal: when the speller detects that the user is looking at one of the five buttons, that button's size increases to indicate that a selection is about to be made. Through such improvements, SSVEP spellers have achieved information transfer rates of up to 300 bits/min in able-bodied users (Rezeika et al., 2018).

Motor-imagery-based spellers, unlike the above two designs, have the advantage that they are not necessarily dependent on any external stimuli. An early example of an imagery-based speller was presented by Blankertz et al. (2007) and named "Hex-o-spell." It is based on two imagined motions: the right hand and the feet. On the screen, six hexagons are arranged around a circle, and an arrow points toward the hexagons. Each hexagon contains five letters, and the first stage of the imagery-based letter selection is to turn the arrow so that it points toward the hexagon containing the desired letter. Every time right-hand motion is imagined, the arrow turns one hexagon to the right; once the arrow is pointed at the correct hexagon, the selection is confirmed using imagined foot motion. In the second stage, the same procedure is used to select among the five letters: moving the arrow to the desired letter one step at a time using hand imagery and confirming the selection using foot imagery. The system achieved an information transfer rate of 2–3 characters/min in able-bodied users, though it was more fatiguing and required more user training than P300- or SSVEP-based spellers (Rezeika et al., 2018). A modified version of its graphical user interface with circles instead of hexagons is shown in Fig. 5.

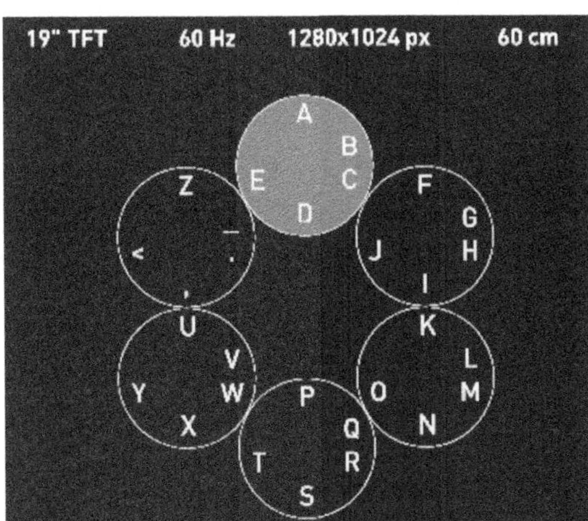

Fig. 5 A modified version of the Hex-o-spell (Blankertz et al., 2007) motor-imagery-based speller. In the first stage of selecting a letter, the user sends motor imagery commands to select one of the six circles. In the second stage, the user sends the same commands to select one of the letters in the previously selected circle. *(From Rezeika, A., Benda, M., Stawicki, P., Gembler, F., Saboor, A., Volosyak, I., 2018. Brain-computer interface spellers: a review. Brain Sci. 8, 57, reused under the Creative Commons Attribution License.)*

2.6 BCI-Triggered Motor Rehabilitation

In motor rehabilitation after stroke, spinal cord injury, traumatic brain injury, or other diseases, patients must perform repetitive, intensive limb exercise to regain their motor functions. Such therapy is increasingly frequently provided by rehabilitation robots that hold the patient's limb and assist in making the desired motion (Lo et al., 2010; Klamroth-Marganska et al., 2014). However, even if the robot provides assistance, the motion should be initiated by the patient, as this allows a tighter coupling between the motor plan in the cortex and its execution through the robot, thus better promoting brain plasticity after the injury (Muralidharan et al., 2011). In patients who still have some residual motion ability, this motion initiation can be detected by a change in limb position (i.e., the robot does not start assisting until the patient has moved their limb at least a little) or by measuring limb EMG, which appears before the actual change in limb position and thus allows a faster robot response (Dipietro et al., 2005). However, these approaches are not feasible for patients who have no residual motion ability. In such severely paralyzed patients, we can instead use a BCI to detect desired motion initiation and have the rehabilitation robot react to it.

BCIs for detection of motion initiation are based on motor imagery: the patient imagines moving the limb that is undergoing rehabilitation, and this imagery is decoded with the same approaches used for, for example, control of mobile robots, then used to trigger a rehabilitation robot that helps carry out the motion. An early clinical demonstration of this approach was performed by Ramos-Murguialday et al. (2013), who divided patients with severe upper limb impairment (no ability to move on their own) into two groups that both participated in ~18 days of training. In the experimental group, patients imagined moving their limb, and a hand-and-arm orthosis then moved the limb in response to detected motor imagery. In the control group, the hand-and-arm orthosis performed the same amount of limb motion in a session, but the motions occurred at random times that had no relation to patient intentions. The experimental group exhibited significantly higher increases in standard scores of functional arm ability, indicating that providing proprioceptive feedback that is contingent upon control of sensorimotor brain activity may improve the beneficial effects of physiotherapy.

Following the Ramos-Murguialday study, several research groups have performed clinical evaluations of BCI-triggered motor rehabilitation, though with mixed results. For example, Ang et al. evaluated robot-aided rehabilitation with and without a BCI using two robotic systems: the

MIT-Manus (Ang et al., 2014a) and the Haptic Knob (Ang et al., 2014b). In both studies, BCI-triggered rehabilitation robots were found to be safe and effective, but no significant intergroup differences were observed between the BCI and non-BCI groups. However, the MIT-Manus study did note that the BCI group exhibited comparable outcome to the non-BCI group even though the number of arm repetitions per exercise session was significantly lower in the BCI group (Ang et al., 2014c). Another recent study found that the outcome of BCI-triggered rehabilitation is correlated with the therapy dose (Young et al., 2015), which suggests that the Ang et al. (2014c) study may have shown negative results due to the difference in dose and that future dose-matched studies may prove the benefits of such BCI-triggered therapy.

Furthermore, several recent technological advancements have the potential to extend the reach of BCI-triggered therapy. For example, Bundy et al. (2017) developed a home-based version of a BCI-triggered rehabilitation robot and showed that using it at home for 12 weeks led to a significant improvement in arm function, demonstrating that such technology does not necessarily need to be limited to rehabilitation hospitals. Furthermore, such BCI-triggered robots have been successfully combined with other types of therapy (Kawakami et al., 2016), showing that the technology does not need to be used on its own, but can become part of a suite of methods and tools used by therapists to achieve optimal rehabilitation outcome. Finally, proof-of-concept systems have been developed that combine EEG with lower limb exoskeletons (López-Larraz et al., 2016; Xu et al., 2014), indicating that this approach could be successfully used for rehabilitation of both upper and lower limbs.

2.7 Adaptive Automation in Cases of Drowsiness and Mental Overload

While the previous sections focused on active BCIs, where the user must actively focus on inputting a command (via SSVEPs, motor imagery, etc.), we now turn our attention to passive BCIs that infer information about the user's mental state without the need for any conscious input (or even awareness) from the user. Specifically, such BCIs can detect undesirable states such as boredom, fatigue/drowsiness, inattention, high stress, and mental overload, allowing a biomechatronic system to either help the user refocus (by, e.g., providing a warning sound) or by taking over part of the task from the user, enabling better overall performance.

Such "adaptive automation" systems were proposed for use with fighter pilots as early as the 1980s and 1990s (Byrne and Parasuraman, 1996), and used classification or regression methods to derive an "operator engagement index" based on the relative power of different frequency bands in the EEG. Adaptive automation was then performed by, for example, activating the autopilot when the human pilot exhibited mental overload. In the 2000s, the general principle of adaptive automation was then extended to many tasks that could result in injury or death due to an inattentive or overwhelmed operator. For example, Wilson and Russell (2003) combined EEG with other physiological responses (heart rate, respiration rate, blink frequency) in order to classify the functional state of US Air Force air traffic control operators during a simulated traffic control task. When discriminating between overload and nonoverload conditions, their classifiers (artificial neural networks and stepwise linear discriminant analysis) achieved accuracies over 90%. The same team later used similar methods to classify the workload level (low or high) in an unmanned aerial vehicle control task, with classification accuracies of 80%–90% (Wilson and Russell, 2007). When high mental workload was detected, the task was modified to make it easier for the operator, resulting in an overall higher percentage of successfully completed tasks.

Adaptive automation is not limited to pilots and military personnel: researchers have frequently used EEG to detect drowsiness, distraction, or stress in car drivers using the same principles. For example, in a recent study by Chuang et al. (2018), driver fatigue was found to result in EEG alpha wave suppression in the occipital cortex as well as increased oxyhemoglobin flow to several parts of the brain (measured using fNIRS) to fight driving fatigue. Although the drivers were still able to successfully complete all tasks, these early physiological markers of fatigue could be used to provide warnings to drivers, for example, by warning them that they should stop and rest soon. In another recent study that focused on driver distraction, participants were asked to drive in a driving simulator while performing different types of secondary tasks (Almahasneh et al., 2014). Distracted driving was primarily reflected in the EEG of the right frontal cortex; however, interestingly, different types of distractions resulted in different EEG responses–for example, math tasks affected the right frontal lobe while decision-making tasks affected the left frontal lobe. This suggests that it may be possible to not only determine whether the driver is distracted, but also to estimate the type (and possibly cause) of distraction. Such information would be beneficial for intelligent cars, which could use it to decide how to most effectively help the driver refocus on the road.

While adaptive automation has the potential to help users avoid negative mental states in critical situations, it is partially limited by the trade-off between accuracy and user-friendliness. Laboratory-grade EEG caps often include 32–64 gelled electrodes for accurate EEG analysis, but we cannot expect car drivers to put on such a system every time they drive at night. Simpler systems with a small number of dry electrodes may be more convenient for users, but would be less accurate, leading to safety and user rejection issues: if the system exhibits too many false positives (e.g., warning sounds when user is not drowsy), the user will simply turn it off; conversely, if the system exhibits too many false negatives (e.g., no warning when user is falling asleep), it will not be able to prevent an accident. At the moment, BCIs for adaptive automation in consumer cars are thus significantly less popular than sensors that either monitor vehicle kinematics (e.g., lane drift) or monitor autonomic nervous system responses through unobtrusive sensors built into the car (e.g., respiration sensors built into the driver's seat (Dziuda et al., 2012)).

2.8 Task Difficulty Adaptation Based on Mental Workload

Task difficulty adaptation is again a passive BCI technology (data obtained without the user's active participation) and can be considered a close relative of the adaptive automation described in the previous section–both applications measure a user's mental state and react to it by changing the behavior of a biomechatronic device. However, the goals of the two are different: while adaptive automation aims to keep the user in a focused mental state to avoid unsafe situations, task difficulty adaptation aims to keep the user appropriately challenged by a task in order to optimize a learning or training process. Such adaptation is based on theories such as flow (Csikszentmihalyi, 1990) and challenge point theory (Guadagnoli and Lee, 2004), which state that optimal engagement and optimal learning/training outcome can be achieved when the user is challenged just below the point of frustration. The goal of the BCI is therefore to estimate the user's workload level and use a form of closed-loop control to keep workload just below the "overload" level while the user is training a task.

One illustrative example of BCI-based difficulty adaptation is in motor rehabilitation: after an injury such as a stroke, patients should exercise intensely to regain their abilities, and should remain focused on the exercise in order to, for example, relearn advanced coordination patterns. If the patient is exercising at a low intensity and is bored, they will not gain much from the exercise; however, if the exercise is very difficult, the patient will become annoyed, lose focus, and not wish to continue. By monitoring the patient's workload level and using it to adapt the exercise difficulty, the BCI-controlled system can achieve optimal rehabilitation outcome. Admittedly, a similar difficulty adaptation could be achieved in a much simpler way by simply monitoring the patient's task success rate and using it as a basis for adaptation. However, this would not capture the patient's internal mental state and would potentially be less reliable, for example, if a patient has a low success rate, it is possible that they are overwhelmed by the task and need an easier one, but it is also possible that they are bored by the task and not putting any effort into it, or that they are trying hard and failing but still enjoying themselves.

Estimation of patient workload for purposes of exercise adaptation in motor rehabilitation was proposed as early as 2007 (Cameirão et al., 2007), and was first implemented using autonomic nervous system responses as workload indicators (Novak et al., 2011), but EEG as a workload indicator was implemented soon afterwards (Novak et al., 2015; George et al., 2012; Park et al., 2015). The closed-loop approach is largely independent of the type of physiological measurement: a rehabilitation robot adapts either its level of assistance or the difficulty of the overall task (e.g., required speed, range of motion) based on the inferred workload. An example a BCI-controlled rehabilitation robot is shown in Fig. 6. However, the main weakness of this technology is its unclear benefit: while some studies have shown that, for example, physiology-based exercise adaptation is more accurate compared to a "ground truth" than simple task-success-based adaptation (Novak et al., 2011), there is so far no evidence that physiology-based adaptation results in better rehabilitation outcome. Thus, adoption of BCI-based adaptation in clinical rehabilitation practice is unlikely until its benefits are more clearly demonstrated.

Aside from motor rehabilitation, several other learning environments could benefit from BCI-based difficulty adaptation. For example, Walter et al. recently developed arithmetic learning software that automatically adapts the difficulty of the presented material based on the learner's EEG (Walter et al., 2017). The EEG-based software was compared to a version that only adapted the difficulty of the material based on the learner's success rate, and the EEG-based version was found to result in a higher learning effect, though the difference was not statistically significant. This presents the same challenge as BCIs for adaptation of rehabilitation difficulty: while the EEG-based system appears to have short-term advantages over a purely

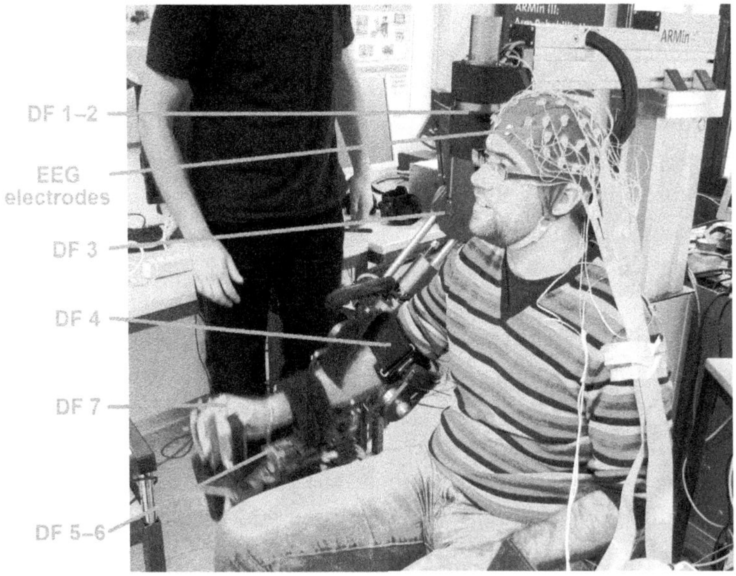

Fig. 6 A person uses a 7-degree-of-freedom rehabilitation robot while a brain-computer interface monitors their mental workload. *DF* = degrees of freedom: DFs 1–3 are in the shoulder (partially obscured by user), DF 4 is in the elbow, DFs 5 and 6 are in the lower arm (lower arm pronation/supination and wrist flexion/extension), and DF 7 is the hand opening/closing module; *EEG* = electroencephalogram. *(From the author's joint research with Prof. José del R. Millán and Dr. Tom Carlson, Ecole Polytechnique Federale de Lausanne, Switzerland.)*

success-rate-based system, it is unclear whether this improvement is large enough to justify the additional complexity and unobtrusiveness. Similar EEG-based prototypes have been developed for, for example, computerized reading tutors (Chang et al., 2013) and serious games that teach fire safety (Ghergulescu and Muntean, 2014), but have also not yet shown clear benefits.

Difficulty adaptation is not limited only to education and training. It can also be used in computer games simply for entertainment: making the game more fun by ensuring that the player is neither bored nor frustrated. An important study in this area was conducted by Chanel et al., who found that player engagement in a game of Tetris can be estimated from EEG with a reasonable accuracy; furthermore, they showed that EEG is a better indicator of engagement than autonomic nervous system responses (Chanel et al., 2011). Ewing et al. (2016) later built on this knowledge to design a BCI that estimated player engagement during Tetris based on frontal and parietal

EEG recordings, then adapted the difficulty of the game based on the engagement estimate. They tested three different EEG-based adaptive Tetris games: a "conservative" system that only adjusted the game speed when the estimated engagement substantially differed from optimal levels, a "liberal" system that adjusted the game speed in response to small deviations from the optimal engagement level, and a moderate system that was essentially a midpoint between the other two. Furthermore, they also tested a Tetris game where participants could manually change the difficulty by saying "increase" or "decrease" out loud. The four versions were tested by 10 participants, with each person trying all four versions. The study unfortunately found no clear advantages of EEG-based over manual adaptation, and participants actually tended to find the manual version to be more immersive. However, it did show that different EEG-based adaptation strategies result in different system behavior, for example, the conservative version tended to increase difficulty more than the liberal one and resulted in higher player alertness. The study thus emphasized the need to not only accurately estimate player engagement using the BCI, but to also intelligently tailor the feedback provided in response to the engagement.

Finally, since most of the BCI-guided examples presented in this section did not demonstrate clear benefits, we end with an example that did not technically use a BCI, but did show a measurable advantage of physiology-guided difficulty adaptation. Liu et al. (2009) measured players' heart rate and EMG during a game of Pong, then used pattern-recognition methods to derive an index of player anxiety from the physiological measurements. The difficulty of the Pong game was adapted based on the physiology-derived index of anxiety, and the adaptation was then compared to adaptation based only on the player's in-game performance. Players found the physiology-based adaptation to result in a more pleasant and more challenging experience than the performance-based one. Thus, it is possible for physiology-based task adaptation to show clear benefits over other adaptation methods, and we remain confident that clearer benefits of BCI-controlled adaptation will be demonstrated in the near future.

2.9 Error-Related Potentials in Biomechatronic Systems

Most of the BCI technologies described in the previous sections essentially use a "fixed" BCI: supervised learning methods are used to train a pattern-recognition algorithm based on previously recorded and labeled data, and the BCI then uses the pattern-recognition algorithm to respond to new data, but does not learn anything from the new data. Thus, even if operating conditions change or the BCI keeps making mistakes, it will not change its previously programmed pattern-recognition algorithms. This puts the onus on the user to learn how to use the BCI effectively, often by trial and error. BCIs that incorporate ERPs, on the other hand, are able to detect that an error has occurred and then take corrective actions. The ERP can be caused either by an error on the part of the user or on the part of the machine, and some studies (though not all) have indicated that larger errors evoke larger ERPs (Gentsch et al., 2009; Spüler and Niethammer, 2015). An excellent, detailed review of ERPs in BCIs is provided by Chavarriaga et al. (2014), and we briefly summarize key developments in this section.

2.9.1 Error Correction

In a first report on the use of ERPs with BCIs, Schalk et al. (2000) demonstrated that, when controlling a cursor with an EEG-based BCI, erroneous control results in an ERP. Since then, several studies of ERPs in response to successful and unsuccessful BCI use have shown that ERPs are relatively stable and occur reliably, allowing BCIs to determine whether the correct desired command was selected based on the user's EEG. Furthermore, the amplitude and waveform of ERPs do not differ significantly between tasks, suggesting that ERP analysis could be independent of the BCI type and the biomechatronic device that it is controlling (Iturrate et al., 2011).

One of the earliest BCIs that used ERPs to correct errors was presented by Millán and Ferrez (2008), who used motor imagery to control a cursor. After each cursor movement, the EEG was checked for ERPs that would indicate an erroneous motion; if one was detected, the cursor was automatically moved back to the previous position. Based on ERP detection, 80% of motions were correctly classified as correct or erroneous, resulting in significantly improved cursor control. An interesting similar concept was presented by Artusi et al. (2011) with a simulated motor-imagery-based BCI: the BCI analyzed the EEG and classified the type of motor imagery, then showed the classification result to the user on the screen before acting on it. If the user exhibited an ERP in response, the classification result was considered erroneous and discarded and the task had to be repeated. Both of these studies showed high potential for ERPs to automatically identify and correct erroneous BCI behavior, though they were conducted with proof-of-concept rather than realistic BCI systems.

Following early proof-of-concept studies, ERP detection was widely implemented in P300 spellers. Essentially, the P300 is used to select a character with approaches such as the matrix-based speller (Section 2.5) and the system shows the selected character to the user, then checks the EEG for an ERP. If an ERP is detected, the character is either immediately deleted (and the P300-based selection process is restarted) or replaced by the second most probable character (Schmidt et al., 2012; Spüler et al., 2012). In able-bodied participants, such error correction has been shown to increase writing speed by $\sim$40% compared to a P300 speller without error correction (Schmidt et al., 2012); furthermore, improvements in writing speed can also be observed in participants with severe impairments such as amyotrophic lateral sclerosis (Spüler et al., 2012). Thus, these studies further validated the potential of ERP-driven error correction in real-world BCIs. Other recent studies have extended this approach to other realistic BCI applications, such as controlling humanoid robots (Salazar-Gomez et al., 2017). In the long term, ERP-driven error correction is likely to become common in a broad range of BCIs, as it does not require any additional hardware (it is based on the EEG) and can significantly improve BCI performance.

2.9.2 Error-Driven Learning

The second possible application of ERPs is to perform error-driven learning, where the underlying algorithms of the BCI are adapted in response to errors (Chavarriaga et al., 2014). For example, Artusi et al. (2011) initially trained a BCI classifier for recognition of fast vs slow motor imagery on a set of previously recorded EEG data. This dataset was then kept in the BCI's memory. When a user interacted with the BCI, incoming EEG was classified as fast or slow motor imagery, and the result was presented to the user on a screen. If no ERP was detected, the classification was considered correct, and the newly recorded EEG signal was added to the dataset in memory together with the classification result. At regular intervals, the motor imagery classifier was then retrained using both the original EEG data and the data obtained from the current user, gradually tailoring the BCI to the current user and increasing its accuracy.

Besides retraining the BCI pattern-recognition algorithms, ERPs can also be used to adapt the behavior of other machines. The user monitors actions taken by an intelligent device; when the device performs the wrong action (e.g., a mobile robot takes the wrong path or a humanoid robot makes the wrong gesture in response to the user), an ERP is detected and the device's control algorithms are automatically updated to reduce the probability of that action being taken in similar future circumstances. A few promising studies in this area have demonstrated that humans generate ERPs in response to erroneous performance of a robotic arm (Kreilinger et al., 2012), a mobile robot (Chavarriaga and Millán, 2010), or a virtual avatar (Pavone et al., 2016) as well as in response to erroneous predictions made by a simulated intelligent car (Zhang et al., 2013), suggesting many potential applications in biomechatronics, for example, identifying when a robotic arm prosthesis has performed an undesired action or identifying when an in-car navigation system has provided the wrong directions to the driver. However, it is still not clear how to respond to ERPs in real-world situations where errors may have multiple possible causes and many possible corrective actions can be taken.

One issue with error-driven learning is that, while ERPs have the potential to detect errors in machine behavior, the ERPs themselves may also be misclassified, for example, a correct BCI action may be misinterpreted as an error. In such cases, error-driven learning will actually increase the probability of future errors by incorrectly retraining the BCI. One possible way to address this would be through probabilistic classifiers: the BCI calculates the probability that an ERP (or lack of ERP) has been detected, and only retrains its algorithms based on this new data if it is sufficiently certain (e.g., above 90%) that it is correct. Such methods have been proposed in the literature (Artusi et al., 2011; Llera et al., 2012), but have primarily been tested with simulated BCIs where prerecorded data are used as a stand-in for actual signal acquisition from a user. Thus, further testing of this approach is needed in natural settings with actual users.

To summarize, BCIs are most commonly used for control of assistive and rehabilitation devices by people with disabilities (e.g., wheelchairs, spellers, prostheses), but can also monitor users' brain activities in a passive fashion and use this information to adapt a mechatronic device–by changing the amount of assistance provided, by changing the difficulty of a task, or by responding to potential errors. Particularly, assistive devices have already been shown to be quite effective, and extensive work is being done to improve the performance and acceptance of BCIs in many biomechatronic applications. However, several challenges do remain, as discussed in the next section.

3 CHALLENGES AND OUTLOOK

In the previous sections, we briefly alluded to some of the challenges facing BCIs in biomechatronic systems. In the next few sections, we will explicitly discuss some of these challenges as well as promising avenues for future research.

3.1 Improving User Friendliness and Resistance to Environmental Conditions

BCIs have long been plagued by a perception of being unwieldy and overly sensitive: in the minds of many researchers, they take a very long time to put on, and their performance is then decimated by even the slightest noise. While this may have been true in the past, BCIs have made enormous strides with regard to user friendliness over the last few years. For example, dry and water-based EEG electrodes have enabled reduced setup time and increased comfort compared to "classic" gelled electrodes, and wireless EEG electrodes have increased signal quality by making BCIs less susceptible to movement artifacts. Furthermore, the use of techniques such as ERPs has allowed BCIs to perform self-correction, increasing their accuracy. However, it is true that BCIs are still inconvenient and error-prone compared to many other technologies (e.g., eye trackers). The situation will doubtlessly improve as some experimental BCI systems become more commonly used, for example, though dry electrodes have achieved promising laboratory results (Guger et al., 2012), they are still relatively uncommon in real-world situations that would benefit from them. Still, new advances in both hardware and software have great potential to improve the robustness of BCIs and could be invented by scientists and engineers in many fields, not just BCI researchers.

3.2 Interindividual Differences

While many BCI studies treat their participant groups as largely homogenous, BCI performance is affected by factors such as personality and cognitive profile (Hammer et al., 2012; Jeunet et al., 2016), motivation (Sheets et al., 2014) and level of experience with the system (Carlson and Millán, 2013). Furthermore, participants with relatively poorly developed brain networks tend to have lower ability to perform motor imagery (Ahn and Jun, 2015), and participants with disabilities frequently exhibit worse BCI performance than able-bodied participants. However, the effects of most of these factors are unclear, and some studies have given conflicting results. For example, a 2012 study by Hammer et al. (2012) found that the accuracy of fine motor skills was a predictor of BCI performance, but a 2014 study by the same research group (Hammer et al., 2014) found that the same variable (measured in the same way) did not reach significance in a somewhat different BCI. As another example, while some studies have found significantly worse BCI performance in participants with disabilities than in able-bodied participants (Spüler et al., 2012), others have found essentially no difference (Leeb et al., 2015), and it is not clear how different pathologies affect performance in different tasks.

Determining the effect of individual characteristics on BCI performance in different tasks is admittedly a daunting task, as it would require multiple studies (due to the need for different tasks) and many participants per study (since the effects of many characteristics would likely be small). The most efficient way to obtain this information may be through a focused review paper that would combine information from many studies to obtain a bigger picture of these effects.

3.3 Training Regimens and User-BCI Coadaptation

BCI performance tends to improve as users train with the system (Lotte et al., 2013; Neuper and Pfurtscheller, 2010). However, this is not a simple dose-response relationship: it is a complex process of the user and machine learning to adapt to each other's idiosyncrasies. Thus, while interacting with a BCI, users will develop their own strategies to compensate for BCI imperfections. For example, in our recent interviews of participants at the Cybathlon 2016 BCI race (Novak et al., 2018), we noted that participants were aware of the delay in detection of motor imagery (up to a second between imagining the motion and the BCI sending a command in response to the detected imagery), and compensated for it by imagining the motion before the command actually needed to be sent. While this led to premature command triggers and consequent penalties in some participants, it was able to improve BCI performance for other participants who were able to master the required prediction process. However, this prediction was not learned instantly: it was part of the BCI training process that, in some participants, involved over a hundred practice races.

As another Cybathlon example, all participants were aware that the actual Cybathlon BCI competition would involve racing in a highly stressful environment with thousands of noisy spectators and that it would not be possible to tailor the BCI to that environment through laboratory training. To make the training more relevant, some participating teams simulated the competition environment in their laboratory using smaller teams of noisy spectators (Novak et al., 2018). Furthermore, after the event, some teams complained about unexpected factors that may have negatively affected their performance, such as increased electromagnetic noise in the environment due to thousands of cellphones and other devices. These examples illustrate two key concepts for future BCI research: BCI training should be optimized for a particular application, and new BCI users should be provided with advice on how to effectively make use of a BCI in a particular application (e.g., how early to perform motor imagery in order to compensate for delays). To the best of the author's knowledge, little systematic research has been done in this direction, and would represent a promising topic for future work.

Furthermore, as emphasized by studies of ERPs for error correction and error-driven learning, the BCI should also adapt to the user. Several strategies for such ERP-driven adaptation have now been proposed and validated, but have largely been limited to adapting the BCI itself. A promising direction that is still in its infancy would be to use ERPs to adapt the behavior of other machines, as demonstrated by Chavarriaga and Millán (2010). This is a much greater challenge than adapting BCIs since it is often unclear how to adapt a machine in response to a detected ERP, for example, we may not be able to determine what specific action caused the ERP or what a more appropriate action would be in that specific situation. Nonetheless, addressing this challenge would greatly broaden the impact of BCIs by creating a new generation of intelligent biomechatronic devices that are responsive to the users' mistakes, preferences, and dislikes.

3.4 Comparison to Other Control Methods

Finally, if BCIs are to achieve widespread adoption, their potential benefits must be made clear to users. While many studies have demonstrated strong benefits of BCIs in applications such as communication, some areas still suffer from unclear usefulness of the technology. One such area is the use of passive BCIs for estimation of mental workload and consequent task difficulty adaptation: while many studies have demonstrated that EEG-based difficulty adaptation achieves better results than performance-based adaptation, it is unclear whether the improvement is sufficient to justify the additional cost, setup time, and inconvenience for the user. This issue has been emphasized by recent reviews of passive BCIs (Brouwer et al., 2015), and is doubly complicated since many studies report only the classification accuracy (e.g., ability to discriminate between high and low workload) of an EEG-based method compared to a performance-based method instead of reporting the effect on the user's enjoyment, learning rate, or other important outcome. The classification accuracy, particularly when calculated offline on prerecorded data, does not necessarily have a clear relationship to BCI performance. For example, studies that experimentally related BCI classification accuracy to user satisfaction by artificially inducing classification errors have found that the relationship is highly nonlinear and occasionally nonmonotonic (van de Laar et al., 2013; McCrea et al., 2017). Furthermore, studies in related fields such as EMG-controlled prosthetics have found that offline classification accuracy does not necessarily correspond to online accuracy, as users will learn to compensate for systematic classification errors and reduce their effect (Jiang et al., 2014; Hargrove et al., 2010).

If possible, BCIs should not only be evaluated with regard to their functional effect (communication speed, enjoyment, rehabilitation outcome, wheelchair navigation speed, etc.), but should also be compared to other control methods that could potentially achieve a better outcome or achieve the same outcome more easily. For example, as SSVEP-based BCIs essentially measure the focus of the user's gaze, their performance could be compared to that of an eye tracker, which measures gaze without the need to attach electrodes to the head. Similarly, EEG-based difficulty adaptation methods could be compared to performance-based adaptation methods, manual adaptation by the user (though this is not recommended by some researchers (Ewing et al., 2016)), or to simple random adaptation. Following a performance analysis, additional cost-benefit analyses could be done to qualitatively or quantitatively compare the different control methods with regard to other factors such as setup time and required user training time. In this way, the potential advantages and disadvantages of BCIs as well as their suitability for different applications could be clearly defined, setting the stage for real-world adoption.

3.5 Outlook

State-of-the-art BCIs have already proven their worth in several assistive biomechatronic systems, and are regularly used by people with severe disabilities who would otherwise not be able to perform everyday activities or even communicate with their loved ones. Furthermore, through the introduction of ERPs into the human-machine interaction process, they are driving the development of a new generation of co-adaptive biomechatronic systems that adapt to the user's preferences, dislikes, and mistakes. While the benefits of BCIs in some applications (e.g., difficulty adaptation) are not yet clear, advances in hardware and software are rapidly increasing both the performance and user friendliness of BCIs, which will undoubtedly lead to their broader adoption in a number of fields.

Furthermore, though most state-of-the-art BCIs are based on noninvasive EEG, implanted electrodes are becoming increasingly accepted and may 1 day lead to the fully seamless human-machine integration that has been predicted by countless science fiction movies.

ACKNOWLEDGMENT

This material is based upon work supported by the National Science Foundation under Grant No. 1717705. Any opinions, findings, and conclusions or recommendations expressed in this material are those of the author and do not necessarily reflect the views of the National Science Foundation.

REFERENCES

Ahn, M., Jun, S.C., 2015. Performance variation in motor imagery brain-computer interface: a brief review. J. Neurosci. Methods 243, 103–110.

Ajiboye, A.B., Willett, F.R., Young, D.R., Memberg, W.D., Murphy, B.A., Miller, J.P., Walter, B.L., Sweet, J.A., Hoyen, H.A., Keith, M.W., Peckham, P.H., Simeral, J.D., Donoghue, J.P., Hochberg, L.R., Kirsch, R.F., 2017. Restoration of reaching and grasping movements through brain-controlled muscle stimulation in a person with tetraplegia: a proof-of-concept demonstration. Lancet 389, 1821–1830.

Almahasneh, H., Chooi, W.T., Kamel, N., Malik, A.S., 2014. Deep in thought while driving: an EEG study on drivers' cognitive distraction. Transport. Res. F: Traffic Psychol. Behav. 26 (PA), 218–226.

Ang, K.K., Chua, K.S.G., Phua, K.S., Wang, C., Chin, Z.Y., Kuah, C.W.K., Low, W., Guan, C., 2014a. A randomized controlled trial of EEG-based motor imagery brain-computer interface robotic rehabilitation for stroke. Clin. EEG Neurosci. 46, 310–320.

Ang, K.K., Guan, C., Phua, K.S., Wang, C., Zhou, L., Tang, K.Y., Ephraim Joseph, G.J., Kuah, C.W.K., Chua, K.S.G., 2014b. Brain-computer interface-based robotic end effector system for wrist and hand rehabilitation: results of a three-armed randomized controlled trial for chronic stroke. Front. Neuroeng. 7.

Antonenko, P., Paas, F., Grabner, R., Gog, T., 2010. Using electroencephalography to measure cognitive load. Educ. Psychol. Rev. 22, 425–438.

Artusi, X., Niazi, I.K., Lucas, M.F., Farina, D., 2011. Performance of a simulated adaptive BCI based on experimental classification of movement-related and error potentials. IEEE J. Emerging Sel. Top. Circuits Syst. 1, 480–488.

Bi, L., Lian, J., Jie, K., Lai, R., Liu, Y., 2014. A speed and direction-based cursor control system with P300 and SSVEP. Biomed. Signal Process. Control 14, 126–133.

Blankertz, B., Krauledat, M., Dornhege, G., Williamson, J., Murray-Smith, R., Klaus, 2007. In: A note on brain actuated spelling with the Berlin brain-computer interface.Universal Access in Human-Computer Interaction. Ambient Interaction. UAHCI 2007, pp. 759–768.

Brouwer, A.-M., Hogervorst, M.A., van Erp, J.B.F., Heffelaar, T., Zimmerman, P.H., Oostenveld, R., 2012. Estimating workload using EEG spectral power and ERPs in the n-back task. J. Neural Eng. 9, 45008.

Brouwer, A.M., Zander, T.O., van Erp, J.B.F., Korteling, J.E., Bronkhorst, A.W., 2015. Using neurophysiological signals that reflect cognitive or affective state: six recommendations to avoid common pitfalls. Front. Neurosci. 9.

Bundy, D.T., Souders, L., Baranyai, K., Leonard, L., Schalk, G., Coker, R., Moran, D.W., Huskey, T., Leuthardt, E.C., 2017. Contralesional brain-computer interface control of a powered exoskeleton for motor recovery in chronic stroke survivors. Stroke 48, 1908–1915.

Byrne, E.A., Parasuraman, R., 1996. Psychophysiology and adaptive automation. Biol. Psychol. 42, 249–268.

Cameirão, M.S., Badia, S.B.i., Mayank, K., Guger, C., PFMJ, V., 2007. In: Physiological responses during performance within a virtual scenario for the rehabilitation of motor deficits.Proceedings of PRESENCE 2007. Barcelona, Spain, pp. 85–88.

Capogrosso, M., Milekovic, T., Borton, D., Wagner, F., Moraud, E.M., Mignardot, J.B., Buse, N., Gandar, J., Barraud, Q., Xing, D., Rey, E., Duis, S., Jianzhong, Y., Ko, W.K.D., Li, Q., Detemple, P., Denison, T., Micera, S., Bezard, E., Bloch, J., Courtine, G., 2016. A brain-spine interface alleviating gait deficits after spinal cord injury in primates. Nature 539, 284–288.

Carlson, T., Millán, J.d.R., 2013. Brain-controlled wheelchairs: a robotic architecture. IEEE Robot. Autom. Mag. 20, 65–73.

Chanel, G., Rebetez, C., Bétrancourt, M., Pun, T., 2011. Emotion assessment from physiological signals for adaptation of game difficulty. IEEE Trans. Syst. Man Cybern. Syst. Hum. 41, 1052–1063.

Chang, K.M., Nelson, J., Pant, U., Mostow, J., 2013. Toward exploiting EEG input in a reading tutor. Int. J. Artif. Intell. Educ. 22, 19–38.

Chao, Z.C., Nagasaka, Y., Fujii, N., 2010. Long-term asynchronous decoding of arm motion using electrocorticographic signals in monkeys. Front. Neuroeng. 3.

Chavarriaga, R., Millán, J.D.R., 2010. Learning from EEG error-related potentials in noninvasive brain-computer interfaces. IEEE Trans. Neural Syst. Rehabil. Eng. 18, 381–388.

Chavarriaga, R., Sobolewski, A., Millán, J.D.R., 2014. Errare machinale est: the use of error-related potentials in brain-machine interfaces. Front. Neurosci. 8.

Cheesborough, J.E., Smith, L.H., Kuiken, T.A., Dumanian, G.A., 2015. Targeted muscle reinnervation and advanced prosthetic arms. Semin. Plast. Surg. 29, 62–72.

Chi, Y.M., Jung, T., Cauwenberghs, G., 2010. Dry-contact and noncontact biopotential electrodes: methodological review. IEEE Rev. Biomed. Eng. 3, 106–119.

Chuang, C.H., Cao, Z., King, J.T., Wu, B.S., Wang, Y.K., Lin, C.T., 2018. Brain electrodynamic and hemodynamic signatures against fatigue during driving. Front. Neurosci. 12, 181.

Collinger, J.L., Wodlinger, B., Downey, J.E., Wang, W., Tyler-Kabara, E.C., Weber, D.J., McMorland, A.J.C., Velliste, M., Boninger, M.L., Schwartz, A.B., 2013. High-performance neuroprosthetic control by an individual with tetraplegia. Lancet 381, 557–564.

Croft, R.J., Barry, R.J., 2000. Removal of ocular artifact from the EEG: a review. Neurophysiol. Clin. 30, 5–19.

Csikszentmihalyi, M., 1990. Flow: The Psychology of Optimal Experience. Harper Perennial, London.

Da Silva, F.L., 2010. EEG: origin and measurement. In: Mulert, C., Lemieux, L. (Eds.), EEG—fMRI: Physiological Basis, Technique, and Applications. Springer, New York, NY, pp. 19–38.

Dipietro, L., Ferraro, M., Palazzolo, J.J., Krebs, H.I., Volpe, B.T., Hogan, N., 2005. Customized interactive robotic treatment for stroke: EMG-triggered therapy. IEEE Trans. Neural Syst. Rehabil. Eng. 13, 325–334.

Duvinage, M., Castermans, T., Petieau, M., Hoellinger, T., Cheron, G., Dutoit, T., 2013. Performance of the Emotiv Epoc headset for P300-based applications. Biomed. Eng. Online. 12.

Dziuda, L., Skibniewski, F.W., Krej, M., Lewandowski, J., 2012. Monitoring respiration and cardiac activity using fiber Bragg grating-based sensor. IEEE Trans. Biomed. Eng. 59, 1934–1942.

Ewing, K.C., Fairclough, S.H., Gilleade, K., 2016. Evaluation of an adaptive game that uses EEG measures validated during the design process as inputs to a biocybernetic loop. Front. Hum. Neurosci. 10.

Farina, D., Jiang, N., Rehbaum, H., Holobar, A., Graimann, B., Dietl, H., Aszmann, O.C., 2014. The extraction of neural information from the surface EMG for the control of upper-limb prostheses: emerging avenues and challenges. IEEE Trans. Neural Syst. Rehabil. Eng. 22, 797–809.

Farwell, L.A., Donchin, E., 1988. Talking off the top of your head: toward a mental prosthesis utilizing event-related brain potentials. Electroencephalogr. Clin. Neurophysiol. 70, 510–523.

Fazli, S., Mehnert, J., Steinbrink, J., Curio, G., Villringer, A., Müller, K.R., Blankertz, B., 2012. Enhanced performance by a hybrid NIRS-EEG brain computer interface. NeuroImage 59, 519–529.

Ferrari, M., Mottola, L., Quaresima, V., 2004. Principles, techniques, and limitations of near infrared spectroscopy. Can. J. Appl. Physiol. 29, 463–487.

Foldes, S.T., Taylor, D.M., 2010. Discreet discrete commands for assistive and neuroprosthetic devices. IEEE Trans. Neural Syst. Rehabil. Eng. 18, 236–244.

Friedrich, E.V.C., Scherer, R., Neuper, C., 2012. The effect of distinct mental strategies on classification performance for brain-computer interfaces. Int. J. Psychophysiol. 84, 86–94.

Gant, K., Guerra, S., Zimmerman, L., Parks, B., Prins, N.W., Prasad, A., 2018. EEG-controlled functional electrical stimulation for hand opening and closing in chronic complete cervical spinal cord injury. Biomed. Phys. Eng. Express. https://doi.org/10.1088/2057-1976/aabb13.

Gentsch, A., Ullsperger, P., Ullsperger, M., 2009. Dissociable medial frontal negativities from a common monitoring system for self- and externally caused failure of goal achievement. NeuroImage 47, 2023–2030.

George, L., Marchal, M., Glondu, L., Lecuyer, A., 2012. In: Combining brain-computer interfaces and haptics: detecting mental workload to adapt haptic assistance.Proceedings of EuroHaptics 2012, pp. 124–135.

Ghergulescu, I., Muntean, C.H., 2014. A novel sensor-based methodology for learner's motivation analysis in game-based learning. Interact. Comput. 26, 305–320.

Girouard, A., Solovey, E.T., Jacob, R.J.K., 2013. Designing a passive brain computer interface using real time classification of functional near-infrared spectroscopy. Int. J. Auton. Adapt. Commun. Syst. 6, 26–44.

Groothuis, J., Ramsey, N.F., Ramakers, G.M.J., Van Der Plasse, G., 2014. Physiological challenges for intracortical electrodes. Brain Stimul. 1–6.

Guadagnoli, M.A., Lee, T.D., 2004. Challenge point: a framework for conceptualizing the effects of various practice conditions in motor learning. J. Mot. Behav. 36, 212–224.

Guger, C., Krausz, G., Allison, B.Z., Edlinger, G., 2012. Comparison of dry and gel based electrodes for P300 brain-computer interfaces. Front. Neurosci. 6.

Hammer, E.M., Halder, S., Blankertz, B., Sannelli, C., Dickhaus, T., Kleih, S., Müller, K.R., Kübler, A., 2012. Psychological predictors of SMR-BCI performance. Biol. Psychol. 89, 80–86.

Hammer, E.M., Kaufmann, T., Kleih, S.C., Blankertz, B., Kübler, A., 2014. Visuo-motor coordination ability predicts performance with brain-computer interfaces controlled by modulation of sensorimotor rhythms (SMR). Front. Hum. Neurosci. 8.

Hargrove, L.J., Scheme, E.J., Englehart, K.B., Hudgins, B.S., 2010. Multiple binary classifications via linear discriminant analysis for improved controllability of a powered prosthesis. IEEE Trans. Neural Syst. Rehabil. Eng. 18, 49–57.

Herrmann, C.S., Munk, M.H., Engel, A.K., 2004. Cognitive functions of gamma-band activity: memory match and utilization. Trends Cogn. Sci. 8, 347–355.

Ho, C.H., Triolo, R.J., Elias, A.L., Kilgore, K.L., DiMarco, A.F., Bogie, K., Vette, A.H., Audu, M.L., Kobetic, R., Chang, S.R., Chan, K.M., Dukelow, S., Bourbeau, D.J., Brose, S.W., Gustafson, K.J., Kiss, Z.H.T., Mushahwar, V.K., 2014. Functional electrical stimulation and spinal cord injury. Phys. Med. Rehabil. Clin. N. Am. 631–654.

Hochberg, L.R., Bacher, D., Jarosiewicz, B., Masse, N.Y., Simeral, J.D., Vogel, J., Haddadin, S., Liu, J., Cash, S.S., van der Smagt, P., Donoghue, J.P., 2012. Reach and grasp by people with tetraplegia using a neurally controlled robotic arm. Nature 485, 372–375.

Hong, K.S., Khan, M.J., 2017. Hybrid brain-computer interface techniques for improved classification accuracy and increased number of commands: a review. Front. Neurorobot. 11, https://doi.org/10.3389/fnbot.2017.00035. Article No. 35.

Horki, P., Solis-Escalante, T., Neuper, C., Müller-Putz, G., 2011. Combined motor imagery and SSVEP based BCI control of a 2 DoF artificial upper limb. Med. Biol. Eng. Comput. 49, 567–577.

Hortal, E., Iáñez, E., Úbeda, A., Perez-Vidal, C., Azorín, J.M., 2015. Combining a brain-machine interface and an electrooculography interface to perform pick and place tasks with a robotic arm. Robot. Auton. Syst. 72, 181–188.

Huppert, T.J., Hoge, R.D., Diamond, S.G., Franceschini, M.A., Boas, D.A., 2006. A temporal comparison of BOLD, ASL, and NIRS hemodynamic responses to motor stimuli in adult humans. NeuroImage 29, 368–382.

Iturrate, I., Montesano, L., Chavarriaga, R., Millán, J.D.R., Minguez, J., 2011. Minimizing calibration time using inter-subject information of single-trial recognition of error potentials in brain-computer interfaces.Proceedings of the Annual International Conference of the IEEE Engineering in Medicine and Biology Society, EMBS, pp. 6369–6372.

Jackson, A.F., Bolger, D.J., 2014. The neurophysiological bases of EEG and EEG measurement: a review for the rest of us. Psychophysiology, 1061–1071.

Jeunet, C., N'Kaoua, B., Lotte, F., 2016. Advances in user-training for mental-imagery-based BCI control: psychological and cognitive factors and their neural correlates. Prog. Brain Res. 228, 3–35.

Jiang, N., Vujaklija, I., Rehbaum, H., Graimann, B., Farina, D., 2014. Is accurate mapping of EMG signals on kinematics needed for precise online myoelectric control? IEEE Trans. Neural Syst. Rehabil. Eng. 22, 549–558.

Kaufmann, T., Schulz, S.M., Grünzinger, C., Kübler, A., 2011. Flashing characters with famous faces improves ERP-based brain-computer interface performance. J. Neural Eng. 8.

Kawakami, M., Fujiwara, T., Ushiba, J., Nishimoto, A., Abe, K., Honaga, K., Nishimura, A., Mizuno, K., Kodama, M., Masakado, Y., Liu, M., 2016. A new therapeutic application of brain-machine interface (BMI) training followed by hybrid assistive neuromuscular dynamic stimulation (HANDS) therapy for patients with severe hemiparetic stroke: a proof of concept study. Restor. Neurol. Neurosci. 34, 789–797.

Khan, M.J., Hong, M.J., Hong, K.-S., 2014. Decoding of four movement directions using hybrid NIRS-EEG brain-computer interface. Front. Hum. Neurosci. 8.

Khushaba, R.N., Kodagoda, S., Lal, S., Dissanayake, G., 2013. Uncorrelated fuzzy neighborhood preserving analysis based feature projection for driver drowsiness recognition. Fuzzy Sets Syst. 221, 90–111.

Klamroth-Marganska, V., Blanco, J., Campen, K., Curt, A., Dietz, V., Ettlin, T., Felder, M., Fellinghauer, B., Guidali, M., Kollmar, A., Luft, A., Nef, T., Schuster-Amft, C., Stahel, W., Riener, R., 2014. Three-dimensional, task-specific robot therapy of the arm after stroke: a multicentre, parallel-group randomised trial. Lancet Neurol. 13, 159–166.

Knudsen, E.B., Moxon, K.A., 2017. Restoration of hindlimb movements after complete spinal cord injury using brain-controlled functional electrical stimulation. Front. Neurosci. 11.

Kreilinger, A., Neuper, C., Müller-Putz, G.R., 2012. Error potential detection during continuous movement of an artificial arm controlled by brain-computer interface. Med. Biol. Eng. Comput. 50, 223–230.

Leeb, R., Tonin, L., Rohm, M., Desideri, L., Carlson, T., Millán, J.D.R., 2015. Towards independence: a BCI telepresence robot for people with severe motor disabilities. Proc. IEEE 103, 969–982.

Lin, Y.P., Wang, Y., Jung, T.P., 2014. Assessing the feasibility of online SSVEP decoding in human walking using a consumer EEG headset. J. Neuroeng. Rehabil. 11.

Liu, C., Agrawal, P., Sarkar, N., Chen, S., 2009. Dynamic difficulty adjustment in computer games through real-time anxiety-based affective feedback. Int. J. Hum. Comput. Interact. 25, 506–529.

Llera, A., Gómez, V., Kappen, H.J., 2012. Adaptive classification on brain-computer interfaces using reinforcement signals. Neural Comput. 24, 2900–2923.

Lloyd-Fox, S., Blasi, A., Elwell, C.E., 2010. Illuminating the developing brain: the past, present and future of functional near infrared spectroscopy. Neurosci. Biobehav. Rev. 269–284.

Lo, A.C., Guarino, P.D., Richards, L.G., Haselkorn, J.K., Wittenberg, G.F., Federman, D.G., Ringer, R.J., Wagner, T.H., Krebs, H.I., Volpe, B.T., Bever, C.T., Bravata, D.M., Duncan, P.W., Corn, B.H., Maffucci, A.D., Nadeau, S.E., Conroy, S.S., Powell, J.M., Huang, G.D., Peduzzi, P., 2010. Robot-assisted therapy for long-term upper-limb impairment after stroke. N. Engl. J. Med. 362, 1772–1783.

Long, J., Li, Y., Wang, H., Yu, T., Pan, J., Li, F., 2012. A hybrid brain computer interface to control the direction and speed of a simulated or real wheelchair. IEEE Trans. Neural Syst. Rehabil. Eng. 20, 720–729.

López-Larraz, E., Trincado-Alonso, F., Rajasekaran, V., Pérez-Nombela, S., Del-Ama, A.J., Aranda, J., Minguez, J., Gil-Agudo, A., Montesano, L., 2016. Control of an ambulatory exoskeleton with a brain-machine interface for spinal cord injury gait rehabilitation. Front. Neurosci. 10.

Lotte, F., Congedo, M., Lécuyer, A., Lamarche, F., Arnaldi, B., 2007. A review of classification algorithms for EEG-based brain-computer interfaces. J. Neural Eng. 4, R1–R13.

Lotte, F., Larrue, F., Mühl, C., 2013. Flaws in current human training protocols for spontaneous Brain-Computer Interfaces: lessons learned from instructional design. Front. Hum. Neurosci. 7, 568.

Lotte, F., Bougrain, L., Cichocki, A., Clerc, M., Congedo, M., Rakotomamonjy, A., Yger, F., 2018. A review of classification algorithms for EEG-based brain-computer interfaces: a 10 year update. J. Neural Eng. 31005.

Ma, J., Zhang, Y., Cichocki, A., Matsuno, F., 2014. A novel EOG/EEG hybrid human–machine interface adopting eye movements and ERPs: application to robot control. IEEE Trans. Biomed. Eng. 62, 876–889.

McCrea, S.M., Geršak, G., Novak, D., 2017. Absolute and relative user perception of classification accuracy in an affective videogame. Interact. Comput. 29, 271–286.

Millán, J., Ferrez, P., 2008. Simultaneous real-time detection of motor imagery and error-related potentials for improved BCI accuracy.Proc 4th Brain-Computer Interface Work Train Course, pp. 197–202.

Millán, J.D.R., Renkens, F., Mouriño, J., Gerstner, W., 2004. Noninvasive brain-actuated control of a mobile robot by human EEG. IEEE Trans. Biomed. Eng. 51, 1026–1033.

Muralidharan, A., Chae, J., Taylor, D.M., 2011. Extracting attempted hand movements from eegs in people with complete hand paralysis following stroke. Front. Neurosci. 5, https://doi.org/10.3389/fnins.2011.00039. Article No. 39.

Naseer, N., Hong, K.-S., 2015. fNIRS-based brain-computer interfaces: a review. Front. Hum. Neurosci. 9.

Neuper, C., Pfurtscheller, G., 2010. Neurofeedback training for BCI control. In: Brain-Computer Interfaces: Revolutionizing Human–Computer Interaction. Springer-Verlag Berlin Heidelberg, pp. 65–78.

Nicolas-Alonso, L.F., Gomez-Gil, J., 2012. Brain computer interfaces, a review. Sensors 12, 1211–1279.

Novak, D., Riener, R., 2015. A survey of sensor fusion methods in wearable robotics. Robot. Auton. Syst. 73, 155–170.

Novak, D., Mihelj, M., Ziherl, J., Olenšek, A., Munih, M., 2011. Psychophysiological measurements in a biocooperative feedback loop for upper extremity rehabilitation. IEEE Trans. Neural Syst. Rehabil. Eng. 19, 400–410.

Novak, D., Mihelj, M., Munih, M., 2012. A survey of methods for data fusion and system adaptation using autonomic nervous system responses in physiological computing. Interact. Comput. 24, 154–172.

Novak, D., Beyeler, B., Omlin, X., Riener, R., 2014. Passive brain-computer interfaces for robot-assisted rehabilitation. In: Brain-Computer Interface Research: A State-of-the-Art Summary. Springer International Publishing, Cham, Switzerland, pp. 73–95.

Novak, D., Beyeler, B., Omlin, X., Riener, R., 2015. Workload estimation in physical human-robot interaction using physiological measurements. Interact. Comput. 27, 616–629.

Novak, D., Sigrist, R., Gerig, N.J., Wyss, D., Bauer, R., Götz, U., Riener, R., 2018. Benchmarking brain-computer interfaces outside the laboratory: the Cybathlon 2016. Front. Neurosci. 11, 756.

Ortner, R., Allison, B.Z., Korisek, G., Gaggl, H., Pfurtscheller, G., 2011. An SSVEP BCI to control a hand orthosis for persons with tetraplegia. IEEE Trans. Neural Syst. Rehabil. Eng. 19, 1–5.

Park, W.N., Kwon, G.H., Kim, D.H., Kim, Y.H., Kim, S.P., Kim, L., 2015. Assessment of cognitive engagement in stroke patients from single-trial EEG during motor rehabilitation. IEEE Trans. Neural Syst. Rehabil. Eng. 23, 351–362.

Pavone, E.F., Tieri, G., Rizza, G., Tidoni, E., Grisoni, L., Aglioti, S.M., 2016. Embodying others in immersive virtual reality: electro-cortical signatures of monitoring the errors in the actions of an avatar seen from a first-person perspective. J. Neurosci. 36, 268–279.

Perdikis, S., Tonin, L., Millán, J.d.R., 2017. Brain racers. IEEE Spectr. 54, 44–51.

Petrov, Y., Nador, J., Hughes, C., Tran, S., Yavuzcetin, O., Sridhar, S., 2014. Ultra-dense EEG sampling results in two-fold increase of functional brain information. NeuroImage 90, 140–145.

Pfurtscheller, G., Müller-Putz, G., Pfurtscheller, J., Rupp, R., 2005. EEG-based asynchronous BCI controls functional electrical stimulation in a tetraplegic patient. EURASIP J. Appl. Signal Process. 19, 3152–3155.

Ramli, R., Arof, H., Ibrahim, F., Mokhtar, N., Idris, M.Y.I., 2015. Using finite state machine and a hybrid of EEG signal and EOG artifacts for an asynchronous wheelchair navigation. Expert Syst. Appl. 42, 2451–2463.

Ramos-Murguialday, A., Broetz, D., Rea, M., Läer, L., Yilmaz, Ö., Brasil, F.L., Liberati, G., Curado, M.R., Garcia-Cossio, E., Vyziotis, A., Cho, W., Agostini, M., Soares, E., Soekadar, S.R., Caria, A., Cohen, L.G., Birbaumer, N., 2013. Brain-machine interface in chronic stroke rehabilitation: a controlled study. Ann. Neurol. 74, 100–108.

Rangayyan, R.M., 2015. Biomedical Signal Analysis, second ed. John Wiley & Sons, Hoboken, NJ.

Rebsamen, B., Guan, C., Zhang, H., Wang, C., Teo, C., Ang, M.H., Burdet, E., 2010. A brain controlled wheelchair to navigate in familiar environments. IEEE Trans. Neural Syst. Rehabil. Eng. 18, 590–598.

Rezeika, A., Benda, M., Stawicki, P., Gembler, F., Saboor, A., Volosyak, I., 2018. Brain-computer interface spellers: a review. Brain Sci. 8, 57.

Riechmann, H., Finke, A., Ritter, H., 2016. Using a cVEP-based brain-computer interface to control a virtual agent. IEEE Trans. Neural Syst. Rehabil. Eng. 24, 692–699.

Salazar-Gomez, A.F., Delpreto, J., Gil, S., Guenther, F.H., Rus, D., 2017. In: Correcting robot mistakes in real time using EEG signals.Proc—IEEE Int Conf Robot Autom, pp. 6570–6577.

Schalk, G., Wolpaw, J.R., McFarland, D.J., Pfurtscheller, G., 2000. EEG-based communication: presence of an error potential. Clin. Neurophysiol. 111, 2138–2144.

Schmidt, N.M., Blankertz, B., Treder, M.S., 2012. Online detection of error-related potentials boosts the performance of mental typewriters. BMC Neurosci. 13.

Sheets, K.E., Ryan, D., Sellers, E.W., 2014. In: The effect of task based motivation on BCI performance: a preliminary outlook.Proceedings of the 6th International Brain-Computer Interface Conference.

Sinclair, C.M., Gasper, M.C., Blum, A.S., 2007. Basic electronics in clinical neurophysiology. In: Blum, A.S., Rutkove, S.B. (Eds.), The Clinical Neurophysiology Primer. Humana Press Inc., New York City, pp. 3–18

Sitaram, R., Zhang, H., Guan, C., Thulasidas, M., Hoshi, Y., Ishikawa, A., Shimizu, K., Birbaumer, N., 2007. Temporal classification of multichannel near-infrared spectroscopy signals of motor imagery for developing a brain-computer interface. NeuroImage 34, 1416–1427.

Soekadar, S.R., Witkowski, M., Gómez, C., Opisso, E., Medina, J., Cortese, M., Cempini, M., Carrozza, M.C., Cohen, L.G., Birbaumer, N., Vitiello, N., 2016. Hybrid EEG/EOG-based brain/neural hand exoskeleton restores fully independent daily living activities after quadriplegia. Sci. Robot. 1, eaag3296.

Spüler, M., Niethammer, C., 2015. Error-related potentials during continuous feedback: using EEG to detect errors of different type and severity. Front. Hum. Neurosci. 9.

Spüler, M., Bensch, M., Kleih, S., Rosenstiel, W., Bogdan, M., Kübler, A., 2012. Online use of error-related potentials in healthy users and people with severe motor impairment increases performance of a P300-BCI. Clin. Neurophysiol. 123, 1328–1337.

Tangermann, M., Müller, K.-R., Aertsen, A., Birbaumer, N., Braun, C., Brunner, C., Leeb, R., Mehrin, C., Miller, K.J., Müller-Putz, G.R., Nolte, G., Pfurtscheller, G., Preissl, H., Schalk, G., Schlögl, A., Vidaurre, C., Waldert, S., Blankertz, B., 2012. Review of the BCI competition IV. Front. Neurosci. 6.

Usakli, A.B., 2010. Improvement of EEG signal acquisition: an electrical aspect for state of the art of front end. Comput. Intell. Neurosci. 2010, 630649.

van de Laar, B., Bos, D.P., Reuderink, B., Poel, M., Nijholt, A., 2013. How much control is enough? Influence of unreliable input on user experience. IEEE Trans. Cybern. 43, 1584–1592.

Vaughan, T.M., Wolpaw, J.R., Donchin, E., 1996. EEG-based communication: prospects and problems. IEEE Trans. Rehabil. Eng. 4, 425–430.

Volosyak, I., Valbuena, D., Malechka, T., Peuscher, J., Gräser, A., 2010. Brain–computer interface using water-based electrodes. J. Neural Eng. 7, 66007.

Walter, C., Rosenstiel, W., Bogdan, M., Gerjets, P., Spüler, M., 2017. Online EEG-based workload adaptation of an arithmetic learning environment. Front. Hum. Neurosci. 11, 286.

Wilson, G.F., Russell, C.A., 2003. Operator functional state classification using multiple psychophysiological features in an air traffic control task. Hum. Factors 45, 381–389.

Wilson, G.F., Russell, C.A., 2007. Performance enhancement in an uninhabited air vehicle task using psychophysiologically determined adaptive aiding. Hum. Factors 49, 1005–1018.

Xu, B., Peng, S., Song, A., Yang, R., Pan, L., 2011. Robot-aided upper-limb rehabilitation based on motor imagery EEG. Int. J. Adv. Robot. Syst. 8, 88–97.

Xu, R., Jiang, N., Mrachacz-Kersting, N., Lin, C., Asin Prieto, G., Moreno, J.C., Pons, J.L., Dremstrup, K., Farina, D., 2014. A closed-loop brain-computer interface triggering an active ankle-foot orthosis for inducing cortical neural plasticity. IEEE Trans. Biomed. Eng. 61, 2092–2101.

Young, B.M., Nigogosyan, Z., Walton, L.M., Remsik, A., Song, J., Nair, V.A., Tyler, M.E., Edwards, D.F., Caldera, K., Sattin, J.A., Williams, J.C., Prabhakaran, V., 2015. Dose-response relationships using brain–computer interface technology impact stroke rehabilitation. Front. Hum. Neurosci. 9.

Zander, T.O., Kothe, C., 2011. Towards passive brain-computer interfaces: applying brain-computer interface technology to human-machine systems in general. J. Neural Eng. 8, 25005.

Zhang, H., Chavarriaga, R., Gheorghe, L., Millan, J.D.R., 2013. In: Inferring driver's turning direction through detection of error related brain activity.Proceedings of the Annual International Conference of the IEEE Engineering in Medicine and Biology Society, EMBS, pp. 2196–2199.

CHAPTER SIX

Upper-Limb Prosthetic Devices

Georgios A. Bertos*[,†,‡], Evangelos G. Papadopoulos*
*National Technical University of Athens, Athens, Greece
†Northwestern University Prosthetics-Orthotics Center, Physical Medicine & Rehabilitation, Feinberg School of Medicine, Chicago, IL, United States
‡Bionic Healthcare, Inc, Chicago, IL, United States

Contents

1 INTRODUCTION

1.1 History

The replacement of a human hand or arm is a truly challenging task. As Aristotle called it, the hand is the "finest tool of all" or the "instrument

Handbook of Biomechatronics
https://doi.org/10.1016/B978-0-12-812539-7.00006-4

of instruments" (Childress, 1980). The complexity of the human hand is evident of its complex anatomy and its dexterity. There are 33 muscles acting on the hand. Of these 33 muscles, 18 are intrinsic and 15 are extrinsic muscles. The human hand also has 27 major bones, and over 20 joint articulations with a total of 27 or more-degrees of freedom (DoF). The arm contributes another 7 DoF. Approximately 1/6 of all the bones and muscles in the human body reside in the two hands. There are complex robotic arms like the MIT/Utah dexterous hand (Jacobsen et al., 1986) which mimic the human hand. The controller of this arm, for example, takes the same space as the space taken by two filing cabinets and is powered by electrical mains. These facts make it impossible to use these robotic arms in replacing the human hand, where power, space, and size are of critical importance for the portable application of prosthetics. Even if the hand is replaced with a simple single-degree-of-freedom (single-DoF) prosthetic hand, as is usually the case nowadays, this still remains a challenging control problem (Weir and Childress, 1996). Technology miniaturization, surgical creativity, and computer science evolution have already enabled the way to multi-degree of freedom (multi-DoF) prosthetic arms.

1.2 How is Success Defined for Upper-Limp Prosthetics?

Childress (1992) presented the following attributes as desirable for prosthesis control:

1. Low mental loading or subconscious control. This means that the person should not put mental effort on how to operate the prosthesis. This should be done subconsciously. In that way a close to the natural way of controlling the human limb is achieved.
2. User friendly or simple to learn to use. In that way the amputee is attracted of learning to use the prosthesis with a minimal effort.
3. Independence in multifunctional control. This means that in multifunctional prostheses, control of one function should not affect the control of any other function.
4. Simultaneous coordinated control of multiple functions (parallel control). This is the ability to coordinate multiple functions of the prosthesis, simultaneously.
5. Direct access and instantaneous response (speed of response). Prosthetic systems should be directly accessible to the user and they should respond immediately.
6. No sacrifice of human functional ability. That is, the prosthesis should be used to supplement, not subtract from, available function.

7. Natural appearance. If possible, the control system should be operated in ways that are esthetically attractive, statically, and dynamically.

In addition, there are several attributes that are desirable to persons responsible for the fitting, maintenance, and modification of a prosthetic controller. Some are:

1. easy to learn; easy and quickly to setup the controller;
2. highly reliable, reproducible;
3. if possible not needed to be fitted in a lab but anywhere; and
4. does not require highly technical skills or high-tech equipment to set-up.

1.2.1 Voice of Customer (Patient)

Peerdeman et al. (2011) after studying acceptance of myoelectric upper-limb prostheses suggest that the integration of sensory information from the prosthesis to the amputee is a gap and should be improved in order to increase user acceptance.

Dudkiewicz et al. (2004) reported that 71% (too high) of upper-limb prosthetic users that participated in a satisfaction study, reported problems with their prosthesis.

Lock et al. (2005) pointed out that for many-degree of freedom (many-DoF) prosthetic arms increased classification accuracy is not correlated to increased usability, meaning that what researchers believe to be a better controller does not lead to better usability results. Therefore, a revisit on the subject has to happen.

Biddiss and Chau (2007) have performed a metasearch study for the last 25 years and reported for pediatric populations rejection rates of 45% for passive and 35% for electric prostheses. For adult populations the rejection rates were 26% for body powered and 23% for electric prostheses. The authors conclude that these high rejection rates make it imperative to investigate further the reasons for abandonment of the prostheses in order to optimize prescription practices and guide the proper design choices (Biddiss and Chau, 2007). Furthermore, not only the rejection causes but also the individual needs of a broad population of amputees should be studied in order to justify personalization or not for the prosthetic process.

Therefore, a more personalized process for prescribing and tailoring the prosthesis to the amputee is needed, which could be facilitated in the future years by technologies like targeted muscle reinnervation (TMR), three-dimensional (3D) printing, and other surgical innovative procedures (see Section 2).

1.2.2 What Would be Ideal?

Amputation is a very traumatic experience. A team of disciplines is needed for the optimal rehabilitation of the patient, namely a team of surgeon, physiatrist, prosthetist, psychologist, occupational therapist, etc. According to Beasley (1981), the only measure of success is: "how well the patient will be reintegrated in normal life." Normal life could be different things for different people according to their priorities and experiences. For example, a teenage girl would prioritize more the cosmesis of the hand: how natural her hand looks and how she is not perceived by others to wear a prosthesis. A farmer might be more interested in using his hand as a tool to perform everyday farming activities. So, where we conclude is that everyone would prefer to have his/her natural hand which is very versatile in functionality. Is this the best we can do? The best we could do is to make the prosthesis to have characteristics that are better than the natural's hand and therefore one might be eager to have a prosthesis better than the natural hand.

1.3 Characteristics of a Prosthesis

1.3.1 Cosmesis

As we mentioned before, cosmesis plays a big role for many amputees. Humans do not want to be different in a negative manner. Cosmetic prosthesis for upper-limb amputees, especially of lower levels (transradial, wrist disarticulation, or finger amputation) is the choice from a lot of amputees. For example, in one study 19 out of 30 upper-limb participants had cosmetic prosthesis (Dudkiewicz et al., 2004). Many amputees, if they can afford it, have a cosmetic prosthesis for social activities, for example, going to a gathering, concert or a festival and have another functional prosthesis at work or at home. It all depends on the character, how social the person is, what age the person is, how the person feels psychologically, and how the environment (family and social group) is treating the person. This is a multivariable subjective situation.

1.3.2 Function: What the Expected Set of Movements Is

Functionality in upper-limb prosthetics is a controversial subject. The human hand is a very delicate, functionally "flexible" instrument which is difficult to replace. By "flexible" we mean ability to perform a broad spectrum of tasks (from piano to heavy lifting). In its entirety, the human hand or arm has a very complicated anatomical structure and control. What matters, is again how that capable "instrument" is used in one's life. That localization is going to dictate what the expected set of movements is for that person. Of course, everyone would desire to have a hand or arm like before, delicate, "flexible" which can perform almost everything. But reality is that technology cannot do that nowadays. Therefore, as designers, we have to think what is the priority for the person who is going to use that hand or arm. What are the tasks that are mostly performed every day? That is, when personalization comes into the picture later on.

The expected set of movements has to do with how many-DoF the prosthesis can do. This has to do with the level of amputation and the needs of the amputee. We have to admit that, the current way of how we manage the prosthesis process, is not asking these questions and more importantly id does not provide to the amputee a prosthesis that is fulfilling his or her personal needs, that is, there is no personalization.

Control Method

As mentioned in Section 1.2, low mental loading which is achieved by subconscious control is a key success factor for upper-limb prostheses. But how subconscious control could be achieved? Subconscious control could be achieved if the prosthetic action is integrated with sensory pathways which inform the user of the state of performance of the task subconsciously.

The control method of many-DoF prostheses, should also be subconscious, actually because of the many-DoFs, controlling many-DoFs simultaneously requires and demands even stronger the control to be subconscious. If not the delays and lack of performance will be augmented and will lead to poor performance.

Performance

As mentioned in Section 1.2, a key factor for success of prosthesis control is the fast response of the controller. Natural delays in humans during reaching from onset of neural command to execution of the task are in the order of a few hundreds of milliseconds (200–300 ms). Therefore, any delays in prosthesis control should not surpass these limits.

Other performance parameters are the weight and enough power for 1 day of regular use (Bertos, 1999).

Higher usability should always be measured. In the past, it was shown that high classification accuracy of many-DoF prostheses does not mean higher usability (Lock et al., 2005).

1.4 Types

1.4.1 Mechanical—Body Powered

The key characteristic of body-powered prostheses is that the amputee senses muscular effort to operate the prosthesis. The development of body-powered prostheses was influenced by development of the aircraft flight technology in the early 19th century and especially of use of the Bowden cable. A Bowden cable consists of an inner core cable that is free to move within a sleeve cable that is fixed in place at either end. Bowden cables are used to mechanically connect the control sticks of the airplane with the airplane's flight surfaces. In that way the pilot feels connected with the control surfaces of the plane, and thus has better control. Bowden cables have also been used in bicycle brakes. Body-powered prostheses have not changed much after their introduction in 1950s. Most of the time they were worn with a harness around the shoulders where one or more Bowden cables are attached. The traditional below-elbow, body-powered prosthesis has a single Bowden cable which runs from the harness to the terminal device (Fig. 1). Opening of the terminal device is achieved by glenohumeral flexion.

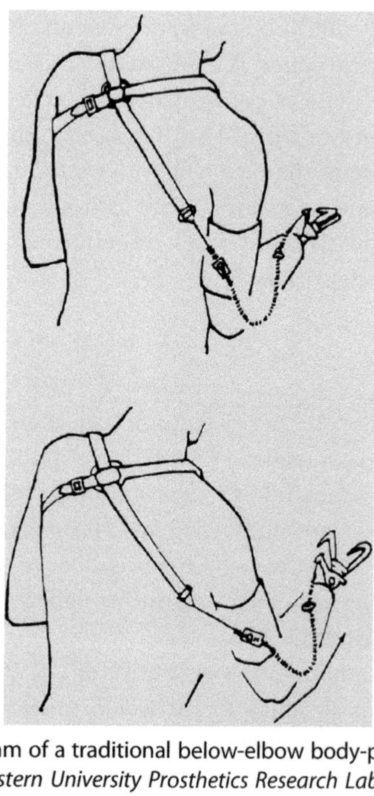

Fig. 1 Schematic diagram of a traditional below-elbow body-powered prosthesis with harness. *(From Northwestern University Prosthetics Research Laboratory (NUPRL).)*

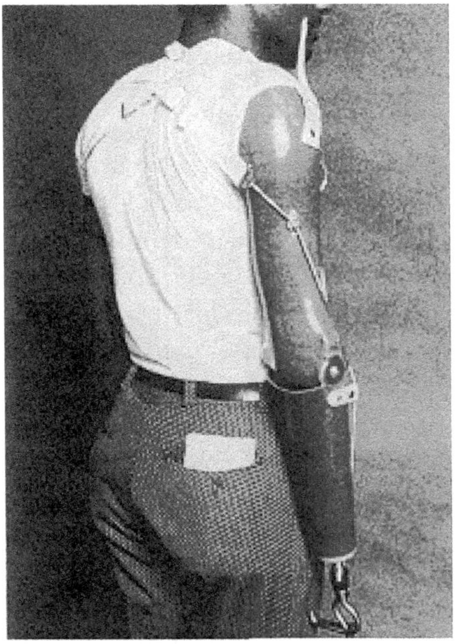

Fig. 2 Above-elbow body-powered prosthesis configuration. The same configuration as the below-elbow body-powered prosthesis (Fig. 1) is used with the addition of a Bowden cable for switching control of elbow flexion-extension to terminal device opening-closing. *(From Northwestern University Prosthetics Research Laboratory (NUPRL).)*

For the case of an above the elbow amputee an additional Bowden cable is used to switch the control of opening and closing to elbow flexion and extension (Fig. 2). The control of both functions is achieved in a serial fashion. The major advantage of body-powered prostheses is that the Bowden cable supplies an interconnection between the amputee and the prosthesis, through which amputees feel that they control an extension of their body. Other advantages include that body-powered prostheses are of low cost, durable, and lightweight. Their disadvantages are that they require a harness to be worn, which is uncomfortable, they have limited range of motion, and all the needed power has to be produced by the muscular system of the amputee. There is no energy enhancement in body-powered systems. An alternative of using a harness and still have body-powered topology is a muscle cineplasty or exteriorized tendons procedure. With these surgical techniques one eliminates the disadvantage of being uncomfortable with the harness and the limited range of motion, but he/she ends up with other disadvantages. These disadvantages include an additional surgery for creating the control sites and needed hygiene for keeping the control sites functional and free of infections (see "Cineplasty" section).

1.4.2 Myoelectric

Myoelectric control systems use muscle electricity as the control method for controlling the prosthesis. Myoelectric control's distinctive characteristic is that it uses electromyogram (EMG) signals from the stump as inputs to control the upper-limb prosthesis. Surface electrodes placed on the skin near a muscle can detect the electricity produced by contracting muscles at the nearby area. The intensity of the EMG signal produced increases as muscle tension increases. The signal is detected from the surface electrodes, amplified, processed, and then used to control the prosthesis (Fig. 3).

Myoelectric control first appeared in the 1940s but it was only until 1970s that it was broadly used in the clinical environment. Today myoelectric control is a favorite but may not be the best way of fitting upper-limb prostheses. It provides open-loop velocity control which is inferior to position control achieved from a position controller like a power-enhanced extended physiological proprioception (EPP) controller. In myoelectric control, the input command signal is proportional to the speed of the prosthesis. Visual feedback is the only feedback to inform the amputee of the state of the prosthesis. The advantage of myoelectric control over power-enhanced EPP control is that myoelectric does not require neither a harness nor a cineplastic surgical procedure. Myoelectric control disadvantage over EPP is that it does not provide proprioceptive sensory feedback. In addition, myoelectric control is velocity control which has been proven to be inferior to position control in positioning tasks (Doubler and Childress, 1984b).

1.4.3 Extended Physiological Proprioception

The problem of control, or how one interfaces the amputee to the prosthetic-mechanical arm is one of the most challenging problems. Today's externally powered systems for upper-limb prostheses use switch or myoelectric controllers implementing open-loop velocity control strategies. Doubler and Childress (1984a,b) have demonstrated that the position control is superior to velocity control in positioning tasks. In addition, prosthesis control techniques that incorporate the body's own proprioceptive sensors and actuators (e.g., body-powered systems) seem to be incorporated easier by prosthetic users and to result in subconscious control (Childress, 1989). Open-loop velocity control implemented by switch or myoelectric control cannot provide feedback. D.C. Simpson, at Edinburgh, in 1969 first

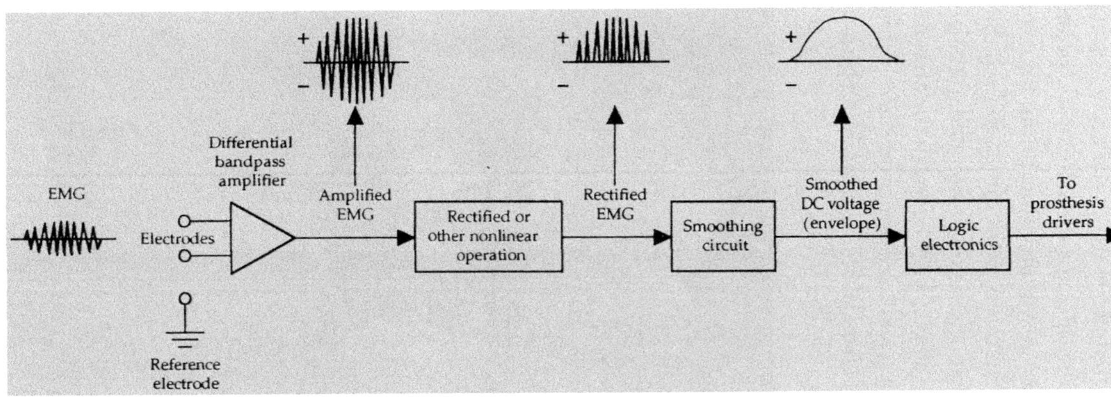

Fig. 3 Schematic diagram of a myoelectric controller used in upper-limb prosthetics. *(From Northwestern University Prosthetics Research Laboratory (NUPRL).)*

realized this and used the Bowden cable technology supplied from the aircraft industry in prosthetics. Simpson (1974) used Bowden cables with pneumatic actuators to build a prosthesis for children (see Fig. 4).

He was the first to coin the phrase "extended physiological proprioception" (EPP) to indicate use of the body's own natural physiological sensors to relate to the operator the state of the prosthetic arm. That is, the operator extends his own proprioception into the prosthesis. The prosthesis becomes an extension of the amputee's self. Simpson used pneumatic technology to enhance the power supplied to the prostheses and at the same time maintained a cable linkage between the amputee and the prosthesis to provide physiological feedback. Of course, the distinctive characteristic of the system was the mechanical linkage providing proprioceptive feedback

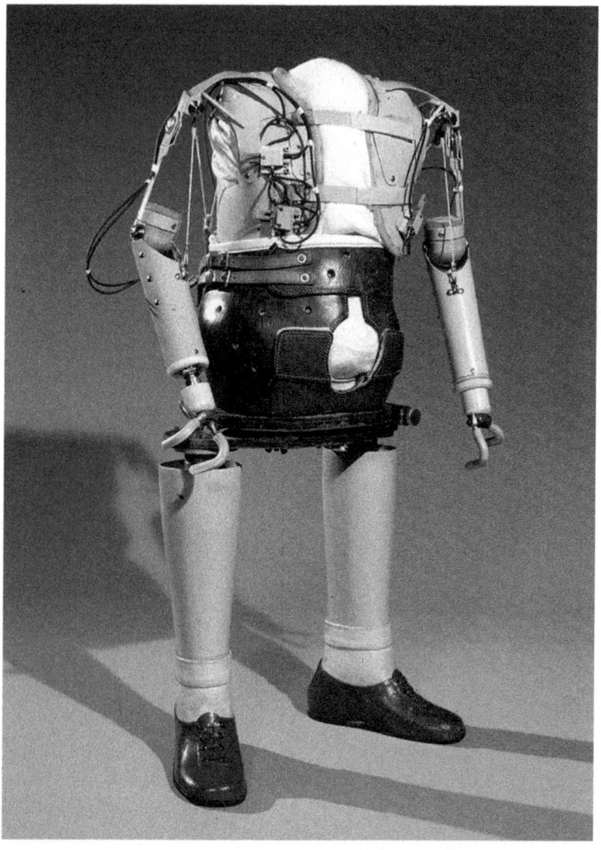

Fig. 4 EPP Prosthesis built by Simpson. *(From https://en.wikipedia.org/wiki/Prosthesis.) From https://commons.wikimedia.org/wiki/File:Artificial_limbs_for_a_thalidomide_child,_1961-1965._(9660575567).jpg*

to the amputee. This is what Simpson clearly saw as the success of the system, and this is why he coined the phrase EPP for it. It can be said that body-powered prostheses are a subset of EPP systems and that is why they were successful in the past. EPP control of externally powered prosthetic joints is similar in concept to power steering of the car. The driver feels, through the handling of the steering wheel, the state of the car's front wheels. Furthermore, the driver cannot "beat" the response of the front wheels because the steering wheel and the front wheels are coupled. This unbeatable coupling provided by a mechanical linkage is the essence of EPP.

The use of tools such as hammers, pens, knives, and racquets illustrate the simple form of EPP. We use these extensions of our body without thinking because we extend our proprioception through these tools. A tennis player does not watch the racquet during swing. The proprioceptive capabilities of the wrist and other joints have been extended to include the racket. The racket becomes a natural extension of the arm. Control of another joint by a physiological joint through EPP is more complex than the EPP control of a simple rigid extension. The artificial joint may be powered by the physiological joint, or it can be externally powered and receiving the control input from the physiological joint. Both can be forms of EPP or not. The prerequisite for the EPP control of another joint is that the two joints, the physiological joint and the artificial joint, must be mechanically interconnected. In that way the physiological and the artificial joint have equivalent kinematics and kinetics. The force, position, and velocity of the artificial joint is transmitted through the Bowden cable and is sensed by the physiological joint and vice versa (Fig. 5). This is an "unbeatable"

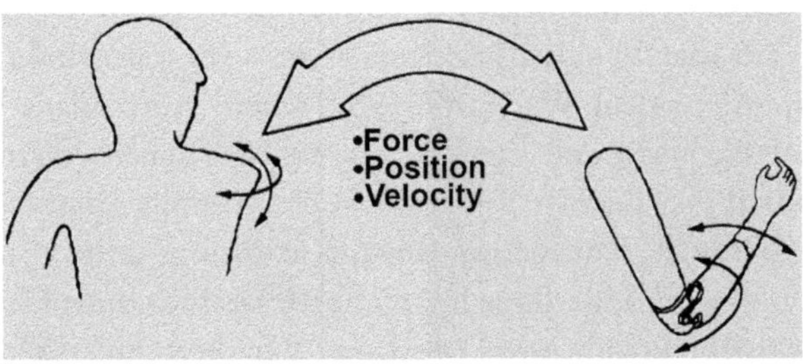

Fig. 5 Diagram showing the "1-1" mapping of the force, position, and velocity between the control site and the prosthesis. This mapping is provided by a rigid mechanical linkage connecting the control site with the prosthesis. Shoulder elevation/depression is associated with elbow flexion/extension. Shoulder protraction/retraction is associated to wrist pronation/supination. *(From Northwestern University Prosthetics Research Laboratory (NUPRL).)*

position servo mechanism. That is, there is a "1-1" mapping between the force, velocity, and position of the control site and the prosthesis.

Cineplasty

As mentioned before, one way of achieving EPP is through means of a harness (see Section 1.4.1). Another way is to create, by means of surgical procedures, direct anchorage sites at the muscle, tendon, or skin of the stump and connect these sites with a Bowden cable to provide a mechanical linkage between the amputee and the prosthesis. Both ways: the harness or the surgical procedures, if used appropriately can provide EPP and its advantages. Both can be used in body-powered topologies or externally powered prostheses. The externally powered prostheses are used when the muscular power of the amputee is not sufficient to provide the necessary energy for his/her daily activities.

Klopsteg and Wilson (1954) argued that it was logical to power an artificial hand or hook by means of voluntary contraction of residuals muscles (cineplasty) rather than by gross body movements like the characteristic shoulder shrug or arm thrust (harness).

The strict definition of cineplasty is any type of surgical procedure which produces some function out of an amputated extremity other than the movement of the extremity itself (Spittler and Fletcher, 1953). Cinematoplasy, kineplasty, cinematization, and cinetization are among the terms used that have the same meaning. The basic idea is that with cineplastic control sites and the attachment to them of a Bowden cable, EPP can be achieved. A "1-1" mapping of position, velocity, and force is shown in Fig. 6.

An Italian from Florence, Giuliano Vanghetti in 1898 (Vanghetti, 1898, 1899a,b, 1900) is generally credited with being the first person to try the idea of using the residual arm muscles to command a prosthesis (Tropea et al., 2017). It is said that Vanghetti conceived his idea following observations of the Italian-Abyssinian War (1896–98), where Abyssinians had cut the right hand at the wrist and the left foot of 800 Italians for punishment. Vanghetti noted that that the forearm muscles in these amputees remained intact and functional. It was from this observation that he conceived of a cineplastic operation to use these intact muscles as the control force for activating the cosmetic prostheses of that time. Vanghetti published 52 ways of connecting muscles and tendons with the prosthesis (Tropea et al., 2017). In 1900, the team of Ceci, Vanghetti, and Redini performed the first cineplastic operation and fitting with a prosthesis (Vanghetti, 1906). They made a skin-lined tendon "loop" motor, using the biceps and triceps. Putti (1917), as professor of orthopedics at Bologna performed a number

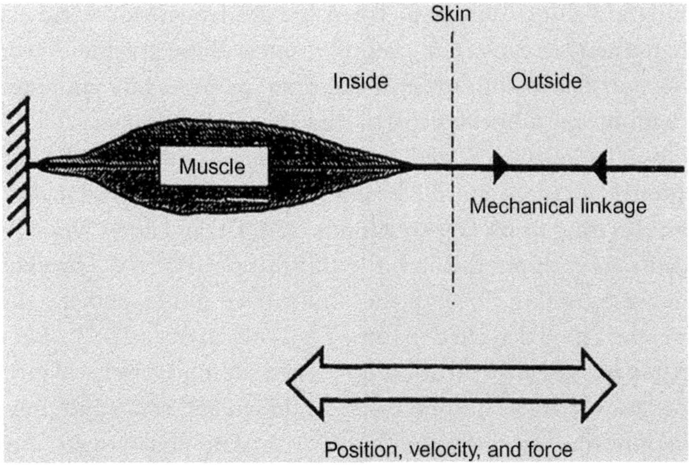

Fig. 6 Schematic representation of the EPP topology when used with a tunnel cineplasty or exteriorized tendons cineplasty. The position, velocity, and force of the prosthetic component is directly correlated with position, velocity, and force of the controlling muscle. *(Drawn by E.C. Grahn; From Northwestern University Prosthetics Research Laboratory (NUPRL).)*

of cineplasty procedures, was the first to suggest no more than two tunnels in a single stump. In Germany, the German surgeon Ernst Ferdinand Sauerbruch (1915, 1916) worked on cineplasty techniques without knowing the existence of the Italian team. Sauerbruch is considered to be the father of the muscle tunnel cineplasty procedure as it is known today. Sauerbruch who was director of surgery at the Greifswald University Hospital in Zurich started working in the field after suggestion of Dr. Stodola, a distinguished Swiss turbine engineering professor. After being Director of Surgery in Zurich (1910–18), Sauerbruch moved to Germany where he was Director of Surgery in Munich (1918–28) and in Berlin (1928–49). Sauerbruch made several contributions in the field. He proposed the use of pairs of agonist and antagonist muscles to control a single-DoF prosthesis, in order to provide more physiological and precise control. He also moved the muscle tunnel from distal to the bone end where Italians had it, to proximal to the bone end. He stressed the necessity of an exercise program for these muscles before and after the cineplasty procedure. He also championed a team organization of physician, prosthetist, surgeon, and technician around the patient. Finally, Sauerbruch's group specially built numerous innovative prostheses for tunnel cineplasty amputees. He was also the first to insist on the importance of performing the tunnel cineplasty only on highly motivated candidates. Only tunnels capable of adequate force and excursion were recommended early in the century, since all the prostheses were body powered. For that reason, most of the time, below-elbow amputees were treated with biceps tunnel cineplasty and above elbow or shoulder disarticulation cases were treated with pectoral tunnel cineplasty. To make the biceps tunnel, incisions are made along three sides of a rectangle (Fig. 7) to permit elevating a skin flap containing skin, fat, and fascia while retaining blood and nerve supply through the fourth side (Fig. 7). The skin flap is rolled into a tube (Fig. 7) with the skin surface inside, and the distal end of the muscle is itself detached to form a smooth surface (Fig. 7). The muscle fibers are separated with a dilator (Fig. 7), not cut transversely, to form a passage through which the skin tube can be drawn and rotated to prevent the pin from pressing on the scar (Fig. 7). A skin graft covers the defect remaining after the skin tube is sutured in place with its ends flared back over the muscle surface (Fig. 7). Petroleum-jelly gauge is placed in the tunnel (Fig. 7) before application of a pressure dressing. An example of a prosthesis suitable for a below-elbow amputee having biceps tunnel cineplasty is shown in Fig. 8. A pin is inserted through the muscle tunnel. A Bowden cable is running from this pin to the prosthesis providing a rigid mechanical linkage, necessary for the implementation of EPP. Contraction of the cineplastized muscle causes the pin to move and the Bowden cable to move and the terminal device to open, or close depending if it is voluntary-opening or voluntary-closing device. For above-elbow amputees, biceps tunnel cineplasty has been used in the past by some surgeons. Also, the pectoralis major muscle has been used by others. The surgical technique involved in constructing the pectoral tunnel parallels that for the case, the base of the skin flap is either toward the axilla or across the lower side. For example prosthesis for an above-elbow amputee having pectoral cineplasty is shown in Fig. 9.

It was only after the World War II that amputation surgery and prosthetics research received proper attention in the United States. This was due the large number of amputated soldiers after the war. In 1945 the Committee on Artificial Limbs of the National Research Council was created, with the goal to organize and execute the needed improvements so that veteran amputees could have access to the best available prostheses. After extensive travel to Europe the Committee's most important finding was the Sauerbruch tunnel cineplasty procedure modified by Lebsche. This procedure is now known in the United States as the Sauerbruch/Lebsche procedure (Fig. 8). Numerous cineplasty procedures were performed after the war in the United States.

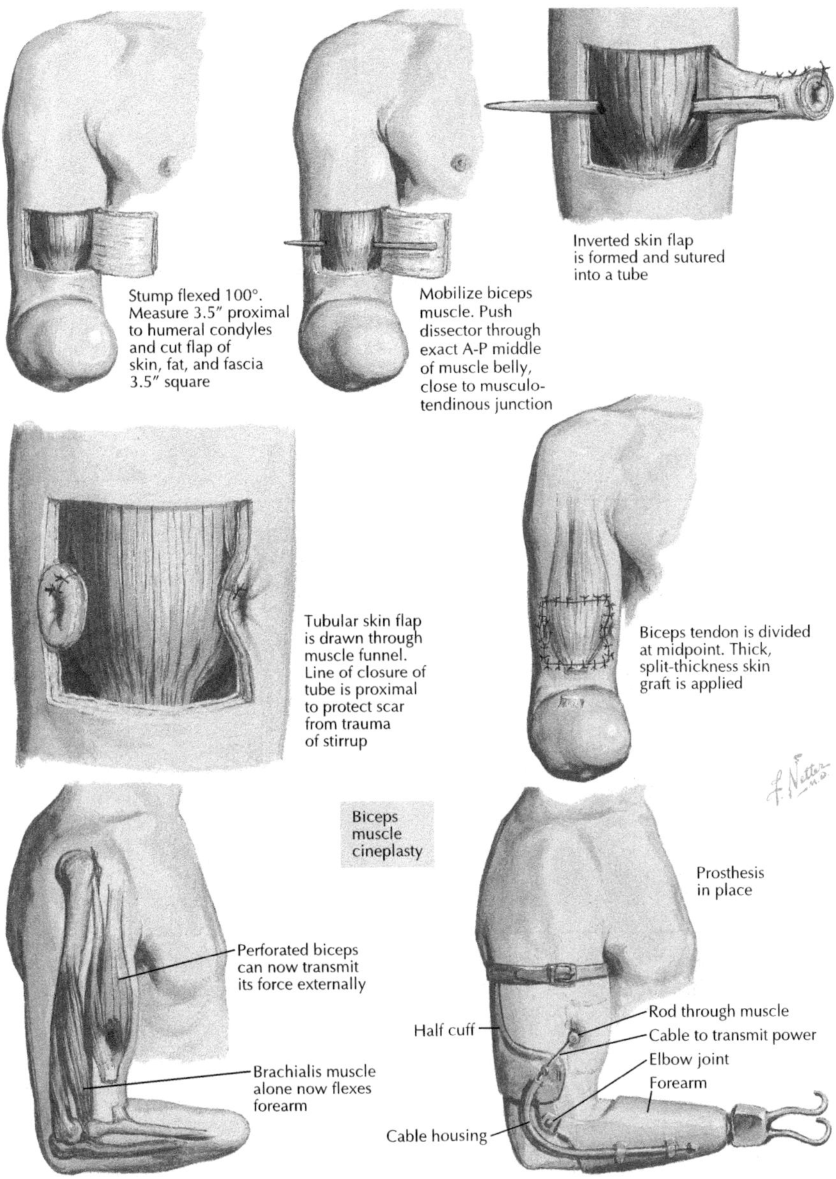

Fig. 7 Construction of the biceps tunnel cineplasty and incorporation of a Bowden-cable prosthesis. *(Image ID: 36724. Used with permission of Elsevier. All rights reserved.)*

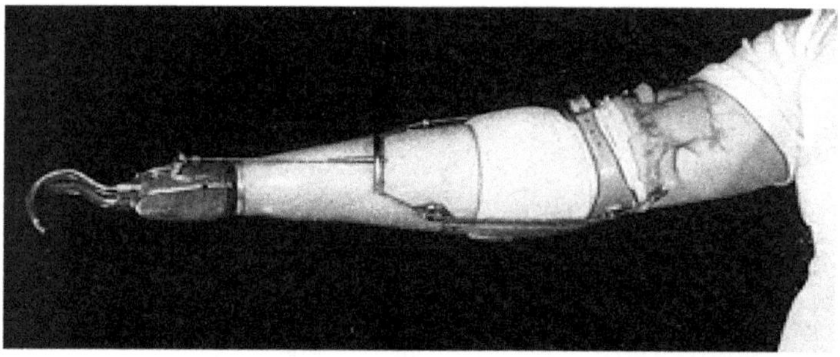

Fig. 8 Example prosthesis for a below-elbow amputee having biceps tunnel cineplasty. *(Reprinted with permission from Klopsteg, P.E., Wilson, P.D., 1954. Human Limbs and Their Substitutes, second ed. National Academy of Sciences, Courtesy of the National Academies Press, Washington, DC.)*

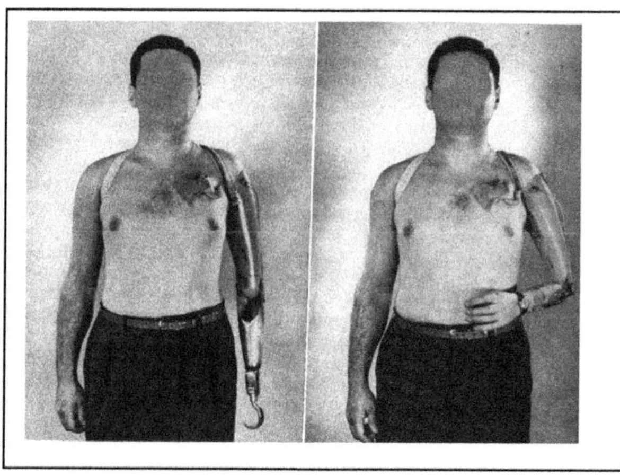

Fig. 9 Example prosthesis for an above-elbow amputee having pectoral tunnel cineplasty. *(From Klopsteg, P.E., Wilson, P.D., 1954. Human Limbs and Their Substitutes, second ed. National Academy of Sciences, Courtesy of the National Academies Press, Washington, DC.)*

Cineplasty was regarded as the best choice for fitting amputated veterans. The retirement of cineplasty's advocate chief surgeons, the rise of myoelectric control and some disadvantages of the procedure led to the decline of the number of cineplasty procedures performed in the United States during the period of 1970 to present.

During the 1960s Dr. Beasley in New York used tendons to create the necessary loops for the cable to be interconnected to. Beasley's "tendon exteriorization" cineplasty uses tendon transfers combined with skin flaps to bring a tendon loop outside of the limb (Beasley and de Bese, 1986). The advantages of this procedure over tunnel cineplasty are:

1. No special procedures for cleaning or ventilating the skin are necessary, thus eliminating a major cause for infectious complications after surgery.
2. The skin of the tendon exteriorization cineplasty has normal cutaneous innervation and optimal circulation.
3. The individual motor units are small, hence the number which can be constructed on a single extremity is limited only by the available innervated skin for use in skin flaps.
4. The system permits selection of either a single muscle or a group of muscles (combined together with a single tendon loop) as the cineplastic motor.
5. Neither muscle excursion nor power is impaired, since dissection occurs only in physiological planes and no significant adhesions result from the surgery.
6. The exteriorized tendons units are esthetically more acceptable than cineplasty.

The tendon exteriorization procedure is presented in Fig. 10 and is as follows: the tendon of an undamaged muscle, or a tendon graft attached to it, is brought up into mobile subcutaneous tissues. The tendon is then enclosed within a proximally based tubed bipedicle skin flap. This skin flap, being proximally based, maintains optimal circulation, sensibility, lymphatic drainage, and remains innervated. The other end of the tendon is looped back on itself or can be attached either to another muscle or anchored to bone.

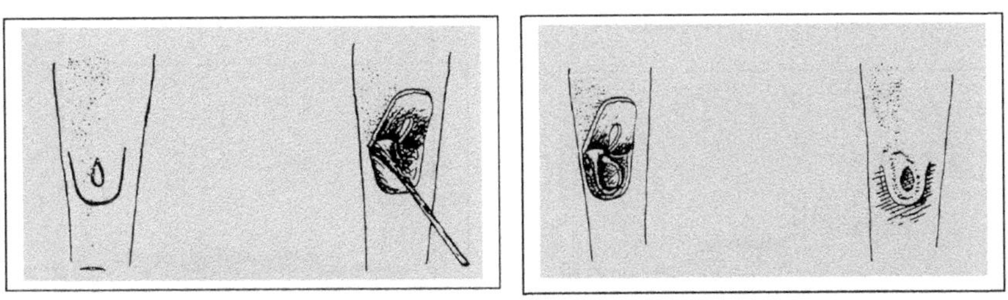

Fig. 10 During tendon exteriorization, the tendon of a selected muscle or a tendon graft substituted for it, is brought above the surface of the body. A tendon loop is formed and enclosed in a proximally based, tubed bipedicle flap, the design of which results in minimal interference with normal cutaneous innervation, vascular supply, and lymphatic drainage. *(From Weir, R.F.f., 1995. Direct Muscle Attachment as a Control Input for a Position-Servo Prosthesis Controller (Ph.D. dissertation). Northwestern University, Evanston, IL.)*

According to Beasley, this tendon exteriorization principle can be applied in a variety of arrangements and the undamaged muscles retain excellent power.

The use of tendon grafts has the advantage of preserving all the force and excursion that the muscle is capable of also of and maintains the Golgi tendon organs and their resultant contributions to the proprioceptive feedback of the muscle. In those instances, where a tendon is not available, the use of artificial tendons attached to the residual muscle and then brought outside the limb in loops has also been suggested.

Besides the many advantages of tendon exteriorization procedure over tunnel cineplasty there is one disadvantage: with the tendon exteriorization procedure the force capability is of the order of 1–1.5 lb.; thus externally powered prostheses must be used. Fig. 11A shows a schematic of the final result of a tendon exteriorization procedure, and Fig. 11B shows a specially modified Otto-Bock hand driven by these exteriorized tendons (Childress et al., 1993).

Since both procedures: tunnel cineplasty and exteriorized tendons are surgical procedures, it is imperative that the surgeon, prosthetist, amputee,

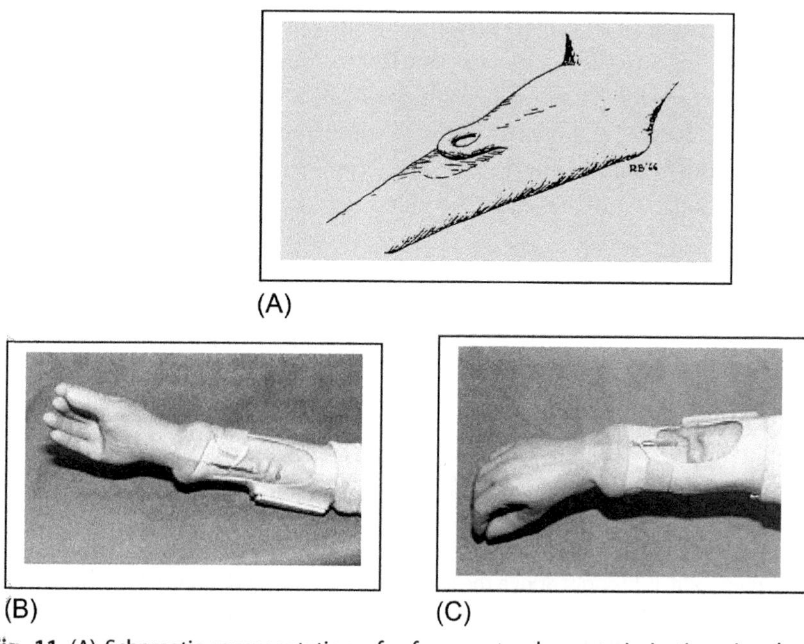

Fig. 11 (A) Schematic representation of a forearm tendon exteriorization cineplasty. (B) Modified Otto-Bock hand to be driven by exteriorized tendons built at Northwestern University Prosthetics Research Laboratory. *((B) Photographs courtesy of Northwestern University Prosthetics Research Laboratory.)*

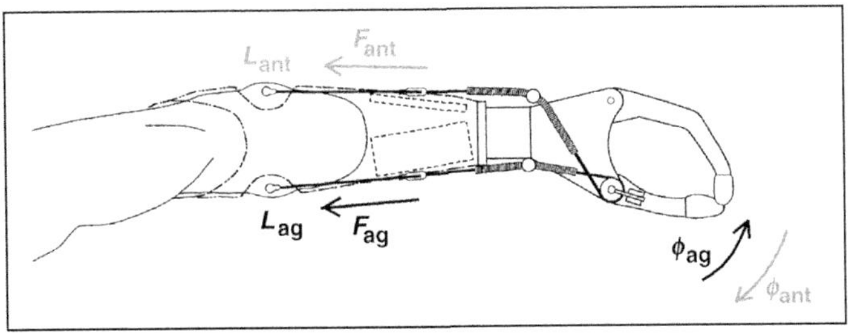

Fig. 12 Bi-directional EPP control using cineplastized forearm flexor and extensor muscles as an agonist/antagonist pair to control opening and closing of an electric hand. During closing, the length of the agonist flexor muscle, L_{ag}, is directly proportional to the angle of closing, F_{ag}. Likewise, during opening, the length of the antagonist extensor muscle, L_{ant}, is directly proportional to the angle of opening, F_{ant}. This control arrangement is analogous to the physiological arrangement for control of natural joint movements. *(From Northwestern University Prosthetics Research Laboratory (NUPRL).)*

engineer, physical therapist, and physician work together as a team in order to have the optimal result.

Fig. 12 shows a schematic of the Classic EPP topology.

A microprocessor-based EPP controller for upper-limb prostheses to be used either for transradial or for transhumeral amputees was developed (Bertos et al., 1997, 1998), eliminating analog electronic problems of the controller developed by Childress et al. (1993).

1.4.4 Many-DoFs

Sequential many-DoF upper-limb prostheses have been used in the 1980s and 1990s especially for high-level amputations since many-DoF arms were needed for those cases. They have been controlling different DoF of the prosthesis sequentially from one control site.

Simultaneous many-DoF upper-limb prostheses were not possible in the past due to the lack of control sites. TMR enabled the creation and miniaturization of sensors and actuators enabled the creation of additional control sites and practical simultaneous many-DoF upper-limb prostheses.

1.5 Technologies That Affect Upper-Limb Prostheses

1.5.1 Materials

The materials for prosthetic devices obviously follow their time. The first materials used were wood and leather. During the Renaissance, materials for prosthetics included iron, steel, copper, and wood (Marshall, 2015).

Modern upper-limb prostheses require a large number of different materials, especially when the prosthesis is active. These may be divided materials based on their function in a prosthetic device, as materials that come into contact with the human body such as sockets, materials employed as a cover of the prostheses (esthetics, feel), and materials employed in the internal construction of the prostheses, such as structural elements. The former are made of biocompatible materials such as fiberglass, thermoplastics, (acrylics, polyester fiber, Perlon), carbon fiber, and Kevlar (Prosthetic, Orthotic Components & Orthopaedic Solutions Catalogue, 2013). For esthetic reasons, a number of foams, such as hard expanded polyurethane foam, are used. However, due to the exceptional strength-to-weight characteristics and quality of superior bio-compatibility, a majority of today's upper-limb prostheses are made from composites with an underlying polymer matrix (Bhuiyan et al., 2015). Still, materials for structural and active components are made of titanium, aluminum, cobalt alloys, or stainless steel (Pandey et al., 2016; Niinomi, 2002). The recent advances in 3D printing added another parameter in the choice of materials, that is, whether they can be applied by a 3D printer. New generations of reasonable cost 3D printers can use composite materials such as carbon fiber, Kevlar, or glass fiber, and therefore can be used in producing functional custom sockets at small cost (Krausz et al., 2016).

When it comes to the design and construction of an active prosthesis that can interact with the environment, then titanium is the best material, as it has high strength, durability, and low density (56% that of steel's), can withstand high and low temperatures and resists corrosion. Therefore, for the same strength, titanium is lighter than steel; however, it is more expensive. On the other hand, aluminum and especially some of its alloys, is also lightweight, is less expensive than titanium, it is easy to form and work with, and is lightweight. Therefore, it can be used in light load and low cost applications. Stainless steel is strong material, of reasonable cost, but heavy. It can be treated to have specific qualities, such as have a hardened surface, and can be machined relatively easy. However, its high density restricts its use to specialized components of active prostheses, such as transmission axles and gears (Bhuiyan et al., 2015).

1.5.2 Control

Upper-limb prostheses replace a missing upper limb of the human body, such as a hand, or an arm. A prosthesis is called active when it includes joints, motors, sensors, a power supply, and the like. From the control point of view, active control of a limb requires two levels of control: (a) the high level, which issues commands to the limb, and receives feedback (bidirectional interface with the brain), and (b) the low-level, which ensures that these commands are followed with the required accuracy.

Although actuation adds the capability for motion and force application, a low-level control system is needed to interpret feedback signals and send appropriate currents to motors so as to achieve the desired motion, or apply the desired force to the environment. In a sense, the control system is required to recreate a force-velocity relationship for the prosthesis similar to that of the missing limb; then the prosthesis is "transparent" and felt as an extension of the patient's body.

To develop a control system, one needs to have some knowledge of the dynamics of the system to be controlled. As this has many-DoF and is described by nonlinear equations of motion, the control system should be nonlinear, too, making its design more involved that for single-DoF linear systems. Prostheses have much in common with robotic arms and exoskeletons (Proietti et al., 2016), therefore control strategies that apply to them can be classified as *position control*, *force control*, and *interaction control*.

In position control, the aim is to ensure that all controlled variables, usually joint angles or displacements, achieve their commanded value at the right time. This must be achieved in the presence of friction, gravity, and even external disturbances, such as loads or impacts, and requires position sensors such as encoders or potentiometers. A large number of control schemes exist such as PID, model-based, LQR, adaptive, etc., (Bhuiyan et al., 2015). However, this type of control is only appropriate when the limb moves in free space.

In force control, the aim is to have the limb apply a desired force or moment to the environment, as for example, when using a hand tool, such as a drill. This type of control requires a force sensor, so that the applied force is available for feedback reasons and appropriate correction by the controller. As it is not possible to control simultaneously the velocity and the force of any body, position, and force controllers are incompatible. To have them both, switching between the two modes would be required. Even then, a delay in switching from position to force control can have undesired effects on the prosthetic and the patient itself.

This problem is addressed by impedance control (Hogan, 1985) and its variants (Calanca et al., 2016), where the aim is to control the relationship between velocity and force, without any switching of controllers. Although this is quite appealing for prostheses applications, the tuning of the control gains is not easy and usually it is task dependent. Recently Variable impedance actuators were introduced, where the controller takes into account the actuator dynamics acting through some physical mechanical compliance, aiming at implementing *variable impedance* according to the task to be performed, (Vanderborght et al., 2013).

An important issue to all control types is the type of command they accept and its interpretation. In the dominant in the early- to mid-1900s classic extended physiological proprioception (Classic EPP), no controller was needed and the connection between the end effector (implement) and the remaining limb was purely mechanical: the prosthetic limb was connected directly to cineplasty sites of residual arm with Bowden Cables (Tavakoli et al., 2017; Klopsteg and Wilson, 1954). As EPP was abandoned in favor of electromechanical prostheses, EMG was used as a high-level command to the active limb.

An EMG-based control system or myoelectric control system, controls the limbs by converting muscle movements to electrical signals allowing the amputees to control the prosthesis more directly (Harvey and Masland, 1941; Jawhar et al., 2011). The EMG signals must be amplified, filtered, and processed to yield root-mean-square (RMS) signals appropriate as control references. However, this introduces undesirable delays in the motion of the prosthetic limb. Unlike to EPP, this type of control does not provide feedback to the patient (proprioception) even if internal feedback is provided to the actuators for low-level control; its implementation requires visual feedback (Cloutier and Yang, 2013). An overview of myoelectric control, and its performance with respect to the characteristics of the ideal myocontroller is presented (Farina et al., 2014). Classic and relatively novel academic methods are described, including techniques for simultaneous and proportional control of multiple-DoF, and the use of individual motor neuron spike trains for direct control. Although myoelectric signals are widely considered as the best available control interface for powered prostheses, many amputees abandon their devices out of frustration due to the lack of precision of the prosthesis' movements (Shehata et al., 2017).

Ideally, a prosthetic device should establish a bidirectional communication between the patient and the prostheses. Currently and in principle, two methods can provide bidirectionality, the Biomechatronic EPP (Mablekos-Alexiou et al., 2015; Moutopoulou et al., 2015) and the direct neural interfaces (invasive or not) (Di Pino et al., 2009; Jerbi et al., 2011). The former represents new topology (Fig. 13) of EPP and aims at elimination of the drawback of cineplasty and Bowden cables, which render the EPP unaesthetic for the user. The core of this concept is based on principles of the field of telerobotics and teleoperation (Yokokohji and Yoshikawa, 1994). In this

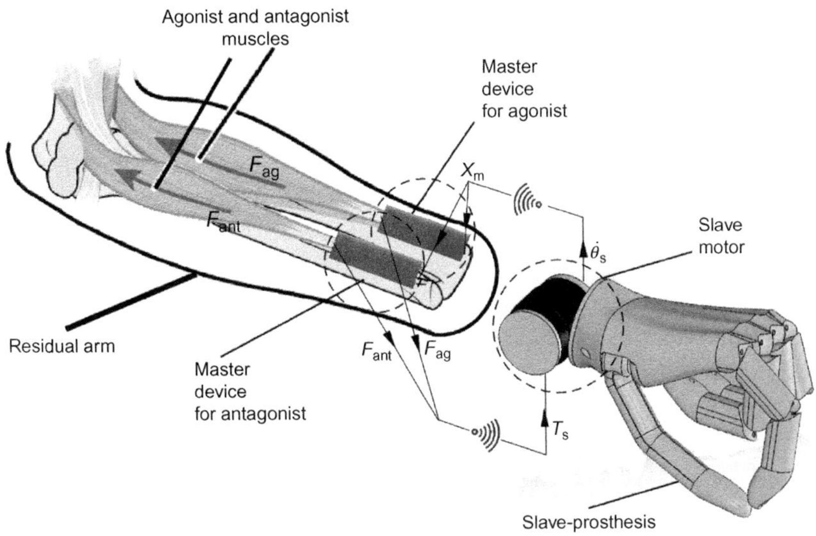

Fig. 13 Proposed control topology of biomechatronic EPP.

topology, a master—slave position-force control scheme is applied, using an implanted lead-screw driven by a DC-motor as the master, and the prosthetic hand as the slave. The implanted lead-screw takes a force command signal from the muscle/tendon attached to. The force command then wirelessly is transmitted to the slave, and a position feedback comes back from the slave to the DC-motor controller, which then moves. As the slave is essentially connected to the muscles, it establishes a bidirectional communication between the patient and the mechatronic device.

Bidirectional alternatives include the direct neural interfaces (invasive or not), often called brain-computer interfaces (BCIs), or more accurately brain-machine interfaces (BMIs) (Di Pino et al., 2009; Jerbi et al., 2011). These correspond to a direct communication path between an enhanced or wired brain and the powered prostheses.

Noninvasive BCI/BMIs have been used to enable high-level control of limbs. A BCI-controlled functional electrical stimulation system to restore upper extremity movements in a person with tetraplegia due to spinal cord injury has been presented (Pfurtscheller et al., 2003). Various neural machine interfaces for voluntary control of externally powered upper-limb prostheses were investigated (Ohnishi et al., 2007; Lebedev and Nicolelis, 2006). The use of electromyographic interfaces and peripheral nerve interfaces for prosthetic control, as well as BMIs suitable for prosthetic control, were examined in detail along with available clinical results (Ohnishi et al., 2007). A recent development in prosthetic hand design employs electroneurographic signals, requiring an interface directly with the peripheral nervous system or the central nervous system to control a prosthetic hand (Cloutier and Yang, 2013).

The current state of the upper-limb prosthetic market, with insights on the accompanying technologies and techniques is presented, along with prominent research solutions (Vujaklija et al., 2016). Moving away from upper-limb cosmetic prostheses, active elbow joints are available today, offering advanced control systems and multiple sensor integration and multijoint articulation. Novel surgical techniques in combination with modern, sophisticated hardware are enabling restoration of dexterous upper-limb functionality.

On the application front, biomechatronic hands provide examples of applied controllers. One of the first robotic hands was the Utah/MIT hand, a tendon operated multi-DoF dexterous robotic hand (Jacobsen et al., 1982, 1984) (Fig. 14). In this hand, a force PD controller was implemented at bandwidth of 50 Hz (Johnston et al., 1996). A biomechatronic optimized design of an anthropomorphic artificial hand for prosthetics and humanoid has been developed (Zollo et al., 2007). Its control system is developed in parallel to its mechanical design and is based on PD controllers with additional terms for compensating its elastic compliance.

Fig. 14 Utah/MIT hand. *(Courtesy of the Computer History Museum.)*

1.5.3 3D Printing

3D printing is an additive manufacturing process that creates a solid physical object from a digital design adding material layer by layer. Although 3D printing technology has been around for >30 years, only recently has become inexpensive. A number of 3D printing technologies and materials exist, with varying cost, and object size, strength, surface, color, etc. Among the technologies, one can identify the following: stereolithography (SLA), digital light processing (DLP), fused deposition modeling (FDM), selective laser sintering (SLS), selective laser melting (SLM), electronic beam melting (EBM), and laminated object manufacturing (LOM) (3D, n.d.). The materials used include glass polyamide, epoxy resin, wax, and metals like titanium, silver, and steel. Among the materials, the most popular is ABS; however, the most promising are composites (strength, lightweight) and metal (strength).

In upper-limb prostheses, three main prostheses parts that can benefit from such technologies are the socket, the arm, and the hand. The benefits of using 3D printed upper-limb devices are many and important: low cost, customization, lightweight.

3D printing will change the fabrication of prosthetic sockets and other limb components drastically. Current generations of 3D printers print composite materials such as carbon fiber, Kevlar, or glass fiber and have the potential to produce fully functional sockets. Latest socket developments are capable of facilitating both implantable and multiple surface electromyography sensors in traditional and osseointegration-based systems (Vujaklija et al., 2016). Many of the open-source hands that are prone to breakage and limited to child sizes can become fully functional at adult sizes. 3D direct laser metal sintering machines are also beginning to be used more in the manufacture of prosthetic components such as artificial fingers and other customizable components (Krausz et al., 2016).

The use of inexpensive, low-end 3D printing technologies for sockets is explored in Herberts et al. (1973). Although 3D printed objects usually are weak and fragile, comfortable prosthetic sockets have been produced and have been used in preliminary fittings with patients.

The first open-source 3D printed hand device was developed in 2012 in South Africa. A charitable organization called Robohand (Fig. 15A), created 3D limb models and uses 3D printers to build lightweight custom arms, hands, and fingers at low cost: \$500–\$2K (Oliker, 2015). The Robohand demonstrated that 3D printing reduces the cost of a prosthetic extremity (Tanaka and Lightdale-Miric, 2016). A large number of available open-

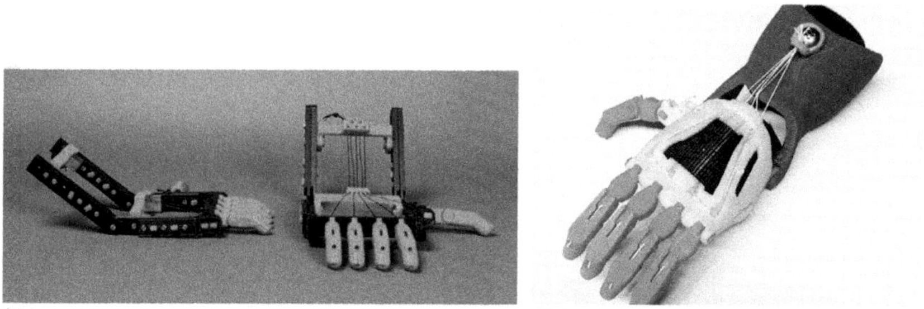

(A) (B)

Fig. 15 Prosthetic hands made by 3D printing techniques. (A) Robohand and (B) Cyborg Beast. *(Part (B): Cyborg Beast by Jorge M. Zuniga Ph.D.)*

source hand models are available today and include the Robohand, Cyborg Beast, Flexy Hand, K-1 Hand, Raptor Reloaded, Second Degree Hand, Osprey Hand, Limbitless Arm, and RIT Arm. These models are available through web sites such as Thingiverse (thingiverse.com), and the NIH 3D Print Exchange (3dprint.nih.gov).

The Cyborg Beast (Fig. 15B) was one of the first projects, which acknowledged the need for a low-cost customizable and prosthesis for children 3–16 years old (Zuniga et al., 2015). The project employed CAD design and 3D printing technology to develop low-cost devices with practical and easy fitting procedures. These body-driven devices are colorful, fun, and provide a general basic functional grasping motion. Although they offer customization and are cheap (~200 euros), they lack any significant functionality. As a result children although initially might feel joy because of the new colorful device in the long term they do not gain any practical benefit (especially children >5–6 years old) in terms of social exclusion and independence in the execution of activities of daily living (ADLs).

Another interesting project is Limbitless, which is the first low-cost customizable myoelectric device (Limbitless, n.d.). Limbitless is 3D printed, low-cost, actuated by an RC servo, which is controlled by an Arduino control board. Its functionality is limited to 1 DoF and therefore its practical significance is very low. Both of these low-cost prosthetic hands are part of greater effort initiated by a community of people who want to assist children with upper-limb deficiencies. The community is called eNable (http://enablingthefuture.org) and provides low-cost customized prosthetic devices similar to Cyborg Beast and Limbitless to children around the world. For a comprehensive review of 3D-printed upper-limb prostheses, see ten Kate et al. (2017).

To reduce the cost of upper-limb myoelectric prostheses and address the limitations of the Robohand, an inexpensive 3D printed prosthesis for patients with transradial limb amputation was developed (Gretsch et al., 2016). The prosthesis is shoulder-controlled and externally powered with an anthropomorphic terminal device. The patient can open and close all five fingers, and move the thumb independently at a cost of US$300. In addition, the device is lightweight, and its size easily scalable. Limitations include low grip strength and decreased durability compared to passive prosthetics.

It is expected that as the cost of 3D printing drops, and as materials become stronger, complex devices such as upper-limb prostheses will benefit from these techniques, leading to customizable, lightweight, easily replaceable, and cost-effective devices (Fig. 15).

1.5.4 Actuators

The actuators are very important elements of prosthetic devices, as they affect motion and interaction forces between the device and the environment. Candidate actuators include DC motors (brushed and brushless), ultrasonic motors, piezoelectric motors, artificial muscles (pneumatic or dielectric electroactive polymer based), shape memory alloys (SMAs), and more.

A large number of factors have to be taken into account for choosing actuators for prosthetic limbs. These include power, power density, voltage, current, torque, torque density, speed, size, weight, precision, hysteresis, repeatability, frequency, efficiency, noise, specific parameters depending on technology, applicability, and cost, and apply both to prosthetics and robotics (Hollerbach et al., 1992; Cura et al., 2003). In many studies, comparisons of actuators based on a number of criteria are presented; however, to adopt some technology for upper-limb prosthesis, one has to include in the comparison, not only the actuator but also the drivers/amplifiers needed, the sensors, and the power source for it. This is because the use of an actuator requires all these subsystems, and all of them have to be embedded in the prosthesis, or carried somehow by the patient. For example, a hydraulic actuator may look very attractive, but when one considers the power supply and the piping needed, then its attractiveness is reduced.

DC motors. Permanent magnet DC motors produce torque due to Lorentz forces acting on their windings. They are produced in miniature sizes of 1 W or even less, and in brushed and brushless forms. As both are low-torque, high speed devices (up to 10–20 krpm), they are used with miniature gearboxes, and are equipped with integrated angle sensors, usually in the form of magnetic or optical encoders. When these are part of a position control system, which regulates the position of a mechanical load precisely, they are called servomotors.

To increase the acceleration, *brushed micromotors* are usually coreless, that is, their windings are glued on a hollow rotor, while the stator carries the permanent magnets. Current commutation is done using a commutator and brushes, which are subject to wear. *Brushless DC micromotors* represent an inversion of the dc brushed motor principle. Here, no brushes are needed, the stator carries the windings, while the rotor carries the permanent magnets. The rotor can be rotating inside the stator, or outside (outrunner motor) for increased torque output, to the expense of higher moment of inertia.

Three versions of a multifunction haptic device that can display touch, pressure, vibration, shear force, and temperature to the skin of an upper extremity amputee have been developed (Kim et al., 2010). For the devices, a number of actuators, such as ultrasonic motors, and electromagnetic motors, were considered. Although ultrasonic motors produced high torque and need no reduction, they had poor frequency response and could not achieve high accelerations for enough time. Therefore, DC brushed micromotors with appropriate gearboxes were selected providing better open-loop frequency response, closed-loop force response, and tapping response in constrained motion.

A 3D printed prosthetic hand for transmetacarpal amputees was developed (Mio et al., 2017). Due to the little space to fit actuators and their associated electronics was actuated by DC micromotors. Four-bar linkage mechanisms were used for the index, middle, ring, and little fingers flexion movements, while a mechanism of cylindrical gears, and worm drive was used for the thumb, all position controlled by local controllers.

A parallel prosthesis aiming to increased force, and reduced weight and size was developed. The prosthesis has four DoF driven by four brushless motors, weighs 1010 g, and can lift 2 kg, while the time for a total excursion of the flexion of the elbow is about 2 s (Escudero et al., 2002).

An overview of past and present artificial hands, developed in the framework of research projects in prosthetics and humanoid robotics is available (Controzzi et al., 2014). Most of them use DC micromotors in conjunction with micromechanisms, for better matching the micromotor to its mechanical load.

Ultrasonic motors. These are electric motors, which produce motion by the mechanical vibration of the stator, placed against the rotor (for rotation) or the slider (for linear displacement). As the mechanical vibrations are in the ultrasonic region, that is, above 20 kHz, they are silent. These motors can be very small in size, they exhibit high power density, high torque and low speed, low moment of inertia, fast response, noiseless operation, self-braking drive, and generate no magnetic fields (Cura et al., 2003; Pons et al., 2002). Its disadvantages include its need for a high-frequency energy source, its short service life due to stator/rotor contact, variations in speed and low efficiency compared to electromagnetic motors, and its requirement for complicated control.

Piezoelectric motors. A piezoelectric or piezo motor is an electric motor, which is based on the change of shape of piezoelectric materials when an electric field is applied. This change of shape, and combined with the stick-slip phenomenon produces mechanical displacements in the form of linear of rotary motion. Compared to dc motors, piezo motors are small and produce large torques, but they are relatively expensive (Da Cunha et al., 2000).

Artificial muscles can be built in principle using pneumatics or dielectric electroactive polymers. This idea is very attractive, because such muscles can fit well in a prosthetic arm, and the load-length curve produced resembles that of the actual limb.

Pneumatic artificial muscles (PAMs) consist of an inflatable inner bladder inside a braided mesh, clamped at the ends, that contracts or extends when supplied with high/low pressure, respectively. As they can only pull, PAMs are applied in agonist and antagonist pairs. This technology was invented in the 1940s and developed in the 1950s as McKibben artificial muscles (Chou and Hannaford, 1996). PAMs are lightweight, fail safe, and compliant. Experimental results indicate that accurate position control is feasible, with power/weight outputs in excess of 1 kW/kg at 200 kPa (Caldwell et al., 1995). However, to operate them, one needs a compressor, which tends to be bulky and noisy, or an external pressurized gas (CO2, air) tank. It also requires solenoid valves, driver electronics, and a battery. Recently, PAMs are of renewed interest due to applications in soft robotics (Greer et al., 2017).

Electroactive polymers were discovered in 1880. They are also known as compliant capacitors, as they have similar behavior to capacitors. These polymers, when stimulated by an electric field, exhibit a change in size or shape; if constrained, they apply large forces (Kim and Tadokoro, 2007). The concept of using dielectric electroactive polymers (EAPs or DEAPs) as artificial muscles was revived recently as it has been demonstrated that some EAPs can exhibit up to a 380% strain (Bar-Cohen, 2001). They have shown great promise due to their low cost, lightweight, simple actuating structure, and good performance in low frequencies with large deformation (Yuan et al., 2016). However to fully deploy their capabilities, DEAPs require very high voltages, of the order of 2 kV DC. Despite the small currents and compact amplifiers, these voltages are not human friendly.

Current DEAP challenges were reviewed with respect to durability, precision control, energy consumption, and anthropomorphic implementation (Biddiss and Chau, 2008). DEAP actuators in powered upper-limb prosthetics is impeded by poor durability and susceptibility to air-borne contaminants, unreliable control owing to viscoelasticity, hysteresis, stress relaxation and creep mechanisms, high voltage requirements, and insufficient stress and strain performance within the confines of anthropomorphic size, weight, and function (Biddiss and Chau, 2008). Although this technology is currently infeasible for upper-limb prosthetics, research continues, aiming at reducing the voltage required and increasing their overall potential (Bar-Cohen et al., 2018).

Shape memory alloys are alloys that convert heat into mechanical displacement through thermo-elastic transformations, passing from martensite to austenite when heated; when cooled, the material returns to austenite. SMAs exhibit shape memory, that is, they return to a predetermined shape when heated (Jani et al., 2014). In practice, this actuator is made by a number of SAM wires in parallel, which can be heated by current passing through the strained wires. Usually the heat is produced by the alloy's own resistance, causing it to contract and return to its original shape, producing large forces.

When these alloys are used in the form of wires, they present a good strength/weight ratio, and high strength/area ratio, rendering this material appropriate for application in upper-limb prostheses. The most common SMA, Nitinol, is composed of nickel and titanium (NiTi). This SMA displays one of the highest work density at 10 J/cm^3, which is 25 times greater than that of electric motors and is able to lift >100 times of its weight. Furthermore, the NiTi SMA is biocompatible, exhibits high wear resistance, and is highly corrosion resistant (Jani et al., 2014).

However, SMAs require high temperatures (up to 100°C) to develop their maximum force and have slow response since it takes time to cool the wires. As their strain is 4%–8.5%, they need either special mechanisms or long lengths to achieve useful displacements. Although recent advancements in SMAs have produced strains of up to 32% using a braided coil design, additional shortcomings including high hysteresis, short service life, and high-energy consumption, still limit their applicability to practical prostheses.

The mechanical design for a five fingered, 20 DoF dexterous hand patterned after human anatomy and actuated by SMA wires strands of 150 μm in diameter, was presented. Two experimental prototypes of a finger were developed, one fabricated by traditional means and another fabricated by rapid prototyping techniques, showing promise for use in prosthetic hands (De Laurentis and Mavroidis, 2002).

1.5.5 MEMS

Microelectromechanical systems (MEMSs) refer to the technology of microscopic devices, particularly those which include moving parts. MEMSs are fabricated using modified semiconductor device fabrication technologies, including molding, plating, wet and dry etching, electro discharge machining, and other similar technologies.

The most common application of MEMS is sensors, such as accelerometers, inertial measurement units (IMUs), magnetic field sensors, microphones, pressure sensors, biosensors (bio-MEMS), and more. MEMS are used in large quantities in modern cars, propelling their proliferation in other areas, including upper-limb prostheses. Here, the MEMSs are mostly used as posture sensors and force/tactile sensors.

The development and preliminary experimental analysis of a soft compliant tactile microsensor with minimum thickness of 2 mm was presented in Beccai et al. (2008). A high shear sensitive 1.4 mm^3 triaxial force microsensor was embedded in a soft, compliant, and flexible packaging. The performance of the sensor was evaluated by static calibration, maximum load tests, noise and dynamic tests, and by focusing on slippage experiments. The experiments showed that the tactile sensor is sufficiently robust for application in artificial hands while sensitive enough for slip event detection.

A tactile sensor designed to measure shear forces for use in robotic and prosthetic hands, where haptic feedback or ability to detect shear forces associated with slip are critical is described and characterized (Tiwana et al., 2011). The sensor employs the principle of differential capacitance to measure the mechanical deflection of the sensor and can be easily mass produced. Sensors with a full-scale displacement range of ± 0.525 mm were produced and the differential capacitance was measured. Shear force transduction was characterized over the range of 0–4 N. A maximum standard deviation of $1.35e-15$ F was measured across the characterized full-scale sensor range of ± 4 N. The sensor output was found to be approximately linear.

A triaxial force sensor was developed with a MEMS as its core component (Sieber et al., 2008). This device allows measuring forces the range of 0–3 N for normal and ± 50 mN for tangential forces with a resolution of

11 bits. Together with a haptic input device, a setup was created allowing palpation and force feeling.

Wearable systems posture sensors for upper body rehabilitation are reviewed (Wang et al., 2017). These include mostly accelerometers and IMUs measuring accelerations and angular velocities, as they yield relatively accurate essential values, are easy to use, and are miniature in size. Similar sensors can be used in upper-limb prostheses to track arm or hand motions, and for safety reasons.

1.5.6 Wireless Power Transfer

Wireless power transfer (WPT) is the transmission of electrical power without wires and is based on technologies using time-varying electric, magnetic, or electromagnetic fields. WPT is useful to power electrical devices where are inconvenient, or not possible, as is the case of body embedded sensors, actuators, and communication devices.

Power can be transferred over short distances (near-field transfer) by alternating magnetic fields and inductive coupling between coils, or by alternating electric fields and capacitive coupling between metal electrodes. Inductive coupling is the most common method of WPT and is used in charging devices such as smart phones, electric shavers, visual prostheses, and implantable medical devices (cardiac pacemakers, cochlear implants) (Sun et al., 2013; Moorey et al., 2014) (Fig. 16). For 20 mm distance

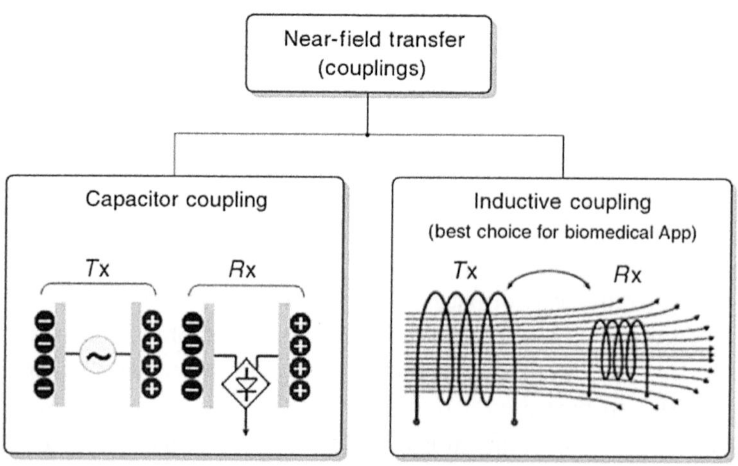

Fig. 16 Capacitive and inductive couplings for WPT. *(From Sun, T.J., Xie, X., Wang, Z.H., 2013. Design challenges of the wireless power transfer for medical microsystems. In: 2013 IEEE International Wireless Symposium (IWS).)*

Table 1 Noncapacitive WPT Options

Parameters	Options Inductive Coupling [11] [12]	RF [6] [13]	Ultrasound [14] [12]
Human safety	Depends on energy transferred	Yes	Yes
Efficiency	73%	48%	21%–35%
Max power	Up to 10 W	<1 W	100 mW
Frequencies	1 kHz–100 MHz	30 kHz–300 GHz	10 kHz–10 MHz

From Moutopoulou, E., Bertos, G.A., Mablekos-Alexiou, A., Papadopoulos, E.G., 2015. Feasibility of a biomechatronic EPP upper-limb prosthesis controller. Conf. Proc. IEEE Eng. Med. Biol. Soc. 2015, 2454–2457. https://doi.org/10.1109/EMBC.2015.7318890.

separation and size of the coil pair, loop diameter, and frequency play a major role in determining WPT performance (Celik and Aydin, 2017).

The noncapacitive WPT couplings, include the inductive, radio frequency (RF), and ultrasound couplings. Of those, the inductive coupling is characterized by high efficiency and power transfer capability and is therefore superior to the other two (Moutopoulou et al., 2015), see Table 1. Also, according to Sun et al. (2013), inductive coupling is considered to be the best choice for biomedical applications.

Candidate biomedical applications include artificial hearts, visual prostheses, ingestible devices (Kim et al., 2014), and upper-limb-embedded biomechatronic devices (Kontogiannopoulos et al., 2018). Implantable neural prosthetic devices typically have power requirements that exceed the capability of reasonably sized implantable batteries. Therefore, transcutaneous magnetic coupling remains the method of choice for powering implanted neural prostheses (Troyk and DeMichele, 2003). A fully wireless EMG recording system that can enable upper-limb prosthesis control while achieving maximum power transfer efficiency through magnetic resonantly coupled (inductive) WPT is described in Bercich et al. (2016). This solution makes notable progress in the efficiency of WPT through loosely coupled inductive links specifically for upper-limb prostheses. As an added benefit of the inductive coupling, data can also be transmitted (Ghovanloo and Najafi, 2004; Troyk and DeMichele, 2003).

For these applications, directivity, system stability, reliability, and efficiency enhancement through the wireless transfer coil design enhancement and operational tunings are required (Kim et al., 2014). Other important parameters include human safety due to a rise in tissue temperature and miniaturization of the relevant electronics (Moutopoulou et al., 2015).

Modeling techniques used in the analysis of WPT systems, with specific emphasis on the approximations that restrict their applicability, aiming at a general modeling technique is given in Moorey et al. (2014). Research in the area of implantable high-power neuroprosthetic devices such as visual prostheses and BCIs focuses on transcutaneous inductive power links formed between a pair of printed spiral coils (PSCs), batch-fabricated using micromachining technology. Optimizing the power efficiency of the wireless link is imperative to minimize the size of the external energy source, heating dissipation in the tissue, and interference with other devices. The theoretical foundation of optimal power transmission efficiency in an inductive link, combined with semiempirical models resulted in two design examples at 1 and 5 MHz, achieving power transmission efficiencies of 41.2% and 85.8%, respectively, at 10-mm spacing (Jow and Ghovanloo, 2007). A method of how to characterize and optimize rectangular coils used in inductive links for general applications is described in Yong-Xi et al. (2011).

2 STATE OF THE ART

2.1 LUKE Arm

The current state of the art for upper-limb prostheses is a many-DoF prosthetic arm named LUKE arm (Fig. 17) which is using myoelectric pattern recognition and TMR-generated control sites. LUKE arm is the previous DEKA arm and it was developed under the DARPA projects

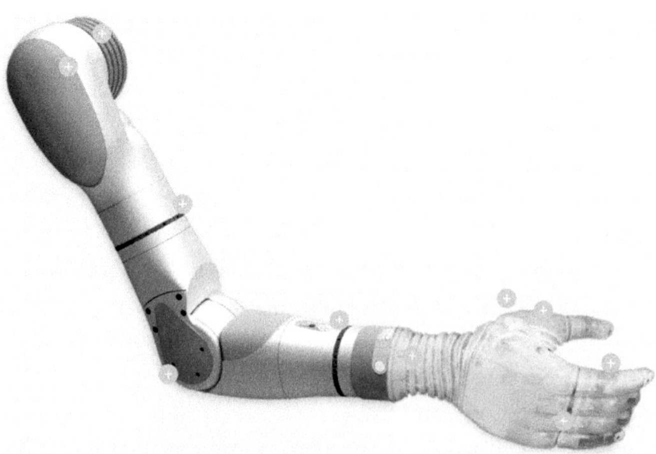

Fig. 17 Mobius Bionics LUKE arm (each of the 10 DoFs is shown as a "+"). *(Photo courtesy Mobius Bionics LLC. Used with permission.)*

Revolutionizing Prosthetics program with collaboration of many academic institutions and DEKA. LUKE arm is commercialized by Mobius Bionics.

The technologies enabling this new generation of upper-limb prosthetics are described further.

2.2 Targeted Muscle Reinnervation

One of the biggest challenges of prosthesis history was that high-level upper-limb amputees did not had much fewer EMG control sites available than needed DoF to control with the prosthesis. TMR is a novel surgical technique which was first introduced in 1995 by Dr. Todd Kuiken of Northwestern University and the former Rehabilitation Institute of Chicago at that time (Kuiken et al., 1995) and solved that problem. It is an extra surgery but its return on investment (ROI), is the creation of additional EMG sites. In 1995, the early experiments of Dr. Kuiken, muscle recovery of hyperreinnervated rat skeletal muscle was found to lead to increased muscle mass and strength compared to the self-reinnervated muscles (Kuiken et al., 1995).

Having positive preliminary animal model results, Dr. Kuiken took his idea to the next level and showed that in an elbow disarticulation amputee, four independent nerve-muscle units were created by anastomosing residual brachial plexus nerves (musculocutaneous, median, radial, and ulnar nerves) of the amputated limb, to dissected and divided aspects of the pectoralis major and minor muscles of the chest. After 5 months, three reinnervations were successful, that is, reinnervated pectoralis muscles were moving at the command of the anastomosed brachial plexus nerves, offering additional EMG sites for prosthesis control, which controlled successfully a three DoF prosthesis: elbow, wrist rotator, and hand (Kuiken et al., 2004; Miller et al., 2008). The muscle can be perceived as the "amplifier" of the nerve, and getting a control signal for prosthesis control from a muscle has less challenges than from a nerve (Childress, 1992; Hijjawi et al., 2006). It was an unexpected discovery that the reinnervated sensory neurons, reinnervated also the skin above the pectoralis muscle, providing sensation of the fingers at the chest (Kuiken et al., 2007; Marasco et al., 2009). It seems that the efferent nerves migrate from the reinnervated site through the muscle and breast tissue and coexist with native chest efferents (Kuiken et al., 2007). During this study, there was an attempt to reinnervate a skin area away from the pectoralis area used for motor control (Kuiken et al., 2007) but without any practical applicable solution for prosthesis control.

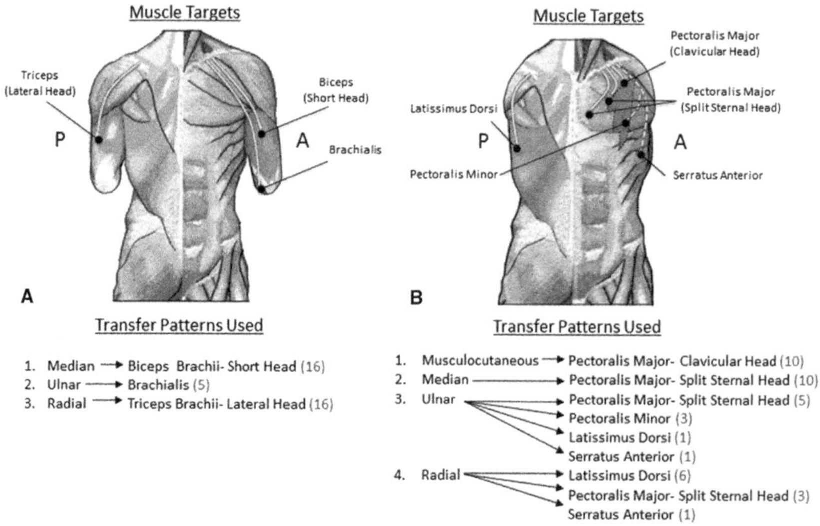

Fig. 18 TMR for (A) transhumeral and (B) shoulder disarticulation amputees. Diagrams illustrate the nerve transfers employed for the (A) transhumeral and (B) shoulder disarticulation procedures. The left side of each image provides a posterior (P) perspective while the right depicts the anterior (A) side. Donor nerves are coapted to the motor nerves of the target muscles via small recipient motor nerve branches. The target muscles are labeled on the diagrams and the yellow lines demonstrate the donor nerves in their transferred positions. The dashed yellow lines indicate nerve transfers that are less frequently used. The parenthetical numbers indicate the frequency with which each transfer was used in this series. *(From Souza, J.M., Cheesborough, J.E., Ko, J.H., Cho, M.S., Kuiken, T.A., Dumanian, G.A., 2014. Targeted muscle reinnervation: a novel approach to postamputation neuroma pain. Clin. Orthop. Relat. Res. 472(10), 2984–2990. https://doi.org/10.1007/s11999-014-3528-7.)*

TMR has been used to create EMG sites for transradial, transhumeral, and shoulder disarticulation (see Fig. 18) upper-limb amputees (Kuiken et al., 2017; Souza et al., 2014). The variation of TMR for transradial amputees involves reinnervation of the median nerve to the flexor digitorum superficialis (FDS) muscle and reinnervation of the ulnar nerve to the flexor carpi ulnaris (FCU) muscle (Kuiken et al., 2017). Another accidental discovery of TMR is that amputees (both transhumeral and shoulder disarticulation) that performed TMRs did not experience any more neuroma or phantom limb pain (Souza et al., 2014).

TMR procedures are well documented at (Kuiken et al., 2017). In addition, TMR training videos for clinicians exist at: https://www.sralab.org/targeted-muscle-reinnervation-training-video.

2.3 Sensing Many-DoFs

Nowadays, the sensing of many EMGs is done via superficial EMG electrodes or intramuscular EMG electrodes as described later in Section 2.6. The following can apply for both.

2.3.1 Artificial Intelligence, Neural Networks, and Pattern Recognition

Pattern recognition for myoelectric control was first introduced (Herberts et al., 1973).

The University of New Brunswick has reinvented and championed since 1990s the pattern recognition paradigm for upper-limb multifunction prosthesis control (Hudgins et al., 1993) by showing that the myoelectric signal is deterministic at the onset of the contraction and that an artificial neural network can be used to classify patterns of the myoelectric signals (Hudgins et al., 1993).

If the physiological musculature exists, then its EMG can be used to control the prosthetic device of an upper-limb amputee, for example, biceps and triceps pair to control a prosthetic elbow (Scheme and Englehart, 2011), leading to intuitive control and low mental loading which are both desirable attributes for prosthesis control (Childress, 1992). In the case that the musculature does not exist, then logical substitutions should be used, for example, using wrist flexor and extensor pair to control a prosthetic hand. The above one muscle pair to one DoF schema becomes impractical when applied for the control of a multi-DoF upper-limb prosthesis, due to co-activation patterns of muscles, deep common muscle activation, and EMG crosstalk (Scheme and Englehart, 2011). One of the proposed solutions of the above problem is the tool of pattern recognition. The idea is simple: a set of myoelectric signals is recorded, for which different features are extracted (amplitude, zero crossings, etc.). During training a set of basic hand postures is performed and their associated features pattern is associated and a classifier is trained. During a real-time hand movement, a classifier algorithm will try to classify the real-time features set to one of the basic classes which is associated to a particular hand posture (from training).

A complete review of the pattern recognition techniques for upper-limp prostheses shows that current pattern recognition success rates are 90%–95%, in a controlled laboratory environment (Scheme and Englehart, 2011). That is, open the hand 9 out of 10 times but one time statistically do not open it. This is not clinically acceptable. On top of this poor clinical performance, there are many practical challenges which impede clinical repeatability and robustness such as: electrode shift from prolonged wearing of prosthetic socket, variation in force levels, and variation in EMG due to recording environment (skin impedance changes over time, motion artifacts, noise, etc.). As accuracy falls below 85%, a multi-DoF prosthesis can become frustrating and usability is poor (Scheme and Englehart, 2011).

In order for surface EMGs and pattern recognition's practical problems to be minimized, frequent training sessions have been proposed. These frequent sessions are not welcome by the amputees which they expect their prosthesis to perform like the natural hand. This makes it even more difficult for pattern recognition controllers to be used clinically.

The main advantage of surface EMG, its noninvasiveness comes with a lot of limitations which might make the application of pattern recognition not practical clinically. The solution to all these problems is the use of IMESs which will reduce or eliminate all the above problems. Use of IMESs will enable simpler pattern recognition classifier performances since they will be close to the "source of truth," that is, the muscles generating the movements instead of the surface of the forearm.

A pattern classifier able to adapt to changes has been shown to decrease the error % of the pattern recognition (Sensinger et al., 2009). Unsupervised (user's intended class is known) pattern recognition for myoelectric control resulted in at least 26% reduction in error and supervised pattern recognition led to smaller reduction in error due to higher uncertainty of correct class (Sensinger et al., 2009).

In addition, Hargrove et al. (2017) have found that pattern recognition as a method for controlling a multi-DoF prosthesis is superior when used for TMR subjects than direct EMG control.

The Osseointegration group in collaboration with Integrum from Sweden has developed and has made available an open source platform called Biopatrec. Biopatrec uses pattern recognition algorithms on acquired from the subject in real-time multi-site EMG signals for the control of a multi-DoF prosthetic arm or hand (Ortiz-Catalan et al., 2013, 2014b).

On the other hand, a competitor of Biopatrec, which is CoApt, LLC, a Chicago-based startup connected to researchers from Shirley Ability Lab—formerly RIC, is commercializing a pattern recognition multi-DoF myoelectric controller with 3-DoFs: one DoF at the hand and the remaining to be at the elbow and wrist (Parker et al., 2006).

2.4 3D Prototyping

3D printers with new materials are becoming available, for example, carbon fiber. In addition, laser sintering machines using metal would provide more durable components (Weir, 2017). This in conjunction with many open-source hand designs (Krausz et al., 2016) can enable rapid manufacturing of the prosthesis components at the prosthetic facility or at home and the prosthesis to be fitted at home by a prosthetist or the amputee prints and takes his/her components of the prosthesis at the prosthetic facility.

2.5 Osseointegration—Osseoperception

Professor Per-Ingvar Brånemark discovered in the 1950s that bone can integrate and coexist "peacefully" with titanium components. He defined osseointegration as "A direct structural and functional connection between ordered living bone and the surface of the load—covering implant" (Branemark et al., 1969). Osseointegration has been proposed for upper-limb prostheses since 1980s after the success of dental implants (Childress, 1997, 1998). The major problem of this technique has been the risk of infection at the skin to implant area (Childress, 1997, 1998). There has been a lot of effort in the past years to optimize implants design, the process and the rehabilitation protocol in order to minimize the risk of infection. In 1999, a treatment protocol called OPRA (Osseointegrated Prostheses for the Rehabilitation of Amputees) was established. Although there is >20 years of experience in transhumeral osseointegration procedures, the orthopedic community still is skeptical of this technique (Tsikandylakis et al., 2014). Results of the first 18 transhumeral patients following the OPRA protocol for upper limb are promising, with a 83% implant survival rate at 5 years and a 38% 5 year incidence of infectious complications of which most of them were not serious and were treated with nonsurgical interventions (Tsikandylakis et al., 2014). Integrum, the company that is commercializing the Osseointegration technology OPRA, was given Humanitarian approval in 2016 from the FDA, to perform 18 Clinical trials for upper-limb amputees in the United States (Li, 2016).

The biggest benefit that Osseointegration provides as a procedure and methodology, other that it eliminates the need of a socket and provides wider range of motion (Fig. 19A), is that there is direct link between the bone, muscles, tendons, receptors, and the prosthesis (Fig. 19). This direct link and engagement provides Osseoperception, the ability of the amputee

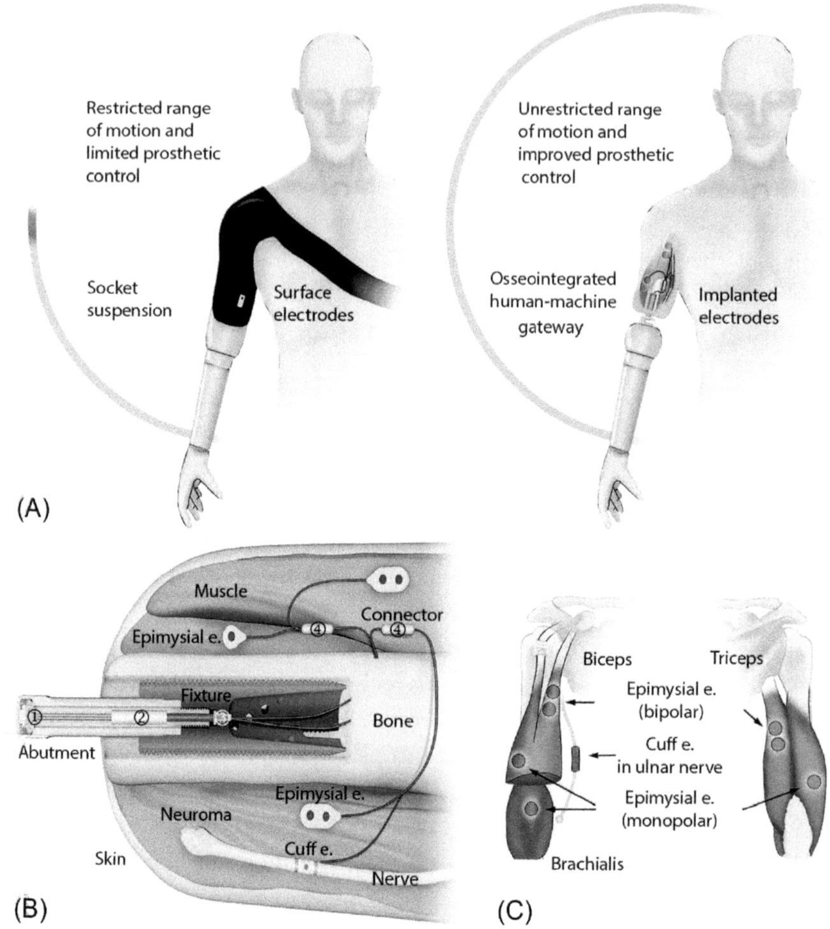

Fig. 19 Upper-limb Osseointegration prosthesis architecture (OHMG). (A) In the conventional socket suspension for high amputations, the adjacent joint is frequently constrained in the range of motion by the socket to provide sufficient suspension. The OHMG eliminates socket-related issues and allows for unrestricted limb motion (see movie S1 downloadable from Ortiz-Catalan et al., 2014a). (B) The prosthetic limb was attached to the abutment, which transferred the load to the bone via the osseointegrated fixture. The abutment screw, which goes through the abutment to the fixture, was designed to maintain the abutment in place. A parallel connector (1) was embedded in the screw's distal end to electrically interface the artificial limb. This connector was electrically linked to a second feedthrough connector (2) embedded in the screw's proximal end. The stack connector (2) interfaced with a pin connector extending from the central sealing component (3), from which leads extended intramedullary and then transcortically to a final connector (4) located in the soft tissue. The leads from the neuromuscular electrodes ("e.") were mated to connector (4). (C) Placement of epimysial and cuff electrodes in the right upper arm. *(From Ortiz-Catalan, M., Hakansson, B., Branemark, R. 2014. An osseointegrated human-machine gateway for long-term sensory feedback and motor control of artificial limbs. Sci. Transl. Med. 6(257), 257re256. https://doi.org/10.1126/scitranslmed.3008933; in order to provide all the details of proposed OHMG platform.)*

to "feel" where his or her prosthesis is without seeing it. The information comes integrative to the amputee by using the remaining afferent (sensory) pathways which are now integrated with the prosthesis and give us an EPP type of control which we know from the past that has advantages over other prosthesis control topologies (Childress, 1997, 1998; Doubler and Childress, 1984a).

For upper-limb amputees, the protocol, and device are part of the Osseointegrated Human-Machine Gateway (OHMG) (Ortiz-Catalan et al., 2014a). The osseointegration implant provides also the gateway or "corridor" for intramuscular EMG electrodes to be placed in the muscles and the wires to come out (Ortiz-Catalan et al., 2013, 2014a). OHMG should be viewed as a platform. Fig. 19 describes all the details of the OHMG platform.

A modified OHMG platform could be used in the future for lower limb Osseointegration prostheses.

As we mentioned before, one of the benefits of all Osseointegrated prostheses is the Osseoperception provided by the receptors and the direct mechanical linkage provided. Therefore, the OHMG, facilitates the integration (and thus "Integrum" is a good name) of the motor and sensory aspects needed for upper-limb prostheses, eliminating the need for wireless interfaces.

All the potential benefits and advantages of osseointegration do not come without problems. The biggest problem of this technique is its long lasting battle with bacteria at the skin interface and its unknown long-term impact on the quality of the bone fixture (Lenneras et al., 2017). Therefore, long-term studies are needed. Radiologically found endosteal bone resorption accompanied with pain at loading might be associated with potential weakness of the bone fixture (Lenneras et al., 2017). Different osseointegration research groups are taking nine different engineering variants of the implant designs and materials in order to achieve a stable mechanical interface between the bone and the implant (Thesleff et al., 2018). The prominent, the ORPA treatment protocol, which involves the traditional surgical technique from Sweden and rehabilitation protocol, involves a threaded titanium abutment screwed into the medullary cavity of the bone (Fig. 20) and a long rehabilitation phase. This treatment protocol has been adapted for transhumeral, transradial, thumb, or finger amputations of the upper limb (Thesleff et al., 2018).

More comparative details on the different surgical techniques and implant systems are given in Thesleff et al. (2018).

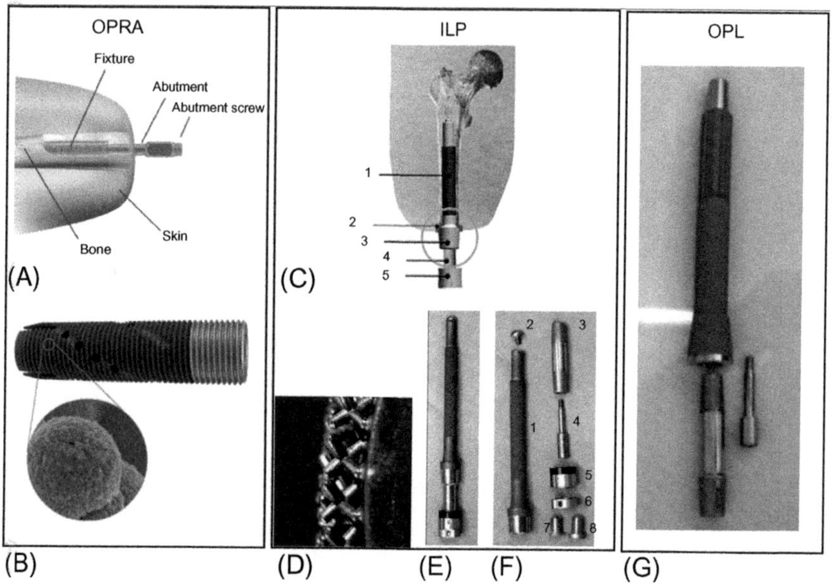

Fig. 20 OPRA. (A) Schematic image of OPRA implant system in an amputated limb; (B) OPRA fixture; the exterior surface in the dark gray region is treated to enhance osseointegration. The lower image shows a close-up of the laser-induced micro structure from the surface treatment; (C) schematic image of the ILP implant system: (1) porous-coated portion of the intramedullary component of the implant system, (2) inner lining, (3) Morse taper, (4) dual cone adapter, (5) knee connecting adapter. The red line indicates the stoma channel; (D) close-up of the spongiosa metal surface to enhance osseointegration and ingrowth; (E) ILP implant system assembled; (F) exploded view of ILP implant system assembly consisting of: (1) intramedullary implant, (2) temporary cover screw, (3) dual cone adapter, (4) safety screw, (5) sleeve, (6) rotating disc (until prosthetist has made final adjustments), (7) final propeller screw, (8) provisional screw; and (G) OPL type-B implant system. *(Copied from Thesleff, A., Branemark, R., Hakansson, B., Ortiz-Catalan, M., 2018. Biomechanical characterisation of bone-anchored implant systems for amputation limb prostheses: a systematic review. Ann. Biomed. Eng. 46(3), 377–391. https://doi.org/10.1007/s10439-017-1976-4.)*

A collaborative effort between Russian and US academic institutions (Shevtsov et al., 2015) involves animal studies with rabbits. In detail, TiO_2 nanotubes along with skin fibroblasts from the rabbit are used as rough (Sul, 2010) coatings on the skin and bone integrated pylon (SBIP), in order to promote less bacteria, better skin interface and better bone ingrowth, with positive preliminary results (Shevtsov et al., 2015). A new approach borrowed from dentistry is the use of porous tantalum trabecular metal (PTTM) collar at the skin abutment interface since it has shown increased skin ingrowth and sealing in dental implants (Bencharit et al., 2014; Deglurkar et al., 2007).

2.6 BIONs and IMESs

2.6.1 Alfred E. Mann Foundation

Alfred E. Mann Foundation was the first one to develop miniature implantable myoelectric sensors (IMESs), the BIONs (latest is a rechargeable battery version BION3) which were intended for broad rehabilitation use, have been used for stimulating the hearing nerve or for intramuscular stimulation for stroke patients (Loeb et al., 2004), see Fig. 21.

2.6.2 IMES

Dr. Weir from Northwestern University Prosthetics Laboratory was the first to use the BIONs made from the Alfred E. Mann foundation, for upper-limb prosthetics use (DeMichele et al., 2008; Schorsch et al., 2008; Troyk et al., 2007; Weir et al., 2009). The BIONs in this case were used as IMES, were not used for stimulating muscles but for picking up the myoelectric activity of the muscle—not via skin surface—but intramuscularly (Fig. 22).

Schematic representation of how IMES, implanted in the muscles of the forearm, communicates via the external coil that is laminated in the prosthetic socket and encircles them when the prosthesis is worn (Fig. 22).

The IMES are injected intramuscularly by the clinician.

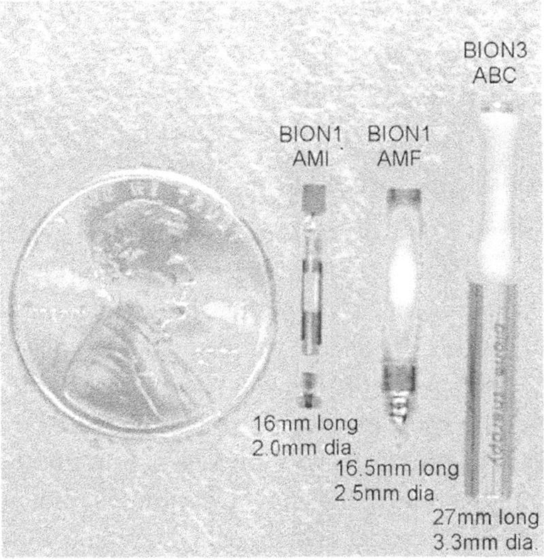

Fig. 21 BIONs. *(From Loeb, G.E., Richmond, F.J., Singh, J., Peck, R.A., Tan, W., Zou, Q., Sachs, N., 2004. RF-powered BIONs for stimulation and sensing. Conf. Proc. IEEE Eng. Med. Biol. Soc. 6, 4182–4185. https://doi.org/10.1109/IEMBS.2004.1404167.)*

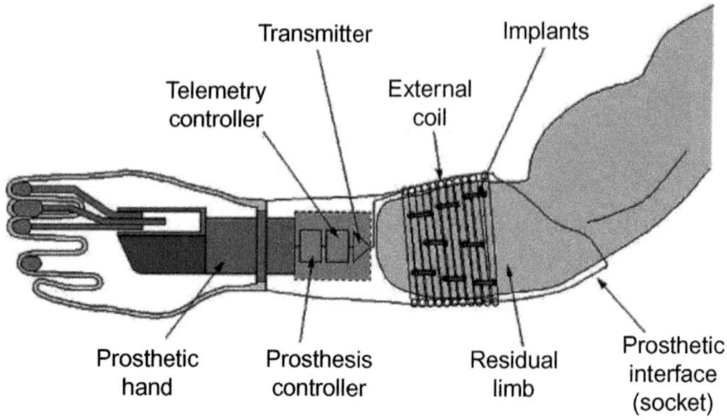

Fig. 22 IMES use for prosthesis control. *(From Weir, R.F., Troyk, P.R., DeMichele, G.A., Kerns, D.A., Schorsch, J.F., Maas, H., 2009. Implantable myoelectric sensors (IMESs) for intramuscular electromyogram recording. IEEE Trans. Biomed. Eng. 56(1), 159–171. https://doi.org/10.1109/TBME.2008.2005942.)*

Variants of the IMES systems for prosthetic use already exist. The Ripple system from Salt Lake City, United States and the MyNode from the Shirley Ryan Ability Lab (formerly known as Rehabilitation Institute of Chicago or RIC) have been developed. The MyoNode (Bercich et al., 2016) has the advantage that is made from off-the-self components.

Even though, these systems have been used in an EMG sensory input paradigm for prosthesis control, there is potential of expanding the paradigm by integrating specific sensory nerve stimulation in order to increase feedback and proprioception in an artificial way. With that holistic paradigm the need for a musculoskeletal model is evident (see Section 2.7.1).

2.7 Neural Feedback Integration

Recently, peripheral nerves have been stimulated by signals connected to touch sensors of prosthetic hands in order to give to the amputees a sense of touch. It is of importance to note that the integration of these sensory signals happens via the Peripheral and Central Nervous Systems, taking advantage of the plasticity of the nervous system, that is, the ability to learn and adapt. This could enhance or complement the widely used myoelectric control of upper-limb prostheses since the lack of proprioceptive feedback is one of its major disadvantages. This breakthrough though makes more evident the need of a model which will determine how the different sensory and motor signals have to coexist as controlling a many-DoF prosthetic hand.

2.7.1 Biomechanics Model

Since the long-term vision for the upper-limb field is a multi-DoF control including both motor pathways (efferent) and sensory (afferent) pathways integration, a good biomechanics model connecting all the control inputs to prosthesis DoFs should be needed in the future. Several attempts have been made but without a satisfactory and stable result up to now. The latest most successful and unsuccessful attempt for the development of such a stable model is made by Blana et al. (2016) and (2017). This is a research area that will be needed in the future in order for multi-DoF prostheses to become functional and available.

2.7.2 Peripheral Nerve Stimulation

The peripheral nerve stimulation that has been integrated with prosthetic arms in the recent years has positioned the upper-limb prostheses to be more integratable with the amputee, since the person "feels" the status of the environment in a natural proprioceptive way.

The existence of peripheral motor and sensory pathways and their satisfactory functionality has been demonstrated for after 2 years after amputation surgery and after any CNS reorganization (Dhillon et al., 2004). Longitudinal intrafascicular electrodes (LIFEs) electrodes implanted in upper-limb amputees connected to force and position sensors of a prosthesis have been used to set grip forces, leading to increased performance and natural integration of the prosthetic arm with the amputee (Dhillon and Horch, 2005).

The group at Case Western Reserve University, a Center for Functional Electrical Stimulation, has developed a cuff-like, flat interface nerve electrode (FINE), which has the advantage that is not penetrating the nerves but on the contrary it forms a surface where the nerves are reshaped upon (Tyler and Durand, 2002). A recent study on two upper-limb amputees that the FINE electrodes are stable, selective with repeatable responses for up to 24 months (Tan et al., 2014) (Fig. 23). Adjusting the average intensity of the stimulation affects the perception area. Adjusting the frequency of the stimulation affects the perception strength (Tan et al., 2014). When these electrodes were connected with force sensors at the tip of a prosthetic hand, increased manipulation performance of delicate objects (cherries) was observed (Tan et al., 2014).

The Biorobotics Institute at Scuole Superiore Sant' Anna (SSSA) used prototype transverse intrafascicular multichannel electrodes (TIME) (Boretius et al., 2010; Stieglitz et al., 2012; Badia et al., 2016) and a

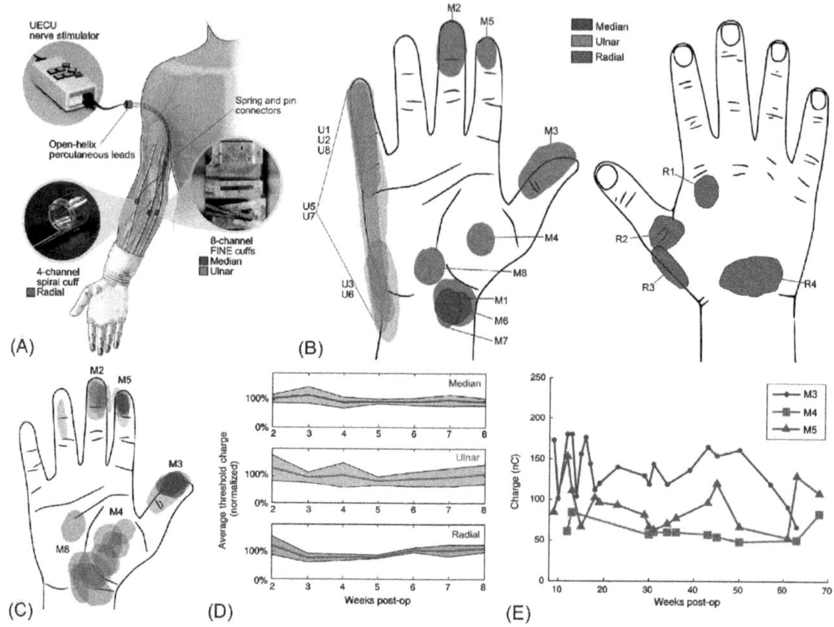

Fig. 23 Stability and selectivity of the FINE electrode. Stability and selectivity of implanted cuff electrodes. (A) We implanted three cuffs with a total of 20 channels in the forearm of subject 1: a four-contact spiral cuff on the radial nerve of the forearm and an eight contact FINE on the median and ulnar nerves. The electrode leads ran subcutaneously to the upper arm and connected to open-helix percutaneous leads via spring-and-pin connectors (27–29). A universal external control unit (UECU, Ardiem Medical) supplied single-channel, charge-balanced, monopolar nerve stimulation. (B) Sensation locations after threshold stimulation at week 3 post-op. Cuff electrodes were highly selective, with each contact producing either a unique location or unique sensation. Here, the letter represents the nerve and the number represents the stimulus channel within the nerve cuff around that nerve. Thus, M3 is the third stimulus channel within the median nerve cuff. Ulnar (U) locations presented the most overlap at threshold, but differentiated in area expansion at suprathreshold responses. The subjects drew the borders around areas of perception. Areas outside the template, for example, M3, represent a small wrap-around of sensation on the digit. (C) Repeated weekly overlapping threshold locations of channels M2, M3, M4, M5, and M8 for weeks 3 through 10 post-op indicated consistent location perception. Locations remained stable for all stimulation waveforms used. (D) Mean normalized threshold charge density for all channels on the median (blue), ulnar (green), and radial (red) cuffs of subject 1 shown as a solid line. Shaded areas indicate the 95% confidence interval. An unbiased, stepwise search determined the threshold. Frequency was a constant 20 Hz. During weeks 2–8, percept thresholds for subject 1 were 95.5 ± 42.5 nC ($n = 59$), 70.7 ± 59.2 nC ($n = 50$), and 40.7 ± 12.4 nC ($n = 24$) for the median, ulnar, and radial nerves, respectively. Linear regression of the threshold stimulation intensity for perception over 8 weeks for every channel was unchanging [18/19, analysis of variance (ANOVA) test, $P \geq 0.067$] or decreasing (1/19, ANOVA, $P = .044$). Subject 2 was also stable ($P \geq .087$) with thresholds of 141 ± 46 nC and 95 ± 47 nC for the median and radial nerves, respectively. (E) Threshold tracking of median channels M3, M4, and M5 to 68 weeks and thereafter showed no significant change in threshold over time ($P = .053$, .587, and .773, respectively). *(From Tan, D.W., Schiefer, M.A., Keith, M.W., Anderson, J.R., Tyler, J., Tyler, D.J., 2014. A neural interface provides long-term stable natural touch perception. Sci. Transl. Med. 6 (257), 257ra138. https://doi.org/10.1126/scitranslmed.3008669.)*

prototype artificial finger (Oddo et al., 2011) with an integrated micro electro-mechanical system (MEMS) sensor to demonstrate restoration of ability to discriminate textural features (Oddo et al., 2016). As shown in Fig. 24, this was achieved mimicking the natural coding using a mechano-neuro-transduction (MNT) process (Oddo et al., 2016).

The same group participated also in the prototype development of an evolution of TIME and LIFE, the self-opening neural interface (SELINE) electrodes (Cutrone et al., 2015), nevertheless a nerve penetrating-through electrode.

The Utah group led by Dr. Normann has been using the utah electrode array (UEA) for cortex applications (like vision restoration) and the utah slant electrode array (USEA) for peripheral nervous system applications (like prosthesis control) (Normann and Fernandez, 2016). The UEA and USEA are commercialized via Blackrock Microsystems, Salt Lake City, UT, United States. The USEA consists of 100, 0.5–1.5 mm long, microneedles, which project out of a $4 \times 4 \times 0.25$ mm thick substrate. A recent study has demonstrated feasibility of the USEA for transradial amputees (Clark et al., 2014; Davis et al., 2016). Nevertheless, these are penetrating electrodes and might exhibit nerve tissue necrosis after long implantation periods and movement artifacts at the periphery (Cutrone et al., 2015).

2.8 Optogenetics

Optogenetics is a powerful neuromodulation method that is using optics (light source) and genetics (modified genes are injected in advance) to monitor activity or excite neural activity in live animals in real time. This new line of research has the potential of eliminating the need for implanted electrodes or other microdevices for stimulating peripheral afferent and efferent nerves with high spatial specificity (Fontaine et al., 2017). This work has been preceded by general applicability research on biomechanics (Towne et al., 2013). Recently, MIT achieved a transdermal optogenetics prototype for ankle activation in mice without the use of any implanted devices for the read in/out (Maimon et al., 2017) (Fig. 25).

2.9 Biomechatronic EPP

Master/slave teleoperation control topology has been used in the Robotics field for many decades. A position-force architecture (Cho et al., 2001; Sheridan, 1992) is proposed (Figs. 26 and 27).

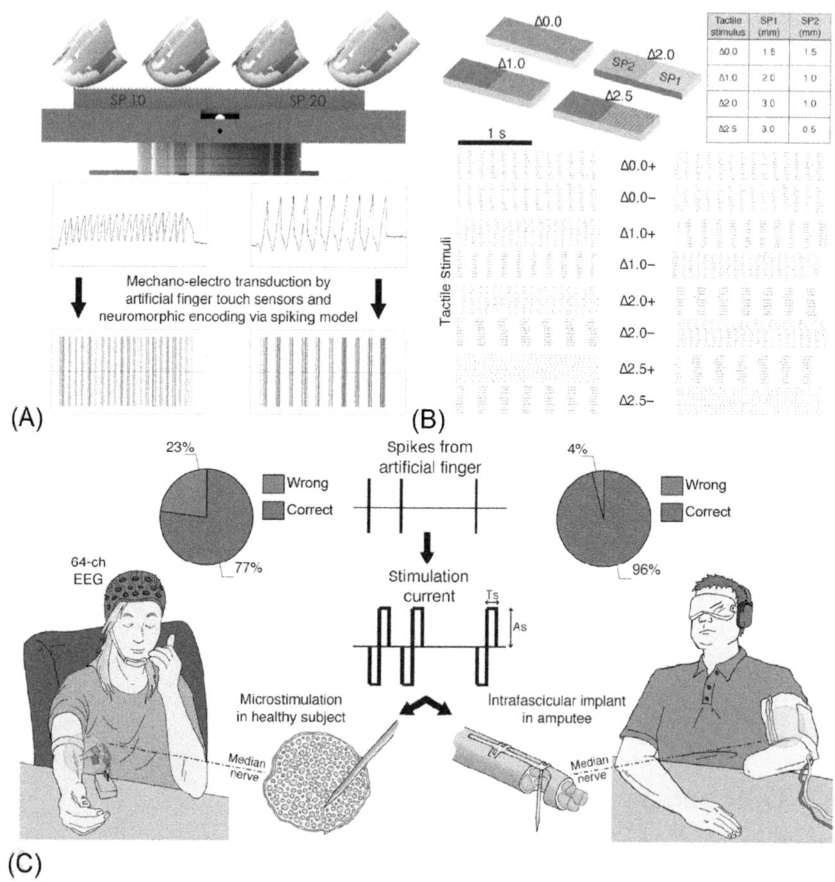

Fig. 24 Restoration of ability to discriminate textural features of a transradial amputee. Experimental setup and performance metrics. (A) Sensorized artificial finger and tactile stimulation platform. (B) Tactile stimuli that were used in the three-alternative forced-choice (3AFC) psychophysical protocol and the raster plot of spike trains that were generated in all sessions with one subject by the artificial finger while the gratings were slid. (C) Setup of percutaneous electrical microstimulation (*left*) and implanted intrafascicular stimulation (*right*) of the median nerve, and discrimination performance during all experimental sessions involving four intact subjects and one transradial upper-limb amputee. Source data of the spike trains that were transduced by the artificial finger while the gratings were indented and slid over have been deposited in Dryad (Oddo et al., 2016). Such spikes were used to trigger the neural stimulator in all the experimental sessions with DAS amputee (raster plot depicted in (B). *(From Oddo, C.M., Raspopovic, S., Artoni, F., Mazzoni, A., Spigler, G., Petrini, F., … Micera, S., 2016. Intraneural stimulation elicits discrimination of textural features by artificial fingertip in intact and amputee humans. elife, 5, e09148. https://doi.org/10.7554/eLife.09148.)*

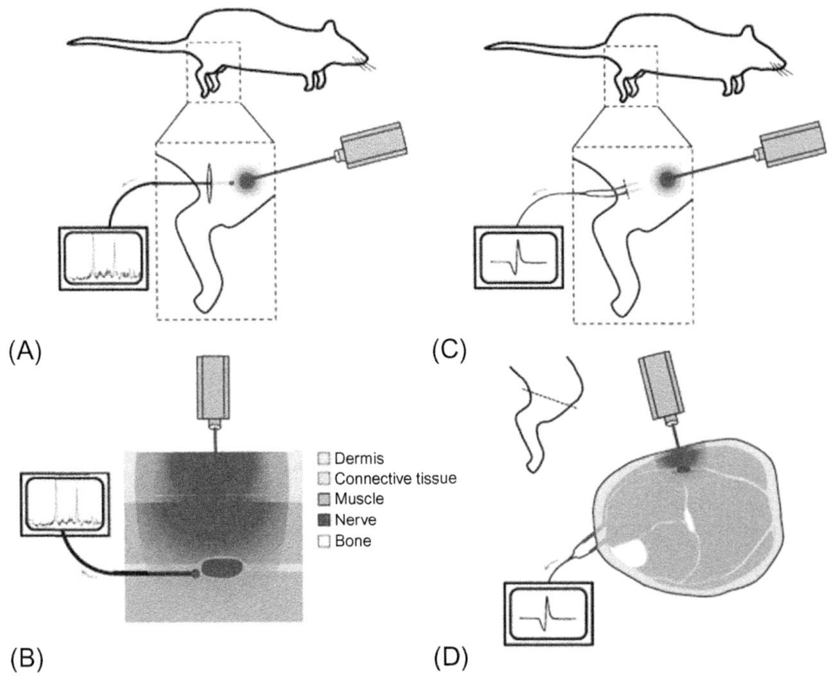

Fig. 25 Transdermal optogenetics read in/out proof of concept. (A) A small ruby sphere connected to a fiber optic is implanted into the rat hind limb via an incision made 1–2 cm proximal to the target measurement location. A 473 nm free-space laser illuminated the ruby sphere through transdermal illumination of the hind limb. Fluorescent emissions from the ruby sphere were collected by a spectrometer via a fiber optic and used to quantify fluence rate. (B) A cross-section of the target measurement location shows the ruby sphere in proximity to the representative common peroneal nerve. (C) Bipolar recording needle electrodes were inserted into the target musculature to record muscle activity in response to transdermal illumination of the nerve. (D) A schematic cross-section of the hind limb depicting connective tissue, musculature, bone, common peroneal nerve, and dermis anatomy. Bipolar recording needle electrodes were used to record muscle activity of both the TA (shown) and GN (not shown) in response to transdermal illumination. Tissue-type legend refers to both (B) and (D) cross sections. *(From Maimon, B.E., Zorzos, A.N., Bendell, R., Harding, A., Fahmi, M., Srinivasan, S., … Herr, H.M., 2017. Transdermal optogenetic peripheral nerve stimulation. J. Neural Eng. 14(3), 034002. https://doi.org/10.1088/1741-2552/aa5e20.)*

Mablekos-Alexiou et al. (2015) and (2016) proposed an evolution topology (Fig. 26) of Classic EPP (Fig. 12) in order to keep the advantages of the classic EPP topology but overcome its disadvantages.

In the proposed topology (Figs. 26 and 27), the amputee via its agonist muscle, sends a force command signal to the controller. Then the controller sends a torque signal to the prosthesis. As a feedback the prosthesis sends a

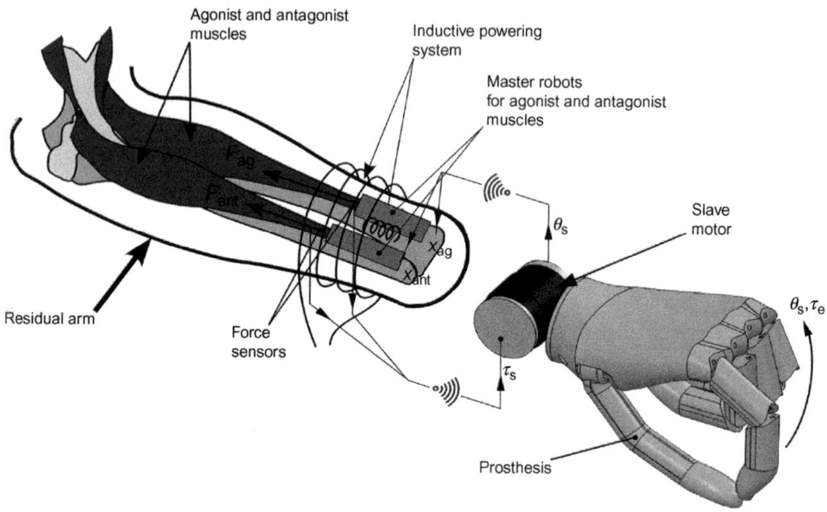

Fig. 26 Biomechatronic EPP topology (Mablekos-Alexiou, 2016).

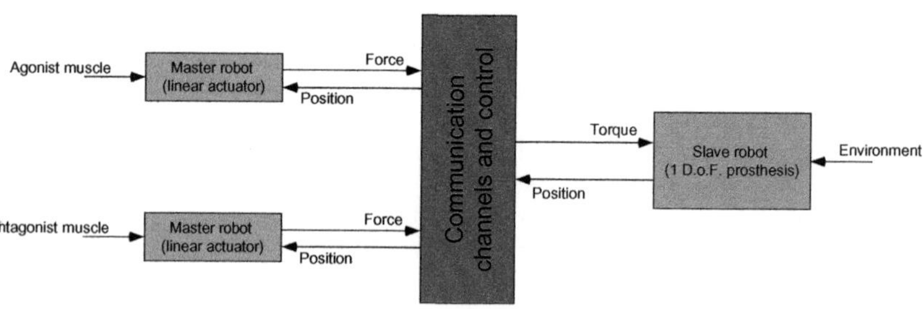

Fig. 27 Master/slave control topology used in the Biomechatronic EPP (Mablekos-Alexiou, 2016).

position signal to the controller and then this is communicated back to the master robot as feedback. The master robots (leadscrews with motors) and other electronics use battery which is charged via inductive coupling as shown in the prosthesis (Fig. 26).

The Biomechatronic EPP topology has been shown (for 1 subject—with a pending research study for 15 subjects) to have equivalent performance with the Classic EPP topology and superior to the myoelectric control (Kontogiannopoulos et al., 2018). Initial thermal and power feasibility analysis (Moutopoulou et al., 2015) is positive. This could be the best building block with inherent subconscious properties that could enable superior upper-limb prostheses in the future.

3 TRENDS FOR THE FUTURE THAT CAN ENABLE BIOMECHATRONICS UPPER-LIMB PROSTHESES

The following are technology trends that could enable the fastest creation of advanced biomechatronic hands and arms.

3.1 Personalization/3D Printing/Fast Prototyping

Modern 3D printing technology can enable personalization where it did not exist before. For example, in order to achieve versatility, one could print tools/extensions of the hand and attach them on a "demand basis." A growing child could print components as it grows. A farmer could print a "tool" and attach it to his modular prosthesis. A new socket could be fabricated on demand at the prosthetic facility or even at the person's home 3D printer.

3D printing could eventually enable what we expressed as ideal at Section 1.2.2, via the versatility of functional endpoint tools that the rapid manufacturing could provide at home.

Another aspect might be customization after surgery. For example, new control sites are created with TMR, the appropriate components for integration of the extra DoF are printed at the hospital and the controller is tuned right there.

3.2 Many-DoFs

Artificial intelligence, pattern recognition (Section 2.3.1) and targeted muscle reinnervation (TMR) (Section 2.5) are enabler technologies which have and will make many-DoF prosthetic arms closer to the ideal or surpass the ideal (see Section 1.2.2) as we perceive now [Interview of Hugh Herr at (Kiss, 2015)]. Independent DoFs do not have to be that independent as described by the "ideal" paradigm (Section 1.2.2), since now the pattern recognition module could identify what the pattern is and decide on what the intended action is. This is a way that was not possible in the past and is certainly another way of achieving the "ideal." The missing functionality now is the subconscious control, still the integrated perception that the human should have for an advanced "ideal" prosthetic arm or hand. Work on sensory neural integration or biomechatronic EPP are technologies which could help address that gap.

3.3 Osseointegration and Osseoperception

Osseointegration is a technique that could help on integrating the prosthesis with the remaining body in a harmonized way with many benefits.

Researchers are trying to resolve practical issues of infections and bone weakening that might happen in a few cases. Not only it can provide to the upper-limb amputee an increased range of motion of the prosthesis (see Fig. 19A and B), it can lead to superior control since via the Osseoperception the amputees feel where the prosthesis is in space without subconsciously.

3.4 EPP and Biomechatronic EPP

EPP as was described at Section 1.4.3 is a paradigm/direction that we have lost over the last decades when prosthetic industry took the path of myoelectric control. We have lost the integrated sensory integration that EPP offers inherently. Biomechatronic EPP (Section 2.9) is a research effort trying to fill in that gap, take out the disadvantages of traditional EPP (cables, harness, and unesthetics) and keep the integrated pathway that tendons and neuromuscular structures in the EPP paradigm provide. Yes, there is surgery that needs to be performed for Biomechatronic EPP but that might happen at the time of amputation.

3.5 Discussion/Realignment

3.5.1 Back to Basics

If we look at the upper-limb prosthetics evolution, we will see that wars have helped progress the state of the art. During World War I, cineplasty was introduced and matured in Italy; during the World War II, EPP, and cineplasty progressed in Germany and United States. During the wars in Middle East with United States, there was substantial progress on TMR which opened the window to many-DoF prosthetic arms.

It is now time, to get back to basics and reflect if we have met the needs of people with amputations. Have we prioritized on their needs? Have we made the process of giving a prosthesis to the amputee, a process that we satisfy his/her needs? Have we defined what the local of practical "ideal" with today's technology is for that amputee?

The technocrats think that what they think will lead to higher usability but (Lock et al., 2005) did not find high correlation between lower classification error and higher usability results. In other words, what researchers think is the "ideal" might not be and might not be usable by users, which is the "ideal." Therefore, we need to get back to basics and define what is "ideal." It seems that the interpretation of the "ideal" for each amputee is subjective and that is the gap process debt that researchers owe to the users.

History Lesson

In the 1920s the airplanes had Bowden cables (a form of EPP) so pilots can "feel" where the flaps are, similar to how car drivers "feel" where the wheels were with the unassisted passive steering. Later on, the modern airplanes use mechatronic joysticks in a master slave topology to convey to the pilot as a feedback the position of the flaps (see Fig. 28).

3.5.1.1 Enable Evolution of Older EPP With Mechatronics

In the 1970s the prosthetic industry took a turn from cineplasty and body-powered prostheses to myoelectric control. The excitement of electronics and the fact myoelectric control would eliminate the need for surgery (cineplasty) or complicated harness and cables for body-powered played a big role in that industry turn? What did we lose though? We lost proprioception, the ability of the human to "feel" the state of the prosthesis subconsciously, by using the remaining afferent (sensory) neural circuitry. This turn maybe was not evident at that time because people substituted with visual feedback or patients did not have the chance to evaluate and choose. It is evident now, though. The lack of proprioception makes many-DoF control more difficult or does not take the current prostheses closer to the "ideal" state for upper-limb prostheses (see Fig. 29). The value of the EPP is shown

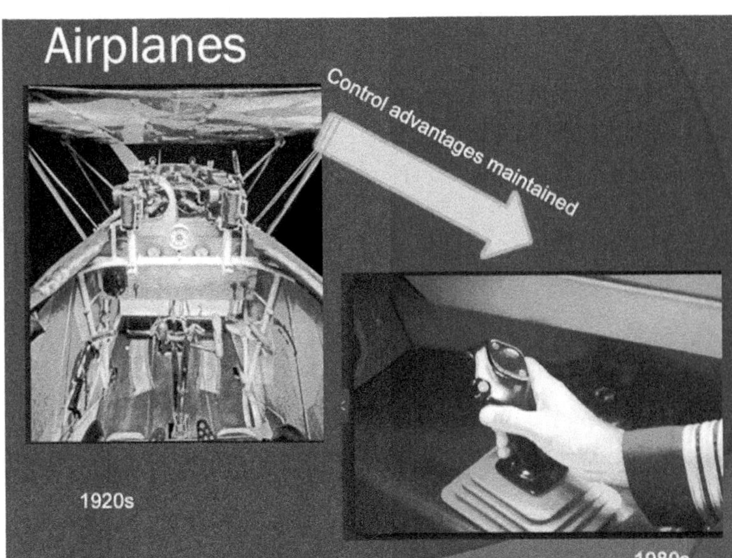

Fig. 28 Airplane's evolution from Bowden cables to modern pilot joysticks.

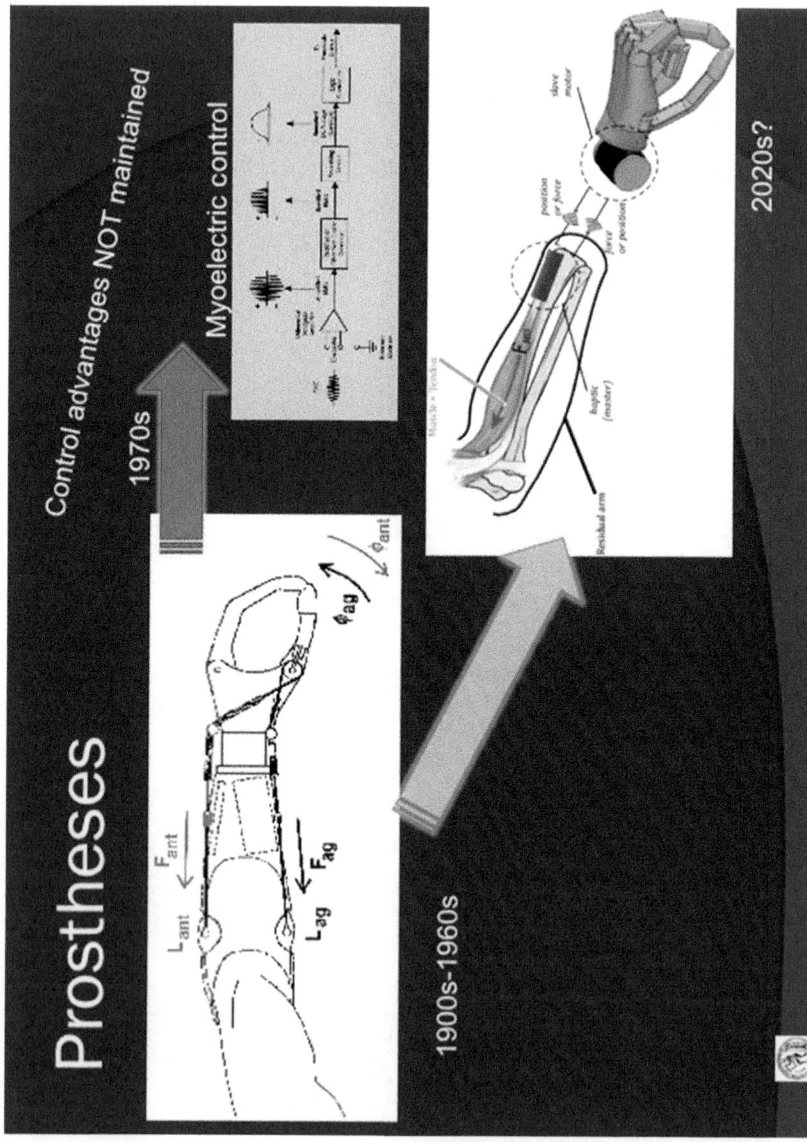

Fig. 29 The evolution of body powered upper-limb prostheses to myoelectric prostheses and the value of the Biomechatronic EPP.

for one DoF but this can be the building block for prosthetic hands or arms for many-DoFs if the implants are miniaturized enough.

Supportive to this direction is the finding that torque control is better than EMG control (Johnson et al., 2017). Also, Doubler and Childress (1984a,b) have shown that traditional EPP position control is of superior quality than velocity or myoelectric control. Recently, Kontogiannopoulos et al. (2018) showed (for one subject, with ongoing study for 15 subjects) that Biomechatronic EPP topology is superior than myoelectric control for 1-DoF prosthesis.

We have to go back to basics: fix the fundamental block of control—use the one of high quality—and then expand to many-DoF prostheses.

AUTHORS' CONTRIBUTIONS

GAB wrote all sections of this chapter except Section 6.5. GAB was also responsible for the structure, content, outline, and review of this chapter. EGP wrote Section 6.5.

REFERENCES

3D. Retrieved from: http://3dprintingfromscratch.com/common/types-of-3d-printers-or-3d-printing-technologies-overview/.

Badia, J., Raspopovic, S., Carpaneto, J., Micera, S., Navarro, X., 2016. Spatial and functional selectivity of peripheral nerve signal recording with the transversal Intrafascicular multichannel electrode (TIME). IEEE Trans. Neural Syst. Rehabil. Eng. 24 (1), 20–27. https://doi.org/10.1109/TNSRE.2015.2440768.

Bar-Cohen, Y., 2001. Electroactive Polymer (EAP) Actuators as Artificial Muscles: Reality, Potential, and Challenges. SPIE Press, Bellingham, WA.

Bar-Cohen, Y., Martin, D., Prillaman, D.L., Taylor, J., Ascione, G., Seacrist, T., … Franzini, G., 2018. Synthetic Muscle Electroactive Polymer (EAP) Based Actuation and Sensing for Prosthetic and Robotic Applications. Paper Presented at the Electroactive Polymer Actuators and Devices (EAPAD) XX.

Beasley, R.W., 1981. General considerations in managing upper limb amputations. Orthop. Clin. North Am. 12 (4), 743–749.

Beasley, R.W., de Bese, G.M., 1986. Upper limb amputations and prostheses. Orthop. Clin. North Am. 17 (3), 395–405.

Beccai, L., Roccella, S., Ascari, L., Valdastri, P., Sieber, A., Carrozza, M.C., Dario, P., 2008. Development and experimental analysis of a soft compliant tactile microsensor for anthropomorphic artificial hand. IEEE-ASME Trans. Mechatron. 13 (2), 158–168. https://doi.org/10.1109/tmech.2008.918483.

Bencharit, S., Byrd, W.C., Altarawneh, S., Hosseini, B., Leong, A., Reside, G., … Offenbacher, S., 2014. Development and applications of porous tantalum trabecular metal-enhanced titanium dental implants. Clin. Implant. Dent. Relat. Res. 16 (6), 817–826. https://doi.org/10.1111/cid.12059.

Bercich, R.A., Wang, Z., Mei, H., Hammer, L.H., Seburn, K.L., Hargrove, L.J., Irazoqui, P.P., 2016. Enhancing the versatility of wireless biopotential acquisition for myoelectric prosthetic control. J. Neural Eng. 13(4), 046012. https://doi.org/10.1088/1741-2560/13/4/046012.

Bertos, Y.A., 1999. A Microprocessor-Based E.P.P. Position Controller for Electric-Powered Upper-Limb Prostheses (Master of Science). Northwestern University, Evanston, IL.

Bertos, Y.A., Heckathorne, C.W., Weir, R.F., Childress, D.S., 1997. In: Microprocessor based E.P.P. position controller for electric-powered upper-limb prostheses. Proceedings of the 19th Annual International Conference of the IEEE Engineering in Medicine and Biology Society, Vol 19, Pts 1–6—Magnificent Milestones and Emerging Opportunities in Medical Engineering (Vol. 19, pp. 2311–2314). IEEE, New York.

Bertos, Y.A., Heckathorne, C.W., Weir, R.F.f., Childress, D.S., 1998. In: A microprocessor based E.P.P. controller for electric-powered prostheses. Paper presented at the International Society of Prosthetists and Orthotists ISPO Conference, Amsterdam, Netherlands, June 1998.

Bhuiyan, M.S., Choudhury, I.A., Dahari, M., 2015. Development of a control system for artificially rehabilitated limbs: a review. Biol. Cybern. 109 (2), 141–162. https://doi.org/10.1007/s00422-014-0635-1.

Biddiss, E.A., Chau, T.T., 2007. Upper limb prosthesis use and abandonment: a survey of the last 25 years. Prosthetics Orthot. Int. 31 (3), 236–257. https://doi.org/10.1080/03093640600994581.

Biddiss, E., Chau, T., 2008. Dielectric elastomers as actuators for upper limb prosthetics: challenges and opportunities. Med. Eng. Phys. 30 (4), 403–418. https://doi.org/10.1016/j.medengphy.2007.05.011.

Blana, D., Kyriacou, T., Lambrecht, J.M., Chadwick, E.K., 2016. Feasibility of using combined EMG and kinematic signals for prosthesis control: a simulation study using a virtual reality environment. J. Electromyogr. Kinesiol. 29, 21–27. https://doi.org/10.1016/j.jelekin.2015.06.010.

Blana, D., Chadwick, E.K., van den Bogert, A.J., Murray, W.M., 2017. Real-time simulation of hand motion for prosthesis control. Comput. Methods Biomech. Biomed. Engin. 20 (5), 540–549. https://doi.org/10.1080/10255842.2016.1255943.

Boretius, T., Badia, J., Pascual-Font, A., Schuettler, M., Navarro, X., Yoshida, K., Stieglitz, T., 2010. A transverse intrafascicular multichannel electrode (TIME) to interface with the peripheral nerve. Biosens. Bioelectron. 26 (1), 62–69. https://doi.org/10.1016/j.bios.2010.05.010.

Branemark, P.I., Adell, R., Breine, U., Hansson, B.O., Lindstrom, J., Ohlsson, A., 1969. Intra-osseous anchorage of dental prostheses. I. Experimental studies. Scand. J. Plast. Reconstr. Surg. 3 (2), 81–100.

Calanca, A., Muradore, R., Fiorini, P., 2016. A review of algorithms for compliant control of stiff and fixed-compliance robots. IEEE-ASME Trans. Mechatron. 21 (2), 613–624. https://doi.org/10.1109/Tmech.2015.2465849.

Caldwell, D.G., Medranocerda, G.A., Goodwin, M., 1995. Control of pneumatic muscle actuators. IEEE Control. Syst. Mag. 15 (1), 40–48. https://doi.org/10.1109/37.341863.

Celik, M.E., Aydin, E., 2017. An efficient inductive coil link design for wireless power transfer to visual prostheses. Acta Phys. Pol. A 132 (3), 535–537. https://doi.org/10.12693/APhysPolA.132.535.

Childress, D.S., 1980. Closed-loop control in prosthetic systems—historical-perspective. Ann. Biomed. Eng. 8 (4–6), 293–303.

Childress, D.S., 1989. Control philosophies for limb prostheses. In: Paul, J. et al., (Eds.), Progress in Bioengineering. Adam Hilger, New York, pp. 210–215.

Childress, D.S., 1992. Control of limb prostheses. In: Bowker, J.W., Michael, J.W. (Eds.), Atlas of Limb Prosthetics, Surgical, Prosthetic, and Rehabilitation Principles. Mosby-Year Book, Inc, St. Louis, MO, pp. 175–199.

Childress, D.S., 1997. The Interfaces between Humans and Limb Replacement Components. Quintessence Publ. Co Inc., Carol Stream, IL

Childress, D.S., 1998. In: Chang, H.K., Zhang, Y.T. (Eds.), Control strategy for upper-limb prostheses. Proceedings of the 20th Annual International Conference of the IEEE Engineering in Medicine and Biology Society, Vol 20, Pts 1–6—Biomedical Engineering Towards the Year 2000 and beyond (Vol. 20, pp. 2273–2275. IEEE, New York.

Childress, D.S., Grahn, E., Weir, R.F.f., Heckathorne, C., Uellendahl, J., 1993. In: Modification of a bock hand for E.P.P. Control by exteriorized tendons.Paper presented at the Proceedings of the 19th Annual Meeting of AAOP.

Cho, H.C., Park, J.H., Kim, K., Park, J.O., 2001. In: Sliding-mode-based impedance controller for bilateral teleoperation under varying time-delay. 2001 IEEE International Conference on Robotics and Automation, Vols I–IV, Proceedings, pp. 1025–1030.

Chou, C.P., Hannaford, B., 1996. Measurement and modeling of McKibben pneumatic artificial muscles. IEEE Trans. Robot. Autom. 12 (1), 90–102. https://doi.org/10.1109/70.481753.

Clark, G.A., Wendelken, S., Page, D.M., Davis, T., Wark, H.A., Normann, R.A., … Hutchinson, D.T., 2014. Using multiple high-count electrode arrays in human median and ulnar nerves to restore sensorimotor function after previous transradial amputation of the hand. Conf. Proc. IEEE Eng. Med. Biol. Soc. 2014, 1977–1980. https://doi.org/10.1109/EMBC.2014.6944001.

Cloutier, A., Yang, J., 2013. Design, control, and sensory feedback of externally powered hand prostheses: a literature review. Crit. Rev. Biomed. Eng. 41 (2), 161–181.

Controzzi, M., Cipriani, C., Carrozza, M.C., 2014. Design of artificial hands: a review. In: Human Hand as an Inspiration for Robot Hand Developments vol. 95, pp. 219–246. https://doi.org/10.1007/978-3-319-03017-3_11.

Cura, V.O., Cunha, F.L., Aguiar, M.L., Cliquet Jr., A., 2003. Study of the different types of actuators and mechanisms for upper limb prostheses. Artif. Organs 27 (6), 507–516.

Cutrone, A., Del Valle, J., Santos, D., Badia, J., Filippeschi, C., Micera, S., … Bossi, S., 2015. A three-dimensional self-opening intraneural peripheral interface (SELINE). J. Neural Eng. 12(1), 016016. https://doi.org/10.1088/1741-2560/12/1/016016.

Da Cunha, F.L., Schneebeli, H.J., Dynnikov, V.I., 2000. Development of anthropomorphic upper limb prostheses with human-like interphalangian and interdigital couplings. Artif. Organs 24 (3), 193–197.

Davis, T.S., Wark, H.A., Hutchinson, D.T., Warren, D.J., O'Neill, K., Scheinblum, T., Greger, B., 2016. Restoring motor control and sensory feedback in people with upper extremity amputations using arrays of 96 microelectrodes implanted in the median and ulnar nerves. J. Neural Eng. 13(3), 036001. https://doi.org/10.1088/1741-2560/13/3/036001.

De Laurentis, K.J., Mavroidis, C., 2002. Mechanical design of a shape memory alloy actuated prosthetic hand. Technol. Health Care 10 (2), 91–106.

Deglurkar, M., Davy, D.T., Stewart, M., Goldberg, V.M., Welter, J.F., 2007. Evaluation of machining methods for trabecular metal implants in a rabbit intramedullary osseointegration model. J. Biomed. Mater. Res. B Appl. Biomater. 80 (2), 528–540. https://doi.org/10.1002/jbm.b.30627.

DeMichele, G.A., Troyk, P.R., Kerns, D., Weir, R.F.F., IEEE, 2008. IMES—Implantable MyoElectric Sensor System: Designing Standardized ASICs. IEEE, New York.

Dhillon, G.S., Horch, K.W., 2005. Direct neural sensory feedback and control of a prosthetic arm. IEEE Trans. Neural Syst. Rehabil. Eng. 13 (4), 468–472. https://doi.org/10.1109/TNSRE.2005.856072.

Dhillon, G.S., Lawrence, S.M., Hutchinson, D.T., Horch, K.W., 2004. Residual function in peripheral nerve stumps of amputees: implications for neural control of artificial limbs. J. Hand. Surg. [Am.] 29 (4), 605–615. discussion 616–618, https://doi.org/10.1016/j.jhsa.2004.02.006.

Di Pino, G., Guglielmelli, E., Rossini, P.M., 2009. Neuroplasticity in amputees: main implications on bidirectional interfacing of cybernetic hand prostheses. Prog. Neurobiol. 88 (2), 114–126. https://doi.org/10.1016/j.pneurobio.2009.03.001.

Doubler, J.A., Childress, D.S., 1984a. An analysis of extended physiological proprioception as a prosthesis-control technique. J. Rehabil. Res. Dev. 21 (1), 5–18.

Doubler, J.A., Childress, D.S., 1984b. Design and evaluation of a prosthesis control system based on the concept of extended physiological proprioception. J. Rehabil. Res. Dev. 21 (1), 19–31.

Dudkiewicz, I., Gabrielov, R., Seiv-Ner, I., Zelig, G., Heim, M., 2004. Evaluation of prosthetic usage in upper limb amputees. Disabil. Rehabil. 26 (1), 60–63. https://doi.org/10.1080/09638280410001645094.

Escudero, A.Z., Alvarez, J., Leija, L., 2002. Development of a parallel myoelectric prosthesis for above elbow replacement.Second Joint EMBS-BMES Conference 2002, Vols. 1–3, Conference Proceedings, pp. 2404–2405. https://doi.org/10.1109/Iembs.2002.1053346.

Farina, D., Jiang, N., Rehbaum, H., Holobar, A., Graimann, B., Dietl, H., Aszmann, O.C., 2014. The extraction of neural information from the surface EMG for the control of upper-limb prostheses: Emerging avenues and challenges. IEEE Trans. Neural Syst. Rehabil. Eng. 22 (4), 797–809. https://doi.org/10.1109/Tnsre.2014.2305111.

Fontaine, A.K., Gibson, E.A., Caldwell, J.H., Weir, R.F., 2017. Optical read-out of neural activity in mammalian peripheral axons: calcium signaling at nodes of Ranvier. Sci. Rep. 7 (1), 4744. https://doi.org/10.1038/s41598-017-03541-y.

Ghovanloo, M., Najafi, K., 2004. A wideband frequency-shift keying wireless link for inductively powered biomedical implants. IEEE Trans. Circuits Syst. I-Regular Papers 51 (12), 2374–2383. https://doi.org/10.1109/Tcsi.2004.838144.

Greer, J.D., Morimoto, T.K., Okamura, A.M., Hawkes, E.W., 2017. Series pneumatic artificial muscles (sPAMs) and application to a soft continuum robot. IEEE Int. Conf. Robot Autom. 2017, 5503–5510. https://doi.org/10.1109/ICRA.2017.7989648.

Gretsch, K.F., Lather, H.D., Peddada, K.V., Deeken, C.R., Wall, L.B., Goldfarb, C.A., 2016. Development of novel 3D-printed robotic prosthetic for transradial amputees. Prosthetics Orthot. Int. 40 (3), 400–403. https://doi.org/10.1177/0309364615579317.

Hargrove, L.J., Miller, L.A., Turner, K., Kuiken, T.A., 2017. Myoelectric pattern recognition outperforms direct control for transhumeral amputees with targeted muscle reinnervation: a randomized clinical. Trial. Sci. Rep. 7 (1), 13840. https://doi.org/10.1038/s41598-017-14386-w.

Harvey, A.M., Masland, R.L., 1941. Actions of curarizing preparations in the human. J. Pharmacol. Exp. Ther. 73 (3), 304–311.

Herberts, P., Almstrom, C., Kadefors, R., Lawrence, P.D., 1973. Hand prosthesis control via myoelectric patterns. Acta Orthop. Scand. 44 (4), 389–409.

Hijjawi, J.B., Kuiken, T.A., Lipschutz, R.D., Miller, L.A., Stubblefield, K.A., Dumanian, G.A., 2006. Improved myoelectric prosthesis control accomplished using multiple nerve transfers. Plast. Reconstr. Surg. 118 (7), 1573–1578. https://doi.org/10.1097/01.prs.0000242487.62487.fb.

Hogan, N., 1985. Impedance control—an approach to manipulation 2. Implementation. J. Dyn. Sys. Meas. Control-Trans. ASME 107 (1), 8–16. https://doi.org/10.1115/1.3140713.

Hollerbach, J., Hunter, I., Ballantyne, J., 1992. A comparative analysis of actuator technologies for robotics. In: Robotics Review 2. MIT Press, Cambridge, MA, pp. 299–342.

Hudgins, B., Parker, P., Scott, R.N., 1993. A new strategy for multifunction myoelectric control. IEEE Trans. Biomed. Eng. 40 (1), 82–94. https://doi.org/10.1109/10.204774.

Jacobsen, S.C., Knutti, D.F., Johnson, R.T., Sears, H.H., 1982. Development of the Utah artificial arm. IEEE Trans. Biomed. Eng. 29 (4), 249–269. https://doi.org/10.1109/TBME.1982.325033.

Jacobsen, S.C., Wood, J.E., Knutti, D.F., Biggers, K.B., 1984. The Utah Mit Dextrous hand—work in progress. Int. J. Robot. Res. 3 (4), 21–50. https://doi.org/10.1177/027836498400300402.

Jacobsen, S.C., Iversen, E.K., Knutti, D.F., Johnson, R.T., Biggers, K.B., 1986. In: Design of the Utah/M.I.T. dextrous hand.Paper Presented at the Proceedings IEEE International Conference on Robotics and Automation, San Francisco, CA, USA, 7–10 April.

Jani, J.M., Leary, M., Subic, A., Gibson, M.A., 2014. A review of shape memory alloy research, applications and opportunities. Mater. Des. 56, 1078–1113. https://doi.org/10.1016/j.matdes.2013.11.084.

Jawhar, I., Mohamed, N., Agrawal, D.P., 2011. Linear wireless sensor networks: classification and applications. J. Netw. Comput. Appl. 34 (5), 1671–1682. https://doi.org/10.1016/j.jnca.2011.05.006.

Jerbi, K., Vidal, J.R., Mattout, J., Maby, E., Lecaignard, F., Ossandon, T., … Bertrand, O., 2011. Inferring hand movement kinematics from MEG, EEG and intracranial EEG: from brain-machine interfaces to motor rehabilitation. IRBM 32 (1), 8–18. https://doi.org/10.1016/j.irbm.2010.12.004.

Johnson, R.E., Kording, K.P., Hargrove, L.J., Sensinger, J.W., 2017. EMG versus torque control of human-machine systems: equalizing control signal variability does not equalize error or uncertainty. IEEE Trans. Neural Syst. Rehabil. Eng. 25 (6), 660–667. https://doi.org/10.1109/TNSRE.2016.2598095.

Johnston, D., Zhang, P., Hollerbach, J., Jacobsen, S., 1996. In: A full tactile sensing suite for dextrous robot hands and use in contact force control.1996 Ieee International Conference on Robotics and Automation, Proceedings, Vols 1–4, pp. 3222–3227.

Jow, U.M., Ghovanloo, M., 2007. Design and optimization of printed spiral coils for efficient transcutaneous inductive power transmission. IEEE Trans. Biomed. Circuits Syst. 1 (3), 193–202. https://doi.org/10.1109/Tbcas.2007.913130.

Kim, K.J., Tadokoro, S., 2007. Electroactive Polymers for Robotic Applications. Springer-Verlag, London. https://www.springer.com/gp/book/9781846283710.

Kim, K., Colgate, J.E., Santos-Munne, J.J., Makhlin, A., Peshkin, M.A., 2010. On the design of miniature haptic devices for upper extremity prosthetics. IEEE-ASME Trans. Mechatron. 15 (1), 27–39. https://doi.org/10.1109/Tmech.2009.2013944.

Kim, J.D., Sun, C., Suh, I.S., 2014. A proposal on wireless power transfer for medical implantable applications based on reviews. 2014 IEEE Wireless Power Transfer Conference (WPTC), pp. 166–169.

Kiss, J., 2015. what if a bionic leg is so good that someone chooses to amputate? In: The Guardian. Retrieved from: https://www.theguardian.com/technology/2015/apr/09/disability-amputees-bionics-hugh-herr-super-prostheses.

Klopsteg, P.E., Wilson, P.D., 1954. Human Limbs and Their Substitutes, second ed. National Academy of Sciences, Courtesy of the National Academies Press, Washington, DC.

Kontogiannopoulos, S., Vangelatos, Z., Bertos, A.G., Papadopoulos, E., 2018. In: A biomechatronic EPP upper-limb prosthesis controller and its performance comparison to other topologies.Paper Presented at the 40th International IEEE Engineering in Medicine and Biology Conference, Honolulu, Hawaii, USA.

Krausz, N.E., Rorrer, R.A., Weir, R.F., 2016. Design and fabrication of a six degree-of-freedom open source hand. IEEE Trans. Neural Syst. Rehabil. Eng. 24 (5), 562–572. https://doi.org/10.1109/TNSRE.2015.2440177.

Kuiken, T.A., Childress, D.S., Rymer, W.Z., 1995. The hyper-reinnervation of rat skeletal-muscle. Brain Res. 676 (1), 113–123.

Kuiken, T.A., Dumanian, G.A., Lipschutz, R.D., Miller, L.A., Stubblefield, K.A., 2004. The use of targeted muscle reinnervation for improved myoelectric prosthesis control in a bilateral shoulder disarticulation amputee. Prosthetics Orthot. Int. 28 (3), 245–253. https://doi.org/10.3109/03093640409167756.

Kuiken, T.A., Marasco, P.D., Lock, B.A., Harden, R.N., Dewald, J.P., 2007. Redirection of cutaneous sensation from the hand to the chest skin of human amputees with targeted reinnervation. Proc. Natl. Acad. Sci. U. S. A. 104 (50), 20061–20066. https://doi.org/10.1073/pnas.0706525104.

Kuiken, T.A., Barlow, A.K., Hargrove, L.J., Dumanian, G.A., 2017. Targeted muscle reinnervation for the upper and lower extremity. Tech. Orthop. 32 (2), 109–116. https://doi.org/10.1097/Bto.0000000000000194.

Lebedev, M.A., Nicolelis, M.A.L., 2006. Brain-machine interfaces: past, present and future. Trends Neurosci. 29 (9), 536–546. https://doi.org/10.1016/j.tins.2006.07.004.

Lenneras, M., Tsikandylakis, G., Trobos, M., Omar, O., Vazirisani, F., Palmquist, A., … Thomsen, P., 2017. The clinical, radiological, microbiological, and molecular profile of the skin-penetration site of transfemoral amputees treated with bone-anchored prostheses. J. Biomed. Mater. Res. A 105 (2), 578–589. https://doi.org/10.1002/jbm.a.35935.

Li, Y., 2016. Osseointegrated Human-Machine Gateway (OHMG) clinical trial identifier NCT03178890. Retrieved from:https://clinicaltrials.gov/ct2/show/NCT03178890.

Limbitless, Retrieved from: http://limbitless-solutions.org.

Lock, B.A., Englehart, K., Hudgins, B., 2005. In: Real-time myoelectric control in a virtual environment to relate usability vs. accuracy.Paper Presented at the MEC '05 Intergrating Prosthetics and Medicine, Proceedings of the 2005 MyoElectric Controls/Powered Prosthetics Symposium, Fredericton, New Brunswick, Canada.

Loeb, G.E., Richmond, F.J., Singh, J., Peck, R.A., Tan, W., Zou, Q., Sachs, N., 2004. RF-powered BIONs for stimulation and sensing. Conf. Proc. IEEE Eng. Med. Biol. Soc. 6, 4182–4185. https://doi.org/10.1109/IEMBS.2004.1404167.

Mablekos-Alexiou, A., 2016. Design and Simulation of a Master Slave Architecture Biomechatronic EPP upper-limb prosthesis. (ΣΧΕΔΙΑΣΜΟΣ ΚΑΙ ΠΡΟΣΟΜΟΙΩΣΗ ΣΥΣΤΗΜΑΤΟΣ ΤΗΛΕΧΕΙΡΙΣΜΟΥ ΠΡΟΣΘΕΤΙΚΟΥ ΑΚΡΟΥ ΜΕ ΙΔΙΟΔΕΚΤΙΚΗ ΑΙΣΘΗΣΗ). (B.Sc.)National Technical University of Athens, Athens.

Mablekos-Alexiou, A., Bertos, G.A., Papadopoulos, E., 2015. In: A biomechatronic extended physiological proprioception (EPP) controller for upper-limb prostheses.2015 IEEE/RSJ International Conference on Intelligent Robots and Systems (Iros), pp. 6173–6178.

Maimon, B.E., Zorzos, A.N., Bendell, R., Harding, A., Fahmi, M., Srinivasan, S., … Herr, H.M., 2017. Transdermal optogenetic peripheral nerve stimulation. J. Neural Eng. 14(3), 034002. https://doi.org/10.1088/1741-2552/aa5e20.

Marasco, P.D., Schultz, A.E., Kuiken, T.A., 2009. Sensory capacity of reinnervated skin after redirection of amputated upper limb nerves to the chest. Brain 132, 1441–1448. https://doi.org/10.1093/brain/awp082.

Marshall, J. 2015 (Producer). (2015-09-21). The History of Prosthetics. UNYQ. Retrieved from: http://unyq.com/the-history-of-prosthetics/.

Miller, L.A., Lipschutz, R.D., Stubblefield, K.A., Lock, B.A., Huang, H., Williams 3rd, T.W., … Kuiken, T.A., 2008. Control of a six degree of freedom prosthetic arm after targeted muscle reinnervation surgery. Arch. Phys. Med. Rehabil. 89 (11), 2057–2065. https://doi.org/10.1016/j.apmr.2008.05.016.

Mio, R., Villegas, B., Ccorimanya, L., Flores, K.M., Salazar, G., Elias, D., 2017. In: Development and assessment of a powered 3D-printed prosthetic hand for transmetacarpal amputees. 2017 3rd International Conference on Control, Automation and Robotics (Iccar), pp. 85–90.

Moorey, C.L., Holderbaum, W., Potter, B., 2014. A review of Modelling techniques used in the analysis of wireless power transfer systems. Electronics 18(2).

Moutopoulou, E., Bertos, G.A., Mablekos-Alexiou, A., Papadopoulos, E.G., 2015. Feasibility of a biomechatronic EPP upper limb prosthesis controller. Conf. Proc. IEEE Eng. Med. Biol. Soc. 2015, 2454–2457. https://doi.org/10.1109/EMBC.2015.7318890.

Nationalshoe, 2013. Prosthetic, Orthotic Components & Orthopaedic Solutions Catalogue 2013. Retrieved from: http://www.nationalshoe.com/pdfs/NS_P_O_2013_Catalogue_LoRes.pdf.

Niinomi, M., 2002. Recent metallic materials for biomedical applications. Metall. Mater. Trans. A-Phys. Metall. Mater. Sci. 33 (3), 477–486. https://doi.org/10.1007/s11661-002-0109-2.

Normann, R.A., Fernandez, E., 2016. Clinical applications of penetrating neural interfaces and Utah electrode array technologies. J. Neural Eng. 13(6), 061003. https://doi.org/10.1088/1741-2560/13/6/061003.

Oddo, C.M., Beccai, L., Wessberg, J., Wasling, H.B., Mattioli, F., Carrozza, M.C., 2011. Roughness encoding in human and biomimetic artificial touch: spatiotemporal frequency modulation and structural anisotropy of fingerprints. Sensors (Basel) 11 (6), 5596–5615. https://doi.org/10.3390/s110605596.

Oddo, C.M., Raspopovic, S., Artoni, F., Mazzoni, A., Spigler, G., Petrini, F., … Micera, S., 2016. Intraneural stimulation elicits discrimination of textural features by artificial fingertip in intact and amputee humans. elife 5, e09148. https://doi.org/10.7554/eLife.09148.

Ohnishi, K., Weir, R.F., Kuiken, T.A., 2007. Neural machine interfaces for controlling multifunctional powered upper-limb prostheses. Expert Rev. Med. Devices 4 (1), 43–53. https://doi.org/10.1586/17434440.4.1.43.

Oliker, A., 2015. 3D printing: Revolutionizing medicine. Am. Q. 8, 46–47 (Spring 2015).

Ortiz-Catalan, M., Branemark, R., Hakansson, B., 2013. BioPatRec: A modular research platform for the control of artificial limbs based on pattern recognition algorithms. Source Code Biol. Med. 8 (1), 11. https://doi.org/10.1186/1751-0473-8-11.

Ortiz-Catalan, M., Hakansson, B., Branemark, R., 2014a. An osseointegrated human-machine gateway for long-term sensory feedback and motor control of artificial limbs. Sci. Transl. Med. 6 (257), 257re256. https://doi.org/10.1126/scitranslmed.3008933.

Ortiz-Catalan, M., Hakansson, B., Branemark, R., 2014b. Real-time and simultaneous control of artificial limbs based on pattern recognition algorithms. IEEE Trans. Neural Syst. Rehabil. Eng. 22 (4), 756–764. https://doi.org/10.1109/TNSRE.2014.2305097.

Pandey, E., Srivastava, K., Gupta, S., Srivastava, S., Mishra, N., 2016. Some biocompatible materials used in medical practices- a review. Int. J. Pharm. Sci. Res. 7 (7), 2748–2755. https://doi.org/10.13040/Ijpsr.0975-8232.7(7).2748-55.

Parker, P., Englehart, K., Hudgins, B., 2006. Myoelectric signal processing for control of powered limb prostheses. J. Electromyogr. Kinesiol. 16 (6), 541–548. https://doi.org/10.1016/j.jelekin.2006.08.006.

Peerdeman, B., Boere, D., Witteveen, H., Huis in 'tVeld, R., Hermens, H., Stramigioli, S., Misra, S., 2011. Myoelectric forearm prostheses: state of the art from a user-centered perspective. J. Rehabil. Res. Dev. 48(6). https://doi.org/10.1682/jrrd.2010.08.0161.

Pfurtscheller, G., Muller, G.R., Pfurtscheller, J., Gerner, H.J., Rupp, R., 2003. 'Thought'—control of functional electrical stimulation to restore hand grasp in a patient with tetraplegia. Neurosci. Lett. 351 (1), 33–36. https://doi.org/10.1016/S0304-3940(03)00947-9.

Pons, J.L., Rodriguez, H., Luyckx, I., Reynaerts, D., Ceres, R., Van Brussel, H., 2002. High torque ultrasonic motors for hand prosthetics: current status and trends. Technol. Health Care 10 (2), 121–133.

Proietti, T., Crocher, V., Roby-Brami, A., Jarrasse, N., 2016. Upper-limb robotic exoskeletons for neurorehabilitation: a review on control strategies. IEEE Rev. Biomed. Eng. 9, 4–14. https://doi.org/10.1109/RBME.2016.2552201.

Putti, V., 1917. Plastiche e Protesi Cinematiche. Chir. d. Org. d. Movimento 1, 419–492.

Sauerbruch, F., 1915. Chirurgische Vorarbiet für eine Wilkürlich Bewegliche Künstliche Hand. Mediziische Klinik 11 (41), 1125–1126.

Sauerbruch, F., 1916. Die Wilkürlich Bewegbare Künstliche Hand. Eine Anleitung für Chirurgen und Techniker, first ed. Julius Springer-Verlag, Berlin.

Scheme, E., Englehart, K., 2011. Electromyogram pattern recognition for control of powered upper-limb prostheses: state of the art and challenges for clinical use. J. Rehabil. Res. Dev. 48 (6), 643–659.

Schorsch, J.F., Weir, R.F.F., IEEE, 2008. Reliability of Implantable MyoElectric Sensors (IMES). IEEE, New York.

Sensinger, J.W., Lock, B.A., Kuiken, T.A., 2009. Adaptive pattern recognition of myoelectric signals: exploration of conceptual framework and practical algorithms. IEEE Trans. Neural Syst. Rehabil. Eng. 17 (3), 270–278. https://doi.org/10.1109/TNSRE.2009.2023282.

Shehata, A.W., Scheme, E.J., Sensinger, J.W., 2017. The effect of myoelectric prosthesis control strategies and feedback level on adaptation rate for a target acquisition task. IEEE Int. Conf. Rehabil. Robot. 2017, 200–204. https://doi.org/10.1109/ICORR.2017.8009246.

Sheridan, T.B., 1992. Telerobotics, Automation, and Human Supervisory Control. MIT Press Cambridge, Cambridge, MA.

Shevtsov, M.A., Yudintceva, N., Blinova, M., Pinaev, G., Galibin, O., Potokin, I., Pitkin, M., 2015. Application of the skin and bone integrated pylon with titanium oxide nanotubes and seeded with dermal fibroblasts. Prosthetics Orthot. Int. 39 (6), 477–486. https://doi.org/10.1177/0309364614550261.

Sieber, A., Valdastri, P., Houston, K., Eder, C., Tonet, O., Menciassi, A., Dario, P., 2008. A novel haptic platform for real time bilateral biomanipulation with a MEMS sensor for triaxial force feedback. Sens. Actuator A Phys. 142 (1), 19–27. https://doi.org/10.1016/j.sna.2007.03.018.

Simpson, D.C., 1974. The choice of control system for the multimovement prosthesis: extended physiological proprioception (EPP). In: Herberts, E.A. (Eds.), The Control of Upper-Extremity Prostheses and Orthoses. Thomas, Springfield, IL, pp. 146–150.

Souza, J.M., Cheesborough, J.E., Ko, J.H., Cho, M.S., Kuiken, T.A., Dumanian, G.A., 2014. Targeted muscle reinnervation: a novel approach to postamputation neuroma pain. Clin. Orthop. Relat. Res. 472 (10), 2984–2990. https://doi.org/10.1007/s11999-014-3528-7.

Spittler, A.W., Fletcher, M.J., 1953. Technique of tunnel cineplastic surgery and prosthetic appliances for cineplasty. In: Edwards, J.W. (Eds.), American Academy of Orthopaedic Surgeons Instructional Course Lectures. vol. 10. Ann Arbor, Michigan, pp. 376–394.

Stieglitz, T., Boretius, T., Navarro, X., Badia, J., Guiraud, D., Divoux, J.L., …Jensen, W., 2012. Development of a neurotechnological system for relieving phantom limb pain using transverse intrafascicular electrodes (TIME). Biomed Tech (Berl) 57 (6), 457–465. https://doi.org/10.1515/bmt-2011-0140.

Sul, Y.T., 2010. Electrochemical growth behavior, surface properties, and enhanced in vivo bone response of TiO2 nanotubes on microstructured surfaces of blasted, screw-shaped titanium implants. Int. J. Nanomedicine 5, 87–100.

Sun, T.J., Xie, X., Wang, Z.H., 2013. In: Design challenges of the wireless power transfer for medical microsystems. 2013 IEEE International Wireless Symposium (IWS).

Tan, D.W., Schiefer, M.A., Keith, M.W., Anderson, J.R., Tyler, J., Tyler, D.J., 2014. A neural interface provides long-term stable natural touch perception. Sci. Transl. Med. 6 (257), 257ra138. https://doi.org/10.1126/scitranslmed.3008669.

Tanaka, K.S., Lightdale-Miric, N., 2016. Advances in 3D-printed pediatric prostheses for upper extremity differences. J. Bone Joint Surg. Am. 98 (15), 1320–1326. https://doi.org/10.2106/Jbjs.15.01212.

Tavakoli, M., Lourenco, J., de Almeida, A.T., 2017. In: 3D Printed endoskeleton with a soft skin for upper-limb body actuated prosthesis. 2017 IEEE 5th Portuguese Meeting on Bioengineering (Enbeng).

ten Kate, J., Smit, G., Breedveld, P., 2017. 3D-printed upper limb prostheses: a review. Disabil. Rehabil. Assist. Technol. 12 (3), 300–314. https://doi.org/10.1080/17483107.2016.1253117.

Thesleff, A., Branemark, R., Hakansson, B., Ortiz-Catalan, M., 2018. Biomechanical characterisation of bone-anchored implant systems for amputation limb prostheses: a systematic review. Ann. Biomed. Eng. 46 (3), 377–391. https://doi.org/10.1007/s10439-017-1976-4.

Tiwana, M.I., Shashank, A., Redmond, S.J., Lovell, N.H., 2011. Characterization of a capacitive tactile shear sensor for application in robotic and upper limb prostheses. Sens. Actuators A Phys. 165 (2), 164–172. https://doi.org/10.1016/j.sna.2010.09.012.

Towne, C., Montgomery, K.L., Iyer, S.M., Deisseroth, K., Delp, S.L., 2013. Optogenetic control of targeted peripheral axons in freely moving animals. PLoS One 8(8), e72691. https://doi.org/10.1371/journal.pone.0072691.

Tropea, P., Mazzoni, A., Micera, S., Corbo, M., 2017. Giuliano Vanghetti and the innovation of "cineplastic operations". Neurology 89 (15), 1627–1632. https://doi.org/10.1212/WNL.0000000000004488.

Troyk, P.R., DeMichele, G.A., 2003. In: Inductively-coupled power and data link for neural prostheses using a class-E oscillator and FSK modulation.Proceedings of the 25th Annual International Conference of the IEEE Engineering in Medicine and Biology Society, Vols 1–4, 25, 3376–3379. https://doi.org/10.1109/Iembs.2003.1280869.

Troyk, P.R., DeMichele, G.A., Kerns, D.A., Weir, R.F., 2007. In: IMES: an implantable myoelectric sensor. 2007 Annual International Conference of the IEEE Engineering in Medicine and Biology Society, Vols 1–16, 1730+. https://doi.org/10.1109/Iembs.2007.4352644.

Tsikandylakis, G., Berlin, O., Branemark, R., 2014. Implant survival, adverse events, and bone remodeling of osseointegrated percutaneous implants for transhumeral amputees. Clin. Orthop. Relat. Res. 472 (10), 2947–2956. https://doi.org/10.1007/s11999-014-3695-6.

Tyler, D.J., Durand, D.M., 2002. Functionally selective peripheral nerve stimulation with a flat interface nerve electrode. IEEE Trans. Neural Syst. Rehabil. Eng. 10 (4), 294–303. https://doi.org/10.1109/TNSRE.2002.806840.

Vanderborght, B., Albu-Schaeffer, A., Bicchi, A., Burdet, E., Caldwell, D.G., Carloni, R., Wolf, S., 2013. Variable impedance actuators: a review. Robot. Auton. Syst. 61 (12), 1601–1614. https://doi.org/10.1016/j.robot.2013.06.009.

Vanghetti, G., 1898. Amputazioni, Disarticolozioni e protesi. Florence, Italy.

Vanghetti, G., 1899a. Plastica dei Monconi a Scopo di Protesi Cinematica. Arch. d. Ortop. 16 (6), 385–410.

Vanghetti, G., 1899b. Plastica dei Monconi a Scopo di Protesi Cinematica. Arch. d. Ortop. 16 (5), 305–324.

Vanghetti, G., 1900. Plastica dei Monconi ed Amputazioni Transitore. Arch. d. Ortop. 17 (5–6), 305–329.

Vanghetti, G., 1906. Plastica e Protesi Cinematiche: Nuova Teoria sulle Amputazioni esulle Protesi. Traversari, Empoli.

Vujaklija, I., Farina, D., Aszmann, O.C., 2016. New developments in prosthetic arm systems. Orthop. Res. Rev. 8, 31–39. https://doi.org/10.2147/Orr.S71468.

Wang, Q., Markopoulos, P., Yu, B., Chen, W., Timmermans, A., 2017. Interactive wearable systems for upper body rehabilitation: a systematic review. J. Neuroeng. Rehabil. 14. https://doi.org/10.1186/s12984-017-0229-y. ARTN 20.

Weir, R.F.f., 2017. Extrapolation of emerging technologies and their long-term implications for myoelectric versus body-powered prostheses. J. Prosthet. Orthot. 29, P63–P74. https://doi.org/10.1097/jpo.0000000000000161.

Weir, R.F.f., Childress, D.S., 1996. Prostheses and Artificial Limbs Encyclopaedia of Applied Physics. pp. 115–140.

Weir, R.F., Troyk, P.R., DeMichele, G.A., Kerns, D.A., Schorsch, J.F., Maas, H., 2009. Implantable myoelectric sensors (IMESs) for intramuscular electromyogram recording. IEEE Trans. Biomed. Eng. 56 (1), 159–171. https://doi.org/10.1109/TBME.2008.2005942.

Yokokohji, Y., Yoshikawa, T., 1994. Bilateral control of master-slave manipulators for ideal kinesthetic coupling - formulation and experiment. IEEE Trans. Robot. Autom. 10 (5), 605–620. https://doi.org/10.1109/70.326566.

Yong-Xi, G., Zhu, D., Jegadeesan, R., 2011. In: Inductive wireless power transmission for implantable devices.Paper Presented at the International Workshop on Antenna Technology (iWAT), Hong Kong, China, 7–9 March 2011.

Yuan, X., Changgeng, S., Yan, G., Zhenghong, Z., 2016. Application review of dielectric electroactive polymers (DEAPs) and piezoelectric materials for vibration energy harvesting. J. Phys. Conf. Ser. 744. https://doi.org/10.1088/1742-6596/744/1/012077.

Zollo, L., Roccella, S., Guglielmelli, E., Carrozza, M.C., Dario, P., 2007. Biomechatronic design and control of an anthropomorphic artificial hand for prosthetic and robotic applications. IEEE-ASME Trans. Mechatron. 12 (4), 418–429. https://doi.org/10.1109/tmech.2007.901936.

Zuniga, J., Katsavelis, D., Peck, J., Stollberg, J., Petrykowski, M., Carson, A., Fernandez, C., 2015. Cyborg beast: a low-cost 3d-printed prosthetic hand for children with upper-limb differences. BMC Res. Notes 8, 10. https://doi.org/10.1186/s13104-015-0971-9.

FURTHER READING

Weir, R.F.f., 1995. Direct Muscle Attachment as a Control Input for a Position-Servo Prosthesis Controller (Ph.D. dissertation). Northwestern University, Evanston, IL.

CHAPTER SEVEN

Lower-Limb Prosthetics

Georgios A. Bertos[*,†,‡], Evangelos G. Papadopoulos[*]
[*]National Technical University of Athens, Athens, Greece
[†]Northwestern University Prosthetics-Orthotics Center, Physical Medicine & Rehabilitation, Feinberg School of Medicine, Chicago, IL, United States
[‡]Bionic Healthcare, Inc, Chicago, IL, United States

Contents

1 HISTORY

Limb amputations have been perceived with fear across civilizations of the past. Partial foot prostheses (great toe of the right foot made of leather and wood) have been identified in mummies of Ancient Egypt dated 15th century

https://doi.org/10.1016/B978-0-12-812539-7.00007-6

BC (Hernigou, 2013). One of the first but very functional lower-limb prostheses was the "peg leg," a wooden leg that was used by military amputees and pirates (Hernigou, 2013). Wooden peg legs have been effective prostheses for thousands of years (Herr et al., 2003). It was not until the 20th century, with the introduction of polymers that the lower-limb prostheses started to be built with plastics. Currently, in lower-limb prosthesis design, we are on the verge of a new era in which embedded microcomputer systems will take over the automatic control of lower-limb states through the automatic control of hydraulic and/or pneumatic-actuated mechanisms. The C-leg of Otto Bock represents the first lower limb of this kind, although a couple of other computer-controlled knees are available at the moment (Michael, 1999). More legs of this kind will emerge. However, the effectiveness and efficiency of these designs will depend on the quality of the algorithm executed, which in turn depends on our understanding of normal and prosthetic walking.

2 HOW IS SUCCESS DEFINED FOR LOWER-LIMB PROSTHETICS?

Success in lower-limb prosthetics is measured by the percentage of the lower-limb amputees using their prostheses, and is functional and happy with them (Webster et al., 2012). Human walking is a basic characteristic of human everyday life. As mentioned in Section 4, able-bodied walking is a repetitious and energy-efficient process. If we were to measure success, we would expect that lower-limb prostheses would enable amputees to walk with similar performance to able-bodied ambulators or outperform it.

Similarly, it is expected that lower-limb prostheses will enable amputees to perform or outperform in other everyday life tasks such as running, jumping, hopping, dancing, ascending and descending stairs, and hiking, the list goes on depending to what is subjectively important to the amputee. Happiness is subjective and personal and the same is true for success in lower-limb prosthetics.

2.1 What Would be Ideal?

The ultimate objective of lower-limb prostheses is to replace the functionality of the natural limb. Since the human legs are well versatile and there is inherent redundancy, they are used for different types of locomotion and activities, that is, jumping, running, dancing, ascending and descending stairs, and walking in an optimal way. For each of these activities, a model (or models or a unified model) that describes that activity is needed in order to intervene and be able to scientifically design a prosthetic "compensatory" device that will compensate for the missing functionality, and together with the remaining limb, will act as a totality close to the natural leg. Out of all these daily activities, walking is the most important one since it is a prerequisite for the rest of the activities; that is, walking is a basic prerequisite for healthy living. Even after surgery, patients need to walk in order to maintain their vascular system without blood clots, which can cause deep vein thrombosis, heart attacks, or strokes. To provide amputees with quality lower-limb prostheses, the functional characteristics of the natural leg during walking must be identified first, that is, a good understanding of the human gait is needed to allow for the development of a good model for it, and have it incorporated these into the prosthesis' design. We believe that some of the functional characteristics of walking are the *shape of the foot*, the *shock absorption of the leg*, and the *"straight leg" nature of walking*. Without understanding these functional characteristics, and without incorporating them into a model of walking, the lower-limb prosthetics will be limited in usefulness and will not achieve the performance and adoption they can.

2.1.1 Adjust to Terrain and Task?

One of the features that would be desirable is that the prosthesis recognizes the terrain conditions and the task that the amputee wants to perform and adjust accordingly (Hansen and Starker, 2017; Major et al., 2018). For example, if the amputee is walking with a walking prosthesis and suddenly he/she wants to run, the prosthesis should recognize his/her intention automatically, and adapt to perform this task accordingly. The same would be desirable for all other tasks, for example, dancing, running, ascending and descending stairs, hopping, etc.

2.1.2 Enable Nonambulatory Amputees (e.g., Bilateral Transfemoral Amputees)

The most challenging population of lower-limb amputees is the bilateral transfemoral amputees. The more proximal the amputation the more difficult it is for an amputee to walk due to the fact that current above the knee prostheses presents significant drawbacks such as limited controllability, and requires significant amount of energy. Duplicate the above drawbacks and add that there are balance problems for the bilateral transfemoral amputee and then it becomes clear why a high percentage of this population is in a wheelchair and not mobile. It would be a measure of success, if prostheses of the future would find ways to enable these amputees to ambulate.

2.1.3 Seamless and Improved Performance

All the different transitions, adjustments and integration of the prosthetic leg should be seamless, that is, if an amputee wants transition from walking to running, the prosthesis should recognize this and adapt accordingly as humans do (Herr et al., 2003). In the same lines, it is recognized that a coordination of all the tasks to be done (i.e., walking, ascending and descending slope, siting, dancing, hopping, etc.) is a need and therefore, the intention of the amputee is important via a high-level controller is of a paramount importance (Windrich et al., 2016). The walking performance of a lower prosthesis of the future should not be inferior to that of able-bodied walking or other locomotive activities. On the contrary, prostheses should be designed to overpass human performance. A good example of this is that amputee runners (called blade runners) with the spring-leaf-like prosthetic feet at (Bròggemann et al., 2008) 100-m sprints records are very close to the able-bodied runners records with regular feet (Bròggemann et al., 2008; List of IPC world records in athletics, 2018).

3 NEEDS/VOICE OF CUSTOMER

We have to consider the real needs of the amputees that lower-limb prostheses have to fill in. This of course depends on the level of amputation. The transtibial amputees can walk acceptably well with the current prostheses. They have needs for cosmesis, increased speed of walking, and skin blisters at the socket interface due to high pressure. In addition, they have issues such as difficulty in ascending and descending stairs, problems with slope walking, and increased energy consumption especially at faster walking speeds (Hansen and Starker, 2017; Windrich et al., 2016).

The transfemoral amputees have needs such as leg controllability, proper shock absorption (shock felt at lower back), increased walking speed, and symmetry with able-bodied side. In addition, they have issues such as difficulty in ascending and descending stairs, problems with slope walking, and increased energy consumption (Windrich et al., 2016). Toe clearance during swing phase is important for transfemoral amputees along with active push off in late stance phase especially for faster walking speeds. In the case of bilateral transfemoral, balance and stability issues are of concern (Hansen and Starker, 2017).

3.1 Stability

Lower-limb prostheses should provide to the amputee stability, that is, prosthetic components which are not going to drive them to instability while ambulating or provide stumbling recovery mechanisms as proposed by Lawson et al. (2010). This is a software monitoring example that can monitor and detect early stumbling and intervene. This becomes more important in the case of bilateral transfemoral amputees where the stability needs are higher due to inherent instabilities. The big picture or theory or biomechanical model is always important. We should note the principle of conservation of angular momentum (Herr et al., 2003) which predicts fairly well the motion of humans during the tasks of standing and walking. It could enable novel prosthetic devices (Herr et al., 2003).

3.2 Walking Speed

Amputees need to be able to achieve the maximum speed they can. Their prosthesis should not be an obstacle on walking as fast as they can. It was proposed in the past that avoidance of high-peak forces and accelerations during gait was the reason that amputees did not achieve the maximum speed they could (Cappozzo, 1991; Gard and Konz, 2003). Gard and Konz (2003) also proposed that providing to the amputee the right shock absorption will be means of improving their walking speed. Walking speed is, therefore, connected to right shock absorption (see Section 3.4).

3.3 Socket Interface Relief of Pressure

One of the sore points that are found to present clinical problems in prosthetics is the socket interface of the prostheses. Shear forces usually create high pressure and blisters, dermatitis, and edema, which make the "symbiosis" of amputees and conventional prostheses difficult (Mak et al., 2001). The most radical solution to this problem is the use of the osseointegration technique (see Section 6.1) where no socket is used (Mak et al., 2001). In lieu of using osseointegration, as mentioned in Mak et al. (2001), the computer-aided design (CAD)/computer-aided manufacturing (CAM) technology can make the socket design and fabrication process more effective and objective and decrease the uncomfortable effects of any nonoptimal socket interface.

3.4 Right Shock Absorption

As mentioned in Hansen and Starker (2017), a big aspect of human walking and therefore, of prosthetic walking is shock absorption. Not having the right shock absorption can cause increased forces due to shock at the hips and spine, which prevents the amputee from achieving high walking speeds. Shock absorption has not been solved as a problem in prosthetic walking (only empirical shock absorber components exist) and the scientific community should pay more attention on resolving it. There have been honest progresses as noted in Section 4. Furthermore, prosthetic shock absorption is even more important in other tasks such as hopping and running.

4 WALKING THEORY

Perry (1992) stated that walking is "controlled falling." Before him, Margaria (1976) said that walking is like an egg rolling end over end (Fig. 1). Saunders et al. (1953) first introduced the compass-gait model (Fig. 2A). Mochon and McMahon (1980) introduced the ballistic walking model, which consists of two stance-phase inverted pendulum legs. Different variations of the model include stance knee flexion, plantar flexion of the stance ankle, and pelvic tilt. Tad McGeer added rockers to the basic compass-gait model and built walking machines that can walk down slight inclines under only the influence of gravity (McGeer, 1990). Coleman and Ruina (1998), Garcia et al. (1998), and Kuo (1999) have also shown that walking can be modeled as an inverted pendulum with rockers. Alexander (1992) added springs to the basic ballistic model.

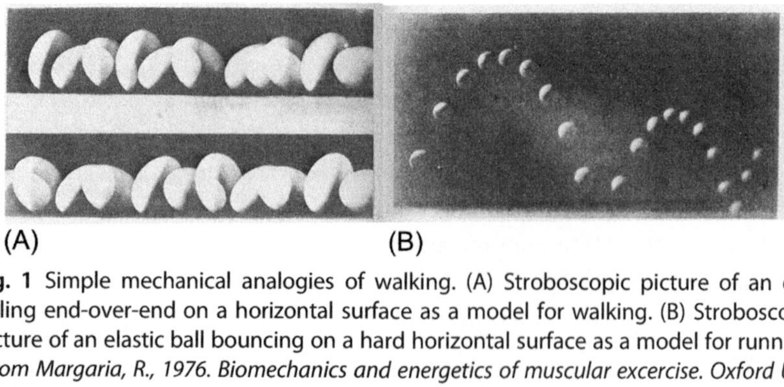

Fig. 1 Simple mechanical analogies of walking. (A) Stroboscopic picture of an egg rolling end-over-end on a horizontal surface as a model for walking. (B) Stroboscopic picture of an elastic ball bouncing on a hard horizontal surface as a model for running. *(From Margaria, R., 1976. Biomechanics and energetics of muscular excercise. Oxford University Press. By permission of Oxford University Press.)*

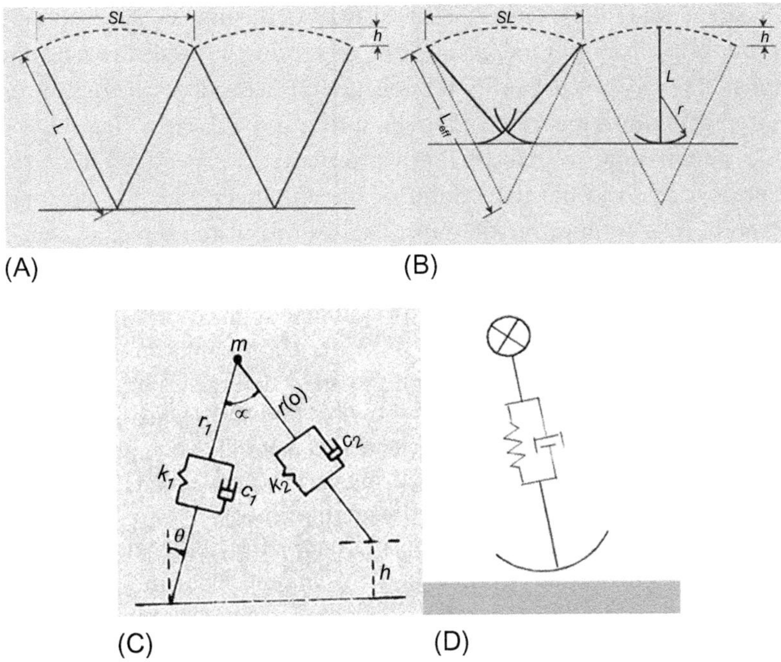

Fig. 2 (A) Compass-gait model, (B) rocker-based inverted pendulum model introduced by Gard and Childress (2000), (C) stiffness-damper model of gait used from Siegler et al. (1982), and (D) stiffness-damper-rocker model of gait used from Gard and Childress (2000).

The Northwestern University Prosthetics Research Laboratory (NUPRL) (Childress, 2002; Gard and Childress, 2002) believed that in order to design the best lower-limb prostheses, a good able-bodied theory of walking should be developed along with amputee deviations. Therefore, the NUPRL developed a theory of walking so we can understand why the body moves the way it moves during walking. An inverted pendulum model with rockers was introduced (Fig. 2B). The characteristics of the rocker are based on the "roll-over shape" which is the equivalent foot/ankle geometry extracted from a person walking during the stance phase (Hansen, 1998; Hansen et al., 2000). The "roll-over shape" functionally lengthens the leg of the inverted pendulum model; thus, it is equivalent to a virtual leg approximately 1.7 times the length of the anatomical leg but without rockers (Fig. 2C). Both models, the leg with rockers or the virtual leg without rockers produce similar center of mass trajectories. The rockers, or equivalently the virtual leg, reduce the peak to peak of the vertical excursion of the center of mass from what it would be otherwise (Gard and Childress, 2000).

Siegler et al. (1982) performed simulations of the human gait with the aid of a simple mechanical model consisting of a spring in parallel to a damping element (Fig. 2C). Gard and Childress (2002) expanded the rocker-based inverted pendulum model by adding a spring and a damper (Fig. 2D).

To design improved lower-limb prostheses, we need to understand the normal gait and the interaction of the amputees with the prostheses, so as to be able to improve the prosthetic gait to match the characteristics of the normal one. Despite many attempts around the world, there is no complete theory of the gait up to now.

Two of the aspects of normal walking we have investigated are the stance-phase knee flexion and pelvic obliquity. We believe that both of these movements provide shock absorption during the early stance phase. Pelvic obliquity was one of the six determinants of gait believed to decrease the vertical excursion of the body center of mass (BCOM) in order to conserve energy (Saunders et al., 1953; Inman et al., 1994, 1981). Using the NUPRL, it was found that the above statement is not true for normal walking (Gard and Childress, 1997a). The peak-to-peak vertical displacement of the center of mass due to pelvic obliquity is not different than the peak-to-peak vertical displacement of the center of mass without pelvic obliquity (Fig. 3A). The conclusion was that pelvic obliquity does not decrease the vertical excursion of the BCOM. Pelvic obliquity is maximum at around the time of contralateral toe-off, being out of phase with the vertical excursion of the BCOM, suggesting that this movement is important for shock absorption in the early stance phase, as suggested by Perry (1992) and Sutherland et al. (1994).

Similar to pelvic obliquity, stance-phase knee flexion during the early stance is one of the six determinants of gait and was believed to lower the vertical excursion of the BCOM in order to conserve energy (Inman et al., 1981, 1994; Saunders et al., 1953). Data show that the effect of the stance-phase knee flexion on the peak-to-peak vertical excursion of the BCOM is negligible (Gard and Childress, 1997a,b, 1999; Fig. 3B). During the stance phase, knee flexion is maximized around the time of contralateral toe-off, and minimized when the knee is nearly fully extended and the trunk reaches its peak vertical displacement during the gait cycle (Fig. 3B). Like pelvic obliquity, stance-phase knee flexion is out of phase with the BCOM vertical displacement due to joint configuration; these results also have been verified by Quesada and Rash (1998).

Pelvic obliquity and stance-phase knee flexion play a critical role in shock absorption during the early stance phase of normal walking. Thus, it might be beneficial for the amputees to incorporate the shock absorption

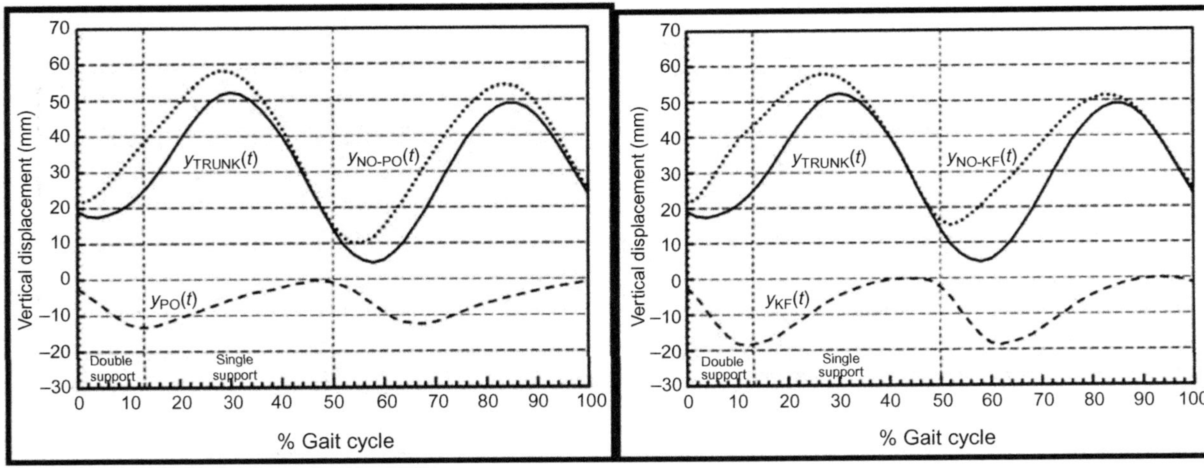

(A) (B)

Fig. 3 Stance-phase knee flexion and pelvic obliquity do not increase vertical BCOM—they introduce a phase shift. (A) The vertical displacement of the body center of mass, $y_{TRUNK}(t)$, the vertical displacement of the body center of mass due to pelvic obliquity, $y_{PO}(t)$, and the vertical displacement of the body center of mass without pelvic obliquity, $y_{NO\text{-}PO}(t)$. (B) The vertical displacement of the body center of mass, $y_{TRUNK}(t)$, the vertical displacement of the body center of mass due to stance-phase knee flexion, $y_{KF}(t)$, and the vertical displacement of the body center of mass without stance-phase knee flexion, $y_{NO\text{-}KF}(t)$ (Gard and Childress, 1997a,b). The peak-to-peak values of $y_{NO\text{-}PO}(t)$ and $y_{NO\text{-}KF}(t)$ are not significantly different from $y_{TRUNK}(t)$.

mechanisms provided by pelvic obliquity and stance-phase knee flexion into the prosthesis design. We might be able to functionally simulate the shock absorption action with devices, which will provide close to normal walking shock absorption effect. We hope that by incorporating the right shock absorption into the prosthesis, the gait will be closer to normal, safer, and more comfortable to the amputee.

Gard and Childress (1997a,b) have introduced an inverted pendulum model with rockers (Fig. 4A). The vertical excursion of the BCOM, h, can be calculated by the constraints imposed by the legs and the "roll-over shape":

$$h = \frac{S_l^2}{8L\rho} \tag{1}$$

where S_l is the step length, L is the anatomical leg length, and ρ is a dimensionless constant approximately equal to 1.7.

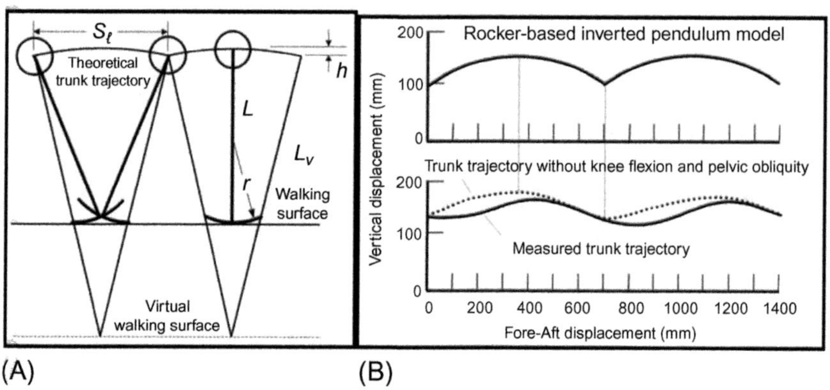

Fig. 4 The rocker-based inverted pendulum of walking. (A) The rocker-based inverted pendulum model of walking. L is the anatomical leg length, r is the foot rocker radius, L_V is the virtual leg length (approximately the height of the subject), S_l is the step length, and h is the vertical excursion of the body center of mass. (B) The model predicts a vertical excursion *(dotted line)* comparable with what we measured for able-bodied ambulators *(solid line)*. The inverted pendulum with rockers model predicts the peak-to-peak vertical excursion of the BCOM. The pelvic obliquity and stance-phase knee flexion introduce a phase shift and provide shock absorption to the system. *((A) Gard, S.A., Childress, D.S., 2001. What determines the vertical displacement of the body during normal walking? J. Prosthet. Orthot. 13(3), 64–67; (B) From Gard, S.A., Childress, D.S., 2000. What determines vertical motion of the body during normal gait. Paper Presented at the 5th Annual Meeting of the Gait and Clinical Movement Analysis Society (GCMAS), Rochester, MN, April 12–15.)*

The vertical excursion of the center of mass is not decreased by pelvic obliquity or stance-phase knee flexion, which we believe provide shock absorption to the system (Fig. 4B) (Gard and Childress, 1997a,b, 2000; Childress & Gard, 1999). Most of the above theoretical results have been confirmed by empirical data (Miff, 2000).

Thus, a shock-absorbing element must be added to the above model (Fig. 4A) in order to stand for the natural shock absorption function provided by the knee flexion, pelvic obliquity, ankle plantar flexion, and the viscoelastic properties of the tissues.

For the above purpose, Bertos (2006) and Bertos et al. (2005) proposed a shock absorption model of walking (Eq. 2, Fig. 5):

$$\frac{y_m(s)}{y_b(s)} = \frac{\frac{B}{M_e}s + \frac{k}{M_e}}{s^2 + \frac{B}{M_e}s + \frac{k}{M_e}} \tag{2}$$

where y_m is the subject's vertical BCOM trajectory, y_b the vertical trajectory of the rocker-based inverted pendulum model, k the stiffness, B the viscous damping, and M_e the effective mass of the body during the stance phase of walking.

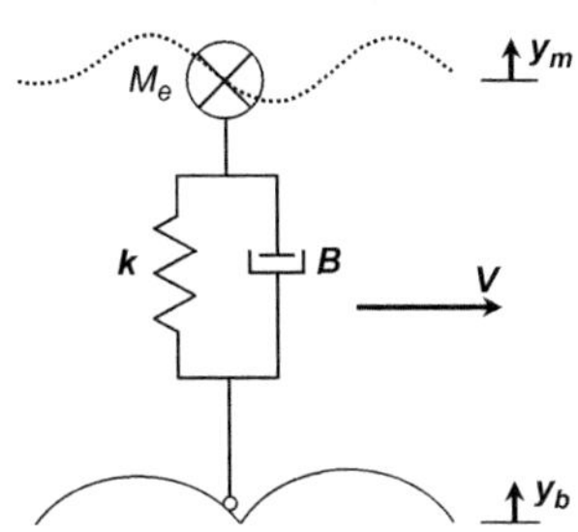

Fig. 5 Shock-absorption model for able-bodied human walking. Able-bodied human walking was modeled with a second-order mechanical vibration system. y_b is the trajectory that a rocker-based inverted pendulum walking with no shock absorption would follow. y_m is the trajectory of the BCOM of one able-bodied walker (which includes any shock absorption effect), M_e the effective mass of the subject during the stance phase of walking, k is the stiffness, B the viscous damping, and v the average forward speed of walking. *(From Bertos, G.A., Childress, D.S., Gard, S.A., 2005. The vertical mechanical impedance of the locomotor system during human walking with applications in rehabilitation. In: IEEE 9th International Conference on Rehabilitation Robotics. IEEE, New York, pp. 380–383.)*

Bertos (2006) and Bertos et al. (2005) identified values (for $n = 7$ subjects) of the dynamic system using steady-state identification techniques. The variance account for VAF% of the model to the data was 85%–95%. The results for one representative subject are shown in Figs. 6–8 (Bertos et al., 2005).

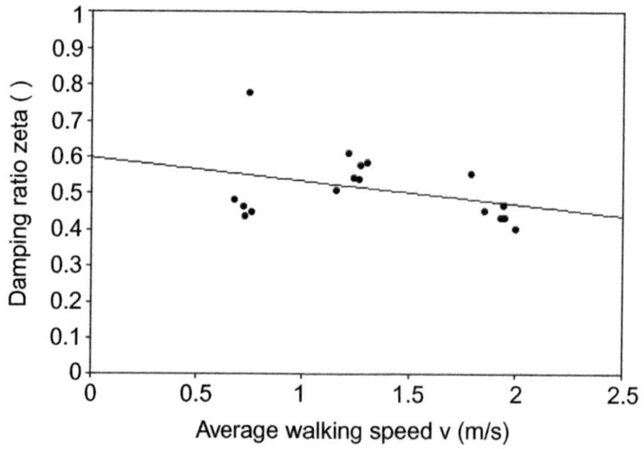

Fig. 6 Estimated damping ratio zeta vs average walking speed *v*. *(From Bertos, G.A., Childress, D.S., Gard, S.A., 2005. The vertical mechanical impedance of the locomotor system during human walking with applications in rehabilitation. In: IEEE 9th International Conference on Rehabilitation Robotics. IEEE, New York, pp. 380–383.)*

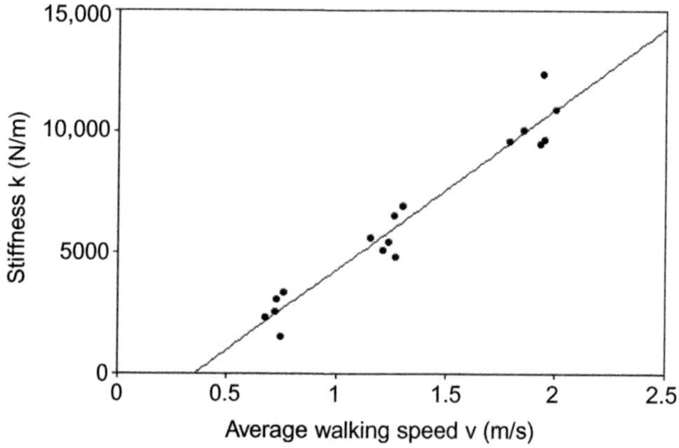

Fig. 7 Estimated stiffness *k* vs average walking speed *v*. *(From Bertos, G.A., Childress, D.S., Gard, S.A., 2005. The vertical mechanical impedance of the locomotor system during human walking with applications in rehabilitation. In: IEEE 9th International Conference on Rehabilitation Robotics. IEEE, New York, pp. 380–383.)*

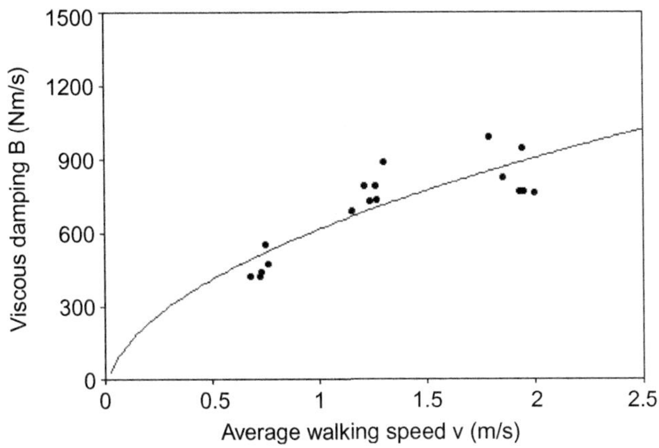

Fig. 8 Estimated viscous damping *B* vs average walking speed *v*. *(From Bertos, G.A., Childress, D.S., Gard, S.A., 2005. The vertical mechanical impedance of the locomotor system during human walking with applications in rehabilitation. In: IEEE 9th International Conference on Rehabilitation Robotics. IEEE, New York, pp. 380–383.)*

The values of the proposed walking model along with estimates of damping ratio ζ, stiffness k, and damping B of other activities of human like bouncing and running are illustrated in Table 1 (Bertos, 2006).

Following this work, Smyrli et al. (2018) performed simulations with a passive biped model with leg compliance and damping, and semicircular feet, and studied its stability as a function of a set of nondimensional parameters.

A unified walking theory model of able-bodied walking is needed for the prostheses of the future to take advantage of nature's optimized evolution. Nevertheless, for unilateral amputees, the prosthetic side must match the unaffected side. In addition, it will generate more symmetry between the two sides, if the prosthesis matches a unified walking model.

4.1 Design Intelligence of Human Legs

A philosophical argumentative quest for where the "design intelligence" of human legs as designed by nature comes from is given by Blickhan et al. (2007). Is it due to structural mechanics? Does the foot serve a purpose for walking? Blickhan et al. (2007) noted: "By placing the foot with its heels and shifting the point of pressure toward the toes, the foot acts like the rim of a wheel" which is in support of the roll-over shape that was introduced by Gard and Childress (1997a,b) in Section 4.

Table 1 Comparison Between Reported Literature of In Vivo Identification of Stiffness and Damping During, Running or Bouncing and Proposed Model's Results During Walking

Author	Task	*k*	*B*	ζ/Method
Greene and McMahon (1979)	Bouncing on a board with knees at a specific angle	37.6 kN/m when the angle θ was 45 degrees	1180 N/(m/s)[a]	0.34 board stiffness
Cavagna (1970)	Bouncing with knees locked and ankles plantar-flexed	34–41 kN/m	570 N/(m/s)	0.18[b] log decrement
Bach et al. (1983)	Bouncing with knees locked and ankles plantar-flexed	31,898 kg/s^2 = 31,898 N/m	3986 kg/s = 406.8 N/(m/s)	0.13[b] log decrement
Zhang et al. (2000)	Standing. Externally induced small amplitude perturbations using a harness-pulley system	28,500 N/m	950 N/(m/s)	0.32[b] random perturbations
McMahon and Greene (1979)	Running	73–117 kN/m	2658–3365 N/(m/s)[a]	0.55 track stiffness
Cavagna (1970)	Bouncing with knees locked and ankles plantar-flexed	34–41 kN/m	570 N/(m/s)	0.18[b] log decrement
Proposed model (Eq. 2)	Walking	Mean = 7 kN/m (range = 2–20 k N/m)	Mean = 800 N/m/s (range = 400–1200 N/m/s)	Mean = 0.577 (range = 0.2–1) sinusoidal analysis and system identification

[a]Values extracted from the values of *k* and ζ assuming an 80 kg subject.
[b]Values extracted from the values of *k* and *B*.

5 ADVANCES IN COMMERCIALLY AVAILABLE LOWER-LIMB PROSTHETICS

5.1 Advances in Shock Absorption Prosthetic Legs

Shock absorption in amputee walking is an important component. Leeuwen et al. (1990) recognized that the absence of the necessary shock absorption in transtibial prostheses might cause proximal joint disease. They claimed that the shock transmitted during prosthetic gait should not be different than the shock transmitted during normal gait. The effectiveness of the cushioned heel as a shock absorber led to its continued use in most prosthetic foot designs (Perry et al., 1997). Edelstein (1988) suggested that all commercially available prosthetic feet provide some degree of shock absorption. The solid ankle cushion heel (SACH) foot has a cushioned heel, which is compressed during heel strike and provides shock absorption, simulating the normal plantar flexion movement of the foot during early stance (Lehmann et al., 1993a,b). Pitkin (1995) has investigated a rolling-joint prosthetic foot/ankle mechanism, which is claimed to incorporate shock absorption, balance, and dorsiflexion functions.

Davies and Holcomb (2001) found significant differences in the heel strike acceleration and heel strike transient amplitude between the prosthetic and the sound side. Poor attenuation in one knee leads to changes in the walking pattern. Amputees have been shown (nonconclusively) to have greater incidence of osteoarthritis than nonamputees; those with transfemoral amputations are three times more likely to exhibit this condition at the hip than a transtibial amputee (Kulkarni et al., 1998). Gitter et al. (1991) compared the gaits of five able-bodied subjects to the gaits of five unilateral transtibial amputees walking with three different feet. They found that regardless of prosthetic foot type, there was a loss of the shock absorption function of the prosthetic-side knee during loading response. Wirta et al. (1991) compared five ankle-foot devices, and found that transtibial amputees preferred walking with those that developed less shock and had greater damping at heel contact. Van Jaarsveld et al. (1990a,b) supported the idea that the reduction of peak accelerations during heel strike was an important aspect of the prosthesis functionality in transtibial amputees. Lehmann et al. (1993a,b) assumed that the shock-absorbing characteristics of prosthetic foot designs are a measure of comfort during gait, and found that the SACH foot attenuated the higher frequency components of acceleration more than the Seattle ankle/lite foot. Snyder et al. (1995) observed excessive stance-phase knee flexion angles in the sound limb of transtibial amputees during gait, which was speculated to be a compensatory mechanism for dynamic deficiencies of the prosthetic side. This may have increased the energy expenditure of walking because of the increased muscular effort by the knee.

Mooney et al. (1995) examined the differences in the ground reaction forces and moments of force during gait between the sound and residual limbs of a transtibial amputee using the Flex Foot's *Re*-Flex vertical shock pylon (VSP) as designed, and with the shock-absorber immobilized; they observed minimal differences between the two testing conditions.

One notable advancement in the area of the shock absorbing legs is the J-leg. It is produced in Canada in very small quantities but has received some positive feedback from the transfemoral (TF) amputees who have tried it. Basically, it includes a spring in the shank, along with a standard locking knee. The knee is locked in extension during the whole gait cycle. When the person wants to sit down, he/she manually unlocks the knee. The floor clearance is facilitated by the design of the foot, which can be considered an end-point "peg-leg." Also, the end-point foot device freely rotates 360 degrees in the transverse plane, which facilitates circumduction of the leg. The spring stiffness is constant, but there different springs are available depending on the body weight. This is an inexpensive leg compared with the new computerized knees of the market. Thus, it might be a good choice for developing economies. However, there is lack of scientific literature on this product. The disadvantage of this leg might be that its appearance is not very cosmetic; but perhaps function is more important than cosmesis.

At the old Leg Laboratory and now Biomechatronics Group of Medial Lab of Massachusetts Institute of Technology, Hugh Herr (2006) has developed an auto-adaptive knee prosthesis for transfemoral amputees, the Rheo-Knee marketed by Össur (Fig. 9). External knee prostheses should move naturally at all locomotory speeds and should perform equally well for all amputees. Using state-of-the-art prosthetic knee technology, a prosthetist must preprogram knee damping values until a knee is comfortable and safe to use. The knee prosthesis should automatically adapt to the amputee without preprogrammed information of any kind from either amputee or prosthetist. With this technology, knee damping is modulated about a single rotary axis using a combination of magnetorheological and frictional effects, and only local sensing of axial force, sagittal plane torque, and knee position are used as control inputs. Early stance damping is automatically adjusted by the controller, using sensory information measured when a patient first walks on

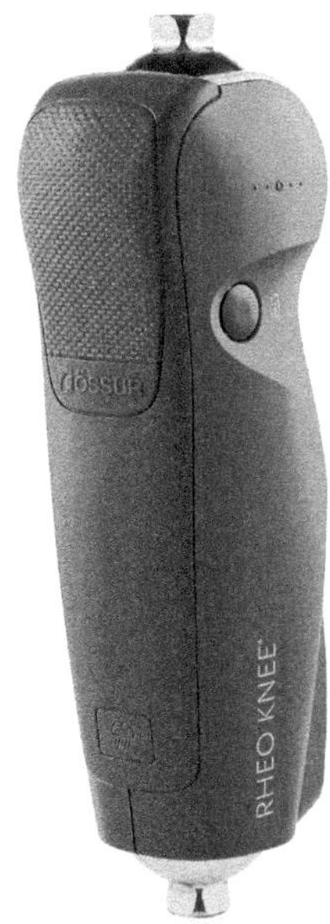

Fig. 9 The Össur RHEO KNEE. *(From www.ossur.com.)*

the prosthesis. With measurements of foot contact time, the controller also estimates forward speed and modulates swing-phase flexion and extension damping profiles to achieve dynamic cosmesis throughout each walking swing phase. The adaptation scheme successfully controls early stance resistance, swing-phase peak flexion angle and extension damping, suggesting that local sensing and computation are all that is required for an amputee to walk in a safe, comfortable, and smooth manner.

The C-Leg 4 (Fig. 10) represents an evolution of the first microprocessor-controlled hydraulic knee with swing and stance-phase control. This innovative knee joint features onboard sensor technology that reads and adapts to the individual's every move. Angles and moments are measured in real-time 50 times per second. Amputees can move on flat terrain at different gait speeds with confidence. Moreover, thanks to the hydraulic stance control feature, which is basically a prevention of buckling, it is easier to tackle slopes, stairs, and other uneven surfaces. The C-leg 4 has also some stance-phase knee flexion.

A prosthetic leg named high intelligence prosthesis (HIP) developed for above-knee amputees by Biedermann Motech (Schwennigen, Germany) uses an array of sensors in the artificial knee component to detect force

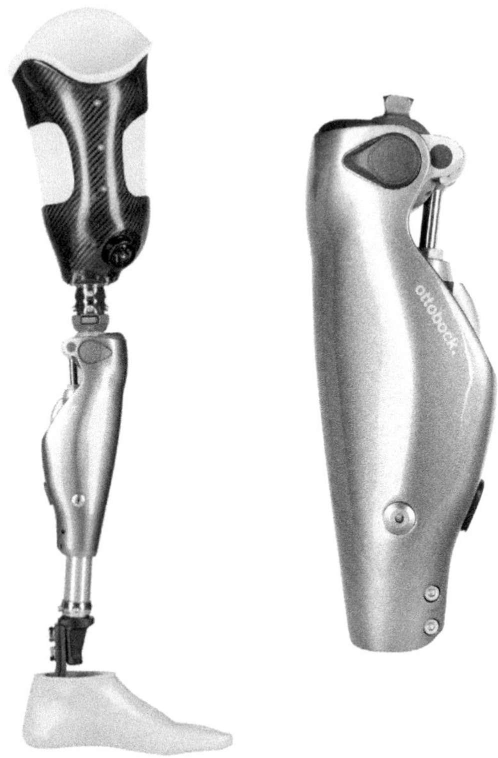

Fig. 10 The C-Leg 4 from Otto-Bock, a microprocessor-controlled knee joint, provides real-time swing phase control and stumble recovery support. *(From http://www.ottobock.com. © by Otto Bock.)*

and moment exerted on the prosthesis and the angular position of the knee joint (Fig. 11). The mechanism also includes a damping device filled with a magnetorheological fluid that can adjust rapidly to changes in external forces. Input from the sensors and software algorithms controls the damping qualities of the device. The fluid, which was developed by Lord Corp. (Cary, North Carolina), (http://www.mrfluid.com), is designed to change consistency—from a fluid to a near-solid state—in response to the strength of a magnetic field applied to it. According to the company, the time required to react to changing forces is 20 times faster than systems that use passive fluids. According to the firm, such results match more closely human neural response times than hydraulic mechanisms with motor-controlled valve systems. The unique characteristics of Lord Rheonetic MR fluid dampers—high controllability, millisecond response time, and velocity-independent force—make this product possible.

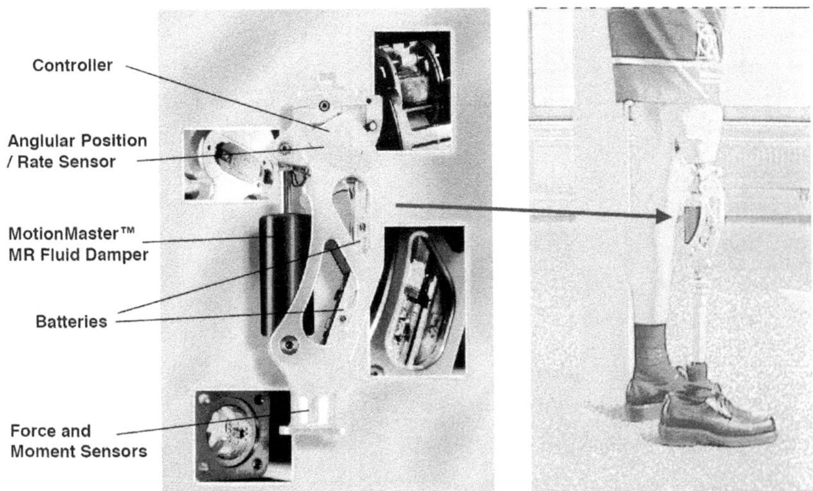

Fig. 11 The high intelligence prosthesis (HIP) developed by Motech is using a magnetorheological fluid damper. *(From David Carlson, J., Matthis, W., Toscano, J.R., 2001. Smart prosthetics based on magnetorheological fluids. In: Proceedings of SPIE.)*

5.2 Knee Shock Absorbers

There is a new class of prosthetic knees that provide simulated stance-phase knee flexion. These include the Otto Bock 3R60 ergonomically balanced stride EBS-PRO-Knee, the Total Knee 2100, and the Endolite ESK+ Knee (Fig. 12). In these, there is a polymer spring with some inherent damping, which provides knee flexion resistance during the early phase of the stance phase (which is obtained through the geometrical setup and the ground reaction force vector and the moments). These knee units appear to flex beginning at heel contact while load is being transferred to the prosthetic limb. The knee extends by the time midstance is reached, similar to physiological knee motion in normal walking. While providing shock absorption like the VSPs, these devices may have better simulated physiological function because they appear to have a period of activation during the gait cycle similar to the normal physiological movement that they are designed to replace. The 3R60 technology of the EBS-PRO-Knee allows up to 15 degrees of cushioned knee flexion and a polymeric spring progressively cushions the increase in loading that occurs as weight is transferred onto the prosthesis. This improvement in knee biomechanics may result in increased comfort during weight bearing and walking. Two hydraulic cylinders–one to influence stance flexion, the other to control the swing phase–offer a more natural gait and a high degree of stability especially noticeable on uneven terrain (Blumentritt et al., 1997).

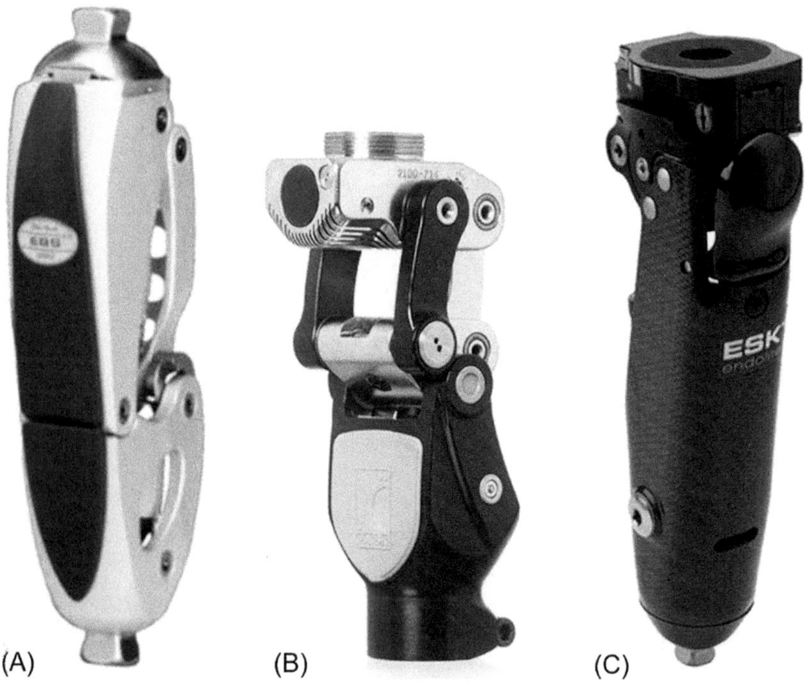

(A) (B) (C)

Fig. 12 Shock absorbing prosthetic knees. (A) EBS-PRO-Knee from Otto Bock using 3R60 technology (ergonomically balanced stride) provides cushioning during stance-phase knee flexion via a small elastomeric bumper, (B) the Total knee 2100 from Össur simulates a stance phase knee flexion, and (C) the Endolite ESK+ with weight activated stance control and Stanceflex provides cushioning at heel strike via a small bumper. *((A) From http://www.ottobock.com. © by Otto Bock; (B) From https://www.ossur.com/prosthetic-solutions/products/dynamic-solutions/total-knee-2100; (C) https://www.endolite.com/products/endolite-esk-with-pspc.)*

5.3 Shock Absorbing Pylons

Miller (1994) and Miller and Childress (1997) analyzed the Flex Foot's *Re*-Flex VSP (Fig. 13). This pylon has a spring-loaded shock absorber, which adds compliance to the prosthesis during activities in which the plantar flexors normally aid, such as going up and down stairs and running.

Gard and Childress (1998) have also investigated the mechanical characteristics of shock absorbing pylons. Using the foot-loading apparatus (FLA) static and dynamic testing was performed on three commercially available shock absorbing pylons: the Flex Foot *Re*-Flex VSP (Fig. 13), Ohio Willow Wood's Stratus Impact Reducing Pylon, and Seattle Limb System's AirStance Pylon.

Fig. 13 The Össur Re-Flex Shock consists of a composite spring in front which provides optimal shock absorption leading to reduced impact on the body. *(From www.ossur.com.)*

The static testing involved slowly loading and unloading the pylon while measuring the linear displacement of the mechanism and the applied force. The stiffness was derived from the force-displacement curves of the data. While the stiffness for the Re-Flex was found to be relatively linear, the stiffness for the AirStance and Stratus were found to be nonlinear. All three pylons showed hysteresis in their force-displacement curves, indicative of energy loss during the load-unload cycle. Dynamic testing involved measuring the response of the pylons to a step of load and unload applied force. The Re-Flex and the Stratus behaved as underdamped second-order systems with overshoot and some small oscillations, whereas the AirStance was overdamped for light loads and underdamped for the heavy ones. The total displacement of the AirStance to the step inputs was one order of magnitude less than that of the other two units, attributable to the much greater stiffness of the AirStance pylon.

Another new pylon is the Endolite TTPRO (Telescopic-Torsion) Pylon (Fig. 14). Perhaps this design is superior to the other pylons because it utilizes helical springs, instead of the elastomer ones used by the Ohio Willow Wood Stratus or the air springs in the Seattle AirStance. In addition, the Endolite unit is considerably less expensive than the Flex Foot's *Re*-Flex VSP. We suspect that the Endolite unit has properties similar to the Re-Flex VSP. The TTPRO pylon primarily behaves as a spring, with very little damping, so shock forces at heel contact during gait are attenuated. Because damping is small, the energy associated with this attenuation is not lost but is stored in the mechanism.

The Ohio Willow's Pathfinder (Fig. 15) is a foot system that integrates both a polycentric ankle and a shock absorber. The design incorporates a lightweight, adjustable pneumatic heel spring in parallel with the toe springs rather than in series as with most shock absorber systems. Therefore, "the

Fig. 14 TTPRO shock absorber from Endolite. *(From https://www.endolite.com/products/ttpro.)*

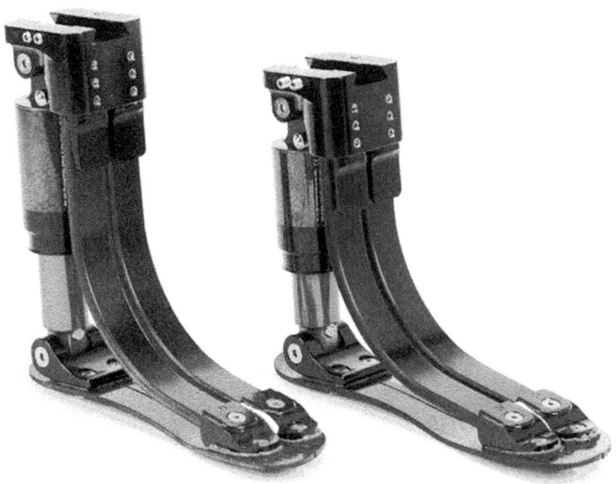

Fig. 15 The Ohio Willow Pathfinder foot system. *(Courtesy of WillowWood.)*

shock absorption action is limited to heel loading and there is no limb shortening problem to address" (Gard, 2002).

5.4 Prosthetic Feet

Versluys et al. (2009) classify the recent timeline of prosthetic feet into three categories: (a) conventional feet, (b) energy storing and returning (ESR) feet, and (c) bionic feet.

The desire of transtibial amputees to participate in sports led to the development of the early ESR feet, which stored energy during early stance by loading a spring with the body weight and then releasing a portion during late stance. The energy lost in the system in the form of friction is high and is dissipated as heat and sound. Early ESR feet include the Seattle foot, the Dynamic Plus foot, the C-Walk, and the Carbon Copy foot.

Advanced ESR feet have better properties than early ESR feet and are shown in Fig. 16.

Hansen et al. (2004) have shown that there is net power generation by the ankle at speeds higher than 1.2 m/s. The need for power generation has led to the design of the so-called "bionic feet," which are active pneumatically or electrically driven feet with objective to generate the above-mentioned net power at the ankle during gait. Different bionic feet have been designed (Fig. 17). This can enable amputees to walk faster and also ascend/descend stairs and walk on slopes.

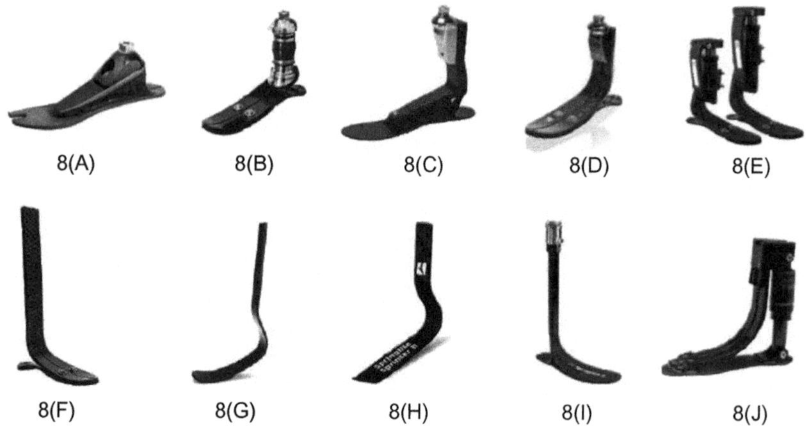

Fig. 16 Advanced ESR feet. (A) Flex-foot Axia. (B) LP-Ceterus. (C) Talux foot. (D) VariFlex. (E) Re-Flex VSP. (F) Modular III. (G) Flex-Sprint. (H) Sprinter. (I) Springlite foot. (J) Pathfinder. *(From Versluys, R., Beyl, P., Van Damme, M., Desomer, A., Van Ham, R., Lefeber, D., 2009. Prosthetic feet: state-of-the-art review and the importance of mimicking human ankle-foot biomechanics. Disabil. Rehabil. Assist. Technol. 4(2), 65–75. https://doi.org/10.1080/17483100802715092.)*

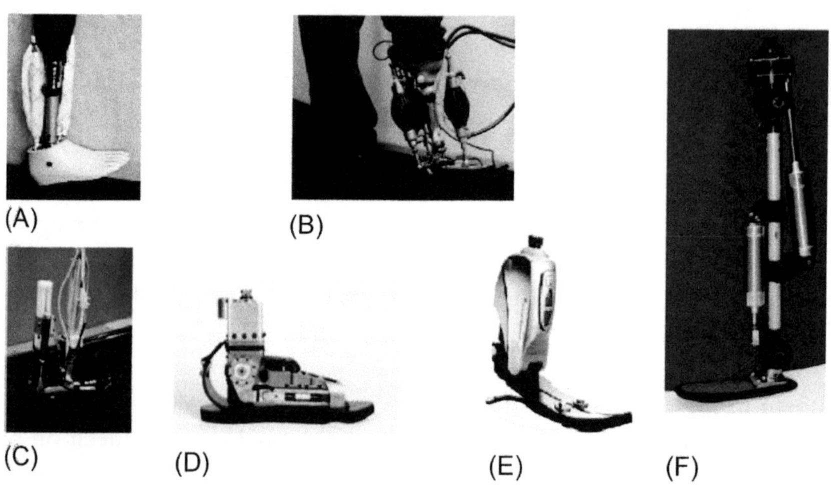

Fig. 17 Bionic feet. (A) TT prosthesis powered by McKibben artificial muscles. (B) TT prosthesis powered by PPAMs. (C) SPARKy. (D) Electrically driven foot of MIT. (E) Proprio foot. (F) Powered transfemoral prosthesis of Vanderbilt University. *(From Versluys, R., Beyl, P., Van Damme, M., Desomer, A., Van Ham, R., Lefeber, D., 2009. Prosthetic feet: state-of-the-art review and the importance of mimicking human ankle-foot biomechanics. Disabil. Rehabil. Assist. Technol. 4(2), 65–75. https://doi.org/10.1080/17483100802715092.)*

6 STATE-OF-THE-ART RESEARCH THREADS AND ENABLING TRENDS

It is interesting to identify current research threads, which are going to enable amputees and shape up novel prostheses in the market.

6.1 Osseointegration

Professor Per-Ingvar Brånemark discovered in the 1950s that bone can integrate and coexist "peacefully" with titanium components. He defined osseointegration as "A direct structural and functional connection between ordered living bone and the surface of the load-covering implant" (Branemark et al., 1969). Osseointegration has been proposed as an alternative technique for prostheses since the 1980s, after the success of dental implants (Childress, 1997, 1998). The major problem of this technique has been the risk of infection at the skin of the implant area (Childress, 1997, 1998). There has been a lot of effort in the past years to optimize implants design, the process, and the rehabilitation protocol in order to minimize the risk of infection. In 1999, a treatment protocol called osseointegrated prostheses for the rehabilitation of amputees (OPRA) was established (Li and Branemark, 2017). The first bilateral transfemoral fitted with osseointegrated prostheses is shown in Fig. 18. Although there is more than 20 years of experience in transfemoral osseointegration procedures, the orthopedic community still is skeptical of this technique (Frossard et al., 2013; Nebergall et al., 2012; Vertriest et al., 2015, 2017).

The biggest benefit that osseointegration provides as a procedure and methodology, other than that it eliminates the need of a socket and provides wider range of motion, is that there is direct link between the bone, muscles, tendons, receptors, and the prosthesis. This direct link and engagement provides osseoperception, the ability of the amputee to "feel" where his or her prosthesis is without seeing it. The information comes integrative to the amputee by using the remaining afferent (sensory) pathways which are now integrated with the prosthesis and give us an extended physiological proprioception (EPP) type of control, which even for lower-limb prostheses is beneficial (e.g., amputee "feels" when foot touches the ground). This leads to increased controllability of the prosthesis and improved balance for the bilateral amputees. It is true that ossoeintegration might be the only feasible option/hope for high-level bilateral transfemoral amputees to ambulate.

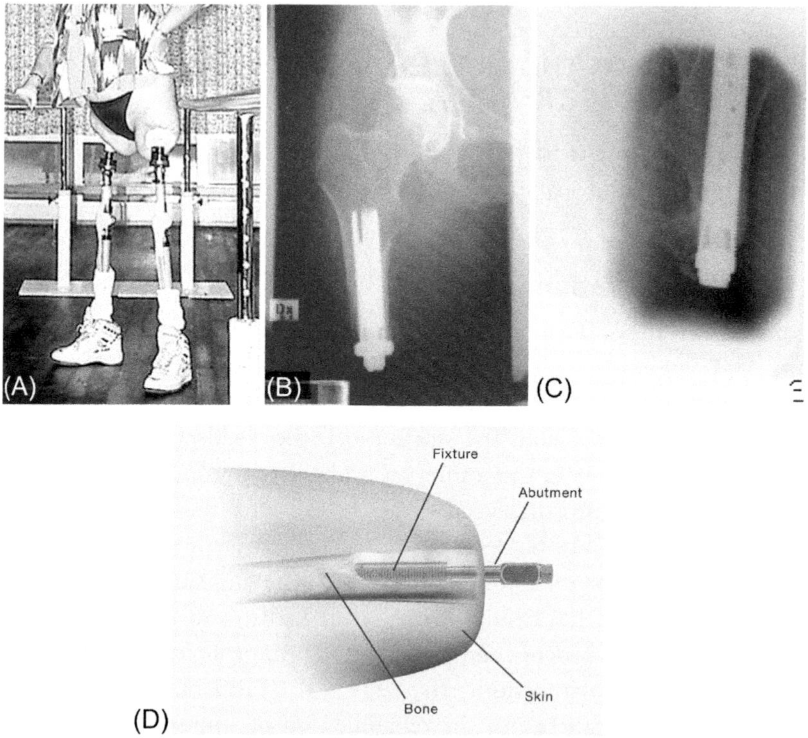

Fig. 18 The first extremity osseointegration patient who was operated in 1990. The patient can stand up and walk with crutches (A). At that time, the implants were of modular design with a distal collar (B and C). (D) The basic implant design of the OPRA implant system. Three major components, the fixture, the abutment, and the abutment screw are used. *(From Li, Y., Branemark, R., 2017. Osseointegrated prostheses for rehabilitation following amputation: the pioneering Swedish model. Unfallchirurg, 120(4), 285–292. https://doi.org/10.1007/s00113-017-0331-4.)*

Integrum, the company that is commercializing the osseointegration technology OPRA, was given humanitarian approval from the FDA to perform 18 clinical trials (Li, 2016) for lower-limb amputees in the United States, in 2015.

A metadata analysis (Hagberg et al., 2014) has shown that the risk of superficial infections is acceptable, and can be treated with the use of oral antibiotics. Results of the first 18 patients following the OPRA protocol were promising, with a 94% success rate at the 2-year follow-up and good quality of life (Hagberg et al., 2008).

"Endosteal bone resorption could be an alarming radiological sign regarding the fixation and future survival of the fixture especially when combined with pain at loading" meaning that the survival of the fixture could be predicted by bone resorption with an X-ray (Lenneras et al., 2017).

All the potential benefits of osseointegration do not come with issues and problems to solve. The biggest problem of this technique is its long-lasting battle with bacteria at the skin interface and its unknown long-term impact on the quality of the bone fixture (Lenneras et al., 2017). Therefore, long-term studies are needed. Radiologically found endosteal bone resorption accompanied with pain at loading might be associated with potential weakness of the bone fixture (Lenneras et al., 2017). Different osseointegration research groups are taking engineering variants of the implant designs and materials in order to achieve a stable mechanical interface between the bone and the implant.

6.2 Inexpensive/Easy and Automated Fabrication

Effective prosthetic feet can be easily and cheaply fabricated as shown by Adamczyk et al. (2006), Adamczyk and Kuo (2013), Hansen and Childress (2000), and Sam et al. (2000, 2004). The key is to know the existing science behind the design of prosthetic feet as dictated by the rollover theory of walking (see Section 4) and use inexpensive fabrication methods and materials for the developing countries. This led to the design of the NUPRL foot. Another modular inexpensively fabricated leg that was intended for the developing countries was the Center of International Rehabilitation (CIR) leg.

The CAD/CAM has been around for at least three decades for prosthetics. Its biggest value is that a scan of the residual limb can be taken by different technologies (e.g., laser), then the CAD model of the personalized and pressure-relieved socket can be generated by the CAD software, and finally the socket is fabricated in minutes by the CAM techniques. Commercially available solutions exist in the market (CAD/CAM SYSTEMS, 2018).

6.3 Targeted Muscle Reinnervation

The targeted muscle reinnervation (TMR) has also been performed for persons with transfemoral amputations (Kuiken et al., 2017a). As reported by Hargrove et al. (2013), a TMR procedure was performed on a 31-year old during knee disarticulation amputation due to a motorcycle accident. The nerve transfer is shown in Fig. 19. The sciatic nerve was split into its tibial and common peroneal branches. The tibial nerve branch was sewn over the motor area of the semitendinosus, and the peroneal nerve branch

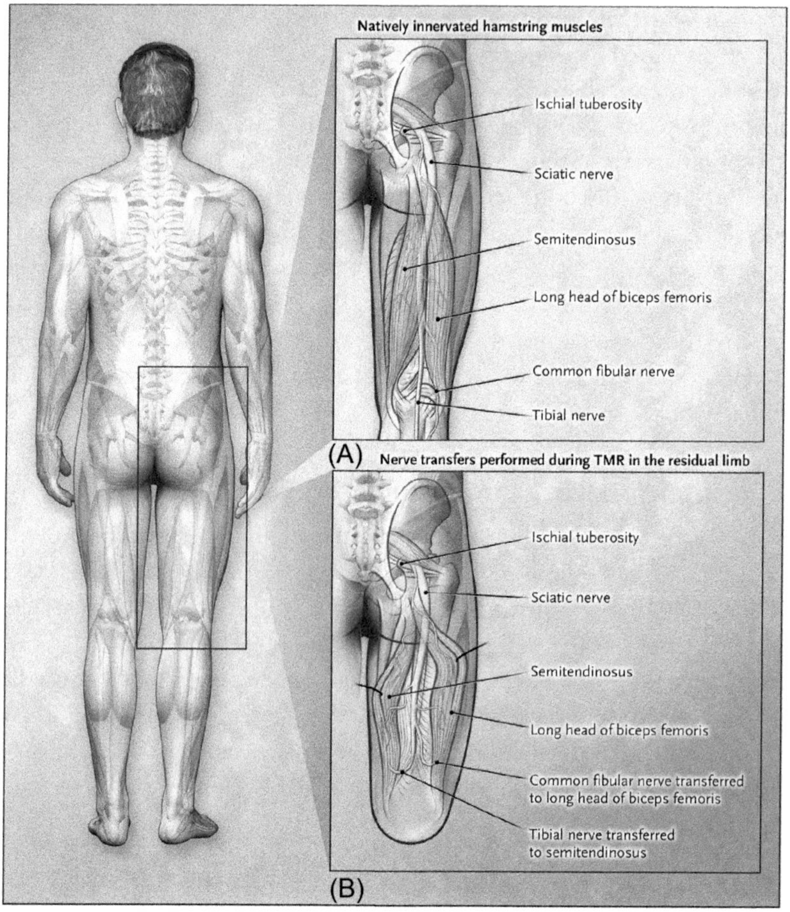

Fig. 19 Natively innervated and surgically reinnervated residual thigh muscles. Posterior views of the anatomy of the upper thigh show natively innervated hamstring muscles (Panel A) and nerve transfers performed during targeted muscle reinnervation (TMR) surgery (panel B) in the residual limb. *(From Hargrove, L.J., Simon, A.M., Young, A.J., Lipschutz, R.D., Finucane, S.B., Smith, D.G., Kuiken, T.A., 2013. Robotic leg control with EMG decoding in an amputee with nerve transfers. N. Engl. J. Med. 369(13), 1237–1242. https://doi.org/10.1056/NEJMoa1300126.)*

was sewn over the motor area of the long head of the biceps femoris. After a few months, the TMR was successful and the amputee could voluntarily activate the motor areas that were reinnervated, leaving extra control sites for prosthesis control, which were used together with prosthesis sensors to a pattern recognition algorithm for controlling different states of prosthesis (walking, ascending and descending the stairs, ramps, and reposition the leg while seated).

Kuiken et al. (2017b) have proposed a novel intramedullary residual limb-lengthening device which is less invasive than Ilizarov's apparatus lengthening technique and which could be used for the management of the lower-limb amputee.

6.4 Micromechatronic Devices

Windrich et al. (2016) summarized all the active lower-limb prostheses under research development (Table 2). The active nature of these prostheses is at different levels as the column "Type" of Table 2 shows. For above the knee (A/K) amputees, the emphasis is on actuators (hydraulic, magnetorheological, and electromechanical) that provide the right "resistance" or impedance at the knee joint during the different phases of the gait cycle (i.e., stance-phase knee flexion and appropriate swing leg resistance based on step frequency during the swing phase). The variable impedance scheme using magnetorheological liquid developed at the MIT and now marketed via Ossur is an example of this category (see Fig. 9).

For the below the knee (B/K) amputees, there are different areas of research as far as actuation systems is concerned: (a) series elastic actuators (e.g., BIOM ankle, Vanderbilt transtibial prosthesis) of variable impedance, (b) higher-level controllers which enable seamless transitions across different states of ambulation (walking on slopes, sitting, etc.), and (c) pneumatic artificial muscles (PAM) application on actuating a BK prosthesis at the Vrije Universiteit in Brussels, Belgium (Versluys et al., 2008).

There is research done on coordination of the prosthetic knee and prosthetic ankle for AK amputees, like the Vanderbilt transfemoral prosthesis (Sup et al., 2009b) (described in Section 6.4.2) and the Cyberleg α-prototype of the Vrije Universiteit, Brussels, Belgium (Flynn et al., 2013; Geeroms et al., 2013).

There is also newer development at the Northwestern University concerning the higher-level coordination controller and synergies of the tasks, described in Section 6.5.

We are presenting below, representative work from the abovementioned newer developments.

It is important to note that there are three startups at the domain of active prostheses: SpringActive, BionX (now OttoBock), and Freedom Innovations of the Netherlands.

As noted in Windrich et al. (2016), there are ambiguous reports on results of active prostheses. Some show increase of walking speed and decrease of metabolic energy (Herr and Grabowski, 2012), but reality is that is new technology and further studies have to be performed in order to quantify the effect.

Table 2 Overview of Different Prostheses

Type	Name of Prosthesis, Institute, Country	Year
A/K	*Agonist-antagonist active knee prosthesis,* Massachusetts Institute of Technology, United States	2008
A/K	University of Sakarya, Adapazari, Turkey	2008
A/K	*Waterloo Active Prosthetic Knee,* University of Waterloo, Canada	2008
A/K	Hebei University of Technology, China	2010
A/K	ETH Zurich, Switzerland	2011
A/K	The University of Alabama, United States	2011
A/K	Department of Mechanical and Aeronautical Engineering, United States	2012
A/K	University of Rhode Island, United States	2012
B/K	*Bionic ankle-foot prosthesis,* Massachusetts Institute of Technology, United States	2006
B/K	*SPARKy,* Arizona State University, United States	2008
B/K	*IPAM (intelligent Prosthesis using Artificial Muscles),* Vrije Universiteit Brussel, Belgium	2008
B/K	Vrije Universiteit Brussel, Belgium	2009
B/K	*PANTOE 1,* Peking University, China	2010
B/K	Marquette University, Milwaukee, United States	2010
B/K	Kanazawa Institute of Technology, Ishikawa, Japan	2011
B/K	*AMP-foot 2.0,* Vrije Universiteit Brussel, Belgium	2012
B/K	*Vanderbilt Transtibial Prosthesis,* Vanderbilt University, United States	2013
A/K + B/K	*Vanderbilt Transfemoral Prosthesis,* Vanderbilt University, United States	2009
A/K + B/K	University of Brasília, Brazil	2009
A/K + B/K	*SmartLeg,* University of Mostar, Bosnia and Herzegovina	2011
A/K + B/K	*Cyberleg alpha,* Vrije Universiteit Brussel, Belgium	2013

The prosthesis are classified as above-knee (A/K), below-knee (B/K) and combined knee-and-ankle prosthesis (A/K + B/K).
From Windrich, M., Grimmer, M., Christ, O., Rinderknecht, S., Beckerle, P. (2016). Active lower limb prosthetics: a systematic review of design issues and solutions. Biomed. Eng. Online, 15(Suppl. 3), 140. https://doi.org/10.1186/s12938-016-0284-9.

6.4.1 BIOM Ankle MIT

Hugh Herr from the MIT Biomechatronics Group of Media Lab has developed a mechatronic prosthetic ankle (BIOM) that enables amputees to walk, run, dance, and climb (Au et al., 2007, 2008; Eilenberg et al., 2010; Rouse et al., 2015), Fig. 20. BIOM was using a series elastic actuator and magnetorheological fluid technology for adjustable damping. This

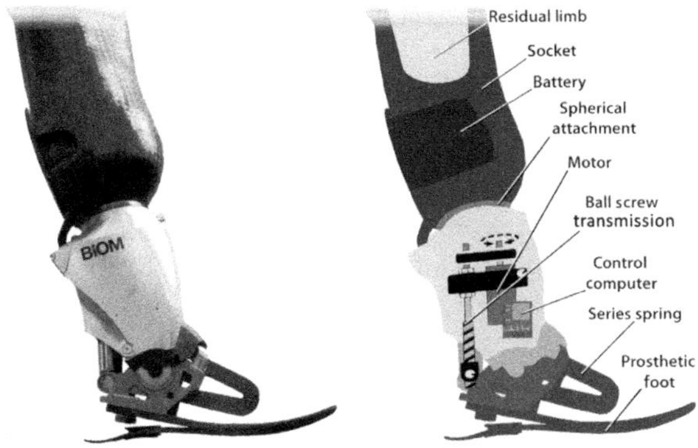

Fig. 20 Schematic of bionic dancing prosthesis. Bionic ankle prosthesis shown (left) with major components highlighted (right). Note the location of the battery in the distal prosthetic socket. *(From Rouse, E.J., Villagaray-Carski, N.C., Emerson, R.W., Herr, H.M. (2015). Design and testing of a bionic dancing prosthesis. PLoS One, 10(8), e0135148. https://doi.org/10.1371/journal.pone.0135148.)*

biomechatronic ankle was patented under US2007/0043449 A1 (Herr et al., 2007) (Fig. 20).

BIOM was sold by the Iwalk, Inc. which later became BionX Medical Technologies, Inc. The Rheo knee was also sold by the Iwalk, Inc. and licensed to Össur. BionX Medical Technologies, Inc. was acquired by Ottobock in March 2017 for \$77 M.

6.4.2 New Active Leg (Vanderbilt)

There has been a recent trend to develop active lower-limb prostheses, especially ankles that will store and/or dissipate energy, and also generate net power during a gait cycle. Examples of these devices are the new leg developed by Varol et al. (2010) and Spanias et al. (2018, 2016b). As stated in Varol et al. (2010), the present transfemoral prostheses can store or dissipate energy but cannot generate net power over a gait cycle. Transfemoral amputees could expend up to 60% more metabolic energy than able-bodied ambulators (Waters et al., 1976). It has also been stated by Hansen et al. (2004) that there is net power generation by the ankle at speeds higher than 1.2 m/s. It is, therefore, justified to develop net power generation prosthesis, since it can improve the metabolic cost of amputee ambulation especially at higher walking speeds. Fig. 21 shows the Vanderbilt prosthesis, which

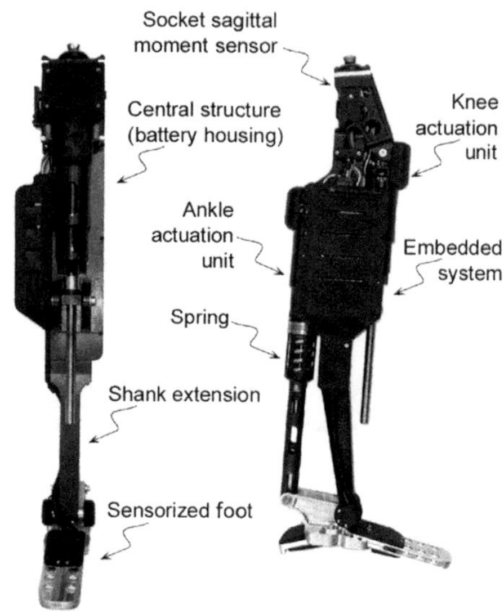

Fig. 21 Self-contained powered knee and ankle transfemoral prosthesis. *(From Varol, H.A., Sup, F., Goldfarb, M., 2010. Multiclass real-time intent recognition of a powered lower limb prosthesis. IEEE Trans. Biomed. Eng. 57(3), 542–551. https://doi.org/10.1109/TBME.2009.2034734.)*

provides net energy during walking, and also identifies the intended uses that the amputee wants to perform. The controller state chart and classification results of the intended states are shown in Figs. 22 and 23 respectively.

6.5 Artificial Intelligence—Pattern Recognition—Machine Learning—Synergies

A new research thread has been developed aiming to use pattern recognition and machine learning on identifying the intentions of the amputee (i.e., ascend or descend stairs, walking). This is in the right direction with the seamless need of transition between intended states that we have stated in Sections 2.1.1 and 2.1.3. An example of this research is the work done by Simon et al. (2016, 2017), Spanias et al. (2014, 2016a,b, 2017, 2018), and Woodward et al. (2016). Using a third generation powered knee-ankle prosthesis designed by the Vanderbilt University (Lawson et al., 2010; Sup et al., 2009a), inputs from different sensors (mechanical sensor data including axial load, ankle and knee angles, velocities, and EMGs) were used to trigger transitions between the following states: stand, walk on level ground, ascend/descend a ramp, and ascend/descend stairs. An overview of the adaptive algorithm is shown in Fig. 24.

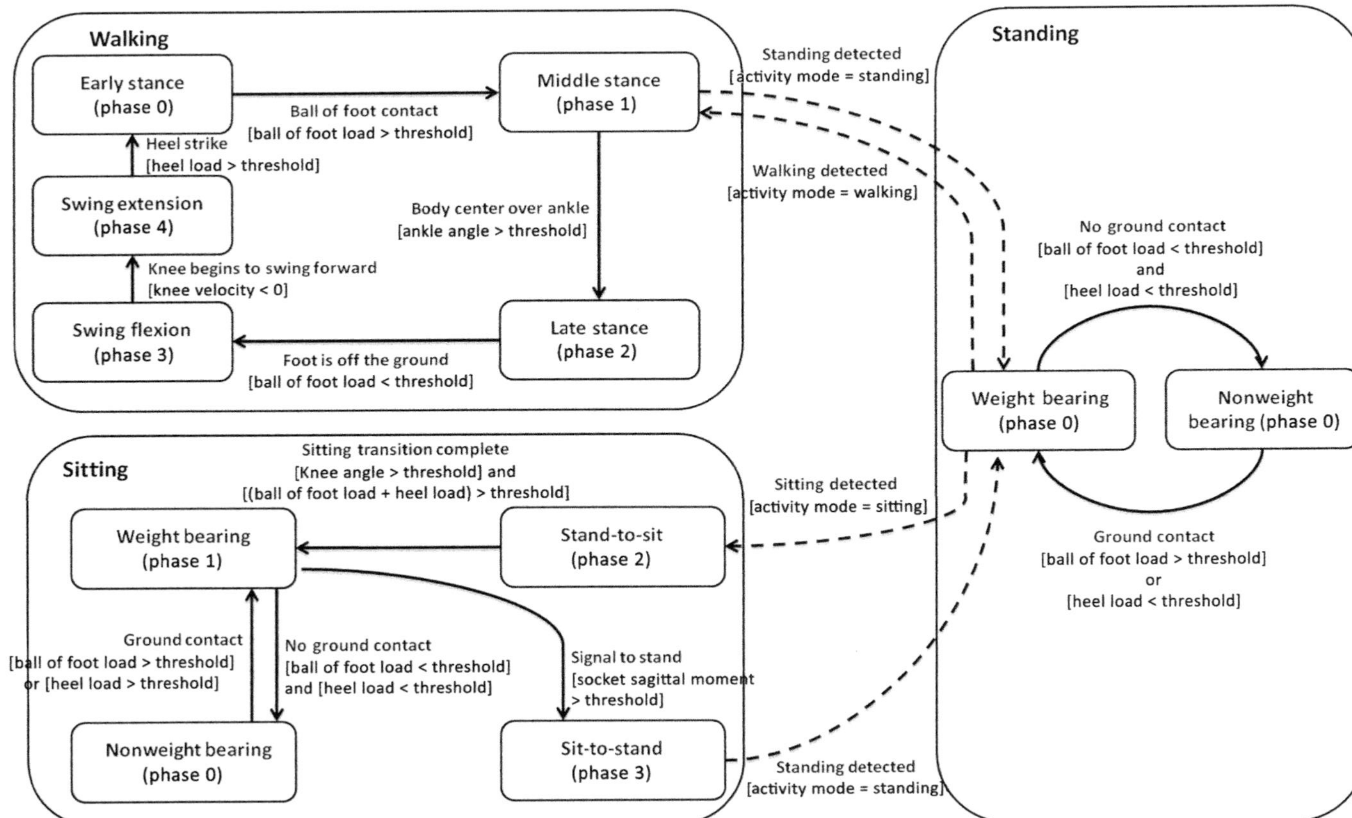

Fig. 22 State chart depicting the phase transitions for standing, walking, and sitting modes. *(From Varol, H.A., Sup, F., Goldfarb, M., 2010. Multiclass real-time intent recognition of a powered lower limb prosthesis. IEEE Trans. Biomed. Eng. 57(3), 542–551. https://doi.org/10.1109/TBME.2009.2034734.)*

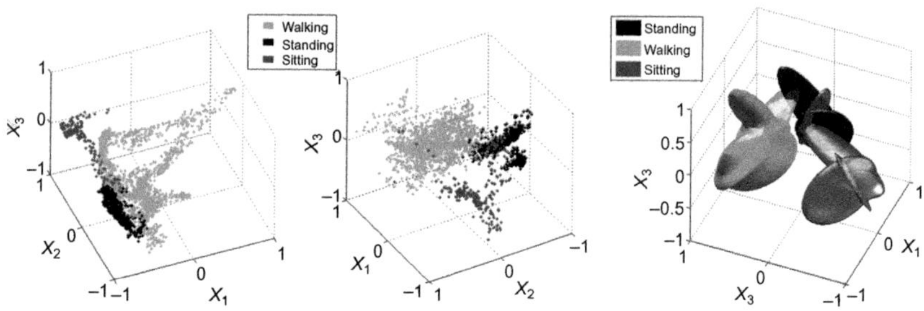

(A) (B)

Fig. 23 (A) PCA (left) and LDA (right) dimension reduced features extracted from 200 sample-long frames. (B) GMMs surface plots for standing, walking, and sitting showing the portions of the feature space, where probability density function is greater than 0.05, for 3D LDA reduced data. *(From Varol, H.A., Sup, F., Goldfarb, M., 2010. Multiclass real-time intent recognition of a powered lower limb prosthesis. IEEE Trans. Biomed. Eng. 57(3), 542–551. https://doi.org/10.1109/TBME.2009.2034734.)*

7 DISCUSSION/REALIGNMENT

As discussed in this chapter, basic walking especially for transtibial amputees can be achieved in a satisfactory degree by conventional prostheses. When we start seeking sports performance or ability to walk on slope, ascend and descend stairs, dancing, jumping, etc., then we are looking at more specialized and advanced prostheses.

Current cutting-edge technologies such as pattern recognition, TMR, osseointegration, and active prostheses are going to enable the unification all the necessary ambulatory tasks to be satisfied and executed seamlessly by a single prosthesis, while advancing performance. In particular, osseointegration can be an enabler technique for bilateral transfemoral amputees.

Further clinical studies are needed in order to quantify the effect of active prostheses on walking speed and metabolic energy.

Attention has to be paid to make sure needs and amputee voice of customer (VOC) are considered when investing in new research threads.

AUTHORS' CONTRIBUTIONS

GAB was responsible for the outline, the structure, and the content of the chapter. GAB wrote all sections. EGP reviewed the chapter.

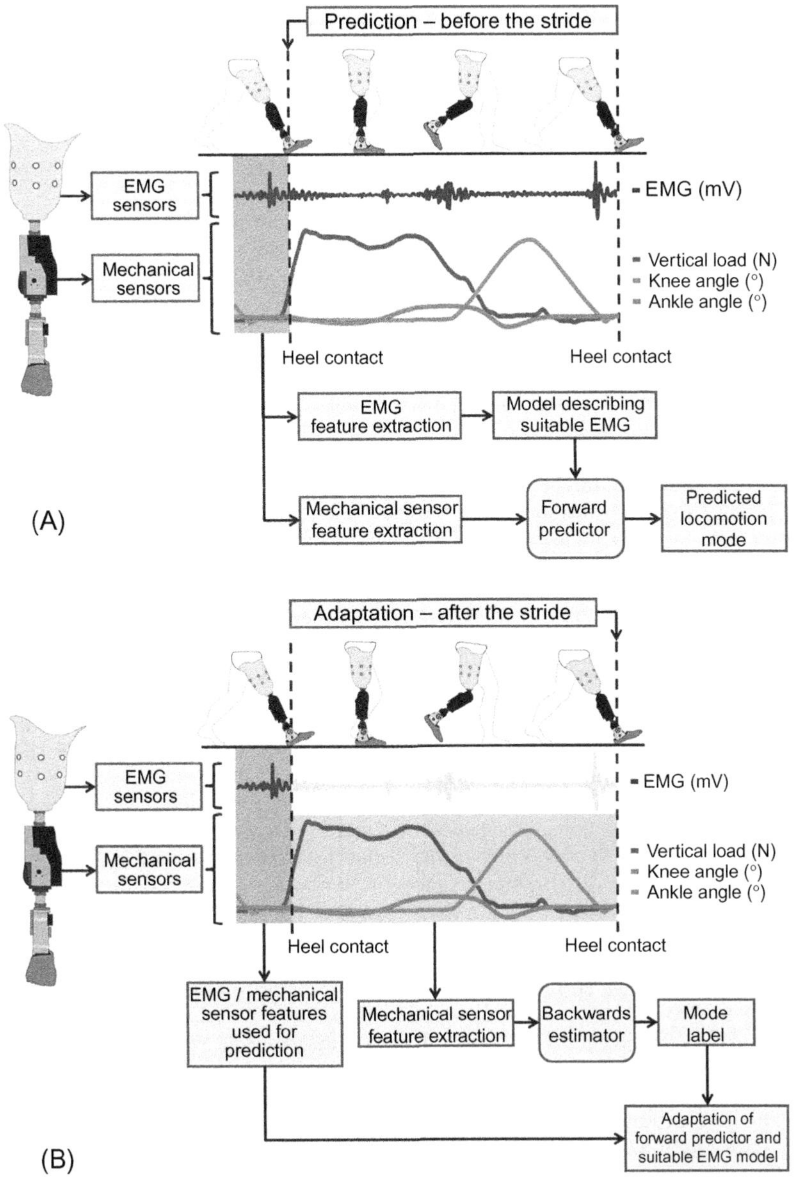

Fig. 24 Overview of the adaptive algorithm. Components include forward prediction (A) and backwards estimation (B). In forward prediction, features are extracted from EMG data and mechanical sensor data acquired before the stride (*red* window) and classified by the forward predictor, which then transitions the prosthesis to the predicted mode. The forward predictor determines whether to use EMG in making its prediction by comparing the EMG feature vector to a model describing suitable EMG data. In backwards estimation, we wait until the users complete their stride with the prosthesis and then classify the acquired mechanical sensor data (*blue* window) as one of the modes of the prosthesis. This provides a label for the pattern of data used for prediction, which is then used to adapt the parameters of the forward predictor and the model describing suitable EMG data. *(From Spanias, J.A., Simon, A.M., Finucane, S.B., Perreault, E.J., Hargrove, L.J., 2018. Online adaptive neural control of a robotic lower limb prosthesis. J. Neural Eng. 15(1), 016015. https://doi.org/10.1088/1741-2552/aa92a8.)*

References

Adamczyk, P.G., Kuo, A.D., 2013. Mechanical and energetic consequences of rolling foot shape in human walking. J. Exp. Biol. 216 (14), 2722–2731. https://doi.org/10.1242/jeb.082347.

Adamczyk, P.G., Collins, S.H., Kuo, A.D., 2006. The advantages of a rolling foot in human walking. J. Exp. Biol. 209 (20), 3953–3963. https://doi.org/10.1242/jeb.02455.

Alexander, R.M., 1992. A model of bipedal locomotion on compliant legs. Philos. Trans. R. Soc. Lond. B Biol. Sci. 338 (1284), 189–198.

Au, S.K., Herr, H., Weber, J., Martinez-Villalpando, E.C., 2007. Powered ankle-foot prosthesis for the improvement of amputee ambulation. Conf. Proc. IEEE Eng. Med. Biol. Soc. 2007, 3020–3026. https://doi.org/10.1109/IEMBS.2007.4352965.

Au, S., Berniker, M., Herr, H., 2008. Powered ankle-foot prosthesis to assist level-ground and stair-descent gaits. Neural Netw. 21 (4), 654–666. https://doi.org/10.1016/j.neunet.2008.03.006.

Bach, T.M., Chapman, A.E., Calvert, T.W., 1983. Mechanical resonance of the human body during voluntary oscillations about the ankle joint. J. Biomech. 16 (1), 85–90.

Bertos, G.A., 2006. Identification of the Mechanical Impedance of the Human Locomotor System and Quantification of Shock Absorption Characteristics: With Applications in Prosthetics (PhD). Northwestern University, Evanston, IL.

Bertos, G.A., Childress, D.S., Gard, S.A., 2005. The vertical mechanical impedance of the locomotor system during human walking with applications in rehabilitation. In: IEEE 9th International Conference on Rehabilitation Robotics. IEEE, New York, pp. 380–383.

Blickhan, R., Seyfarth, A., Geyer, H., Grimmer, S., Wagner, H., Gunther, M., 2007. Intelligence by mechanics. Philos. Trans. R. Soc. A Math. Phys. Eng. Sci. 365 (1850), 199–220. https://doi.org/10.1098/rsta.2006.1911.

Blumentritt, S., Scherer, H.W., Wellershaus, U., Michael, J.W., 1997. Design principles, biomechanical data and clinical experience with a polycentric knee offering controlled stance phase knee flexion: a preliminary report. J. Prosthet. Orthot. 9 (1), 18–24.

Branemark, P.I., Adell, R., Breine, U., Hansson, B.O., Lindstrom, J., Ohlsson, A., 1969. Intra-osseous anchorage of dental prostheses. I. Experimental studies. Scand. J. Plast. Reconstr. Surg. 3 (2), 81–100.

Bròggemann, G., Arampatzis, A., Emrich, F., Potthast, W., 2008. Biomechanics of double transtibial amputee sprinting using dedicated sprinting prostheses. Sports Technol. 1 (4–5), 220–227.

CAD/CAM SYSTEMS, 2018. Retrieved from: https://opedge.com/ProductDirectory/CADCAM_Systems.

Cappozzo, A., 1991. The mechanics of human walking. In: Patla, A. (Ed.), Adaptability of Human Gait Implications for the Control of Locomotion. Elsevier Science Publishers, North-Holland, pp. 167–186.

Cavagna, G., 1970. Elastic bounce of the body. J. Appl. Physiol. 29, 279–282.

Childress, D.S., 1997. The Interfaces Between Humans and Limb Replacement Components. Quintessence Publ Co Inc., Carol Stream, IL.

Childress, D.S., 1998. Control Strategy for Upper-Limb Prostheses. In: Chang, H.K., Zhang, Y.T. (Eds.), Proceedings of the 20th Annual International Conference of the IEEE Engineering in Medicine and Biology Society, vol. 20, Pts 1–6—Biomedical Engineering Towards the Year 2000 and Beyond, vol. 20. IEEE, New York, pp. 2273–2275.

Childress, D.S., 2002. Peak of the Vertical Ground Reaction Force is Proportional to Square of Walking Speed. personal communication.

Childress, D.S., Gard, S.A., 1999. In: Vertical Movement of the Trunk in Human Walking. Paper Presented at the XVIIth Congress of the International Society of Biomechanics (ISB), Calgary, Canada, August 8–13th.

Coleman, M., Ruina, A., 1998. An uncontrolled walking toy that cannot stand still. Phys. Rev. Lett. 80 (16), 3658–3661.

Davies, T.C., Holcomb, S., 2001. Measuring shock absorption of lower limb amputees. In: RESNA 2001, June 22–26, Reno, Nevada.

Edelstein, J.E., 1988. Prosthetic feet: state of the art. Phys. Ther. 68 (12), 1874–1881.

Eilenberg, M.F., Geyer, H., Herr, H., 2010. Control of a powered ankle-foot prosthesis based on a neuromuscular model. IEEE Trans. Neural Syst. Rehabil. Eng. 18 (2), 164–173. https://doi.org/10.1109/TNSRE.2009.2039620.

Flynn, L., Geeroms, J., Jimenez-Fabian, R., Vanderborght, B., Vitiello, N., Lefeber, D., 2013. Ankle-knee prosthesis with powered ankle and energy transfer development of the CYBERLEGs alpha-prototype. In: Neurotechnix: Proceedings of the International Congress on Neurotechnology, Electronics and Informatics, pp. 224–228. https://doi.org/10.5220/0004664702240228.

Frossard, L., Haggstrom, E., Hagberg, K., Branemark, R., 2013. Load applied on bone-anchored transfemoral prosthesis: characterization of a prosthesis-a pilot study. J. Rehabil. Res. Dev. 50 (5), 619–634.

Garcia, M., Ruina, A., Chatterjee, A., Coleman, M., 1998. The simplest walking model: stability, complexity, and scaling. ASME J. Biomech. Eng. 120 (2), 281–288.

Gard, S.A., 2002. Stiffness and damper in a rocker-based inverted pendulum model (Personal Communication).

Gard, S.A., Childress, D.S., 1997a. The effect of pelvic list on the vertical displacement of the trunk during normal walking. Gait Posture 5 (3), 233–238.

Gard, S.A., Childress, D.S., 1997b. The effect of stance-phase knee flexion on the vertical displacement of the trunk during normal walking. In: Paper Presented at The 21st Annual Meeting of the American Society of Biomechanics, Clemson University, Clemson, SC, September 24–27.

Gard, S.A., Childress, D.S., 1998. Mechanical characterization of vertical shock-absorbing pylons for lower-limb prostheses. In: Paper Presented at the 1st Annual Meeting of the VA Rehabilitation Research & Development Service, Washington, DC, October 1–3.

Gard, S.A., Childress, D.S., 1999. The influence of stance-phase knee flexion on the vertical displacement of the trunk during normal walking. Arch. Phys. Med. Rehabil. 80 (1), 26–32.

Gard, S.A., Childress, D.S., 2000. What determines vertical motion of the body during normal gait. In: Paper Presented at the 5th Annual Meeting of the Gait and Clinical Movement Analysis Society (GCMAS), Rochester, MN, April 12–15.

Gard, S.A., Childress, D.S., 2002. Stiffness and Damper in a Rocker-Based Inverted Pendulum Model. personal communication.

Gard, S.A., Konz, R.J., 2003. The effect of a shock-absorbing pylon on the gait of persons with unilateral transtibial amputation. J. Rehabil. Res. Dev. 40 (2), 109–124.

Geeroms, J., Flynn, L., Jimenez-Fabian, R., Vanderborght, B., Lefeber, D., 2013. Ankle-knee prosthesis with powered ankle and energy transfer for CYBERLEGs alpha-prototype. IEEE Int. Conf. Rehabil. Robot. 2013, 6650352. https://doi.org/10.1109/ICORR.2013.6650352.

Gitter, A., Czerniecki, J.M., DeGroot, D.M., 1991. Biomechanical analysis of the influence of prosthetic feet on below-knee amputee walking. Am. J. Phys. Med. Rehabil. 70 (3), 142–148.

Greene, P.R., McMahon, T.A., 1979. Reflex stiffness of man's anti-gravity muscles during kneebends while carrying extra weights. J. Biomech. 12, 881–891.

Hagberg, K., Branemark, R., Gunterberg, B., Rydevik, B., 2008. Osseointegrated trans-femoral amputation prostheses: Prospective results of general and condition-specific quality of life in 18 patients at 2-year follow-up. Prosthetics Orthot. Int. 32 (1), 29–41. https://doi.org/10.1080/03093640701553922.

Hagberg, K., Hansson, E., Branemark, R., 2014. Outcome of percutaneous osseointegrated prostheses for patients with unilateral transfemoral amputation at two-year follow-up. Arch. Phys. Med. Rehabil. 95 (11), 2120–2127. https://doi.org/10.1016/j.apmr.2014.07.009.

Hansen, A.H., 1998. Influence of Prosthetic Foot Mechanics on Alignment of Trans-tibial Prostheses. (Master's thesis). Northwestern University, Evanston, IL.

Hansen, A.H., Childress, D.S., 2000. Roll-over shapes of the human foot/ankle complex. In: Enderle, J.D. (Ed.), Proceedings of the 22nd Annual International Conference of the IEEE Engineering in Medicine and Biology Society, Vols 1–4 (vol. 22). IEEE, New York, pp. 828–830.

Hansen, A., Starker, F., 2017. Prosthetic foot principles and their influence on gait. In: Müller, B., Wolf, S.I. (Eds.), Handbook of Human Motion. Springer International Publishing AG.

Hansen, A.H., Gard, S.A., Childress, D.S., 2000. The determination of foot/ankle roll-over shape: clinical and research applications. In: Paper Presented at the IEEE-EMBS/Shriner's Hospital Workshop on Pediatric Gait: A New Millenium in Clinical Care and Motion Analysis Technology, Chicago, IL, July 21st.

Hansen, A.H., Childress, D.S., Miff, S.C., Gard, S.A., Mesplay, K.P., 2004. The human ankle during walking: implications for design of biomimetic ankle prostheses. J. Biomech. 37 (10), 1467–1474. https://doi.org/10.1016/j.jbiomech.2004.01.017.

Hargrove, L.J., Simon, A.M., Young, A.J., Lipschutz, R.D., Finucane, S.B., Smith, D.G., Kuiken, T.A., 2013. Robotic leg control with EMG decoding in an amputee with nerve transfers. N. Engl. J. Med. 369 (13), 1237–1242. https://doi.org/10.1056/NEJMoa1300126.

Hernigou, P., 2013. Ambroise pare IV: the early history of artificial limbs (from robotic to prostheses). Int. Orthop. 37 (6), 1195–1197. https://doi.org/10.1007/s00264-013-1884-7.

Herr, H., 2006. (Producer). http://www.ai.mit.edu/people/hherr/Research.html.

Herr, H.M., Grabowski, A.M., 2012. Bionic ankle-foot prosthesis normalizes walking gait for persons with leg amputation. Proc. Biol. Sci. 279 (1728), 457–464. https://doi.org/10.1098/rspb.2011.1194.

Herr, H., Whiteley, G., Childress, D., 2003. Cyborg technology—biomimetic orthotic and prosthetic technology. In: Bar-Cohen, Y., Breazeal, C. (Eds.), Biologically Inspired Intelligent Robots. SPIE Press, pp. 103–144.

Hugh M. Herr, Samuel K. Au, Peter Dilworth, & Daniel Joseph Paluska. (2007). USA Patent No. US2007/0043449 A1. USPTO.

Inman, V.T., Ralston, R.J., Todd, F., 1981. Human Walking. Williams & Wilkins, Baltimore, MD.

Inman, V.T., Ralston, H.J., Todd, F., 1994. Human locomotion. In: Rose, J., Gamble, J.G. (Eds.), Human Walking. Williams & Wilkins, Baltimore, MD, pp. 1–22 (Chapter 1).

Kuiken, T.A., Barlow, A.K., Hargrove, L.J., Dumanian, G.A., 2017a. Targeted muscle Reinnervation for the upper and lower extremity. Tech. Orthop. 32 (2), 109–116. https://doi.org/10.1097/Bto.0000000000000194.

Kuiken, T.A., Butler, B.A., Sharkey, T., Ivy, A.D., Li, D., Peabody, T.D., 2017b. Novel intramedullary device for lengthening transfemoral residual limbs. J. Orthop. Surg. Res. 1253. https://doi.org/10.1186/s13018-017-0553-8.

Kulkarni, J., Adams, J., Thomas, E., Silman, A., 1998. Association between amputation, arthritis and osteopenia in British male war veterans with major lower limb amputations. Clin. Rehabil. 12 (4), 348–353.

Kuo, A.D., 1999. Stabilization of lateral motion in passive dynamic walking. Int. J. Robot. Res. 18 (9), 917–930. https://doi.org/10.1177/02783649922066655.

Lawson, B.E., Atakan Varol, H., Sup, F., Goldfarb, M., 2010. Stumble detection and classification for an intelligent transfemoral prosthesis. Conf. Proc. IEEE Eng. Med. Biol. Soc. 2010, 511–514. https://doi.org/10.1109/IEMBS.2010.5626021.

Leeuwen, J.L., Van Speth, L.A.W.M., Daanen, H.A.M., 1990. Shock absorption of below-knee prostheses: a comparison between the SACH and the multiflex foot. J. Biomech. 23 (5), 441–446.

Lehmann, J.F., Price, R., Boswell-Bessette, S., Dralle, A., Questad, K., 1993a. Comprehensive analysis of dynamic elastic response feet: Seattle ankle/lite foot versus SACH foot. Arch. Phys. Med. Rehabil. 74 (8), 853–861.

Lehmann, J.F., Price, R., Boswell-Bessette, S., Dralle, A., Questad, K., deLateur, B.J., 1993b. Comprehensive analysis of energy storing prosthetic feet: flex foot and Seattle foot versus standard SACH foot. Arch. Phys. Med. Rehabil. 74 (11), 1225–1231.

Lenneras, M., Tsikandylakis, G., Trobos, M., Omar, O., Vazirisani, F., Palmquist, A., … Thomsen, P., 2017. The clinical, radiological, microbiological, and molecular profile of the skin-penetration site of transfemoral amputees treated with bone-anchored prostheses. J. Biomed. Mater. Res. A 105 (2), 578–589. https://doi.org/10.1002/jbm.a.35935.

Li, Y., 2016. Osseointegrated Human-Machine Gateway (OHMG) clinical trial identifier NCT03178890. Retrieved from: https://clinicaltrials.gov/ct2/show/NCT03178890.

Li, Y., Branemark, R., 2017. Osseointegrated prostheses for rehabilitation following amputation: the pioneering Swedish model. Unfallchirurg 120 (4), 285–292. https://doi.org/10.1007/s00113-017-0331-4.

List of IPC world records in athletics, 2018. Retrieved from: https://en.wikipedia.org/wiki/List_of_IPC_world_records_in_athletics.

Major, M.J., Serba, C.K., Chen, X.L., Reimold, N., Ndubuisi-Obi, F., Gordon, K.E., 2018. Proactive locomotor adjustments are specific to perturbation uncertainty in below-knee prosthesis users. Sci. Rep. 8. 1863. https://doi.org/10.1038/s41598-018-20207-5.

Mak, A.F., Zhang, M., Boone, D.A., 2001. State-of-the-art research in lower-limb prosthetic biomechanics-socket interface: a review. J. Rehabil. Res. Dev. 38 (2), 161–174.

Margaria, R., 1976. Biomechanics and energetics of muscular excercise. Oxford University Press.

McGeer, T., 1990. Passive dynamic walking. Int. J. Rob. Res. 9 (2), 62–82.

McMahon, T.A., Greene, P.R., 1979. The influence of track compliance on running. J. Biomech. 12, 893–904.

Michael, J.W., 1999. Modern prosthetic knee mechanisms. Clin. Orthop. Relat. Res. 361, 39–47.

Miff, S., 2000. The Effects of Step Length, Cadence, and Walking Speed on Gait Kinematics, Kinetics and Energetics (MS Thesis). Northwestern University, Evanston, IL.

Miller, L.A., 1994. A Biomechanical Analysis of a Vertical Compliance Shock Pylon for a Below-Knee Amputee System. (Master's Thesis). Northwestern University, Evanston, IL.

Miller, L.A., Childress, D.S., 1997. Analysis of a vertical compliance prosthetic foot. J. Rehabil. Res. Dev. 34 (1), 52–57.

Mochon, S., McMahon, T.A., 1980. Ballistic walking. J. Biomech. 13 (1), 49–57.

Mooney, J., Hill, S., Supan, T., Barth, D., 1995. Comparison of floor reaction and rotational forces in the gait of a transtibial amputee using a *re*-flex VSP* flex foot design: a pilot study. In: Paper Presented at the Proceedings of the 21st Annual Meeting & Scientific Symposium of the AAOP (American Academy of Orthotists and Prosthetists), New Orleans, March 21–25.

Nebergall, A., Bragdon, C., Antonellis, A., Karrholm, J., Branemark, R., Malchau, H., 2012. Stable fixation of an osseointegated implant system for above-the-knee amputees: titel

RSA and radiographic evaluation of migration and bone remodeling in 55 cases. Acta Orthop. 83 (2), 121–128. https://doi.org/10.3109/17453674.2012.678799.

Perry, J., 1992. Gait Analysis Normal and Pathological Function. SLACK, New Jersey.

Perry, J., Boyd, L.A., Rao, S.S., Mulroy, S.J., 1997. Prosthetic weight acceptance mechanics in transtibial amputees wearing the single Axis, Seattle lite, and flex foot. IEEE Trans. Rehabil. Eng. 5 (5), 283–289.

Pitkin, M.R., 1995. Mechanical outcomes of a rolling-joint prosthetic foot and its performance in the dorsiflexion phase of transtibial amputee gait. J. Prosthet. Orthot. 7 (4), 114–123.

Quesada, P.M., Rash, G.S., 1998. Simulation of walking without stance phase knee flexion. Gait Posture 7 (2), 151–152.

Rouse, E.J., Villagaray-Carski, N.C., Emerson, R.W., Herr, H.M., 2015. Design and testing of a bionic dancing prosthesis. PLoS One. 10(8). e0135148. https://doi.org/10.1371/journal.pone.0135148.

Sam, M., Hansen, A.H., Childress, D.S., 2000. Mechanical characterization of prosthetic feet using a prosthetic foot loading apparatus. In: Enderle, J.D. (Ed.), Proceedings of the 22nd Annual International Conference of the IEEE Engineering in Medicine and Biology Society, vols, 1–4 (vol. 22). Ieee, New York, pp. 1968–1971.

Sam, M., Hansen, A.H., Childress, D.S., 2004. Characterisation of prosthetic feet used in low-income countries. Prosthetics Orthot. Int. 28 (2), 132–140.

Saunders, J.B., Inman, V.T., Eberhart, H.D., 1953. The major determinants in normal and pathological gait. J. Bone Joint Surg. 35 (A3), 543–558.

Siegler, S., Seliktar, R., Hyman, W., 1982. Simulation of human gait with the aid of a simple mechanical model. J. Biomech. 15 (6), 415–425.

Simon, A.M., Spanias, J.A., Ingraham, K.A., Hargrove, L.J., 2016. Delaying ambulation mode transitions in a powered knee-ankle prosthesis. Conf. Proc. IEEE Eng. Med. Biol. Soc. 2016, 5079–5082. https://doi.org/10.1109/EMBC.2016.7591869.

Simon, A.M., Ingraham, K.A., Spanias, J.A., Young, A.J., Finucane, S.B., Halsne, E.G., Hargrove, L.J., 2017. Delaying ambulation mode transition decisions improves accuracy of a flexible control system for powered knee-ankle prosthesis. IEEE Trans. Neural Syst. Rehabil. Eng. 25 (8), 1164–1171. https://doi.org/10.1109/TNSRE.2016.2613020.

Smyrli, A., Bertos, G., Papadopoulos, E., 2018. Efficient stabilization of zero-slope walking for bipedal robots following their passive fixed-point trajectories. In: Proc. IEEE International Conference on Robotics and Automation (ICRA '18), Brisbane, Australia, May 21–25, pp. 5733–5738.

Snyder, R.D., Powers, C.M., Fontaine, C., Perry, J., 1995. The effect of five prosthetic feet on the gait and loading of the sound limb in dysvascular below-knee amputees. J. Rehabil. Res. Dev. 32 (4), 309–315.

Spanias, J.A., Perreault, E.J., Hargrove, L.J., 2014. A strategy for labeling data for the neural adaptation of a powered lower limb prosthesis. Conf. Proc. IEEE Eng. Med. Biol. Soc. 2014, 3090–3093. https://doi.org/10.1109/EMBC.2014.6944276.

Spanias, J.A., Perreault, E.J., Hargrove, L.J., 2016a. Detection of and compensation for EMG disturbances for powered lower limb prosthesis control. IEEE Trans. Neural Syst. Rehabil. Eng. 24 (2), 226–234. https://doi.org/10.1109/TNSRE.2015.2413393.

Spanias, J.A., Simon, A.M., Perreault, E.J., Hargrove, L.J., 2016b. Preliminary results for an adaptive pattern recognition system for novel users using a powered lower limb prosthesis. Conf. Proc. IEEE Eng. Med. Biol. Soc. 2016, 5083–5086. https://doi.org/10.1109/EMBC.2016.7591870.

Spanias, J.A., Simon, A.M., Hargrove, L.J., 2017. Across-user adaptation for a powered lower limb prosthesis. IEEE Int. Conf. Rehabil. Robot. 2017, 1580–1583. https://doi.org/10.1109/ICORR.2017.8009473.

Spanias, J.A., Simon, A.M., Finucane, S.B., Perreault, E.J., Hargrove, L.J., 2018. Online adaptive neural control of a robotic lower limb prosthesis. J. Neural Eng. 15(1). 016015. https://doi.org/10.1088/1741-2552/aa92a8.

Sup, F., Varol, H.A., Mitchell, J., Withrow, T.J., Goldfarb, M., 2009a. Preliminary evaluations of a self-contained anthropomorphic transfemoral prosthesis. IEEE ASME Trans. Mechatron. 14 (6), 667–676. https://doi.org/10.1109/TMECH.2009.2032688.

Sup, F., Varol, H.A., Mitchell, J., Withrow, T.J., Goldfarb, M., 2009b. Self-contained powered knee and ankle prosthesis: initial evaluation on a transfemoral amputee. IEEE Int. Conf. Rehabil. Robot. 2009, 638–644. https://doi.org/10.1109/ICORR.2009.5209625.

Sutherland, D.H., Kaufman, K.R., Moitoza, J.R., 1994. Human locomotion. In: Rose, J., Gamble, J.G. (Eds.), Human Walking. In: vol. 2. Williams & Wilkins, Baltimore, MD, pp. 23–44.

Van Jaarsveld, H.W., Grootenboer, H.J., De Vries, J., 1990a. Accelerations due to impact at heel strike using below-knee prosthesis. Prosthetics Orthot. Int. 14 (2), 63–66.

Van Jaarsveld, H.W., Grootenboer, H.J., de Vries, J., Koopman, H.F., 1990b. Stiffness and hysteresis properties of some prosthetic feet. Prosthetics Orthot. Int. 14 (3), 117–124.

Varol, H.A., Sup, F., Goldfarb, M., 2010. Multiclass real-time intent recognition of a powered lower limb prosthesis. IEEE Trans. Biomed. Eng. 57 (3), 542–551. https://doi.org/10.1109/TBME.2009.2034734.

Versluys, R., Desomer, A., Lenaerts, G., Van Damme, M., Beyl, P., Van der Perre, G., … Lefeber, D., 2008. A pneumatically powered below-knee prosthesis: design specifications and first experiments with an amputee. In: 2nd IEEE RAS & EMBS International Conference on Biomedical Robotics and Biomechatronics (Biorob 2008), Vols 1 and 2, p. 372. https://doi.org/10.1109/Biorob.2008.4762842.

Versluys, R., Beyl, P., Van Damme, M., Desomer, A., Van Ham, R., Lefeber, D., 2009. Prosthetic feet: state-of-the-art review and the importance of mimicking human ankle-foot biomechanics. Disabil. Rehabil. Assist. Technol. 4 (2), 65–75. https://doi.org/10.1080/17483100802715092.

Vertriest, S., Coorevits, P., Hagberg, K., Branemark, R., Haggstrom, E., Vanderstraeten, G., Frossard, L., 2015. Static load bearing exercises of individuals with transfemoral amputation fitted with an osseointegrated implant: reliability of kinetic data. IEEE Trans. Neural Syst. Rehabil. Eng. 23 (3), 423–430. https://doi.org/10.1109/TNSRE.2014.2337956.

Vertriest, S., Coorevits, P., Hagberg, K., Branemark, R., Haggstrom, E.E., Vanderstraeten, G., Frossard, L.A., 2017. Static load bearing exercises of individuals with transfemoral amputation fitted with an osseointegrated implant: Loading compliance. Prosthetics Orthot. Int. 41 (4), 393–401. https://doi.org/10.1177/0309364616640949.

Waters, R.L., Perry, J., Antonelli, D., Hislop, H., 1976. Energy cost of walking of amputees: the influence of level of amputation. J. Bone Joint Surg. Am. 58 (1), 42–46.

Webster, J.B., Hakimi, K.N., Williams, R.M., Turner, A.P., Norvell, D.C., Czerniecki, J.M., 2012. Prosthetic fitting, use, and satisfaction following lower-limb amputation: a prospective study. J. Rehabil. Res. Dev. 49 (10), 1493–1504. https://doi.org/10.1682/Jrrd.2012.01.0001.

Windrich, M., Grimmer, M., Christ, O., Rinderknecht, S., Beckerle, P., 2016. Active lower limb prosthetics: a systematic review of design issues and solutions. Biomed. Eng. Online 15 (Suppl. 3), 140. https://doi.org/10.1186/s12938-016-0284-9.

Wirta, R.W., Mason, R., Calvo, K., Golbranson, F.L., 1991. Effect on gait using various prosthetic ankle-foot devices. J. Rehabil. Res. Dev. 28 (2), 13–24.

Woodward, R.B., Spanias, J.A., Hargrove, L.J., 2016. User intent prediction with a scaled conjugate gradient trained artificial neural network for lower limb amputees using a powered prosthesis. Conf. Proc. IEEE Eng. Med. Biol. Soc. 2016, 6405–6408. https://doi.org/10.1109/EMBC.2016.7592194.

Zhang, L.Q., Xu, D., Makhsous, M., Lin, F., 2000. Stiffness and viscous damping of the human leg. In: 24th Annual Meeting, American Society of Biomechanics (Chicago, IL).

CHAPTER EIGHT

Upper and Lower Extremity Exoskeletons

Andres F. Ruiz-Olaya*, Alberto Lopez-Delis[†], Adson Ferreira da Rocha[‡]

*Faculty of Electronics and Biomedical Engineering, Antonio Nariño University, Bogotá, Colombia
[†]Medical Biophysics Center, University of Oriente, Santiago de Cuba, Cuba
[‡]Biomedical Engineering Program, University of Brasilia, Brasilia, Brazil

Contents

1 CONCEPTS AND FUNDAMENTALS OF EXOSKELETONS

1.1 Definitions

In biology, exoskeleton is a kind of external covering on an animal to protect or support it, for example, the shell of a crab. In the engineering field, the concept of the exoskeleton-type system is an extension of the exoskeleton in

Handbook of Biomechatronics
https://doi.org/10.1016/B978-0-12-812539-7.00011-8

biology, which refers to systems that expand or augment a person's physical abilities (Kazerooni, 2008). For instance, those devices help a person lift or carry heavier loads, run faster, and jump higher. Table 1 shows an analogy between the biological exoskeleton and the exoskeleton system in the engineering field, and their potential applications. Upper and lower exoskeletons could offer humans the kind of protection, support, enhancement, and sensing which they afford in nature.

Exoskeletons have segments and joints that correspond to some extend to those of the human body (Fig. 1). Those devices can be seen as a technology to extend, complement, substitute, or enhance the human function and capability or to empower the human limb where it is worn out (Maciejasz et al., 2014). There is a one-to-one correspondence between human anatomical joints and the robot joints or sets of joints. This kinematic compliance is a key aspect in achieving ergonomic human-robot interfaces. Taking into account that humans and exoskeletons are in close physical interaction, there is an effective transfer of power between the human and the robot (Ruiz et al., 2008).

1.2 Classification and Applications of Exoskeletons

There are several classifications for exoskeletons. According to the principle of action, they could be divided into active and passive exoskeletons. Active devices use an external power source, whereas the mechanics of the passive exoskeletons relies on kinetic energy and human strength (Pons, 2008).

Exoskeletons can also be classified according to the human limb onto which the external framework couple to the human body. Thus, exoskeletons can be classified in upper-limb (either including or excluding the hand), lower-limb, and full-body exoskeletons. Upper-limb exoskeletons enhance the manipulation function, and normally include the shoulder, elbow, and wrist articulations. A number of investigations devoted to the application of the exoskeletons for the upper limbs suggest a wide scope of possible usage. Lower-limb exoskeletons provide support, stability, and mobility (locomotion).

Applications of exoskeletons include power amplifier, telemanipulation, rehabilitation and motor training, virtual reality, and haptics (Ruiz et al., 2008).

1.2.1 Power Amplifier

The main purpose of a robotic exoskeleton in this application is to amplify the physical capacities of a human. As a result, the person provides control signals to the exoskeleton, while the device delivers mechanical power in

Table 1 Analogy Between the Biological Exoskeleton and Exoskeletons in the Engineering Field

	The Biological Exoskeleton		Exoskeleton in the Engineering Field	
Function		**Example**		**Application/Example**
Support	Supporting the body of the invertebrates	• Because molluscs have a soft body, they are more fragile • It is also more difficult for them to support their body in terrestrial environments or to attach to substrates in aquatic habitats	Supporting physically disabled patient or walking assistance	• Rehabilitation engineering for the human motor system
Enhancement	Enhancing the power of animals	• Ingrowths of the arthropod exoskeleton known as apodemes serve as attachment sites for muscles • Similar to tendons, apodemes can stretch to store elastic energy for jumping, notably in locusts	Strengthening the human operator	• Power amplification
Protection	Protecting the animal's body	• The shell of a crab	Protecting the human operator	• External armor for soldier, rescue devices, safe manipulation in risky environments
Sensing	Obtaining the information, sensorium	• The spider's rigid exoskeleton readily conducts vibrations, transmits mechanical stress that may be caused by substance vibrations, by gravity, or by the spiders's own movement	Interface of human operator and the environment to acquire information	• Telemanipulation • Virtual reality • Entertainment

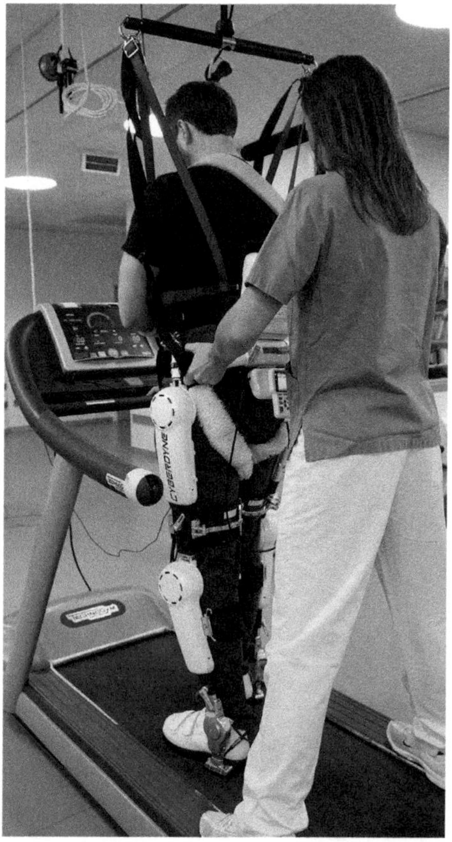

Fig. 1 Exoskeletons have segments and joints that correspond to some extend to those of the human body. Open Access article with unrestricted use permission (Nilsson et al., 2014).

order to accomplish a particular task. Thus, exoskeletons are currently under development for enhancement of human motor performance in the military (Zoss et al., 2006), and for industrial applications (de Looze et al., 2015).

1.2.2 Telemanipulation

This application comprises the set of technologies that enable tasks to be executed remotely. A robotic exoskeleton acts as a master device in a teleoperation system. In bilateral control mode, it allows the operator to control a remote robotic arm (slave). Interaction forces between the remote robot arm and its environment are fed back to the master and applied by the exoskeleton to the human arm.

1.2.3 Rehabilitation and Motor Training

The rehabilitation field is a key application domain for the development of exoskeletons, to help disabled people with difficulties in moving (Kiguchi et al., 2004; Colombo, 2001).For rehabilitation applications, the exoskeleton permits to assist in several active and passive therapies. Thus, the device emulates and replicates movements and exercises that a physiotherapist executes when working with a patient. The exoskeleton should be able to replicate with a patient the movements performed with a therapist during the treatment.

There are a significant number of papers published concerning robotic exoskeleton in therapy, and about their effectiveness in the functional recovery after stroke (Kwakkel et al., 2008; Brokaw et al., 2013; Milot et al., 2013; Chang and Kim, 2013). Some studies conclude that robot-aided therapy can elicit improvements in arm function that are distinct from the conventional therapy and supplements conventional methods to improve outcomes (Brokaw et al., 2013).

Most of the reports in the literature using robotic exoskeletons in therapy focus on treatment of poststroke paralysis of the upper and lower limbs (Louie and Eng, 2016). Other works concern using exoskeletons for rehabilitation after cerebrospinal traumas, for multiple sclerosis, for tremor treatment, and for compensation of grasping function of the hand (Maciejasz et al., 2014).

1.2.4 Virtual Reality and Haptics

In this application, exoskeletons aim to exert a reactive force on the user while they are using a virtual reality (VR) headset. It is the opposite of exoskeletons for teleoperation, which are used to extract kinematics data from the user. It could be used for gaming, motor rehabilitation, and training.

1.3 The Role of Biomechatronics in Exoskeletons

Exoskeletons are biomechatronic devices that interact with the human body. The human motor control system (HMCS) can be modeled as in Fig. 2A (Lobo-Prat et al., 2014). The HMCS consists of a mechanical structure, the plant, which represents the skeleton and passive tissues; the actuators, which represent the muscles; and a controller, which represents the central nervous system and receives sensory feedback from the physiological sensors. An artificial motor control system (AMCS) such as exoskeletons

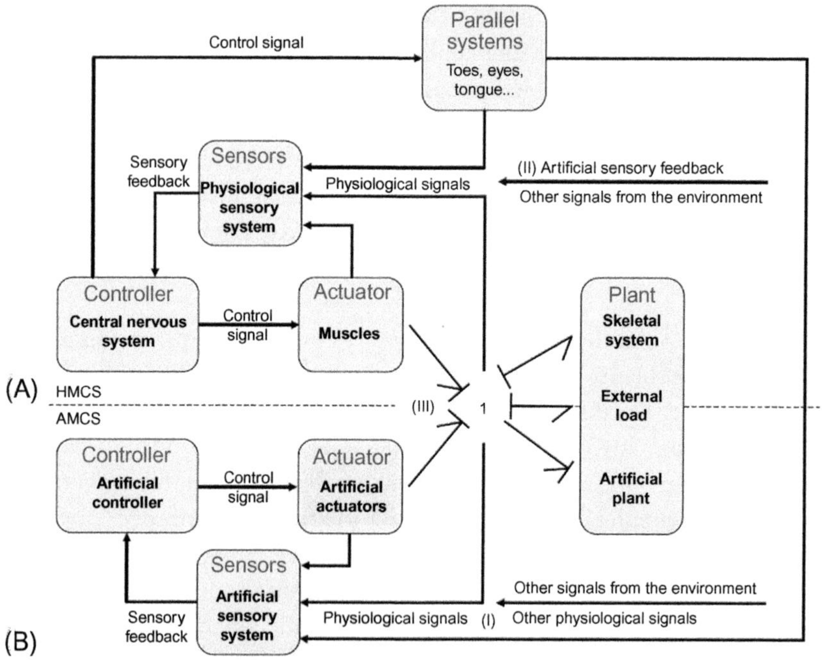

Fig. 2 Schematic block diagram of the human motor control systems (A) in parallel with the artificial movement control system (B) (i.e., exoskeleton). Three kinds of interactions between the HMCS and AMCS can be distinguished: (I) detection of the motion intention of the user; (II) provision of feedback to the user regarding the state of the AMCS, the HMCS or the environment; and (III) exchange of mechanical power between plants. Open Access article with unrestricted use permission (Lobo-Prat et al., 2014).

work in parallel to the HMCS and can be modeled with the same components as the HMCS: a plant representing the mechanical structure and passive elements, such as springs or dampers, and an artificial controller that receives the data measured from the sensors and generates control signals to operate the actuators (Fig. 2B; Lobo-Prat et al., 2014).

Several aspects of biomechatronics could be incorporated while developing exoskeletons. First, bioinspiration could be extended in the development of mechatronic systems, for example, the development of bioinspired mechatronic components, that is, structure, actuators, and control architectures. Second, exoskeletons permit information exchanging between the device and humans. Thus, bioelectric signals (EMG, EEG, EOG) could be used to control the exoskeleton and permit and more "natural interaction," without using external pushbuttons, joysticks, or other elements.

Finally, exoskeletons work close to human body, thus the following aspects must be considered:

(a) the external framework should replicate the structure of the upper or lower limb;
(b) the device should be lightweight, strong, and safe;
(c) there must be a possibility of changing the elements to permit the exoskeleton structure be length adaptable; and
(d) exoskeletons should perform a range of movements required to accomplish the activity or function.

One of the biggest challenges for robotic exoskeletons that interface with persons closely is to assure the safety of the user. It is important to establish a safety guideline appropriate for elderly and disabled human users and to develop and integrate both mechanical and electrical safety systems in exoskeletons. To meet stringent standards, redundant safety mechanisms must be in place.

2 A BRIEF HISTORY OF EXOSKELETON RESEARCH

The first mention of a device resembling an exoskeleton was Yang's running aid (Yagn, 1890) patented in 1890. It was a simple bow/leaf-spring operating parallel to the legs, whose function is to augment running and jumping. Each leg spring was engaged during the foot contact to effectively transfer the body's weight to the ground and to reduce the forces borne by the stance leg. During the aerial phase, the parallel leg spring was designed to disengage in order to allow the biological leg to freely flex and to enable the foot to clear the ground.

The studies on wearable equipment have been going on for more than 50 years by military institutions, private companies, and research groups in several countries. The main components for the development of exoskeleton robots (XoRs) include mechanism design technology, human intent measurement technology, and human-robot control technology. For the successful development of robotic exoskeleton systems, designers should take into consideration the field of application, the purpose of power support, and to which part of the body the robot would give support. In the late 1960s, the General Electric company, funded by military institutions of the United States of America (USA), developed and tested what the researchers called a body amplifier prototype based on a master-slave system named

Hardiman ("Human Augmentation Research and Development Investigation"). The prototype remained incomplete at the time of its termination (Kazerooni et al., 1968). General Electric Co. developed the concept of human-amplifiers through the Hardiman project from 1966 to 1971. The Hardiman concept was a robotic master/slave configuration in which two overlapping exoskeletons were implemented. The inner one was set to follow human motion while the outer one implemented an hydraulically empowered version of the motion performed by the inner exoskeleton (Kazerooni, 1990).

Other research projects were conducted in *Serbia* in the 1970s (Hristic and Vukobratovic, 1973), and at the Massachusetts Institute Technology (MIT) in the 1980s (Seireg and Grundman, 1981). However, few studies were carried on during the next 20 years because of fundamental technological limitations, especially in control hardware. At the end of the 20th century, with the rapid progress in computer science, as well as control and drive technologies, the Defense Advanced Research Projects Agency (DARPA), an agency of the Department of Defense of the USA, started new efforts in the development of exoskeletons (Garcia et al., 2002). This renewed interest in the United States led other groups and institutes in other countries (including Japan, Russia, the United Kingdom, Germany, Korea, and Singapore) to start their own projects (Li et al., 2014). Many results have been published since the beginning of the 21st century, as well as several reviews discussing the state-of-the-art and future perspectives (Dollar and Hugh, 2008; Yang et al., 2008; Kazerooni, 2008).

Rehabilitation and functional compensation are very important potential applications of exoskeletons and wearable robotics. Worldwide, an estimated 185 million people use wheelchairs and other functional assistance devices daily. Furthermore, almost 20% of the world's population is now aged over 65 years, and this proportion may exceed 35% until 2050. The assistive rehabilitation exoskeletons have the potential for use in many applications, and a very important use in the near future should be in the rehabilitation of upper and lower limbs.

2.1 Upper Extremity Exoskeletons

The primary applications of upper-limb exoskeletons were originally teleoperation and power amplification. An example is the ESA human arm exoskeleton for Space Robotics Telepresence (Schiele and Van der Helm, 2006), developed as a human-machine interface for master-slave robotic teleoperation with force feedback. Later, exoskeleton applications were considered for the rehabilitation and assistance of disabled or elderly people, for example, upper- and lower-limb orthoses. Assist of upper-limb motion is important in daily activities, and several kinds of upper-limb exoskeletons have been proposed (Maciejasz et al., 2014) in order to improve the quality of life of physically weak persons. Usually, the movable range of human shoulder is 180 degrees in flexion, 60 degrees in extension, 180 degrees in abduction, 75 degrees in adduction, 100–110 degrees in internal rotation, and 80–90 degrees in external rotation. The limitation of the movable range of forearm pronation-supination motion is 50–80 degrees in pronation and 80–90 degrees in supination, and the elbow flexion-extension motion is 145 degrees in flexion and −5 degrees in extension, see Fig. 3. Those exoskeletons are controlled to assist the upper-limb motion

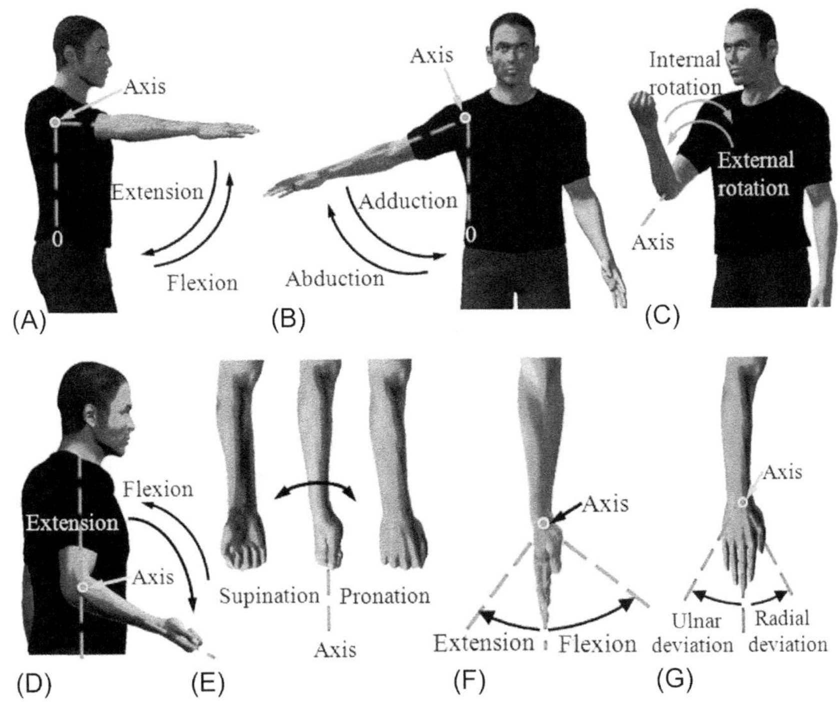

Fig. 3 Upper-limb motions: (A) shoulder flexion/extension, (B) shoulder abduction/adduction, (C) shoulder internal/external rotation, (D) elbow flexion/extension, (E) forearm supination/pronation, (F) wrist flexion/extension, and (G) wrist ulnar/radial deviation. *(From Gopura, R.A.R.C., Bandara, D.S.V., Kiguchi, K., Mann, G.K.I., 2016. Developments in hardware systems of active upper-limb exoskeleton robots: a review. Robot. Auton. Syst. 75, 203–220, with permission from Elsevier.)*

of the user in accordance with the user's motion intention by monitoring the electromyographic (EMG) signals of certain muscles involved in the upper-limb motion. At least five degrees of freedom (DOF) must be provided assuming that the location of the rotation center of the shoulder joint of the exoskeleton is the same as that of the user. As a matter of fact, more DOF are required to assist all upper-limb motion, since the human shoulder complex, which consists of the scapula, clavicle, and humerus, moves conjointly, providing seven DOF for upper-limb motion (Zatsiorsky, 1998).

An example of upper-limb exoskeleton is the wearable orthosis for tremor assessment and suppression (WOTAS) device, which was presented within the framework of the DRIFTS project as a promising solution for patients who cannot use medication to suppress the tremor (Manto et al., 2003). WOTAS exhibits three DOF corresponding to elbow flexion-extension, forearm pronation-supination and wrist flexion-extension, while restricting adduction-abduction movements of the wrist (Fig. 4). The ARMin system is a rehabilitation exoskeleton with six DOF designed to

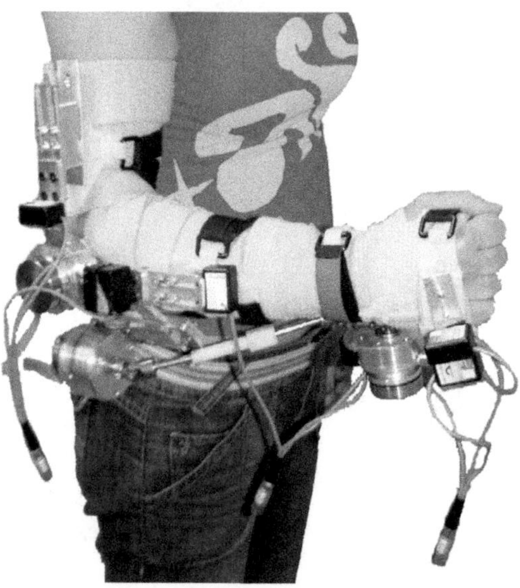

Fig. 4 WOTAS final version for control of human upper-limb three movements control: flexion-extension elbow, flexion-extension wrist, and pronation-supination forearm. *(From Rocon, E., Ruiz, A.F., Belda-Lois, J.M., Moreno, J.C., Pons, J.L., Raya, R., Ceres, R., 2008. Diseño, desarrollo y validación de dispositivo robótico para la supresión del temblor patológico. Revista Iberoamericana de Automática e Informática Industrial RIAI, 5(2), 79–92, with permission from Elsevier.)*

enable training for specific activities of daily living (Nef et al., 2006). Kousidou et al. (2006) have incorporated the Salford arm into the Rehabilitation Laboratory System for virtual rehabilitation of complex three-dimensional trajectories in the workspace. Carignan et al. described a prototype with five DOF exoskeleton systems currently under development that focuses on shoulder rehabilitation (Carignan et al., 2005). A four DOF power-assist exoskeleton (Kiguchi, 2007), which assists shoulder vertical and horizontal flexion/extension motion, elbow flexion/extension motion, and forearm pronation/supination motion, has been developed as an example of the effective EMG-based control method for the activation on exoskeletons according the user's motion intention. The effectiveness of the power-assist exoskeleton is verified by the experiment. It mainly consists of four main links, an upper-arm holder, a wrist holder, four DC motors, the shoulder mechanism of the moving center of rotation, the mechanism for shoulder inner/outer rotation motion assist, an elbow joint, a wrist force sensor, and driving wires.

Other devices for upper-limb rehabilitation, labeled as coaching devices, do not generate any forces but provide specific feedback (Maciejasz et al., 2014). These devices serve as input interfaces for interaction with therapeutic games in VR, using video-based motion recognition (Sanchez et al., 2004), ArmeoSpring from Hocoma AG (Gijbels et al., 2011).

Some systems for rehabilitation of fingers or hands have even higher numbers of DOF. Examples include the system proposed by Hasegawa et al., with 11 DOF (Hasegawa et al., 2008) and the hand exoskeleton developed at the Technical University of Berlin with 20 DOF (Fleischer et al., 2009). The sEMG signals from the contralateral healthy limb have also been used to control movements of the affected limb (Li et al., 2006). This method has also been implemented in the Bi-Manu-Track system (Hesse et al., 2003), in the ARMOR exoskeleton (Mayr et al., 2008), and in the device proposed by Kawasaki et al. (2007). The use of the other limb to control the affected one is especially useful during rehabilitation after stroke.

2.2 Lower Extremity Exoskeletons

In the past years, several companies have been developing lower-body robotic exoskeletons that allow paraplegics or wheelchair users to stand and walk and even climb stairs. These robotic devices use battery-powered electric motors to actuate hip and knee joints and sometimes also the ankle joints, and are controlled by motion or signals from sensors and microcomputers. BLEEX was the first load-carrying and energetically autonomous exoskeleton (Zoss et al., 2006). With an anthropomorphic design, BLEEX has left and right three-segment legs, being analogous to the human thigh, shank, and foot. Each leg has seven DOFs: hip flexion/extension (f/e) and abduction/adduction (a/a), knee f/e, and ankle dorsi/plantar flexion (d/p). The hip presents intra/extra rotation, and the ankle, inversion/eversion—a/a. ReWalk Robotics, formerly ARGO Medical Technologies, offers two products: the ReWalk Rehabilitation, launched in 2011, and the ReWalk Personal, which became available internationally in 2012. The ReWalk was developed by Dr. Amit Goffer, an Israeli scientist who became quadriplegic after an accident in 1997. It consists of a metal brace that supports the legs and part of the upper body, electric motors that supply movement of the hips, knees and ankles, a tilt sensor, and a backpack that contains a computer and a power supply (Esquenazi et al., 2012). An allied product device is produced by Rex Bionics: the REX Rehab and REX Personal. The REX was designed specifically for users with high levels of mobility impairment, including paraplegic and quadriplegic users, and allows them to navigate stairs and ramps safely. In contrast to the ReWalk, it does not require crutches or a walking frame to provide stability. The device is powered by DC motors and it is controlled by a simple keypad and joystick (Bogue, 2015). The Indego Powered Leg Orthosis prototype presented at OTWorld (2014), in Leipzig, Germany from research at Vanderbilt University, is a battery-powered, lower-body exoskeleton that provides up to 4h of use and weights 26lbs (12kg). The exoskeleton uses gyroscopes and other inertial sensors that allow it to mirror natural human movement; LED indicators and a wireless software interface provide control over parameters such as stride length and step frequency. Cyberdyne, a spin-off from the University of Tsukuba, developed the HAL (full-body exoskeleton) units, mainly used for nonmilitary applications, such as nursing and assisting the disabled in waking. The system was certified by Underwriters Laboratories to ISO13485 with the international quality standard for medical devices and by the global safety certificate. HAL uses sensors on the user's skin for detecting myoelectric signals for estimating his or her intended motion (Bogue, 2015). Based on these signals, servo motors try to produce the same torque as that caused by the contraction of human muscle, synchronizing the movement of the exoskeleton with the intention of the user. The controller of HAL uses battery-powered small PCs that were equipped with wireless network cards, and located in the back of the exoskeleton. Fig. 5 presents some of lower-limb exoskeleton developed by research groups.

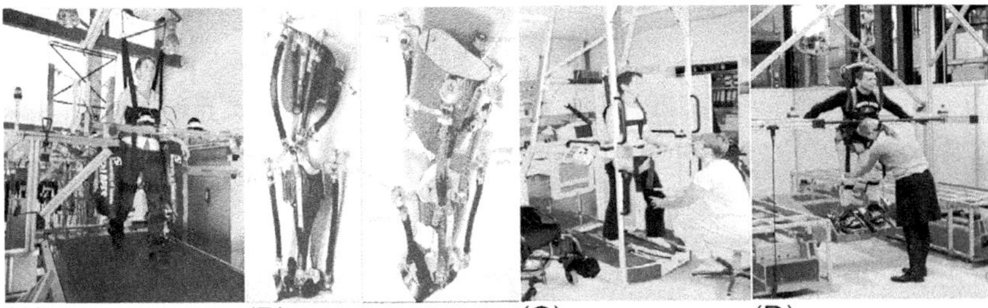

Fig. 5 Lower-limb exoskeleton with permission of open access articles with unrestricted use permission (Fleerkotte et al., 2014; Sawicki and Ferris, 2009; Schmidt et al., 2007). (A) LOPEZ, (B) KAFO, (C) gait trainer GTI, and (D) haptic walker.

In the ATLAS project, an active orthosis has been developed for gait assistance, in particular among children suffering from quadriplegia (Merodio et al., 2012). The first prototype of the ATLAS exoskeleton provided active motion for the hip and knee f/e, with the ankle f/e underactuated, by connecting to a linkage between the thigh and shank. The aim of the designers is that the device could to be a completely autonomous assistive orthosis, in which the user only supply locomotion maneuver triggers, for example, start and stop, stand up, and sit down. The IHMC Mobility Assist Exoskeleton presented in Kwa et al. (2009) has three actuated DOFs on the hip a/a and f/e and knee f/e, and two passive DOFs on the hip rotation and ankle d/p. The anthropomorphic design and its series of elastic actuators (SEA), enable the IHMC exoskeleton to work in different modes, like zero assistance mode, performance augmentation mode, and gait rehabilitation mode. The wearable walking helper (WWH) is a wearable gravity-compensating hip-knee (HK) exoskeleton developed to assist the locomotion activities of disabled and elderly people (Nakamura et al., 2005). The assistive torques provided by the WWH are proportional to the torques calculated based on an approximated human body model, user's postures and motions. Experiments with a subject standing up and sitting down showed a reduction of EMG activities at the rectus femoris, thus proving efficacy to the WWH as an antigravity exoskeleton. The XoR prototype has been developed for the postural control of elderly people and persons with mobility disability (Hyon et al., 2011). The XoR is implemented with a hybrid driving concept combining pneumatic artificial muscles (PAMs) and electric motors—the former acts as a gravity balancer while the latter acts as a dynamic compensator. The user's posture is defined by joint angles and ground reaction forces while the motion intention is estimated based on EMG signals. The hip-knee-ankle-foot (HKAF) lower-limb exoskeleton presented by He and Kiguchi (2007) has been designed to assist the movements of physically weak people. It consists of one passive DOF for the ankle d/p, and two active DOFs for the hip and knee f/e joints.

The desired assistances for hip and knee movements are estimated through an EMG-based neurofuzzy controller, from eight muscles on the thigh. LOPES is the first application of adaptive oscillators on a lower-limb assistive exoskeleton (Ronsse et al., 2011). It is based on a trunk-hip-knee frame-based treadmill-mounted exoskeletons with actuated hip f/e, a/a, and knee f/e. The RoboKnee is a knee exoskeleton designed to assist the wearer during stairs climbing and squatting with heavy loads (Pratt et al., 2004). RoboKnee, consists of a thigh and a shank brace, jointed on the knee and connected by a linear SEA joint.

The exoskeletons presented above, have been designed to assist different kinds of human lower-limb movements, such as supporting heavy loads, ground-level walking, sit/stand transitions, squatting, ascending and descending stairs, and even running. The subjects used in the studies are also diversified, including elderly people, healthy people, people with muscular weakness, people with lower-limb disability, or totally lost lower-limb functions (Tingfang et al., 2015).

Even though the kinematics and kinetics characteristics of lower-limb joints greatly differ in each kind of locomotion, in the control process, the exoskeletons are usually divided into a series of phases: detection and prediction of these phases are based on the exoskeleton sensory systems, which are fundamental for the control strategy. Due to the complexity in evaluating user's psychological effort, in the reported examples, there are only few works involving these indexes. In addition, due to the current prototypical nature of human augmentation/assistance devices, safety and dependability factors have been poorly dealt with. In the examples above, very few validations including two or more continuous tasks, and at the current level they rely on state machines and vocal commands which do not facilitate switching between tasks, thus interrupting the user's movements (Tingfang et al., 2015).

Generally, there are two main issues associated with the strategy for developing assistive technology for upper and lower extremities, with respect to the mutual interactions: the physical interaction, that is, the mechanical power transfer, and a cognitive interaction, for information exchange (Pons, 2010). These two issues affect each other: a consistent and effective mechanical power transfer is fundamental for the comfort of the wearer, and for the efficiency of the exoskeleton, since it must rely on correct kinematics and kinetics information. On the other hand, the physical interaction is mostly related to the low-level controller of the robotic system—for example, bandwidth of the system, motor own dynamics, performances, and characteristics of the power supply and of the actuation elements (leverages, springs, pneumatic chambers) (Pons, 2010).

3 DESIGN AND IMPLEMENTATION OF EXOSKELETONS

3.1 Kinematics and Dynamics of Exoskeletons

Kinematics can be defined as the branch of mechanics dealing with the description of the motion of bodies or fluids without reference to the forces producing the motion (Pons, 2008). When referring to multi-body, jointed mechanisms as in the case of robots, and more specifically exoskeletons, kinematics deals with analysis of the motion of each robot link with respect to a reference frame (Pons, 2008). Dynamics is the part of classical mechanics that studies objects in motion and the causes of this motion, for example, forces (Pons, 2008). When considering multibody, jointed mechanisms like wearable robots, dynamics deals with the analysis of movement in specific in a configuration and working space as a function of internal forces (e.g., torque at each joint actuator) and external forces (e.g., interaction force with the environment) (Pons, 2008).

The kinematics involves an analytical description of motion as a function of time and the nonlinear relationship between robot end effector position as well as the orientation and robot configuration (Pons, 2008). The mobility, M, of a robot composed of a number of links is defined as the number of independent parameters, $\vec{q}$, required to fully specify the position of every link (Pons, 2008). A particular robot configuration is a vector of realizable values, q_i, $i=1, \ldots, n$, for the independent parameters at time t. The redundancy of a robot is an indicator of the number of available robot configurations for a particular position of the end effector position. High redundancy makes control complex but improves dexterity (Pons, 2008).

From the explanation above, there may be a forward and an inverse relationship between a robot position and orientation and its configuration. The forward kinematics problem deals with the specification of robot position and orientation, as a function of robot configuration. The inverse kinematics involves the determination of robot configuration as a function of robot position and orientation (Pons, 2008). Any type of coordinates system—Cartesian, $\{x, y, z\}$, cylindrical,$\{r, \theta, z\}$, or spherical, $\{\rho, \varphi, \theta\}$ can be used to fully determine robot position. The selection of a particular coordinate system depends on the kinematic structure of the robot (Pons, 2008).

In robotics, the preference is to describe the position and orientation in a more compact form based on the translation and rotation of the coordinate frame. A rotation matrix, R, is a transformation matrix that, when multiplied by a vector, has the effect of changing the direction of the vector but not its magnitude. A rotation is an orthonormal transformation in which the opposite rotation is represented by the transposed version of the original matrix, $R^{-1} = R^T$ (Pons, 2008). The analysis of kinematics of robots is usually based on homogeneous transformation matrices. In jointed, multilink mechanisms like robots, the relative motion of links around a joint can be simply described by homogeneous transformation matrices. Finding the form of the forward kinematic problem for a robot can be approached by the Denavit-Hartenberg (D-H) convention. The D-H convention establishes an algorithm for assigning a set of coordinate systems which are related through translation and rotation transformations. The transformation between successive coordinate systems takes into account the particular kinematics of robot joints, as shown in Eq. (1) (the general form of the transformation matrix between two consecutive coordinate systems) (Pons, 2008).

$$T_{i-1}^{i} = \begin{bmatrix} \cos\theta_i & -\cos\alpha_i \sin\theta_i & \sin\alpha_i \sin\theta_i & \alpha_i \cos\theta_i \\ \sin\theta_i & \cos\alpha_i \theta_i & -\sin\alpha_i \cos\theta_i & \alpha_i \sin\theta_i \\ 0 & \sin\alpha_i & \cos\alpha_i & d_i \\ 0 & 0 & 0 & 1 \end{bmatrix} \quad (1)$$

When considering multibody, jointed mechanisms like wearable robots, dynamics deals with the analysis of movement in a configuration, and working space as a function of internal forces (e.g., torque at each joint actuator) and external forces (e.g., force interactions with the environment) (Pons, 2008). Two instances of the relationship between force and movement can be identified: the forward dynamics problem and the inverse dynamics problem. The forward dynamic model expresses the evolution of joint and working coordinates as a function of the force and torque involved. The Inverse dynamics model describes forces and torques as a function of the evolution of joint coordinates in time (Pons, 2008).

In robotics, Newtonian and Lagrangian mechanics are used to derive the dynamic model of a robot. The Newton-Euler formulation is based on a description of mechanics in vector functions, while the Lagrange-Euler formulation is based on scalar functions, as shown in Eqs. (2), (3), respectively (Barrientos et al., 1997).

$$\sum F = m\dot{v} \quad \sum T = I \times w + w \times (I \times w) \tag{2}$$

where F is the force, m is the mass, and $\dot{v}$ the linear acceleration of the link. In the other equation, T is the torque, I is the inertial matrix, and w the angular velocities of the link.

$$\frac{d}{dt}\frac{\partial \zeta}{\partial \dot{q}_i} - \frac{\partial \zeta}{\partial q_i} = \tau \quad \zeta = k - u \tag{3}$$

where:

q_i, generalized coordinates (in this case articulates)
τ, vector of forces and applied pairs in the
ζ, Lagrangian function
k, kinetic energy
u, potential energy

In addition, equations of motion are equations that describe the behavior of a system as a function of time (Barrientos et al., 1997). Robot dynamics can be represented by linear state-space variable equations. On the other hand, the interaction between different links is described by nonlinear differential equations. It is possible to use the state-space formulation for control design, in either the nonlinear or the linear forms (Barrientos et al., 1997).

3.2 Human Factors and Biomechanics

The exoskeleton should be anthropomorphic and ergonomic, not only in shape but also in function. The exoskeleton should be analogous to the human limbs in the case of joint positions and distribution of DOF. In most of cases, designing the kinematics of an exoskeleton generally consists on trying to replicate human limb kinematics. This approximation has a major disadvantage due to the fact that it is impossible to replicate human kinematics with a mechanic structure and conventional robotic joints (Scott and Winter, 1993). Also, morphology varies among people and several models of human kinematics in the biomechanics literature differ in some aspects, due to the complex geometry, redundancy, and DOF of the human limbs.

Furthermore, unpredictability of joint axes locations and body segment sizes, for instance, can disturb interaction between an exoskeleton and the human operator, depending on the exoskeleton's kinematic design.

This especially applies to exoskeletons that are wearable and kinematically equivalent to the human arm. Typical biomechanical effects that cannot easily be captured within a human arm model used for exoskeleton development include (Schiele and van der Helm, 2006):

- The intersubject variability of human limb link parameters (Denavit-Hartenberg parameters such as length of bones, distances between rotation axes, orientations of rotation axes).
- The variability within an individual subject of joint centers of rotation during movement. This can cause misalignments in the joints axes of exoskeleton and human joints.
- The intersubject variability of body segment dimensions: mass, size, volume, and so forth.

The unavoidable kinematic incompatibility between the robot and the human limb can cause several problems, such as unwanted reaction forces in the human joints, shear forces, and additional pressure at the attachment points. A key aspect of human-exoskeleton interaction relies on an adequate transmission of mechanical power generated by exoskeleton to the human body. Transmitting power from the device to the human body is challenging because biological tissues and interfaces deform and displace when forces are applied, absorbing power. Thus, a part of the mechanical power generated by the exoskeleton is not used for the enhancement of human motor performance, but is absorbed in compression of soft tissues, or lost to unwanted effects (i.e., skin/tissue stretch and slippage of the exoskeleton with respect to the skin).

Effective ways for the exoskeletons to transmit mechanical power to the body are essential. Exoskeletons interact with the human body by means of multiple physical contact points, frequently using a wide physical interface such as a cuff or an orthosis to smoothly transmit the loads to the user. The human-robot physical interface should be designed to provide a safe and comfortable interaction, while transmitting the torque/force to the human body. Conventional "shell and strap" style attachments are found on most of the developed exoskeletons in the literature. These systems consist of a rigid (or semirigid) shell with one or more strap-style fasteners and padding for subject comfort.

3.3 Technologies in Exoskeletons

Robotic exoskeletons involve sensors, actuators, mechanical structures, algorithms, and control strategies capable of acquiring information to execute a motor function. A key feature of exoskeletons is the direct interaction between human and device. This aspect could be divided into cognitive human-robot interaction (cHRI) and physical human-robot interaction (pHRI). cHRI relates to how the user controls the exoskeleton. pHRI relates to the application of controlled forces between human and exoskeleton.

Interaction of exoskeleton with the user involves three main modules: sense, decision, and execution (Knaepen et al., 2014). Developing robotic exoskeletons relates to including technologies to accomplish function of each module.

The sense module acquires the information data from the human operator as well as device sensors. The decision module interprets the sensing information and organizes the activities in the whole system. The execution module is responsible for the actuation, providing mechanical power.

Acquiring information from the human operator for cHRI could be implemented using bioelectric signals such as the electromyogram (EMG), which evaluates and records physiologic properties of muscles; electroencephalogram (EEG, which monitors brain waves), and electrooculogram (EOG, which monitors eye movements). On the other hand, pHRI involves acquiring kinematics and kinetic information. A critical aspect while designing exoskeletons relates to measuring of the interaction forces between the device and the user's limbs, which can be used to assess the performance of the user in executing a task (e.g., the level of effort spent by a patient in completing a therapy). A common way to measure interaction force/torque is to adapt a force sensor between the cuff and the exoskeleton link, which provide accurate measurements. Table 2 shows several sensor technologies to implement cHRI and pHRI.

There are several actuator technologies that have been used to provide mechanical power for exoskeletons, which include pneumatic, hydraulic, and electric actuators (Gopura et al., 2016). Pneumatic and hydraulic actuators have good power-to-weight ratio but unfortunately they don't have much precision, and it is difficult to implement accurate positional control with them due to their nonlinear behavior. Electric actuators are the most used element in the literature for powered exoskeletons, because they could be controlled with high precision; however, the power-to-weight ratio is not so good (Gopura et al., 2016). The series-elastic actuator (SEA) is a kind of actuator that implements a continuously variable transmission between a motor and a series-elastic element, used to power exoskeletons (Veneman, 2007). SEA actuators have been used in a number of exoskeletons because of the inherent compliance.

Table 2 Sensor Technologies to Implement cHRI and pHRI

	Signals to Acquire	**Sensor Technology**
pHRI (physical human-robot interaction)	Kinematic Information	Potentiometer, encoder, electrogoniometer, accelerometer, gyroscopes, IMU
	Kinetic Information	Strain gage, piezoresistive sensor, force/torque sensor
cHRI (cognitive human-robot interaction)	Muscle activity information	Electromyography
	Brain activity information	Electroencephalography
	Ocular movement information	Electrooculography

A special type of pneumatic actuator, called PAMs or McKibben-type actuators are often used in several exoskeletons (Ramos and Meggiolaro, 2014). Such actuators consist of an internal bladder surrounded by braided mesh shell with flexible, but nonextensible, threads. The bladder is pressurized, and the actuator increases its diameter and shortens according to its volume, thus providing tension at its ends.

When selecting actuators for an exoskeleton, it is required to define an appropriate location. Thus, the actuators could be located close to the joints that are actuated. This configuration simplifies power transmission by using direct drives on joint. However, it increases the weight of the distal part of the exoskeleton and the inertia makes it more difficult to control the overall system. On the other hand, locating the actuators in the part that remains constrained reduces the weight and inertia of the distal part. However, a mechanical power transmission mechanism is required. This complicates the mechanical structure and may lead to difficulties with control due to friction.

Energy efficiency is a major problem for robotic exoskeletons. Those systems require considerable energy to accelerate and decelerate the limbs and to dynamically support the body mass against gravity. Supplying power to such devices for several hours is well beyond the capabilities of current battery technology. Currently, there are multiple efforts to develop efficient power sources for exoskeleton aimed to enable ambulatory applications. Lithium polymer batteries, with a specially formed dry polymer, currently offer the advantage of unrestricted shape and can therefore be thinner than the lithium-ion-based design. They provide double the energy density of lithium-ion batteries.

3.4 Control for Exoskeletons

The exoskeleton control system can be categorized according to the model system, the physical parameters, the hierarchy, and the usage. These considerations lead to different control schemes (Anam and Al-Jumaily, 2012).

According to the model-based control system, the control strategy for the skeleton can be divided into two types: the dynamic model and the muscle model-based control (Anam and Al-Jumaily, 2012). The dynamic exoskeleton model is derived through modeling the human body as rigid links joined together by joints (bones). This model is formed from combination of inertial, gravitational, Coriolis, and centrifugal effects (Anam and Al-Jumaily, 2012). The dynamic model can be obtained through three ways: the mathematical model, the system identification, and the artificial intelligent method (Anam and Al-Jumaily, 2012):

- The mathematical model is obtained by modeling the exoskeleton theoretically based on physical characteristics of the system (Anam and Al-Jumaily, 2012).
- The system identification method is based in parameters estimation. In the BLEEX exoskeleton researchers have implemented the least-squares method for swing-phase control (Ghan et al., 2006). The least square is utilized to estimate the parameter of the dynamic model.
- Based on the pairs of input-output data. Aguirre-Ollinger et al. also employed the recursive least square method to estimate the dynamic model parameters of one DOF lower exoskeleton (Aguirre-Ollinger et al., 2007).
- The use of an artificial intelligence method to allow solution many nonlinear problems has attracted some researchers to employ in the dynamic model identification. Xiuxia et al. (2008) used the wavelet neural network to identify the dynamic model of exoskeleton. They implemented the wavelet neural network in the virtual joint torque control as inverse dynamic model.

The muscle models have been used in the exoskeleton control schemes. Unlike the dynamic model, the muscle model predicts the muscle forces deployed by the muscles of the human limb joint as a function of muscle neural activities and the joint kinematics (Anam and Al-Jumaily, 2012).

The input is the EMG signals and the output is force estimation. The muscle model can be obtained by using the parametric and nonparametric muscle model. The parametric muscle model is commonly implemented using the Hill-based muscle model (Anam and Al-Jumaily, 2012). This model can be regarded as the biological mechanics of the musculoskeletal limb model and it is composed of three elements: a contractile element (CE), a series element (SE), and a parallel element (PE). The Hill-based model generates the output as the function of EMG activity and the muscle length. The nonparametric muscle model does not need information of muscle and joint dynamics (Anam and Al-Jumaily, 2012).

Based on the physical parameters, the exoskeleton control system can be classified into position, torque-force, and force interaction controllers (Anam and Al-Jumaily, 2012). The position control scheme is commonly utilized to make sure the exoskeleton joints turn in a desired angle. The control system based on torque-force controller is generally applied in the low-level controller; meanwhile, the high-level controller is the impedance controller which controls the interaction force between human and the exoskeleton (Anam and Al-Jumaily, 2012). The main goal of torque/force controller is to provide proper help for the users in performing a task so that the force of human-exoskeleton interaction goes to zero. The impedance controller is an extension of position control and it does not only control the position and the force but also control a relation and an interaction between the exoskeleton and the human body, the output of the impedance model is the force that becomes the reference force for the force-torque controller. This interaction is applied as the high-level controller; its main goal is to provide proper help for the users in performing a task so that the force of human-exoskeleton interaction goes to zero. The interaction force can be controlled by either the impedance controller or the admittance controller (Anam and Al-Jumaily, 2012). The basic characteristic of the impedance controller is that it accepts position and produces force. While, the admittance controller is the opposite of the impedance controller; it accepts the force and yields the position (Anam and Al-Jumaily, 2012).

From the hierarchy point of view, the exoskeleton control system can be grouped into three levels, which they are task-level, high-level, and low-level controllers (Anam and Al-Jumaily, 2012). The task-level controller is the highest level controller whose function is based on the task designed. The high-level controller is responsible for controlling the force of human-exoskeleton interaction based on the information from the task-level controller. In the low-level controller, which is the lowest level, the duty is to control the position or force of the exoskeleton joints. This controller interacts directly with the exoskeleton (Anam and Al-Jumaily, 2012).

In the usage-based control systems, the exoskeleton control system can also be categorized according to the sort of applications such as the VR controller, the tele-operation controller, and the gait controller (Anam and Al-Jumaily, 2012). This controller has been applied in the most upper-limb exoskeletons for use on VR controllers; for example: in performing therapy exercises, it guides and helps the patient to carry on the tasks such as a virtual object reaching, an object moving by virtual hand, a ball game, a labyrinth game, a virtual wall painting, and a reaching and motion constrain task (Anam and Al-Jumaily, 2012). In those applications, the exoskeletons are considered as haptic devices.

4 EXOSKELETONS: CHALLENGES AND TRENDS

4.1 Applications

Trends of exoskeletons can be split up into two different applications: medical and nonmedical. Medical applications focus on enhancing or recovering human motor function for a wide range of patients several neuromotor disabilities. On the other hand, nonmedical applications focus on the industrial, military, and entertainment fields.

4.1.1 Medical Applications

Rehabilitation applications are one of most dynamic fields for exoskeletons, which are designed to assist paralyzed patients, and they should be able to respond to any command control made by the patient. This must be based on a precise control of the mechanical interaction with the patient's limb (Ruiz et al., 2008). Furthermore, with other applications, more than assisting the movement, the goal is to help the patient recover his/her sensorimotor capability. Brain computer interface systems promise to enhance application for sensorimotor and neuromotor rehabilitation of patients integrating user commands directly from brain. Recently, a study of a mind-control exoskeleton that permits to a patients group regain sensation and move previously paralyzed muscles was done by Donati et al. (2016).

The biological actuators of human body muscles can be used instead of external actuators. For this purpose, a controlled electrical stimulation of the muscles leading to their contraction can be applied (Doucet et al., 2012). The electrical stimulation can be used to generate muscle contraction in otherwise paralyzed limbs to produce functions such as grasping, walking, and standing. This electrical stimulation is known as functional electrical stimulation (FES). Hybrid systems incorporating exoskeletons with other technologies such as FES have been reported in the literature (del-Alma et al., 2014). Hybrid actuation and control have a considerable potential for walking rehabilitation, with adequately control strategies of hybrid systems that command FES and robotic controllers.

4.1.2 Nonmedical Applications

Currently, wearable robotics designed to be part in an industrial setting is the fastest growing field of exoskeleton research. Exoskeletons for industry and the workplace offer three main advantages: reduction in work-related injuries, saving billions of dollars in medical fees, sick leave, and lawsuits. Exoskeleton has lowered worker fatigue, leading to increased worker alertness, productivity, and work quality. It has the ability to keep quality and experienced personnel past their physical prime in the work force longer.

In addition of using exoskeletons for the human motor performance of soldiers, the military are looking to build VR simulators for troop training (e.g., firing a cannon). Exoskeletons have been envisioned to support this kind of training.

Other future applications of exoskeletons focus on gaming. There are commercial organizations aimed to develop a full exoskeleton that is suspended in the air and provide the appropriate resistance to make the user feel they are walking, swimming, or interacting with objects. For example, to swing a virtual axe, the player will have to feel resistance at the hands via a glove-type exoskeleton. Currently, gaming exoskeletons do not aim to simulate entire objects but just their effects. If a gamer is playing a first-person shooter then a vest could compress to simulate the player being hit.

4.2 Technologies

Technologies are in most instances the limiting factor in developing new exoskeletons. Exoskeletons for portable and ambulatory applications are limited in the literature, one of the reasons being a lack of enabling technologies. Ambulatory scenarios require miniaturized, robust, and energetically efficient technologies, for example, control, sensors, and actuators. Challenges and trends of technologies for exoskeletons can be split up into a generic categorization applicable for any mechatronic system: a signal domain (e.g., controllers, sensors); energy domain that includes the source of energy and the conversion into mechanical power that is applied through the system; and mechanical domain that involves how mechanical power is transported and how the different joints are supported (e.g., cables, linkages, transmission).

4.2.1 Signal Domain

In order to enhance cHRI of users controlling exoskeletons using bioelectric signals, multiple-source signal fusion is an emerging approach. Signal fusion permits that multimodal signals be combined to provide sufficient information for motion intention decoding.

Thus, sEMG plays an important role in the control of exoskeletons taking into account its relative ease of acquisition and abundant content of neural information; however, implementation of the EMG-based pattern recognition algorithms is not easy to be accomplished due to some difficulties, such as EMG signals are time varying and highly nonlinear. Furthermore, the activity level of each muscle for a certain motion is different between each person. A trend for EMG-based controlled exoskeleton relies on using non-EMG signals that are combined with sEMG signals to realize a more precise extraction of motor commands. Furthermore, acquisition of information using high-density sensors array provide more information to improve control.

4.2.2 Energy Domain

In several applications, the exoskeleton must be able to generate high forces to sustain, assist, and/or perturb the motor capabilities of the user. Thus, taking into account of current actuator technologies with characteristic of size, weight, and torque, it is limited to power multiple joints.

A trend for actuator technologies is muscle-like actuators, which are built using soft materials that have good properties, and they behave like human muscles. Most of them are made of elastomers, including silicon and rubber, and so they are inherently safe. This technology enables the development of "soft exoskeletons" (Majidi, 2014). Active polymers appear promising, being thin, lightweight, compliant and able to perform both sensing and actuation. However, fundamental enhancements would be required for the feasible use in exoskeletons. Similar to shape-memory alloys, forces are generally low and take time to build up (i.e., low bandwidths), which results in the need for large stacked configurations (Villoslada et al., 2015).

4.2.3 Mechanical Domain

An exoskeleton must be able to interact with the human body, a very complex kinematic structure that includes multiple DOF. Exoskeletons must have a large number of active joints, each with a wide range of motion to be able to follow as well as to assist movements within a large workspace. For rehabilitation applications, the majority of existing exoskeletons cannot be widely used by patients with limited functions of the upper and lower limbs because they are heavy, dependent on external power supply, and expensive (Ruiz et al., 2006).

Smaller, lighter actuators with gearboxes could generate sufficient forces; however, gearboxes add friction to the system, reducing overall dynamic performance. The power transmission technologies with high transmission efficiency and minimum friction are required so the exoskeleton systems can be more efficient. Moreover, the back-drivability of the transmission is also essential for these systems to eliminate possible discomfort to the user.

According to appearance, most of existing robotic exoskeletons often cause discomfort to the user, especially when they wear it for daily activities. Thus, a challenge for new exoskeletons is to improve esthetics.

4.3 Exoskeleton Design

In the field of healthcare, field exploration, and cooperative human assistance, robots and machines must become increasingly less rigid and specialized and instead approach the mechanical compliance and versatility of materials and organisms found in nature. As with their natural counterparts, this next generation of robots must be elastically soft and capable of safely interacting with humans or navigating through tightly constrained environments (Majidi, 2014). This is the choice of Soft robots, which are primarily composed of easily deformable matter such as fluids, gels, and elastomers that match the elastic and rheological properties of biological tissue and organs. A soft robot must adapt its shape and locomotion strategy for a broad range of tasks, obstacles, and environmental conditions (Majidi, 2014). This emerging class of elastically soft, versatile, and biologically inspired machines represents an exciting and highly interdisciplinary paradigm in engineering that could revolutionize the role of robotics in health care, field exploration, and cooperative human assistance. The most immediate application of emerging soft robot technologies will be in the domain of human motor assistance and co-robotics. For example, a soft active ankle-foot orthotic (AFO) could help prevent foot dragging for patients that suffer gait abnormalities such as drop foot (Majidi, 2014).

The Harvard's soft exosuit team provided a first proof-of-concept results showing that its wearable robot could lower energy expenditure in healthy people walking with a load on their back (Panizzolo et al., 2016). Lightweight exosuits exoskeleton are a new class of soft robots that combine classical robotic design and control principles with functional apparel to increase the wearer's strength, balance, and endurance (Panizzolo et al., 2016). Soft Exosuits offer a new way to assist the elderly in maintaining or restoring their gait, in rehabilitating children and adults with movement disorders due to stroke, multiple sclerosis, and Parkinson's disease, or to ease the physical burden of soldiers, firefighters, paramedics, farmers, and others whose jobs require them to carry extremely heavy loads (Panizzolo et al., 2016). For decades, engineers have built exoskeletons that use rigid links in parallel with the biological anatomy to increase strength and endurance in wearers, and to protect them from injury and physical stress. A number of systems have been developed that show strong commercial potential, for example, in helping spinal-cord injury patients walk, or enabling soldiers carry heavy loads. However, rigid exoskeletons often fail to allow the wearer to perform his or her natural joint movements, are generally heavy and can hence cause fatigue.

Wyss Institute researchers are pursuing a new paradigm (Panizzolo et al., 2016): the use of soft clothing-like "exosuits." An exosuit (Fig. 6) does not contain any rigid elements, so the wearer's bone structure must sustain all the compressive forces normally encountered by the body—plus the forces generated by the exosuit. The soft exosuits translating small amounts of force, applied by mechanical actuators in the suit at the right time into effective motions. In addition to soft exosuits that enhance the functionality of lower extremities, ongoing work at the Wyss is also developing prototypes that improve mobility of the upper extremities. The Wyss Institute is collaborating with ReWalk Robotics, Ltd., to accelerate the development of the Institute's lightweight, wearable soft exosuit technologies for assisting people with lower-limb disabilities (Panizzolo et al., 2016). The agreement with ReWalk will help speed the design of assistive exosuits that could help patients suffering from stroke and multiple sclerosis to regain mobility.

During its development, the soft exosuit has inspired the innovation of entirely new forms of functional textiles, flexible power systems, soft sensors, and control strategies that integrate the suit and its wearer in ways that mimic the natural biomechanics of the human musculoskeletal system (Majidi, 2014; Panizzolo et al., 2016). Study coauthor and biomechanics expert Ken Holt, Ph.D., P.T., Associate Professor at Boston University's Department of Physical Therapy and Athletic Training, has worked alongside

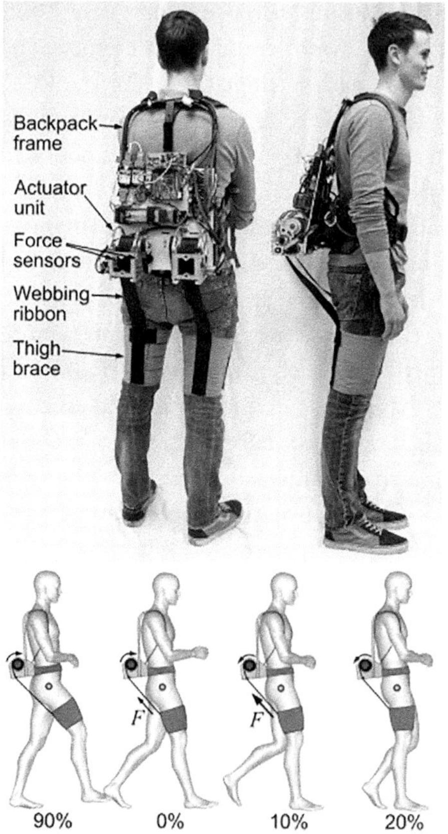

Fig. 6 Top, exosuit to assist hip extension. Two actuator units are mounted on a backpack frame, and connected to cloth thigh braces with webbing. Bottom, overview of device operation. Starting at 90% in the gait cycle, which extends from one heelstrike to the next, the actuator units retract the webbing. *(From Asbeck, A.T., Schmidt, K., Walsh, C.J., 2015. Soft exosuit for hip assistance. Robot. Auton. Syst. 73, 102–110, with permission from Elsevier.)*

Walsh and the team since the beginning of the project and has helped the team grow their expertise in running protocols to evaluate the effect of the exosuit on wearers (Panizzolo et al., 2016).

The soft exosuit team's researchers found that wearers significantly adapted their gait with increasing levels of assistance. The changes were most significant at the ankle joint but also at the hip as the exosuit included straps coupling the assistance from the back of the lower legs to the front of the hip in a beneficial manner. Other studies had reported that there can be an energy transfer between the ankle and other joints (Panizzolo et al., 2016).

4.4 Control for Exoskeleton

Despite decades of research relating to multifunctional myoelectric control, there is much to be made before myoelectric control can effectively be integrated into daily life commercial applications. From the ability to extract proper muscle activity information for most potential users, the future of simultaneous multifunctional control applications relies on producing reliable control schemes utilizing robust representations of muscle synergies (Ison and Artemiadis, 2014). The future directions should lie in three main areas: the development of real-time control applications and standardized metrics to compare performance across differing techniques and improve the surface electromyography recordings through high-density surface EMG (HDsEMG) and the development of hybrid prediction and learning schemes for user-friendly control (Ison and Artemiadis, 2014).

The motor learning-based control schemes train a motor system to develop and refine synergies associated with system dynamics of a specific mapping function relating sEMG inputs with control outputs. The user learns the system dynamics via feedback while interacting with the control interface. This scheme consistently reports significant learning while achieving good performance metrics (Ison and Artemiadis, 2014). These metrics are generally specific to the given task and are difficult to compare to other control methods implemented in real time. For real-time implementation and testing of control schemes, it is necessary to standardize metrics that can compare performance and efficiencies across different schemes, including comparisons between pattern recognition and motor learning (Ison and Artemiadis, 2014).

Advancements in recording technology have made HDsEMG electrodes a viable option for myoelectric controllers. The high-density electrodes provide a more complete set of information to allow for richer processing and more robust control schemes (Ison and Artemiadis, 2014). From a macroscale view, HDsEMG provides opportunities to describe two-dimensional distributions of muscle activity as well as intensity, compensating for electrode shift and cross talk. In addition, provides redundancy in signals such that they can be subsets that allow for more efficient estimation without losing control performance (Ison and Artemiadis, 2014).

Attempts to reduce the training phases have been made in classification schemes using adaptive learning and pretrained models. The hybrid approach, which aims to use natural population-wide approaches in order to develop new forms of synergies, may be the key to efficient, user-friendly simultaneous multifunctional control that would be widely accepted by users (Ison and Artemiadis, 2014). Pattern recognition-based schemes commonly require an initial adaptation to the outputs when controlling the device in real time. Gibson et al demonstrated that a control scheme trained on a variety of users can extract the low-level population-wide synergies and provide good performance in offline analysis, and better performance in real-time given visual feedback (Ison and Artemiadis, 2014). Recent implementations of simple control schemes based on extracted synergies have shown robust performance compared to more complex classifiers. These results suggest that intuitive, user-independent control schemes can be developed to provide user-friendly, low-level control without requiring an intense training phase from the user (Ison and Artemiadis, 2014).

Similarly, recent trends and attempts in developing electroencephalography (EEG)-based control methods have shown the potential of this area in the modern bio-robotics field. A new approach of combining both control methods, which use the advantages, and diminish the disadvantages, of each system might therefore be a promising approach (Lalitharatne et al., 2013). In this case, EEG signals can be used to compensate for insufficient information in the EMG signals. Numerous examples such as wheelchairs, prosthetics, exoskeletons/orthoses (Lalitharatne et al., 2013) show the effectiveness of EMG-based control methods. However, these EMG-based control approaches used alone have some disadvantages that depend on the user and on the application. In cases where the user cannot generate sufficient muscle signals, EMG-based methods are not useful for movement intention detection. For example, a person who has a totally paralyzed upper limb may not be able to use a device such as an exoskeleton due to the difficulty of getting control signals from the muscles of the paralyzed limb. In this case also, EEG can be used to compensate for the missing EMG signals (Lalitharatne et al., 2013). Even if all required muscles for EMG are available, EEG can still be used to remove the effect of fatigue or undesired tremor. Applications of the hybrid approaches may vary from a simple game control application for an able-bodied person through to a prosthetic arm and exoskeleton control application for an amputee or motor disabilities person.

Technology is one of the limiting factors for hybrid EEG-EMG-based control approaches. High-density EEG systems can provide a lot of details, but it is sometimes not practical to use such systems when they cover the whole head of the user, as the user may feel uncomfortable (Lalitharatne et al., 2013). Compact and low-weight designs for EEG and EMG data measuring systems need to be introduced, in order to allow use when users need to move around. Nevertheless, there are many issues yet to improve the effectiveness of the hybrid EEG-EMG methods for use in bio-robotic applications (Lalitharatne et al., 2013). It is important for future studies to present quantitative performance evaluations for such hybrid EEG-EMG approaches, in order to demonstrate their effectiveness in comparison to other control methods.

5 CONCLUSION

The capacity of applying dynamic forces to the body and specifically to upper and lower limbs opens the application field of exoskeletons. Those devices are designed to enhance the human motor performance by the external framework, in the military, in the industry, and for medical applications. Nowadays, the exoskeleton systems are forging ahead with high integration using other emerging technologies including VR, haptics, videogames, and soft robotics, among others. However, several challenges remain and there are some design constraints in the development of exoskeletons. When developing portable exoskeletons, a tradeoff between power and weight must be considered. Specifically, advances in actuation and energy storage technologies, intelligent power management, and mechanical design are required before seeing exoskeletons widespread use.

REFERENCES

Aguirre-Ollinger, G., Colgate, J.E., Peshkin, M.A., Goswami, A., 2007. In: Active-impedance control of a lower-limb assistive exoskeleton.Rehabilitation Robotics, ICORR 2007, IEEE 10th International Conference, pp. 188–195.

Anam, K., Al-Jumaily, A.A., 2012. Active exoskeleton control systems: state of the art. Procedia Eng. 41, 988–994.

Barrientos, A., Peñín, L.F., Balaguer, C., Aracil, R., 1997. Fundamentos de robótica. vol. 256. McGraw-Hill, Madrid, Spain.

Bogue, R., 2015. Robotic exoskeletons: a review of recent progress. Ind. Robot Int. J. 42 (1), 5–10.

Brokaw, E.B., Nichols, D., Holley, R.J., Lum, P.S., 2013. Robotic therapy provides a stimulus for upper limb motor recovery after stroke that is complementary to and distinct from conventional therapy. Neurorehabil. Neural Repair 28 (4), 367–376.

Carignan, C., Liszka, M., Roderick, S., 2005. In: Proc. IEEE Int. Conf. on Advanced Robotics (ICAR), Seattle (Ed.), Design of an exoskeleton with scapula motion for shoulder rehabilitation. pp. 524–531.

Chang, W.H., Kim, Y.H., 2013. Robot-assisted therapy in stroke rehabilitation. J. Stroke 15 (3), 174–181.

Colombo, G., 2001. Driven gait orthosis for improvement of locomotor training in paraplegic patients. Spinal Cord 39, 252–255.

de Looze, M.P., Bosch, T., Krause, F., Stadler, K.S., O'Sullivan, L.W., 2015. Exoskeletons for industrial application and their potential effects on physical work load. Ergonomics 59 (5), 671–681.

del-Alma, A.J., Gil-Agudo, A., Pons, J.L., Moreno, J.C., 2014. Hybrid FES-robot cooperative control of ambulatory gait rehabilitation exoskeleton. J. Neuroeng. Rehab. 11 (27), 1–15.

Dollar, A.M., Hugh, H., 2008. Lower extremity exoskeletons and active orthoses: challenges and state-of-the-art. IEEE Trans. Robot. 24 (1), 144–158.

Donati, A.R.C., Shokur, S., Morya, E., et al., 2016. Long-term training with brain-machine interfaces induces partial neurological recovery in paraplegic patients. Sci. Rep. 6 (30383), 1–16.

Doucet, B.M., Lam, A., Griffin, L., 2012. Neuromuscular electrical stimulation for skeletal muscle function. Yale J. Biol. Med. 85 (2), 201–215.

Esquenazi, A., Talaty, M., Packel, A., Saulino, M., 2012. The rewalk powered exoskeleton to restore ambulatory function to individuals with thoraciclevel motor-complete spinal cord injury. Am. J. Phys. Med. Rehab. 91, 911–921.

Fleerkotte, B.M., Koopman, B., Buurke, J.H., van Asseldonk, E.H., van der Kooij, H., Rietman, J.S., 2014. The effect of impedance-controlled robotic gait training on walking ability and quality in individuals with chronic incomplete spinal cord injury: an explorative study. J. Neuroeng. Rehab. 11 (26), 1–15.

Fleischer, C., Kondak, K., Wege, A., Kossyk, I., 2009. In: Research on exoskeletons at the TU Berlin.Proc. German Workshop on Robotics, 9–10 June, Braunschweig, Germany.

Garcia, E., Sater, J.M., Main, J., 2002. Exoskeletons for human performance augmentation(EHPA): a program summary. J. Robot. Soc. Japan 20 (8), 44–48.

Ghan, J., Steger, R., Kazerooni, H., 2006. Control and system identification for the Berkeley lower extremity exoskeleton (BLEEX). Adv. Robot. 20 (9), 989–1014.

Gijbels, D., Lamers, I., Kerkhofs, L., Alders, G., Knippenberg, E., Feys, P., 2011. The Armeo Spring as training tool to improve upper limb functionality in multiple sclerosis: a pilot study. J. Neuroeng. Rehab. 8 (5), 1–8.

Gopura, R.A.R.C., Bandara, D.S.V., Kiguchi, K., Mann, G.K.I., 2016. Developments in hardware systems of active upper-limb exoskeleton robots: a review. Robot. Auton. Syst. 75, 203–220.

Hasegawa, Y., Mikami, Y., Watanabe, K., Sankai, Y., 2008. In: Five-fingered assistive hand with mechanical compliance of human finger.IEEE International Conference Robotics and Automation (ICRA), Pasadena, CA, pp. 718–724.

He, H., Kiguchi, K., 2007. A study on EMG-based control of exoskeleton robots for human lower-limb motion assist.Information Technology Applications in Biomedicine, ITAB 2007, 6th International Special Topic Conference, pp. 292–295.

Hesse, S., Schulte-Tigges, G., Konrad, M., Bardeleben, A., Werner, C., 2003. Robot-assisted arm trainer for the passive and active practice of bilateral forearm and wrist movements in hemi paretic subjects. Arch. Phys. Med. Rehabil. 84 (6), 915–920.

Hristic, D., Vukobratovic, M., 1973. In: Development of active aids for handicapped.Proc. III International Conference on Biomedical Engineering. Sorrento, Italy, pp. 123–129.

Hyon, S., Morimoto, J., Matsubara, T., Noda, T., Kawato, M., 2011. In: XoR: hybrid drive exoskeleton robot that can balance.Intelligent Robots and Systems, IROS, IEEE/RSJ International Conference, pp. 3975–3981.

Ison, M., Artemiadis, P., 2014. The role of muscle synergies in myoelectric control: trends and challenges for simultaneous multifunction control. J. Neural Eng. 11(5), 051001.

Kawasaki, H., Ito, S., Ishigure, Y., Nishimoto, Y., Aoki, T., Mouri, T., Sakaeda, H., Abe, M., 2007. In: Development of a hand motion assist robot for rehabilitation therapy by patient self-motion control.Proc. IEEE 10th International Conference on Rehabilitation Robotics(ICORR). Noordwijk, Netherlands, pp. 234–240.

Kazerooni, H., 1990. Human–robot interaction via the transfer of power and information signals. IEEE Trans. Syst. Man Cyber 20 (2), 450–463.

Kazerooni, H., 2008. In: A review of the exoskeleton and human augmentation technology. ASME Dynamic Systems and Control Conference, 20–22 October, Michigan, USA, pp. 1549–1557.

Kazerooni, H., Steger, R., Hung, L., 1968. Hardiman I prototype project, special interim study. In: General electric report, pp. S-68–1060.

Kiguchi, K., 2007. Active exoskeletons for upper-limb motion assist. Int. J. Humanoid Robot. 4 (03), 607–624.

Kiguchi, K., Tanaka, T., Fukuda, T., 2004. Neuro-fuzzy control of a robotic exoskeleton with EMG signals. IEEE Trans. Fuzzy Syst. 12, 481–490.

Knaepen, K., Beyl, P., Duerinck, S., Hagman, F., Lefeber, D., Meeusen, R., 2014. Human-robot interaction: kinematics and muscle activity inside a powered compliant knee exoskeleton. IEEE Trans. Neural. Syst. Rehabil. Eng. 22 (6), 1128–1137.

Kousidou, S., Tsagarakis, N., Caldwell, D.G., Smith, C., 2006. In: Assistive exoskeleton for task based physiotherapy in 3-dimensional space.Biomedical Robotics and Biomechatronics, BioRob 2006, The First IEEE/RAS-EMBS International Conference, pp. 266–271.

Kwa, H.K., Noorden, J.H., Missel, M., Craig, T., Pratt, J.E., Neuhaus, P.D., 2009. In: Development of the IHMC mobility assist exoskeleton.Robotics and Automation, ICRA'09, IEEE International Conference on, IEEE, pp. 2556–2562.

Kwakkel, G., Kollen, B.J., Krebs, H.I., 2008. Effects of robot-assisted therapy on upper limb recovery after stroke: a systematic review. Neurorehabil. Neural Repair 22 (2), 111–121.

Lalitharatne, T.D., Teramoto, K., Hayashi, Y., Kiguchi, K., 2013. Towards hybrid EEG-EMG-based control approaches to be used in bio-robotics applications: current status, challenges and future directions. Paladyn J. Behav. Robot. 4 (2), 147–154.

Li, Q., Wang, D., Du, Z., Song, Y., Sun, L., 2006. In: sEMG based control for 5 DOF upper limb rehabilitation robot system.Proc. IEEE International Conference on Robotics and Biomimetics (ROBIO), Kunming, China, pp. 1305–1310.

Li, Z., Xie, H., Li, W., Yao, Z., 2014. Proceeding of human exoskeleton technology and discussions on future research. Chinese J. Mech. Eng. 27 (3), 437–447.

Lobo-Prat, J., Kooren, P.N., Stienen, A.H., Herder, J.L., Koopman, B.F., Veltink, P.H., 2014. Non-invasive control interfaces for intention detection in active movement-assistive devices. J. Neuroeng. Rehabil. 11 (168), 1–22.

Louie, D.R., Eng, J.J., 2016. Powered robotic exoskeletons in post-stroke rehabilitation of gait: a scoping review. J. Neuroeng. Rehabil. 13 (53), 1–10.

Maciejasz, P., Eschweiler, J., Gerlach-Hahn, K., Jansen-Troy, A., Leonhardt, S., 2014. A survey on robotic devices for upper limb rehabilitation. J. Neuroeng. Rehabil. 11 (3), 1–29.

Majidi, C., 2014. Soft robotics: a perspective—current trends and prospects for the future. Soft Robot. 1 (1), 5–11.

Manto, M., Topping, M., Soede, M., Sanchez-Lacuesta, J.J., Harwin, W., Pons, J.L., Willimas, J., Skaarup, S., Normie, L., 2003. Dynamically responsive intervention for tremor suppression. IEEE Eng. Med. Biol. Mag. 22 (3), 120–132.

Mayr, A., Kofler, M., Saltuari, L., 2008. ARMOR: an electromechanical robot for upper limb training following stroke. A prospective randomized controlled pilot study. Handchir. Mikrochir. Plast. Chir. 40, 66–73.

Merodio, D.S., Soto, M.C., Arevalo, J.C., Armada, E.G., 2012. Control motion approach of a lower limb orthosis to reduce energy consumption. Int. J. Adv. Robot. Syst. 9, 1–8.

Milot, M.H., Spencer, S.J., Chan, V., Allington, J.P., Klein, J., Chou, C., Bobrow, J.E., Cramer, S.C., Reinkensmeyer, D.V., 2013. A crossover pilot study evaluating the functional outcomes of two different types of robotic movement training in chronic stroke survivors using the arm exoskeleton BONES. J. Neuroeng. Rehabil. 10 (112), 1–12.

Nakamura, T., Saito, K., Wang, Z., Kosuge, K., 2005. In: Realizing model-based wearable antigravity muscles support with dynamics terms.Intelligent Robots and Systems, IROS 2005, IEEE/RSJ International Conference, pp. 2694–2699.

Nef, T., Mihelj, M., Colombo, G., Riener, R., 2006. In: ARMin-robot for rehabilitation of the upper extremities.Robotics and Automation, ICRA 2006, Proceedings 2006 IEEE International Conference, pp. 3152–3157.

Nilsson, A., Vreede, K.S., Häglund, V., Kawamoto, H., Sankai, Y., Borg, J., 2014. Gait training early after stroke with a new exoskeleton—the hybrid assistive limb: a study of safety and feasibility. J. Neuroeng. Rehabil. 11 (92), 1–11.

Panizzolo, F.A., Galiana, I., Asbeck, A.T., Siviy, C., Schmidt, K., Holt, K.G., Walsh, C.J., 2016. A biologically-inspired multi-joint soft exosuit that can reduce the energy cost of loaded walking. J. Neuroeng. Rehabil. 13 (43), 1–14.

Pons, J.L., 2008. Wearable Robots: Biomechatronic Exoskeletons. John Wiley & Sons, New Jersey, USA.

Pons, J.L., 2010. Rehabilitation exoskeletal robotics. IEEE Eng. Med. Biol. Mag. 29 (3), 57–63.

Pratt, J.E., Krupp, B.T., Morse, C.J., Collins, S.H., 2004. In: The roboknee: an exoskeleton for enhancing strength and endurance during walking.Robotics and Automation, Proceedings ICRA'04, IEEE International Conference on vol. 3, IEEE, pp. 2430–2435.

Ramos, J.L., Meggiolaro, M.A., 2014. In: Use of surface electromyography for human amplification using an exoskeleton driven by artificial pneumatic muscles. 5th IEEE/RAS-EMBS International Conference on Biomedical Robotics and Biomechatronics (BioRob 2014), 12–15 August, Sao Paulo, Brazil,

Ronsse, R., Lenzi, T., Vitiello, N., Koopman, B., Van Asseldonk, E., De Rossi, S.M.M., Van den Kieboom, J., Van der Kooij, H., Carrozza, M.C., Ijspeert, A.J., 2011. Oscillator-based assistance of cyclical movements: model-based and model free approaches. Med. Biol. Eng. Comput. 49, 1173–1185.

Ruiz, A.F., Forner-Cordero, A., Rocon, E., Pons, J.L., 2006. In: Exoskeletons for rehabilitation and motor control.The First IEEE/RAS-EMBS International Conference on Biomedical Robotics and Biomechatronics.

Ruiz, A.F., Rocon, E., Raya, R., Pons, J.L., 2008. In: Coupled control of human-exoskeleton systems: an adaptative process.Conference on Human System Interactions.

Sanchez, R., Reinkensmeyer, D., Shah, P., Liu, J., Rao, S., Smith, R., Cramer, S., Rahman, T., Bobrow, J., 2004. Monitoring functional arm movement for home-based therapy after stroke.Conference Proceedings IEEE Engineering Medical Biology Society, San Francisco, CA, vol. 7, pp. 4787–4790.

Sawicki, G.S., Ferris, D.P., 2009. A pneumatically powered knee-ankle-foot orthosis (KAFO) with myoelectric activation and inhibition. J. Neuroeng. Rehabil. 6 (23), 1–16.

Schiele, A., van der Helm, F.C.T., 2006. Kinematic design to improve ergonomics in human machine interaction. IEEE Trans. Neural Syst. Rehabil. Eng. 14 (4), 456–469.

Schmidt, H., Werner, C., Bernhardt, R., Hesse, S., Krüger, J., 2007. Gait rehabilitation machines based on programmable footplates. J. Neuroeng. Rehabil 4 (2), 1–7.

Scott, S.H., Winter, D.A., 1993. Biomechanical model of the human foot: kinematics and kinetics during the stance phase of walking. J. Biomech. 26, 1091–1104.

Seireg, A., Grundmann, J., 1981. Design of a Multitask Exoskeletal Walking Device for Paraplegics. Biomechanics of Medical Devices Inc, New York, pp. 569–644.

Tingfang, Y., Marco, C., Calogero, M.O., Nicola., V., 2015. Review of assistive strategies in powered lower-limb orthoses and exoskeletons. Robot. Auton. Syst. 64, 120–136.

Veneman, J.F., 2007. Design and evaluation of the gait rehabilitation robot LOPES. PhD thesis, University of Twente, Enschede, The Netherlands.

Villoslada, A., Flores, A., Copaci, D., Blanco, D., Moreno, L., 2015. High-displacement flexible shape memory alloy actuator for soft wearable robots. Robot. Auton. Syst. 73, 91–101.

Xiuxia, Y., Gui, L., Zhiyong, Y., Wenjin, G., 2008. Lower extreme carrying exoskeleton robot adative control using wavelet neural networks. In: Natural Computation, ICNC'08, Fourth International Conference, pp. 399–403.

Yagn, N., 1890. Apparatus for Facilitating Walking, Running, and Jumping. U.S. Patents 420 179 and 438 830.

Yang, C., Zhang, J., Chen, Y., et al., 2008. A review of exoskeleton-type systems and their key technologies. Proc. Inst. Mech. Eng. C 222 (8), 1599–1612.

Zatsiorsky, V.M., 1998. Kinematics of Human Motion. Human Kinetics, Urbana Champaign, IL.

Zoss, A.B., Kazerooni, H., Chu, A., 2006. Biomechanical design of the Berkeley lower extremity exoskeleton (BLEEX). IEEE/ASME Trans. Mechatron. 11 (2), 128–138.

FURTHER READING

Asbeck, A.T., Schmidt, K., Walsh, C.J., 2015. Soft exosuit for hip assistance. Robot. Auton. Syst. 73, 102–110.

Perry, J.C., 2006. Design and development of a 7 degree-of-freedom powered exoskeleton for the upper limb. PhD dissertation, University of Washington, Seattle, WA.

Rocon, E., Ruiz, A.F., Belda-Lois, J.M., Moreno, J.C., Pons, J.L., Raya, R., Ceres, R., 2008. Diseño, desarrollo y validación de dispositivo robótico para la supresión del temblor patológico. Revista Iberoamericana de Automática e Informática Industrial RIAI 5 (2), 79–92.

Sergeyev, A., Alaraje, N., Seidel, C., Carlson, Z., Breda, B., 2013. In: Design of a pneumatically powered-wearable exoskeleton with biomimetic support and actuation.Proc. Aerospace Conference, 2013 IEEE, 2–9 March 2013.

CHAPTER NINE

Upper Extremity Rehabilitation Robots: A Survey

Borna Ghannadi, Reza Sharif Razavian, John McPhee
Systems Design Engineering, University of Waterloo, Waterloo, ON, Canada

Contents

1 INTRODUCTION

Upper extremity movement defects are caused by different sources such as upper limb component injuries and surgeries, overuse (Skirven et al., 2011), stroke, traumatic brain injury, spinal cord injury, motoneuron defects, and neurological diseases such as cerebral palsy and Parkinson's disease (Maciejasz et al., 2014). Most of these defects need sessions of physical therapy to improve joint range of motion (ROM), strengthen muscles (Skirven et al., 2011), restore functional capabilities, and resolve impairments (Maciejasz et al., 2014).

Stroke causes longstanding impairments, and it has a noticeable risk factor in older adults. One-sixth of people worldwide will experience stroke in

Handbook of Biomechatronics
https://doi.org/10.1016/B978-0-12-812539-7.00012-X

their lifetime; 15 million people are suffering from stroke every year. Following these trends, it is estimated that 23 million stroke cases will happen in 2030 (Mendis, 2013). Thus, procedures to rehabilitate this long-term disability are essential (Broeren et al., 2004; Oujamaa et al., 2009; Turolla et al., 2013; Hatem et al., 2016). Studies have reported that following a stroke upper extremity motor defects have the highest prevalence among movement disorders (Bansil et al., 2012; Mehrholz et al., 2012). Therefore, rehabilitation approaches for upper extremity motor control and function recovery are of importance. Consequently, this chapter will focus on upper extremity movement disorders in poststroke patients.

Neurological complications of stroke are various (Fulk et al., 2014) and need to be considered in rehabilitation therapy. Some of these complications are:

1. *Hemispheric behavioral differences*: Stroke patients may show different behaviors in doing a task. Those with right hemiplegia have difficulty accomplishing consecutive tasks; these patients may need some assistance in their therapy. On the other hand, patients with left hemiplegia have task perception problems, and they overestimate their abilities. Fluctuations in doing a task are common among them. To address the wrong perception, safety issues should be considered carefully.
2. *Perceptual dysfunction*: It is common among left hemiplegia patients, and can be revealed as one of these symptoms: body scheme, spatial relation, and agnosia. The body scheme is the difficulty in realizing the relationship between body parts. The spatial relation is having trouble in perceiving the relationship between body and other objects. The agnosia is the problem in distinguishing incoming information, which can be visual, auditory, or tactile.
3. *Osteoporosis and fracture risk*: Because of the lack of physical activity, these patients may get osteoporosis. Osteoporosis is a bone disease for which the mass of bone will decrease and cause fractures.

There are two main types of training for stroke rehabilitation: *unilateral* and *bilateral* (Wu et al., 2013). Unilateral training is a therapy for the single impaired limb. *Constraint-induced therapy*, which is an intensive use of the impaired limb while constraining the unaffected limb, is a kind of unilateral training therapy. Taking into account bimanual daily activities like hand washing, the idea of getting more help from undamaged neural pathways, and case-dependent use of unilateral training, has led to bilateral training theory. Bilateral training is used for symmetric, asymmetric, and complementary movements of both impaired and unimpaired limbs

(Stoykov and Corcos, 2009). In symmetric movements of the upper extremity, arms are moved in the same way. In asymmetric movements, arm movements are opposing. In complementary movements, both arms are performing a combinatory task.

Although unilateral and bilateral training approaches are different, they are pursuing the same goal. Recent studies (Wu et al., 2013; van Delden et al., 2013) have stated that there are no significant outcomes that can make one method of training superior to the other. The procedures of these training methods are developed by motor learning theories. These theories are sometimes contradicting and are not fully determined; some of the available ones are (Brewer et al., 2007; Muratori et al., 2013; Hatem et al., 2016):

- *Implicit or explicit learning*: Implicit learning is unconscious during indirect task execution, while explicit learning is directed. *Bobath concept training* can be defined as an implicit learning exercise; it facilitates voluntary movement by handling specific points of the patient's body.
- *Massed or variable practice*: Massed practice (*repetitive task training*) is repetitive single task accomplishment, while variable practices (*task-oriented training* and *goal-directed training*) are for training multiple tasks. In the task-oriented (*task-specific*) training, a real-life practice is provided to reacquire a specific skill. The goal-directed (*client-centered*) training is a type of task-specific training in which the practice is defined based on the directed goals of the patient and therapist.
- *Feedback distortion or assistance*: Feedback distortion is magnifying movement errors instead of assisting the patient to reduce the errors.
- *Real-world practice*: This can be done by virtual reality methods that are enhanced by visual, auditory, or tactile feedback.

Although it has been found that therapy is effective in the treatment of movement disorders, therapy hours per patient have decreased because of economic burdens (Reinkensmeyer et al., 2002). Studies have shown that comprehensive and optimal stroke care can decrease the associated costs significantly (Krueger et al., 2012; Blacquiere et al., 2017). This optimal care can be achieved by implementing new technologies. That is why the design and development of biomechatronic devices (i.e., rehabilitation robots) have gained more importance.

To show the need for rehabilitation robots, we should survey the goals of therapy (Reinkensmeyer, 2009; Richards and Malouin, 2015; Hatem et al., 2016):

- *Increase activity*: It is done by the use of Thera-bands, pegboards, and blocks in conventional therapy.

- *Provide intense repetitive and engaging exercises*: This is the best practice guideline for therapy (Richards and Malouin, 2015). Conventionally, a therapist's labor and enthusiasm plays an important role in providing these exercises. However, there is a high interest in applying less therapist labor-intensive modes, and this is in contrast to conventional therapy approach for providing intense repetitive exercises.
- *Provide assistance*: Conventionally is accomplished with the help of therapists, splints, and arm-supports.
- *Improve assessment*: Traditionally is achieved by force gauges, goniometers, and timers.
- *Provide feedback*: This can be visual, auditory, or tactile.

Considering these goals and their effectiveness, increasing physically impaired patient population (Maciejasz et al., 2014), the limited number of therapists and decreased therapy hours because of economic issues (Reinkensmeyer et al., 2002), and versatile features offered by robotic devices justify the employment of rehabilitation robots in therapy sessions. These features include automation and versatility in procedures and assessments while applying intense repetitive and engaging exercises (Reinkensmeyer, 2009).

There are various reviews of upper extremity rehabilitation robots (devices) in the literature (Hesse et al., 2003a; Brewer et al., 2007; Brochard et al., 2010; Lo and Xie, 2012; Maciejasz et al., 2014; Babaiasl et al., 2016; Proietti et al., 2016; Brackenridge et al., 2016; Gopura et al., 2016; Huang et al., 2017), and there are different classifications for these robots. However, there is a lack of comprehensive classification of these robots. In this study, we thoroughly categorize these robots for different contexts. Here we use upper extremity rehabilitation robots and upper extremity rehabilitation devices interchangeably. It is worth noting that upper extremity rehabilitation devices include passive and active robots. Mechanical, and visual and auditory feedback devices are part of passive robots.

In the next sections, classification of these robots based on different approaches is discussed. Next, a proper planning for rehabilitation is presented. Finally, recent developments and research opportunities in the field of upper extremity rehabilitation robots are reviewed and conclusions are made.

2 CLASSIFICATION BY MECHANICAL DESIGN

The mechanical design of upper extremity rehabilitation robotic systems can be classified as *manipulanda* or *exoskeletons* (Maciejasz et al., 2014).

Manipulanda are end-effector-based robots that have a simple structure and control algorithms. Thus, it is hard to perform special movements of a distinct joint using these robots. Another design issue in these robots is that the end-effector at most can provide 6 degree-of-freedom (DOF). Hence, the number of anatomical movements should not exceed 6; otherwise, it will cause redundancy, which may be unsafe.

These devices can be composed of multiple robots (*multirobot manipulandum* in Fig. 1) such as "iPAM" (Jackson et al., 2007, 2013) and "REHAROB" (Fazekas et al., 2006), which are dual-robot manipulanda. However, generally, these devices are a single robot (*single-robot manipulandum* in Fig. 1). The "InMotion Arm" (which is the commercial version of "MIT-MANUS" (Krebs et al., 1998)), "HapticMaster" (Van der Linde and Lammertse, 2003), and "ReoGo" (from Motorika Medical Inc.) are some examples of single-robot manipulanda.

It is worth noting that some of these devices are connected to the body segments by cables (*cable-based devices*), and in some references, cable-based

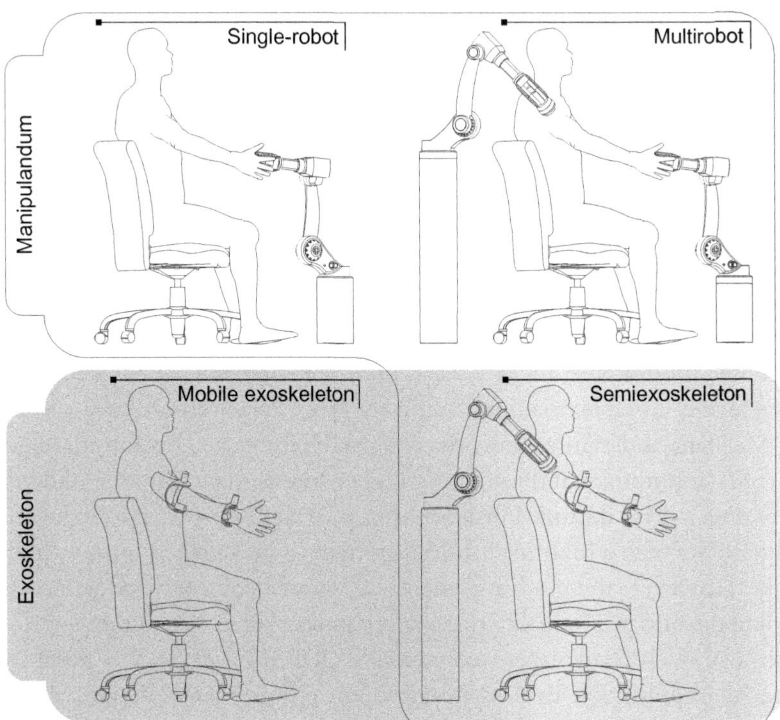

Fig. 1 Mechanical classification of upper extremity rehabilitation robots.

devices are categorized separately. Nonetheless, we have considered them as a type of manipulanda (*cable-based manipulanda*). Cable-based manipulanda can also be categorized as single-robot and multirobot. "DIEGO" (from Tyromotion GmbH) and "MariBot" (Rosati et al., 2005) are examples of cable-based single-robot manipulanda. "GENTLE/S" (Loureiro et al., 2003), which is an integrated "HapticMaster" with a cable-based mechanism, is a type of cable-based multirobot manipulanda.

Exoskeletons can provide movements to particular joints (see Fig. 1), and the number of anatomical movements can exceed 6. Nonetheless, increasing the number of moving parts increases the number of device modules, so the system setup becomes difficult. Moreover, since the shoulder has a variable joint center, the mechanical design and control algorithms become more complicated. Mostly these robots are combined with weight supporting devices or manipulanda (*semiexoskeleton* in Fig. 1). "ArmeoPower" (which is based on "ARMin III" (Nef et al., 2009)) and "ArmeoSpring" (which is based on "T-WREX" (Sanchez et al., 2004)) are commercial semiexoskeletons (Proietti et al., 2016; Maciejasz et al., 2014). If exoskeletons are not connected to any external mechanism, they will be mobile (*mobile exoskeleton* in Fig. 1). "CyberGrasp" (Adamovich et al., 2009) and "RUPERT" (Balasubramanian et al., 2008) are examples of these devices.

Manipulanda are most often used for training nonmobile gross movements (e.g., reaching task); on the other hand, exoskeletons are perfect for training mobile or joint-specific movements (i.e., perform specific movements of distinct body joints, e.g., grasping task). Manipulanda usually enjoy lower cost margins than exoskeletons as well as less complicated setups and shorter patient-preparation time for therapy. The selection of one of these two different devices highly depends on the level of the patient's disability; for example, in early stages of stroke when the patient is more vulnerable and unstable, manipulandum training seems to be a safer choice.

Mechanical design of these devices can be improved by considering the patient's ergonomics and removing higher transformation ratios using efficient direct-drive motors. Furthermore, exoskeletons benefit from the use of lighter parts with a high mechanical strength to be attached to the patient's body. However, these advancements are limited by the production cost; finding the best price-quality trade-off requires proper design methodology, such as model-based system engineering (MBSE). MBSE is a designated modeling application that supports system requirements, design, analysis, verification, and validation of conceptual designs throughout the development and lifecycle phases.

3 CLASSIFICATION BY TRAINING

Based on Brewer et al. (2007), these robotic systems can be categorized by training approaches. Accordingly, these robots are classified as either unilateral or bilateral trainers. Unilateral trainers compromise repetitive practice of a single arm, while bilateral trainers perform bimanual therapy. Compared with unilateral trainers, there are a limited number of bilateral devices available in the literature (Sheng et al., 2016). Both the classes of trainers can provide gross and/or fine motor movements.

In gross motor movements, massed practice with explicit learning is accomplished. Gross motor movement is an established method of therapy used in various rehabilitation robots. Unilateral trainers, such as "MIT-MANUS" (Krebs et al., 1998), "GENTLE/S" (Loureiro et al., 2003), "MariBot" (Rosati et al., 2005), "ARM Guide" (Kahn et al., 2006), and "ARMin" (Nef et al., 2007), and bilateral trainer "MIME" (Burgar et al., 2000) are used for gross motor movements.

Fine motor movements are mostly related to hand and wrist rehabilitation. This method can be used for increasing ROM or regulation of motor tasks like independent movements of fingers. Unilateral trainers, such as "Hand Mentor" (Koeneman et al., 2004), "HEXORR" (Schabowsky et al., 2010), "HandTutor" (Carmeli et al., 2011), "Amadeo" (Sale et al., 2012), and "VAEDA glove" (Thielbar et al., 2017), and bilateral trainer "Bi-Manu-Track" (Hesse et al., 2003b) provide fine motor movements.

Some rehabilitation robots can be used for both gross and fine motor movements. "RUPERT" (Sugar et al., 2007), the single arm "CADEN-7" (also known as "EXO-UL7") (Perry et al., 2007; Simkins et al., 2013), "ARMin III" (Nef et al., 2009), and "Universal Haptic Drive" (Oblak et al., 2010) are unilateral trainers of this type. The double arm "EXO-UL7" (Rosen and Perry, 2007; Simkins et al., 2013) is a bilateral trainer that provides both gross and fine motor movements.

Together with the above tasks, some robots have additional features such as real-world practice (Patton et al., 2004), functional electrical stimulation (FES) (Hu and Tong, 2014), electromyography (EMG) (Rahman et al., 2015), electroencephalogram (EEG) (Fok et al., 2011), gravity compensation (Stienen et al., 2007; Moubarak et al., 2010), feedback distortion (Brewer et al., 2008), telerehabilitation (Ivanova et al., 2015), and progress assessment (e.g., "KINARM" is used for motor function assessments (Coderre et al., 2010; Mostafavi et al., 2015)).

As discussed in Section 1, there is no significant advantage that can make one method of training superior to the other. Both unilateral and bilateral trainers are pursuing the same goal, and their selection depends on the patient's condition and his/her level of disability. Hence, plotting a general guideline for the selection of a suitable trainer is a complicated and cumbersome procedure, and it is case-dependent. For example, in early stages of stroke, a unilateral trainer who provides gross movements is a generally preferable choice. In the next stages, this training can be combined with real-world practice. For fine movements, if exoskeletons are not affordable, FES can be used instead. Finally, to quantify functional activities of the subject, bio-feedback features (EMG and EEG) can be used.

4 CLASSIFICATION BY FORM OF REHABILITATION

Upper extremity rehabilitation robots can support daily activities and are designed for home or clinical use (Maciejasz et al., 2014). The target population for most of these robotic systems is poststroke patients, for whom these robots can be active, passive, haptic, or coaching devices.

Active devices provide active/passive assistance therapy. In passive mode, the robot moves the patient's limb without any muscular activity of the passive patient, while in active mode the patient is active during training. Most upper extremity rehabilitation robots are active devices (Maciejasz et al., 2014). In contrast to active devices, *passive devices* perform passive resistance therapy. These devices are used to provide different types of muscle strengthening exercises including isometric, isotonic, isokinetic, and iso-contractile. "Biodex System 4 Pro" is used for isokinetic exercises (Cvjetkovic et al., 2015), "MEM-MRB" is an isokinetic and iso-contractile exercise machine (Oda et al., 2009), and "PLEMO" (Kikuchi et al., 2007) and "WOTAS" (Rocon et al., 2007) are other examples of passive devices.

In addition to active and passive devices, there are some devices that do not explicitly assist or resist the patient's movement; instead they are used for real-world practice. *Haptic devices* transfer tactile sensing to the patient. They do not assist or resist movement, but they provide real-world practice by incorporating haptic feedback while a patient is manipulating virtual objects in the simulated environment (i.e., virtual reality). There are various examples of virtual reality in rehabilitation research in which actuated feedback is implemented (Todorov et al., 1997; Prisco et al., 1998; Jack et al., 2001; Sveistrup, 2004). In Johnson et al. (2004) and Wamsley et al. (2017), gaming steering wheels are used to generate force feedback for poststroke upper extremity rehabilitation. "Handreha" is a hand-wrist haptic device that is used for hemiplegic children rehabilitation (Bouri et al., 2013). *Coaching devices* coach the individual by providing real-world practice via visual or auditory feedback. For example, "T-WREX" monitors functional arm movements during a home-therapy (Sanchez et al., 2004), and "DIEGO" (from Tyromotion GmbH) with active gravity compensation and "Microsoft Kinect" are used in virtual rehabilitation (Tseng et al., 2014).

Once again, selection of a suitable form of rehabilitation depends on the patient's condition and his/her level of disability. Recommending a general guideline for this selection requires significant years of experience with movement disorder therapy. Studies have shown that assisted therapy with active devices is prevalant for most rehabilitation procedures, and other forms of rehabilitation can be achieved by means of these active devices if needed (Maciejasz et al., 2014).

5 CLASSIFICATION BY CONTROL SCENARIOS

Human arm motions are controlled by the biological feed-forward and feedback control commands of the central nervous system (CNS) (Mehrabi et al., 2017). The feed-forward commands are predicted using an internal model of the arm. Feedback commands are corrective commands generated by the assessment of movements by sensory organs. Any electronic controller that can maintain these characteristics might be advantageous in rehabilitation robotics.

For exerting therapy approaches by upper extremity rehabilitation robots, different control algorithms are utilized. The control inputs are dynamic measurements such as force and torque signals, kinematic displacement and velocity signals, and triggers such as switches and EMG signals. Their feedbacks to the user are tactile, visual, auditory, or electrical (FES). The control strategies for these robots are categorized as (Maciejasz et al., 2014; Proietti et al., 2016) high- and low-level control scenarios. High-level control scenarios help to stimulate motor plasticity, and low-level control scenarios are used to implement high-level scenarios. These control scenarios with their subcategories are summarized in Fig. 2.

5.1 High-Level Control Scenarios

As shown in Fig. 2, there are three high-level control scenarios (Marchal-Crespo and Reinkensmeyer, 2009; Maciejasz et al., 2014; Proietti et al., 2016), which are assistive, resistive, and corrective control.

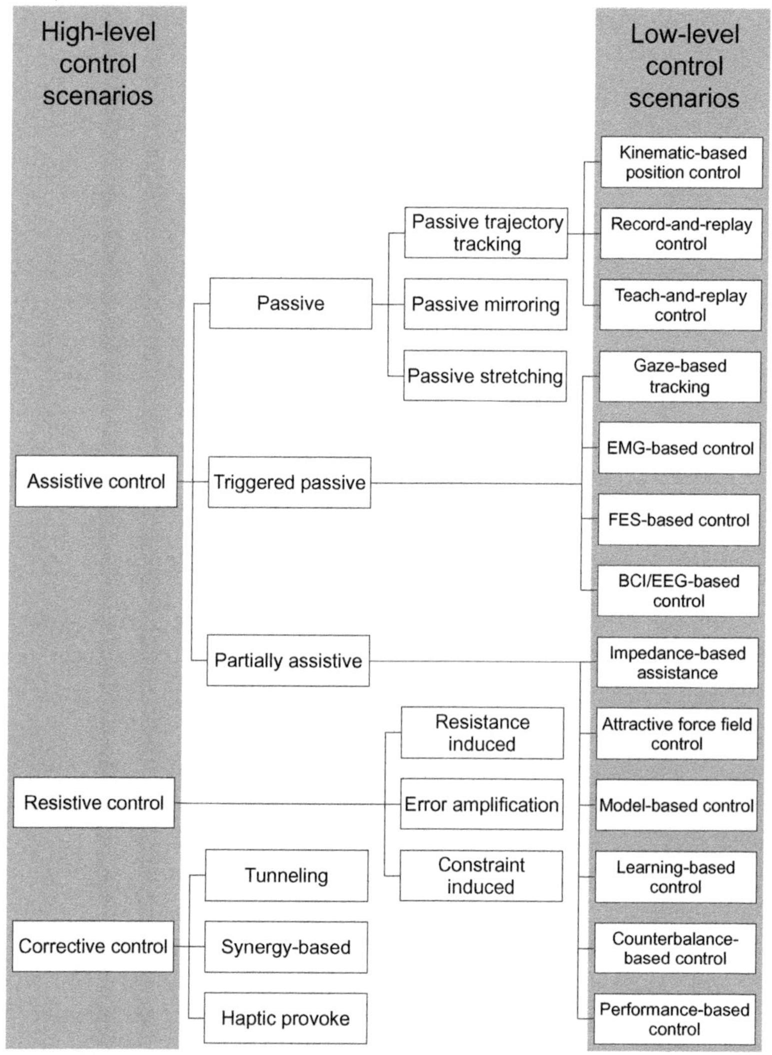

Fig. 2 Different control scenarios in rehabilitation robotics.

In assistive control, the robot helps the patient's movements using passive, triggered passive, or partially assistive control.

In passive control, the device tries to constrain the patient's hand to the desired track. This track can be defined in different ways. If it is a reference tracking control, then it is called *passive trajectory tracking*. This trajectory can be achieved by *kinematic-based position control*, where the tracking is done on a smooth trajectory (Krebs et al., 2003; Johnson et al., 2006; Brewer et al., 2006; Amirabdollahian et al., 2007; Wolbrecht et al., 2007; Rosati et al., 2007; Loureiro and Harwin, 2007; Montagner et al., 2007; Erol and Sarkar, 2007) that is determined by the "minimum-jerk" hypothesis (Flash and Hogan, 1985). The reference trajectory can be obtained from unimpaired volunteers in the so-called "record-and-replay" method (Kousidou et al., 2007; Staubli et al., 2009), or it can be generated by the therapist guidance, which is called "teach-and-replay" (Pignolo et al., 2012). If the desired trajectory is a path followed by the unimpaired limb, it is called *passive mirroring*, which is based on bilateral training (Pignolo et al., 2012; Guo et al., 2013). Finally, in the *passive stretching*, the limbs are coordinated by measuring the angle-resistance torque relation (Ren et al., 2013).

In triggered passive control, the device uses biosignals as control inputs, but this triggering may cause slacking in which the patient does not show any effort and waits for the robot assistance. These controllers are gaze-based tracking (Loconsole et al., 2011; Novak and Riener, 2013), EMG-based (Crow et al., 1989; Dipietro et al., 2005; Stein et al., 2007; Choi and Kim, 2007; Duc et al., 2008; Cesqui et al., 2013; Loconsole et al., 2014; Rahman et al., 2015; Leonardis et al., 2015; Elbagoury and Vladareanu, 2016), FES-based (Hu and Tong, 2014; Kapadia et al., 2014), and brain-computer interface (BCI)-based (which also includes EEG-based controllers) (Fok et al., 2011; Frisoli et al., 2012; Sakurada et al., 2013; Venkatakrishnan et al., 2014; Dremstrup et al., 2014; Brauchle et al., 2015; Barsotti et al., 2015).

Partially assistive control is implemented by various methods (see Fig. 2). In *impedance-based assistance*, different variations of impedance and admittance controls are used to control the rehabilitation robot (Reinkensmeyer et al., 2000; Colombo et al., 2005; Kahn et al., 2006; Gupta and O'Malley, 2006; Carignan et al., 2009; Culmer et al., 2010; Tsai et al., 2010; Miller and Rosen, 2010; Yu et al., 2011). In *attractive force field control*, some types of manipulability ellipsoid are used to apply force in specific directions (Kim et al., 2013; Yamashita, 2014). If a musculoskeletal upper extremity model is used to implement a model-based assistive control in an exoskeleton, it is a *model-based assistance* (Ding et al., 2008, 2010). If the adaption to the performance index is done from trial to trial, it is called *learning-based control*. Offline adaptive (Balasubramanian et al., 2008; Wolbrecht et al., 2008; Pérez-Rodríguez et al., 2014; Proietti et al., 2015) and artificial intelligence (AI) (Hernández Arieta et al., 2007) controls are among this type of control structure. In *counterbalance-based control*, the device applies active/passive counterbalance to the patient limb for gravity compensation (Sanchez et al., 2006; Sukal et al., 2006; Stienen et al., 2007; Montagner et al., 2007; Jackson et al., 2007; Mihelj et al., 2007). Lastly, if the robotic system tracks the performance of the patient using an error-based strategy and adapts some features for assistance, this is *performance-based adaptive control* (Kahn et al., 2004; Krebs et al., 2003; Riener et al., 2005).

As in Fig. 2, there are different methods to implement resistive (challenge-based) control. In *resistance induced* therapy, the robot resists patient's movements (Morris et al., 2004; Patten et al., 2006). In *error amplification* (feedback distortion) therapy, the robot amplifies kinematic (Patton et al., 2006a,b), visual (Wei et al., 2005; Brewer et al., 2006; Patton et al., 2006b), or tactile errors (Liu et al., 2017). Finally, sometimes constraint-induced therapy is used in resistive robotic control (Johnson et al., 2003; Shaw et al., 2005).

Corrective control is a kind of time-independent assistive control, in which the assistance is done when there are large tracking, coordination, or skill errors. This can be achieved by *tunneling*, in which an impedance-based control is applied at the boundaries of a wider trajectory (Guidali et al., 2011; Klamroth-Marganska et al., 2014; Mao et al., 2015). *Coordination* (synergy-based) control prevents large coordination errors between joints during a rehabilitation task (Guidali et al., 2009; Brokaw et al., 2011; Crocher et al., 2012). Finally, *haptic provoke* is used for providing real-world experience based on gaming control schemes (Burdea, 2003; Patton et al., 2004; Broeren et al., 2006; Yeh et al., 2013).

It was mentioned in Section 1 that optimal care is of great importance for rehabilitation robotics. This optimal care can be achieved only if the robot has an understanding of the coupled human-robot rehabilitation system. Thus, one major stream of recent studies is dedicated to the improvement of triggered passive control methods, which will be discussed in Section 7: Recent developments and research opportunities. Patient preparation is the downside for the direct use of biosignals (triggered passive control); however, partially assistive controllers use internal bio-inspired models of the patients to make decisions. Consequently, another major stream of recent research is focused on partially assistive control methods since these devices can assist the patients using some helpful bio-inspired information. Later in Section 7, recent developments and research opportunities, some of these developments, will be discussed.

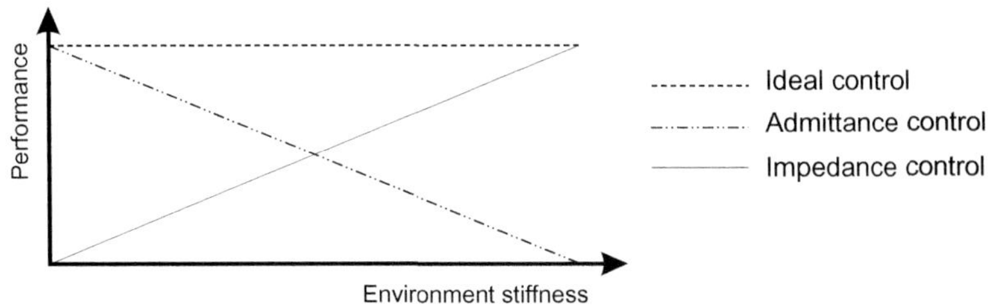

Fig. 3 Qualitative performance of impedance and admittance controllers in different environments.

5.2 Low-Level Control Scenarios

In robotic rehabilitation, since the human body is interacting with the mechatronic device, safety issues in the design of appropriate control strategies are very important. Conventional position or force control approaches (because of poor dynamic interaction modeling) are not safe enough to be implemented in these devices (Hogan, 1985). Therefore, modified control approaches like impedance and admittance control are used. In impedance control, the position of the impaired limb is measured, and appropriate force is applied (i.e., it is a force control with a position feedback), while in admittance control the applied force by the impaired limb is measured and the corresponding movement is imposed (i.e., it is a position control with a force feedback). Use of these methods is design and task specific. Impedance control has a poor accuracy; however, it becomes more stable by increasing the environment stiffness (see Fig. 3, which is adopted from Ott et al., 2010). On the other hand, as in Fig. 3, admittance control in stiff environments is not stable, while it has a good accuracy in less stiff environments. Implementing admittance control needs high transmission ratios to be considered in the mechanical design, while impedance control works well with direct drives (i.e., it is efficient for a light-weight back-drivable robot) (Ott et al., 2010; Proietti et al., 2016).

6 REHABILITATION PLANNING

Since rehabilitation robots are in contact with the human body, proper planning for rehabilitation needs design and decisions that consider the patient. The goal of the human-robot interaction (HRI) field is the design, development, and assessment of *human-centered* products (Goodrich and Schultz, 2007; Louie et al., 2017). HRI research in upper extremity robotic rehabilitation dates back to the 1990s (Van der Loos et al., 1999).

The interaction term in HRI for rehabilitation robots can be categorized into two levels: physical and social. Mechanical upper extremity devices have physical interactions, while noncontact upper extremity devices such as "Microsoft Kinect" are considered to have social interaction. Most active rehabilitation robots that provide different types of visual and auditory feedback have physical-social interaction.

To study HRI in rehabilitation robotics, one should consider HRI parameters: interaction arrangement, user interface, ability level, learning and adaption, exterior design, and therapy time (Louie et al., 2017). In robotic rehabilitation, *interaction arrangement* includes single-robot and single-user, single-robot and multiple-user, and multiple-robot and single-user; this arrangement can help to find the required mechanical design. Robot *user interface* can be auditory, tactile, or visual; the type of training can be distinguished by the user interface. *Ability level* indicates the robot's ability to perform a task, and this factor can have 10 levels varying from no-assistance to independent control modes; these levels indicate the form of rehabilitation. Regarding *learning and adaption*, both robot and user should learn and adapt to each other's performances, and this can motivate the type of control scenario. *Therapy time* is each rehabilitation session's duration, and it is important to consider patient fatigue in control scenario selection.

In addition to the HRI parameters, HRI metrics including user acceptance, user participation, user accompaniment, and user safety should be considered. These metrics are used for postprocessing the results of a rehabilitation task with a robot. User acceptance indicates how much the user is satisfied with the robot, user participation shows how long the user is engaged in the robotic rehabilitation task, user accompaniment evaluates how often the user is accompanying the robotic task (learning and adaption), and the robot's reliability is assessed by user safety (which is ensured by limiting the robot's ROM, kinetic variables, and motor torques).

To have a systematic and human-centered approach for optimal mechanical design, these HRI metrics and parameters should be included in the system requirements of the MBSE design process.

7 RECENT DEVELOPMENTS AND RESEARCH OPPORTUNITIES

In previous sections, we categorized upper extremity rehabilitation robots by mechanical design, type of training, form of rehabilitation, and control scenarios. In this section, we focus on recent advancements in the control strategies for upper extremity rehabilitation robots with different mechanical designs, including single- and multirobot manipulanda, mobile exoskeletons, and semiexoskeletons.

7.1 BCI-Based Strategies for Control and Rehabilitation

Methods for recording electrical (e.g., EEG) or magnetic fields (e.g., functional magnetic resonance imaging (fMRI) and functional near-infrared spectroscopy (fNIRS)) are used to monitor brain activities. Studies have shown that the intention to perform a specific physical activity generates consistent EEG patterns in BCI (Liu et al., 2012; Xu et al., 2014). BCI may recover brain plasticity and motor function by means of focused attention on and guidance of activation patterns of brain signals (Daly and Huggins, 2015; Yao et al., 2017). This feature motivates the application of BCI in rehabilitation robotics. Recent advancements in real-time signal processing, identification of new brain signal patterns, widespread acceptance of BCI, and less-satisfactory intense rehabilitation methods have increased the interest in BCI deployment.

BCI-based rehabilitation studies (Ang and Guan, 2015, 2017; Ang et al., 2015) at the Nanyang Technological University (Singapore) have led to well-established results in the use of BCI for rehabilitation robots. In Ang et al. (2015), they used the "MIT-MANUS" (single-robot manipulandum) with their proposed EEG-based motor imagery BCI (BCI-MANUS therapy) and compared the rehabilitation results with MANUS therapy.

In the MANUS therapy, poststroke subjects performed self-paced voluntary reaching movements. The robot assisted the subject if there were no detectable movements from them after a 2-second interval. Prior to the BCI-MANUS therapy, the robot was calibrated based on the recorded EEG signals when the subject was asked to imagine a voluntary reaching movement while the robot's end-effector was locked in its position. Then, in the BCI-MANUS therapy, the subject was asked to imagine voluntary reaching movements with minimal voluntary movements. Based on the trained subject-specific motor imagery results, the robot manipulated the subject's arm toward the target.

Results of the study showed that the BCI-MANUS therapy is more effective than the MANUS therapy. Furthermore, despite the reduced number of repetitions (i.e., less intensity) in the BCI-MANUS, it results in motor gains similar to more intense robotic therapy. Although BCI-based rehabilitation has been successful in laboratory-based studies, it needs more clinical trials. Currently, available BCIs can improve motor function, if they are applied in a larger number of therapy sessions (Mrachacz-Kersting et al., 2016). Further developments of this system depend on our knowledge of motor recovery and skill learning, involved motor centers, and intervention mechanisms. Discoveries in these areas will lead to more reliable clinical BCI-based therapy (Daly and Huggins, 2015).

7.2 FES-Based Strategies for Control and Rehabilitation

With FES, a series of electrical pulses are applied to the skeletal muscles of the affected limb to compensate for the loss of voluntary neural commands. It is possible to modulate the amount of force produced in the muscles by controlling either the electrical current or pulse-width of the stimulation (Sharif Razavian et al., 2018). FES has been shown to be an effective therapy program in restoring hand function in severe chronic stroke patients (Kapadia et al., 2014; Thrasher et al., 2008). Due to the complexity of the FES control, the combination of robotic and FES therapy paradigms has been proposed (Hu and Tong, 2014; Kapadia et al., 2014). In such setups, the robot is usually used to resist the motion while "guiding" the patient's limb, while FES is the main driver of the affected limb. Therefore, a robotic controller is needed to allow for such interactive movement.

Combination of FES with an upper extremity stroke rehabilitation robot is an ongoing research, which is mostly focused on its possibility (Hu and Tong, 2014; Kapadia et al., 2014). Recently, at the University of Leeds (United Kingdom), a proof of concept study on the feasibility of this combination has been performed (O'Connor et al., 2015). In this study, "iPAM" (double-robot manipulandum) was used to assist active reaching of a subject, and "Odstock Pace" (neuromuscular electrical stimulator) was assisting and restoring grasp in the subject. In a big picture view, if "Odstock Pace" is viewed as an exoskeleton, this system can be considered as a semiexoskeleton (see Fig. 1), which is used for *reach-and-grasp* arm movement.

The objective of this study was to enable natural prehension (reach-and-grasp) instead of over-imposed therapy, which is achieved by separate reaching and grasping exercises. "iPAM" provides arm reaching (shoulder and elbow motion) from a target to another target; once the hand is close to the reaching target, "Odstock Pace" is triggered by "iPAM" and it stimulates forearm muscles to open the patient's hand. The results of this study proved the possibility of combining FES with an upper extremity rehabilitation robot.

The effectiveness of the FES therapy seems to be tied to the simultaneous activation of sensory and motor pathways in the nervous system, which coupled with the associated mental effort may increase the neuroplasticity (Daly and Wolpaw, 2008). Therefore, the use of EEG in the detection of motor imagery and proper timing of FES signals is proposed as a possible solution to further improve the therapy outcome (Marquez-Chin et al., 2016).

7.3 EMG-Based Strategies for Control and Rehabilitation

EMG signals are used to evaluate the amount of muscle activity during a specific task. If upper extremity rehabilitation robots target deficits in muscle activations, their therapy will be more beneficial. The best way to capture muscle activation patterns is to use bio-feedback (i.e., EMG) signals. In a study by the Rehabilitation Institute of Chicago (RIC, United States), a special voice and EMG-driven mobile exoskeleton (called "VAEDA glove") for hand rehabilitation has been developed (Thielbar et al., 2017). Compared to other hand rehabilitation robots, the "VAEDA glove" is advantageous since it allows for practice of functional task.

Poststroke patients were divided into two groups: (1) with rehabilitation robot therapy (VAEDA) and (2) traditional fine-motor rehabilitation therapy (No-VAEDA). The therapy was focused on *grasp-and-release* tasks. In VAEDA therapy, the voice commands triggered the movement and the EMG command drove the actuators. Results of this study showed that the patients with VAEDA therapy could achieve better performances in physiotherapy assessments.

Despite the satisfactory outcomes of EMG-based rehabilitation, it is not suitable for performing complex movements. The success of EMG-based methods highly depends on how well muscle synergies and activation patterns are identified. The learning algorithm which is used to relate muscle activations to physical activities plays an important role in the establishment of better EMG-based rehabilitation. Advancements in deep learning will provide a platform for EMG-based therapy in complex activities.

7.4 Model-Based Strategies for Control and Rehabilitation

Best design practices demand a proper understanding of the whole system, which for this case consists of a human body interacting with a rehabilitation robot. This interaction will affect rehabilitation procedures; however, there is a lack of studies considering human body interaction with the rehabilitation robot. In Ding et al. (2010), a musculoskeletal upper extremity model (without including muscle dynamics) was used to implement a model-based assistive controller for a full-body rehabilitation exoskeleton.

At the University of Zurich (Switzerland), model-based arm weight compensation is used inside the controller for "ARMin V" (semiexoskeleton). The results of this study showed that with active model-based gravity compensation, the patient's effort will drop significantly.

The biological control structure of the CNS can be represented by an nonlinear model predictive control (NMPC) with receding horizon. In the NMPC, a forward dynamics model is used to generate gross optimal movements, and feedback information is used for fine-tuning. NMPC is used in a variety of applications in biomechanics (Mehrabi et al., 2017) and automotive control (Maitland and McPhee, 2018). Recent progress in the development of NMPC motivated researchers at the University of Waterloo (Ontario, Canada) to control a rehabilitation robot using NMPC with a nonlinear dynamic HRI model (Ghannadi et al., 2017). In this research, the HRI model was confined within an NMPC of the single-robot manipulandum (which is designed and developed by the Toronto Rehabilitation Institute (TRI) and Quanser Consulting Inc.). The proposed controller used a musculoskeletal model of the upper extremity to predict human movements and muscle activations (Mehrabi et al., 2017), thereby providing optimal assistance to the patient. In this study, the controller successfully predicts the muscular activations in model-in-the-loop simulations.

Model-based strategies for rehabilitation are more appealing than the triggered-passive methods since they do not require patient preparation for sensor attachment. However, the models should be identified within an acceptable accuracy to ascertain the validity of bio-inspired information. This accuracy should be achieved with a proper parameter identification procedure that is done with the use of bio-sensors in pretests with the robot. Thus, having a systematic approach for pretests and developing powerful tools for parameter identification is a key element in the success of these methods.

8 CONCLUSION

In this chapter, a review of upper extremity rehabilitation robots was presented, considering their mechanical design, type of training, form of rehabilitation, and control scenarios. Then, recent enhancements in the field of rehabilitation robotics were introduced.

In the human body, the arm motion is controlled by the CNS, so controllers that have any characteristics of the CNS might be advantageous for rehabilitation robotics. Since triggered passive controllers are dealing with biosignals, they can provide powerful tools for rehabilitation by inclusion of biological feedback. Thus, recent developments in rehabilitation robotics are mostly focused on leveraging these type of controllers to improve the quality of *biologically plausible* therapy. Furthermore, model-based controllers (e.g., NMPC) can also provide *biomechanically plausible* tools for rehabilitation; consequently, some studies in recent years have been focused on this idea.

Traditional physical therapies suffer from various inadequacies (Jorgensen et al., 1995; Ifejika-Jones and Barrett, 2011) and may result in significant financial burdens from costly therapy sessions (Dong et al., 2006; Krebs and Hogan, 2012). It is important to continue advancing rehabilitation robots, supported by innovative motor learning scenarios (Brewer et al., 2007; Cano-de-la Cuerda et al., 2015) and the optimization of mechatronic design and control algorithms, since they can result in effective in-home rehabilitation and patient care (Dong et al., 2006; Poli et al., 2013). Furthermore, these interactive and friendly robots can provide variations in delivering therapy (building on new achievements in motor learning studies) (Brewer et al., 2007; Reinkensmeyer, 2009), and meaningful restoration of functional activities (Krebs and Volpe, 2013). In conclusion, we fully expect that more progress will be made in the near future to improve the design and control of rehabilitation robots for providing biologically plausible *autonomous* therapy.

GLOSSARY

AI Artificial intelligence
BCI Brain-computer interface
CNS Central nervous system
DOF Degree-of-freedom
EEG Electroencephalogram
EMG Electromyography
FES Functional electrical stimulation
fMRI Functional magnetic resonance imaging
fNIRS Functional near-infrared spectroscopy
HRI Human-robot interaction
MBSE Model-based system engineering
NMPC Nonlinear model predictive control
ROM Range of motion
TRI Toronto Rehabilitation Institute

REFERENCES

Adamovich, S.V., Fluet, G.G., Mathai, A., Qiu, Q., Lewis, J., Merians, A.S., 2009. Design of a complex virtual reality simulation to train finger motion for persons with hemiparesis: a proof of concept study. J. Neuroeng. Rehabil. 1743-00036 (1), 28. https://doi.org/10.1186/1743-0003-6-28. Available from: http://jneuroengrehab.biomedcentral.com/articles/10.1186/1743-0003-6-28.

Amirabdollahian, F., Loureiro, R., Gradwell, E., Collin, C., Harwin, W., Johnson, G., 2007. Multivariate analysis of the Fugl-Meyer outcome measures assessing the effectiveness of GENTLE/S robot-mediated stroke therapy. J. Neuroeng. Rehabil. 1743-00034 (1), 4. https://doi.org/10.1186/1743-0003-4-4. Available from: http://www.springerlink.com/index/10.3758/BF03203630.

Ang, K.K., Guan, C., 2015. Brain-computer interface for neurorehabilitation of upper limb after stroke. Proc. IEEE 103 (6), 944–953. https://doi.org/10.1109/JPROC.2015.2415800. Available from: http://ieeexplore.ieee.org/document/7105815/.

Ang, K.K., Guan, C., 2017. EEG-based strategies to detect motor imagery for control and rehabilitation. IEEE Trans. Neural Syst. Rehabil. Eng. 25 (4), 392–401. https://doi.org/10.1109/TNSRE.2016.2646763. Available from: http://ieeexplore.ieee.org/document/7802578/.

Ang, K.K., Chua, K.S.G., Phua, K.S., Wang, C., Chin, Z.Y., Kuah, C.W.K., Low, W., Guan, C., 2015. A randomized controlled trial of EEG-based motor imagery brain-computer interface robotic rehabilitation for stroke. Clin. EEG Neurosci. 46 (4), 310–320. https://doi.org/10.1177/1550059414522229. Available from: http://journals.sagepub.com/doi/10.1177/1550059414522229.

Babaiasl, M., Mahdioun, S.H., Jaryani, P., Yazdani, M., 2016. A review of technological and clinical aspects of robot-aided rehabilitation of upper-extremity after stroke. Disabil. Rehabil. Assist. Technol. 1748-310711 (4), 263–280. https://doi.org/10.3109/17483107.2014.1002539. Available from: http://www.tandfonline.com/doi/full/10.3109/17483107.2014.1002539.

Balasubramanian, S., Wei, R., Perez, M., Shepard, B., Koeneman, E., Koeneman, J., He, J., 2008. RUPERT: an exoskeleton robot for assisting rehabilitation of arm functions. In: 2008 International Conference on Virtual Rehabilitation (ICVR). IEEE, pp. 163–167. Available from: http://ieeexplore.ieee.org/document/4625154/.

Bansil, S., Prakash, N., Kaye, J., Wrigley, S., Manata, C., Stevens-Haas, C., Kurlan, R., 2012. Movement disorders after stroke in adults: a review. Tremor Other Hyperkinet. Mov. 2160-82882, 1–7. Available from: http://www.pubmedcentral.nih.gov/articlerender.fcgi?artid=3570045&tool=pmcentrez&rendertype=abstract.

Barsotti, M., Leonardis, D., Loconsole, C., Solazzi, M., Sotgiu, E., Procopio, C., Chisari, C., Bergamasco, M., Frisoli, A., 2015. A full upper limb robotic exoskeleton for reaching and grasping rehabilitation triggered by MI-BCI. In: 2015 IEEE International Conference on Rehabilitation Robotics (ICORR). IEEE, pp. 49–54. Available from: http://ieeexplore.ieee.org/document/7281174/.

Blacquiere, D., Lindsay, M.P., Foley, N., Taralson, C., Alcock, S., Balg, C., Bhogal, S., Cole, J., Eustace, M., Gallagher, P., Ghanem, A., Hoechsmann, A., Hunter, G., Khan, K., Marrero, A., Moses, B., Rayner, K., Samis, A., Smitko, E., Vibe, M., Gubitz, G., Dowlatshahi, D., Phillips, S., Silver, F.L., 2017. Canadian stroke best practice recommendations: telestroke best practice guidelines update 2017. Int. J. Stroke 1747-493012 (8), 886–895. https://doi.org/10.1177/1747493017706239. Available from: http://journals.sagepub.com/doi/10.1177/1747493017706239.

Bouri, M., Baur, C., Clavel, R., Zedka, M., Newman, C.J., 2013. "Handreha": a new hand and wrist haptic device for hemiplegic children. In: The Sixth International Conference on Advances in Computer-Human Interactions (ACHI)pp. 286–292. Available from: https://www.thinkmind.org/index.php?view=article&articleid=achi_2013_11_20_20392.

Brackenridge, J., Bradnam, L.V., Lennon, S., Costi, J.J., Hobbs, D.A., 2016. A review of rehabilitation devices to promote upper limb function following stroke. Neurosci. Biomed. Eng. 221338524 (1), 25–42. https://doi.org/10.2174/2213385204666160303220102. Available from: http://www.eurekaselect.com/openurl/content.php?genre=article&issn=2213-3852&volume=4&issue=1&spage=25http://www.ingentaconnect.com/content/ben/nbe/2016/00000004/00000001/art00006.

Brauchle, D., Vukelić, M., Bauer, R., Gharabaghi, A., 2015. Brain state-dependent robotic reaching movement with a multi-joint arm exoskeleton: combining brain-machine interfacing and robotic rehabilitation. Front. Hum. Neurosci. 1662-51619, 564. https://doi.org/10.3389/fnhum.2015.00564. Available from: http://journal.frontiersin.org/Article/10.3389/fnhum.2015.00564/abstracthttp://www.pubmedcentral.nih.gov/articlerender.fcgi?artid=PMC4607784http://www.ncbi.nlm.nih.gov/pubmed/26528168.

Brewer, B.R., Klatzky, R., Matsuoka, Y., 2006. Initial therapeutic results of visual feedback manipulation in robotic rehabilitation. In: 2006 International Workshop on Virtual RehabilitationIEEE, pp. 160–166. Available from: http://ieeexplore.ieee.org/lpdocs/epic03/wrapper.htm?arnumber=1707546http://ieeexplore.ieee.org/document/1707546/.

Brewer, B.R., McDowell, S.K., Worthen-Chaudhari, L.C., 2007. Poststroke upper extremity rehabilitation: a review of robotic systems and clinical results. Top. Stroke Rehabil. 1074-935714 (6), 22–44. https://doi.org/10.1310/tsr1406-22. Available from: http://www.ncbi.nlm.nih.gov/pubmed/18174114.

Brewer, B.R., Klatzky, R., Matsuoka, Y., 2008. Visual feedback distortion in a robotic environment for hand rehabilitation. Brain Res. Bull. 0361923075 (6), 804–813. https://doi.org/10.1016/j.brainresbull.2008.01.006. Available from: http://linkinghub.elsevier.com/retrieve/pii/S0361923008000099. http://www.ncbi.nlm.nih.gov/pubmed/18394527.

Brochard, S., Robertson, J., Médée, B., Rémy-Néris, O., 2010. What's new in new technologies for upper extremity rehabilitation? Curr. Opin. Neurol. 1350-754023, 683–687. https://doi.org/10.1097/WCO.0b013e32833f61ce.

Broeren, J., Rydmark, M., Sunnerhagen, K.S., 2004. Virtual reality and haptics as a training device for movement rehabilitation after stroke: a single-case study. Arch. Phys. Med. Rehabil. 0003999385, 1247–1250. https://doi.org/10.1016/j.apmr.2003.09.020.

Broeren, J., Dixon, M., Sunnerhagen, K.S., Rydmark, M., 2006. Rehabilitation after stroke using virtual reality, haptics (force feedback) and telemedicine. Stud. Health Technol. Inform. 0926-9630124, 51–56. Available from: http://www.ncbi.nlm.nih.gov/pubmed/17108503.

Brokaw, E.B., Murray, T., Nef, T., Lum, P.S., 2011. Retraining of interjoint arm coordination after stroke using robot-assisted time-independent functional training. J. Rehabil. Res. Dev. 1938-135248 (4), 299–316. Available from: http://www.ncbi.nlm.nih.gov/pubmed/21674385.

Burdea, G.C., 2003. Virtual rehabilitation-benefits and challenges. Methods Inf. Med. 0026-127042 (5), 519–523. https://doi.org/10.1267/METH03050519. Available from: http://www.ncbi.nlm.nih.gov/pubmed/14654886.

Burgar, C.G., Lum, P.S., Shor, P.C., Machiel Van der Loos, H.F., 2000. Development of robots for rehabilitation therapy: the Palo Alto VA/Stanford experience. J. Rehabil. Res. Dev. 37 (6), 663–673. Available from: http://www.ncbi.nlm.nih.gov/pubmed/11321002.

Cano-de-la Cuerda, R., Molero-Sánchez, A., Carratalá-Tejada, M., Alguacil-Diego, I.M., Molina-Rueda, F., Miangolarra-Page, J.C., Torricelli, D., 2015. Theories and control models and motor learning: clinical applications in neurorehabilitation. Neurología (English Edition) 2173580830 (1), 32–41. https://doi.org/10.1016/j.nrleng.2011.12.012. Available from: http://linkinghub.elsevier.com/retrieve/pii/S2173580814001424. http://www.sciencedirect.com/science/article/pii/S2173580814001424.

Carignan, C., Tang, J., Roderick, S., 2009. Development of an exoskeleton haptic interface for virtual task training. In: 2009 IEEE/RSJ International Conference on Intelligent Robots and Systems (IROS). IEEE, pp. 3697–3702. Available from: http://ieeexplore.ieee.org/document/5354834/.

Carmeli, E., Peleg, S., Bartur, G., Elbo, E., Vatine, J.-J., 2011. HandTutor enhanced hand rehabilitation after stroke—a pilot study. Physiother. Res. Int. 16 (4), 191–200. https://doi.org/10.1002/pri.485. Available from: http://www.ncbi.nlm.nih.gov/pubmed/20740477.

Cesqui, B., Tropea, P., Micera, S., Krebs, H., 2013. EMG-based pattern recognition approach in post stroke robot-aided rehabilitation: a feasibility study. J. Neuroeng. Rehabil. 1743-000310 (1), 75. https://doi.org/10.1186/1743-0003-10-75. Available from: http://jneuroengrehab.biomedcentral.com/articles/10.1186/1743-0003-10-75.

Choi, C., Kim, J., 2007. A real-time EMG-based assistive computer interface for the upper limb disabled. In: 2007 IEEE 10th International Conference on Rehabilitation Robotics (ICORR). IEEE, pp. 459–462. Available from: http://ieeexplore.ieee.org/document/4428465/.

Coderre, A.M., Amr Abou Zeid, A.A., Dukelow, S.P., Demmer, M.J., Moore, K.D., Demers, M.J., Bretzke, H., Herter, T.M., Glasgow, J.I., Norman, K.E., Bagg, S.D., Scott, S.H., 2010. Assessment of upper-limb sensorimotor function of subacute stroke patients using visually guided reaching. Neurorehabil. Neural Repair 24 (6), 528–541. https://doi.org/10.1177/1545968309356091. Available from: http://www.ncbi.nlm.nih.gov/pubmed/20233965.

Colombo, R., Pisano, F., Micera, S., Mazzone, A., Delconte, C., Carrozza, M.C., Dario, P., Minuco, G., 2005. Robotic techniques for upper limb evaluation and rehabilitation of stroke patients. IEEE Trans. Neural Syst. Rehabil. Eng. 1534-432013 (3), 311–324. https://doi.org/10.1109/TNSRE.2005.848352. Available from: http://www.ncbi.nlm.nih.gov/pubmed/16200755.

Crocher, V., Sahbani, A., Robertson, J., Roby-Brami, A., Morel, G., 2012. Constraining upper limb synergies of hemiparetic patients using a robotic exoskeleton in the perspective of neuro-rehabilitation. IEEE Trans. Neural Syst. Rehabil. Eng. 1534-432020 (3), 247–257. https://doi.org/10.1109/TNSRE.2012.2190522. Available from: http://ieeexplore.ieee.org/document/6177268/. http://www.ncbi.nlm.nih.gov/pubmed/22481836.

Crow, J.L., Lincoln, N.B., Nouri, F.M., Weerdt, W.D., 1989. The effectiveness of EMG biofeedback in the treatment of arm function after stroke. Int. Disabil. Stud. 0259-914711 (4), 155–160. https://doi.org/10.3109/03790798909166667. Available from: http://www.tandfonline.com/doi/full/10.3109/03790798909166667.

Culmer, P.R., Jackson, A.E., Makower, S., Richardson, R., Cozens, J.A., Levesley, M.C., Bhakta, B.B., 2010. A control strategy for upper limb robotic rehabilitation with a dual robot system. IEEE/ASME Trans. Mechatron. 1083-443515 (4), 575–585. https://doi.org/10.1109/TMECH.2009.2030796. Available from: http://ieeexplore.ieee.org/document/5263023/.

Cvjetkovic, D.D., Bijeljac, S., Palija, S., Talic, G., Radulovic, T.N., Kosanovic, M.G., Manojlovic, S., 2015. Isokinetic testing in evaluation rehabilitation outcome after ACL reconstruction. Med. Arch. (Sarajevo, Bosnia and Herzegovina) 69 (1), 21–23. https://doi.org/10.5455/medarh.2015.69.21-23. Available from: http://www.ncbi.nlm.nih.gov/pubmed/25870471.

Daly, J.J., Huggins, J.E., 2015. Brain-computer interface: current and emerging rehabilitation applications. Arch. Phys. Med. Rehabil. 0003999396 (3), S1–S7. https://doi.org/10.1016/j.apmr.2015.01.007. Available from: http://linkinghub.elsevier.com/retrieve/pii/S0003999315000209. http://www.ncbi.nlm.nih.gov/pubmed/25721542.

Daly, J.J., Wolpaw, J.R., 2008. Brain-computer interfaces in neurological rehabilitation. Lancet Neurol. 147444227 (11), 1032–1043. https://doi.org/10.1016/S1474-4422(08)70223-0. Available from: http://www.sciencedirect.com/science/article/pii/S1474442208702230?via%3Dihub.

Ding, M., Ueda, J., Ogasawara, T., 2008. Pinpointed muscle force control using a power-assisting device: system configuration and experiment. In: Proceedings of the 2nd Biennial IEEE/RAS-EMBS International Conference on Biomedical Robotics and Biomechatronics, BioRob 2008. IEEE, pp. 181–186. Available from: http://ieeexplore.ieee.org/document/4762829/.

Ding, M., Hirasawa, K., Kurita, Y., Takemura, H., Takamatsu, J., Mizoguchi, H., Ogasawara, T., 2010. Pinpointed muscle force control in consideration of human motion and external force. In: 2010 IEEE International Conference on Robotics and Biomimetics (ROBIO). IEEE, pp. 739–744. Available from: http://ieeexplore.ieee.org/document/5723418/.

Dipietro, L., Ferraro, M., Palazzolo, J.J., Krebs, H.I., Volpe, B.T., Hogan, N., 2005. Customized interactive robotic treatment for stroke: EMG-triggered therapy. IEEE Trans. Neural Syst. Rehabil. Eng. 1534-432013 (3), 325–334. https://doi.org/10.1109/TNSRE.2005.850423. Available from: http://www.pubmedcentral.nih.gov/articlerender.fcgi?artid=PMC2752646. http://www.ncbi.nlm.nih.gov/pubmed/16200756.

Dong, S., Lu, K.-Q., Sun, J.Q., Rudolph, K., 2006. Smart rehabilitation devices: part II—adaptive motion control. J. Intell. Mater. Syst. Struct. 1045-389X17 (7), 555–561. https://doi.org/10.1177/1045389X06059076. Available from: http://www.pubmedcentral.nih.gov/articlerender.fcgi?artid=2424262&tool=pmcentrez&rendertype=abstract.

Dremstrup, K., Niazi, I.K., Jochumsen, M., Jiang, N., Mrachacz-Kersting, N., Farina, D., 2014. Rehabilitation using a brain computer interface based on movement related cortical potentials—a review. In: Roa Romero, L.M. (Ed.), XIII Mediterranean Conference on Medical and Biological Engineering and Computing 2013. Springer International Publishing, Cham, pp. 1659–1662.

Duc, D.M., Kazuhiko, T., Takanori, M., 2008. EMG-moment model of human arm for rehabilitation robot system. In: 2008 10th International Conference on Control, Automation, Robotics and Vision. IEEE, pp. 190–195. Available from: http://ieeexplore.ieee.org/document/4795515/.

Elbagoury, B.M., Vladareanu, L., 2016. A hybrid real-time EMG intelligent rehabilitation robot motions control based on Kalman filter, support vector machines and particle swarm optimization. In: 2016 10th International Conference on Software, Knowledge, Information Management & Applications (SKIMA). IEEE, pp. 439–444. Available from: http://ieeexplore.ieee.org/document/7916262/.

Erol, D., Sarkar, N., 2007. Intelligent control for robotic rehabilitation after stroke. J. Intell. Robot. Syst. 0921-029650 (4), 341–360. https://doi.org/10.1007/s10846-007-9169-2. Available from: http://link.springer.com/10.1007/s10846-007-9169-2.

Fazekas, G., Horvath, M., Toth, A., 2006. A novel robot training system designed to supplement upper limb physiotherapy of patients with spastic hemiparesis. Int. J. Rehabil. Res. 29 (29) Available from: https://insights.ovid.com/pubmed?pmid=16900048.

Flash, T., Hogan, N., 1985. The coordination of arm movements: an experimentally confirmed mathematical model. J. Neurosci. 0270-64745 (7), 1688–1703. Available from: https://doi.org/4020415. http://www.ncbi.nlm.nih.gov/pubmed/4020415. http://www.jneurosci.org/cgi/content/abstract/5/7/1688.

Fok, S., Schwartz, R., Wronkiewicz, M., Holmes, C., Zhang, J., Somers, T., Bundy, D., Leuthardt, E., 2011. An EEG-based brain computer interface for rehabilitation and restoration of hand control following stroke using ipsilateral cortical physiology. In: 2011

Annual International Conference of the IEEE Engineering in Medicine and Biology Society. IEEE, pp. 6277–6280. Available from: http://ieeexplore.ieee.org/document/6091549/.

Frisoli, A., Loconsole, C., Leonardis, D., Banno, F., Barsotti, M., Chisari, C., Bergamasco, M., 2012. A new gaze-BCI-driven control of an upper limb exoskeleton for rehabilitation in real-world tasks. IEEE Trans. Syst. Man Cybern. Part C Appl. Rev. 1094-697742 (6), 1169–1179. https://doi.org/10.1109/TSMCC.2012.2226444. Available from: http://ieeexplore.ieee.org/document/6392463/.

Fulk, G., O'sullivan, S.B., Schmitz, T.J., 2014. Physical Rehabilitation, sixth ed. F.A. Davis Company. ISBN 978-0-8036-2579-2.

Ghannadi, B., Mehrabi, N., Sharif Razavian, R., McPhee, J., 2017. Nonlinear model predictive control of an upper extremity rehabilitation robot using a two-dimensional human-robot interaction model. In: IEEE/RSJ International Conference on Intelligent Robots and Systems (IROS). IEEE, Vancouver, British Columbia, Canada, pp. 502–507. Available from: http://ieeexplore.ieee.org/document/8202200/.

Goodrich, M.A., Schultz, A.C., 2007. Human-robot interaction: a survey. Found. Trends Hum. Comput. Interact. 1551-39551 (3), 203–275. https://doi.org/10.1561/1100000005. Available from: http://www.nowpublishers.com/article/Details/HCI-005.

Gopura, R.A.R.C., Bandara, D.S.V., Kiguchi, K., Mann, G.K.I., 2016. Developments in hardware systems of active upper-limb exoskeleton robots: a review. Robot. Auton. Syst. 0921889075, 203–220. https://doi.org/10.1016/j.robot.2015.10.001. Available from: http://www.sciencedirect.com/science/article/pii/S0921889015002274.

Guidali, M., Schmiedeskamp, M., Klamroth, V., Riener, R., 2009. Assessment and training of synergies with an arm rehabilitation robot. In: 2009 IEEE International Conference on Rehabilitation Robotics (ICORR). IEEE, pp. 772–776. Available from: http://ieeexplore.ieee.org/document/5209516/.

Guidali, M., Duschau-Wicke, A., Broggi, S., Klamroth-Marganska, V., Nef, T., Riener, R., 2011. A robotic system to train activities of daily living in a virtual environment. Med. Biol. Eng. Comput. 0140-011849 (10), 1213–1223. https://doi.org/10.1007/s11517-011-0809-0. Available from: http://link.springer.com/10.1007/s11517-011-0809-0. http://www.ncbi.nlm.nih.gov/pubmed/21796422.

Guo, S., Zhang, W., Wei, W., Guo, J., Ji, Y., Wang, Y., 2013. A kinematic model of an upper limb rehabilitation robot system. In: 2013 IEEE International Conference on Mechatronics and Automation. IEEE, pp. 968–973. Available from: http://ieeexplore.ieee.org/document/6618046/.

Gupta, A., O'Malley, M.K., 2006. Design of a haptic arm exoskeleton for training and rehabilitation. IEEE/ASME Trans. Mechatron. 1083-443511 (3), 280–289. https://doi.org/10.1109/TMECH.2006.875558. Available from: http://ieeexplore.ieee.org/document/1642690/.

Hatem, S.M., Saussez, G., della Faille, M., Prist, V., Zhang, X., Dispa, D., Bleyenheuft, Y., 2016. Rehabilitation of motor function after stroke: a multiple systematic review focused on techniques to stimulate upper extremity recovery. Front. Hum. Neurosci. 1662-516110, 442. https://doi.org/10.3389/fnhum.2016.00442. Available from: http://journal.frontiersin.org/Article/10.3389/fnhum.2016.00442/abstract. http://www.ncbi.nlm.nih.gov/pubmed/27679565.

Hernández Arieta, A., Kato, R., Yu, W., Yokoi, H., 2007. The man-machine interaction: the influence of artificial intelligence on rehabilitation robotics. In: 50 Years of Artificial IntelligenceSpringer, Berlin, Heidelberg, pp. 221–231. Available from: http://link.springer.com/10.1007/978-3-540-77296-5_21.

Hesse, S., Schmidt, H., Werner, C., Bardeleben, A., 2003. Upper and lower extremity robotic devices for rehabilitation and for studying motor control. Curr. Opin. Neurol. 1350-754016, 705–710. https://doi.org/10.1097/00019052-200312000-00010.

Hesse, S., Schulte-Tigges, G., Konrad, M., Bardeleben, A., Werner, C., 2003. Robot-assisted arm trainer for the passive and active practice of bilateral forearm and wrist movements in hemiparetic subjects. Arch. Phys. Med. Rehabil. 0003999384 (6), 915–920. https://doi.org/10.1016/S0003-9993(02)04954-7. Available from: http://www.sciencedirect.com/science/article/pii/S0003999302049547?via%3Dihub.

Hogan, N., 1985. Impedance control: an approach to manipulation: part I—theory. J. Dyn. Syst. Meas. Control. 00220434107 (1), 1. https://doi.org/10.1115/1.3140702. Available from: http://dynamicsystems.asmedigitalcollection.asme.org/article.aspx?articleid=1403621.

Hu, X., Tong, R.K.Y., 2014. FES in rehabilitation robotics. In: 2014 IEEE 19th International Functional Electrical Stimulation Society Annual Conference (IFESS). IEEE, pp. 1–3. Available from: http://ieeexplore.ieee.org/document/7036730/.

Huang, X., Naghdy, F., Naghdy, G., Du, H., Todd, C., 2017. Robot-assisted post-stroke motion rehabilitation in upper extremities: a survey. Int. J. Disabil. Hum. Dev. 16 (3), 233–247. https://doi.org/10.1515/ijdhd-2016-0035. Available from: http://www.degruyter.com/view/j/ijdhd.2017.16.issue-3/ijdhd-2016-0035/ijdhd-2016-0035.xml.

Ifejika-Jones, N.L., Barrett, A.M., 2011. Rehabilitation-emerging technologies, innovative therapies, and future objectives. Neurotherapeutics 193372138 (3), 452–462. https://doi.org/10.1007/s13311-011-0057-x. Available from: http://www.pubmedcentral.nih.gov/articlerender.fcgi?artid=PMC3148149. http://www.ncbi.nlm.nih.gov/pubmed/21706265.

Ivanova, E., Kruger, J., Steingraber, R., Schmid, S., Schmidt, H., Hesse, S., 2015. Design and concept of a haptic robotic telerehabilitation system for upper limb movement training after stroke. In: IEEE 14th International Conference on Rehabilitation Robotics (ICORR). IEEE, pp. 666–671. Available from: http://ieeexplore.ieee.org/document/7281277/.

Jack, D., Boian, R., Merians, A.S., Tremaine, M., Burdea, G.C., Adamovich, S.V., Recce, M., Poizner, H., 2001. Virtual reality-enhanced stroke rehabilitation. IEEE Trans. Neural Syst. Rehabil. Eng. 153443209 (3), 308–318. https://doi.org/10.1109/7333.948460. Available from: http://ieeexplore.ieee.org/document/948460/.

Jackson, A.E., Culmer, P.R., Makower, S.G., Levesley, M.C., Richardson, R.C., Cozens, J.A., Williams, M.M., Bhakta, B.B., 2007. Initial patient testing of iPAM—a robotic system for Stroke rehabilitation. In: IEEE 10th International Conference on Rehabilitation Robotics (ICORR). IEEE, pp. 250–256. Available from: http://ieeexplore.ieee.org/document/4428435/.

Jackson, A.E., Levesley, M.C., Makower, S.G., Cozens, J.A., Bhakta, B.B., 2013. Development of the iPAM MkII system and description of a randomized control trial with acute stroke patients. In: IEEE 13th International Conference on Rehabilitation Robotics (ICORR), vol. 2013. IEEE, pp. 1–6. Available from: http://www.ncbi.nlm.nih.gov/pubmed/24187226.

Johnson, M.J., Van der Loos, H.F.M., Burgar, C.G., Shor, P., Leifer, L.J., 2003. Design and evaluation of Driver's SEAT: a car steering simulation environment for upper limb stroke therapy. Robotica 0263-574721 (1), 13–23. https://doi.org/10.1017/S0263574702004599. Available from: http://www.journals.cambridge.org/abstract_S0263574702004599.

Johnson, M.J., Trickey, M., Brauer, E., Feng, X., 2004. TheraDrive: a new stroke therapy concept for home-based, computer-assisted motivating rehabilitation. In: The 26th Annual International Conference of the IEEE Engineering in Medicine and Biology Society (EMBS), vol. 4. IEEE, pp. 4844–4847. Available from: http://ieeexplore.ieee.org/document/1404340/.

Johnson, M.J., Wisneski, K.J., Anderson, J., Nathan, D., Smith, R.O., 2006. Development of ADLER: the activities of daily living exercise robot. In: Proceedings of the First IEEE/RAS-EMBS International Conference on Biomedical Robotics and Biomechatronics,

2006, BioRob 2006, vol. 2006. IEEE, pp. 881–886. Available from: http://ieeexplore.ieee.org/document/1639202/.

Jorgensen, H.S., Nakayama, H., Raaschou, H.O., Vive-Larsen, J., Stoier, M., Olsen, T.S., 1995. Outcome and time course of recovery in stroke. Part II: time course of recovery. The Copenhagen Stroke Study. Arch. Phys. Med. Rehabil. 0003999376, 406–412. https://doi.org/10.1016/S0003-9993(95)80568-0.

Kahn, L.E., Rymer, W.Z., Reinkensmeyer, D.J., 2004. Adaptive assistance for guided force training in chronic stroke. In: The 26th Annual International Conference of the IEEE Engineering in Medicine and Biology Society, vol. 3. IEEE, pp. 2722–2725. Available from: http://ieeexplore.ieee.org/document/1403780/.

Kahn, L.E., Zygman, M.L., Rymer, W.Z., Reinkensmeyer, D.J., 2006. Robot-assisted reaching exercise promotes arm movement recovery in chronic hemiparetic stroke: a randomized controlled pilot study. J. Neuroeng. Rehabil. 1743-00033, 12. https://doi.org/10.1186/1743-0003-3-12. Available from: http://www.pubmedcentral.nih.gov/articlerender.fcgi?artid=PMC1550245. http://www.ncbi.nlm.nih.gov/pubmed/16790067.

Kapadia, N.M., Nagai, M.K., Zivanovic, V., Bernstein, J., Woodhouse, J., Rumney, P., Popovic, M.R., 2014. Functional electrical stimulation therapy for recovery of reaching and grasping in severe chronic pediatric stroke patients. J. Child Neurol. 0883-073829 (4), 493–499. https://doi.org/10.1177/0883073813484088. Available from: http://journals.sagepub.com/doi/10.1177/0883073813484088. http://www.ncbi.nlm.nih.gov/pubmed/23584687.

Kikuchi, T., Xinghao, H., Fukushima, K., Oda, K., Furusho, J., Inoue, A., 2007. Quasi-3-DOF rehabilitation system for upper limbs: its force-feedback mechanism and software for rehabilitation. In: IEEE 10th International Conference on Rehabilitation Robotics (ICORR). IEEE, pp. 24–27. Available from: http://ieeexplore.ieee.org/document/4428401/.

Kim, H., Miller, L.M., Fedulow, I., Simkins, M., Abrams, G.M., Byl, N., Rosen, J., 2013. Kinematic data analysis for post-stroke patients following bilateral versus unilateral rehabilitation with an upper limb wearable robotic system. IEEE Trans. Neural Syst. Rehabil. Eng. 1534432021 (2), 153–164. https://doi.org/10.1109/TNSRE.2012.2207462. Available from: http://ieeexplore.ieee.org/document/6252060/.

Klamroth-Marganska, V., Blanco, J., Campen, K., Curt, A., Dietz, V., Ettlin, T., Felder, M., Fellinghauer, B., Guidali, M., Kollmar, A., Luft, A., Nef, T., Schuster-Amft, C., Stahel, W., Riener, R., 2014. Three-dimensional, task-specific robot therapy of the arm after stroke: a multicentre, parallel-group randomised trial. Lancet Neurol. 1474442213 (2), 159–166. https://doi.org/10.1016/S1474-4422(13)70305-3. Available from: http://linkinghub.elsevier.com/retrieve/pii/S1474442213703053. http://www.ncbi.nlm.nih.gov/pubmed/24382580.

Koeneman, E.J., Schultz, R.S., Wolf, S.L., Herring, D.E., Koeneman, J.B., 2004. A pneumatic muscle hand therapy device. In: Annual International Conference of the IEEE Engineering in Medicine and Biology Society (EMBS), vol. 4. IEEE, pp. 2711–2713. Available from: http://ieeexplore.ieee.org/document/1403777/.

Kousidou, S., Tsagarakis, N.G., Smith, C., Caldwell, D.G., 2007. Task-orientated biofeedback system for the rehabilitation of the upper limb. In: 2007 IEEE 10th International Conference on Rehabilitation Robotics (ICORR). IEEE, pp. 376–384. Available from: http://ieeexplore.ieee.org/document/4428453/.

Krebs, H.I., Hogan, N., 2012. Robotic therapy: the tipping point. Am. J. Phys. Med. Rehabil. 1537-738591, S290–S297. https://doi.org/10.1097/PHM.0b013e31826bcd80. Available from: http://www.pubmedcentral.nih.gov/articlerender.fcgi?artid=PMC3480667. http://www.ncbi.nlm.nih.gov/pubmed/23080044.

Krebs, H.I., Volpe, B.T., 2013. Rehabilitation robotics. Handb. Clin. Neurol. 0072975211 0, 283–294. https://doi.org/10.1016/B978-0-444-52901-5.00023-X. Available from: http://www.pubmedcentral.nih.gov/articlerender.fcgi?artid=PMC4688009. http://www.ncbi.nlm.nih.gov/pubmed/23312648.

Krebs, H.I., Hogan, N., Aisen, M.L., Volpe, B.T., 1998. Robot-aided neurorehabilitation. IEEE Trans. Rehabil. Eng. 6 (1), 75–87. Available from: http://www.ncbi.nlm.nih.gov/pubmed/9535526.

Krebs, H.I., Palazzolo, J.J., Dipietro, L., Ferraro, M., Krol, J., Rannekleiv, K., Volpe, B.T., Hogan, N., 2003. Rehabilitation robotics: performance-based progressive robot-assisted therapy. Auton. Robot. 0929559315 (1), 7–20. https://doi.org/10.1023/A:1024494031121. Available from: http://link.springer.com/10.1023/A:1024494031121.

Krueger, H., Lindsay, P., Cote, R., Kapral, M.K., Kaczorowski, J., Hill, M.D., 2012. Cost avoidance associated with optimal stroke care in Canada. Stroke 0039249943 (8), 2198–2206. https://doi.org/10.1161/STROKEAHA.111.646091. Available from: http://stroke.ahajournals.org/content/early/2012/05/24/STROKEAHA.111.646091.

Leonardis, D., Barsotti, M., Loconsole, C., Solazzi, M., Troncossi, M., Mazzotti, C., Castelli, V.P., Procopio, C., Lamola, G., Chisari, C., Bergamasco, M., Frisoli, A., 2015. An EMG-controlled robotic hand exoskeleton for bilateral rehabilitation. IEEE Trans. Haptics 1939-14128 (2), 140–151. https://doi.org/10.1109/TOH.2015.2417570. Available from: http://ieeexplore.ieee.org/document/7072553/.

Liu, C., Wang, H., Pu, H., Zhang, Y., Zou, L., 2012. EEG feature extraction and pattern recognition during right and left hands motor imagery in brain-computer interface. In: 5th International Conference on BioMedical Engineering and Informatics. IEEE, pp. 506–510. Available from: http://ieeexplore.ieee.org/document/6513023/.

Liu, Y., Li, C., Ji, L., Bi, S., Zhang, X., Huo, J., Ji, R., 2017. Development and implementation of an end-effector upper limb rehabilitation robot for hemiplegic patients with line and circle tracking training. J. Healthcare Eng. 2040-22952017, 1–11. https://doi.org/10.1155/2017/4931217. Available from: https://www.hindawi.com/journals/jhe/2017/4931217/.

Lo, H.S., Xie, S.Q., 2012. Exoskeleton robots for upper-limb rehabilitation: state of the art and future prospects. Med. Eng. Phys. 1350453334 (3), 261–268. https://doi.org/10.1016/j.medengphy.2011.10.004. Available from: http://www.sciencedirect.com/science/article/pii/S1350453311002694#fig0015.

Loconsole, C., Bartalucci, R., Frisoli, A., Bergamasco, M., 2011. A new gaze-tracking guidance mode for upper limb robot-aided neurorehabilitation. In: 2011 IEEE World Haptics Conference. IEEE, pp. 185–190. Available from: http://ieeexplore.ieee.org/document/5945483/.

Loconsole, C., Dettori, S., Frisoli, A., Avizzano, C.A., Bergamasco, M., 2014. An EMG-based approach for on-line predicted torque control in robotic-assisted rehabilitation. In: 2014 IEEE Haptics Symposium (HAPTICS)IEEE, pp. 181–186. Available from: http://ieeexplore.ieee.org/document/6775452/.

Louie, W.-Y.G., Mohamed, S., Nejat, G., 2017. Human-robot interaction for rehabilitation robots. In: Encarnação, P., Cook, A.M. (Eds.), Robotic Assistive Technologies: Principles and Practice. Taylor & Francis Group, CRC Press, pp. 25–70. Available from: http://www.crcnetbase.com/doi/10.1201/9781315368788-3.

Loureiro, R.C.V., Harwin, W.S., 2007. Reach & grasp therapy: design and control of a 9-DOF robotic neuro-rehabilitation system. In: 2007 IEEE 10th International Conference on Rehabilitation Robotics (ICORR). IEEE, pp. 757–763. Available from: http://ieeexplore.ieee.org/document/4428510/.

Loureiro, R., Amirabdollahian, F., Topping, M., Driessen, B., Harwin, W., 2003. Upper limb robot mediated stroke therapy—GENTLE/s approach. Auton. Robot.

0929559315 (1), 35–51. https://doi.org/10.1023/A:1024436732030. Available from: http://link.springer.com/10.1023/A:1024436732030. http://link.springer.com/article/10.1023/A%3A1024436732030.

Maciejasz, P., Eschweiler, J., Gerlach-Hahn, K., Jansen-Troy, A., Leonhardt, S., 2014. A survey on robotic devices for upper limb rehabilitation. J. Neuroeng. Rehabil. 1743-000311(1), 3. https://doi.org/10.1186/1743-0003-11-3. Available from: http://www.jneuroengrehab.com/content/11/1/3.

Maitland, A., McPhee, J., 2018. Fast NMPC with mixed-integer controls using quasi-translations. In: 6th IFAC Conference on Nonlinear Model Predictive Control. Madison, Wisconsin, USA.

Mao, Y., Jin, X., Gera Dutta, G., Scholz, J.P., Agrawal, S.K., 2015. Human movement training with a cable driven arm exoskeleton (CAREX). IEEE Trans. Neural Syst. Rehabil. Eng. 1534-432023 (1), 84–92. https://doi.org/10.1109/TNSRE.2014.2329018. Available from: http://ieeexplore.ieee.org/document/6826540/. http://www.ncbi.nlm.nih.gov/pubmed/24919202.

Marchal-Crespo, L., Reinkensmeyer, D.J., 2009. Review of control strategies for robotic movement training after neurologic injury. J. Neuroeng. Rehabil. 1743-00036 (1), 20. https://doi.org/10.1186/1743-0003-6-20. Available from: http://www.jneuroengrehab.com/content/6/1/20.

Marquez-Chin, C., Marquis, A., Popovic, M.R., 2016. EEG-triggered functional electrical stimulation therapy for restoring upper limb function in chronic stroke with severe hemiplegia. Case Rep. Neurol. Med. 2016, 1–11. https://doi.org/10.1155/2016/9146213. Available from: https://www.hindawi.com/journals/crinm/2016/9146213/.

Mehrabi, N., Sharif Razavian, R., Ghannadi, B., McPhee, J., 2017. Predictive simulation of reaching moving targets using nonlinear model predictive control. Front. Comput. Neurosci. 1662-518810, 143. https://doi.org/10.3389/fncom.2016.00143. Available from: http://journal.frontiersin.org/article/10.3389/fncom.2016.00143/full.

Mehrholz, J., Hädrich, A., Platz, T., Kugler, J., Pohl, M., 2012. Electromechanical and robot-assisted arm training for improving generic activities of daily living, arm function, and arm muscle strength after stroke. In: Jan, M. (Ed.), Cochrane Database of Systematic Reviews. In: vol. 6. John Wiley & Sons, Ltd, Chichester, p. CD006876. Available from: http://doi.wiley.com/10.1002/14651858.CD006876.pub3. http://onlinelibrary.wiley.com/doi/10.1002/14651858.CD006876.pub3/pdf/standard. http://www.ncbi.nlm.nih.gov/pubmed/22696362.

Mendis, S., 2013. Stroke disability and rehabilitation of stroke: World Health Organization perspective. Int. J. Stroke 8 (1), 3–4. https://doi.org/10.1111/j.1747-4949.2012.00969.x. Available from:. http://www.ncbi.nlm.nih.gov/pubmed/23280261.

Mihelj, M., Nef, T., Riener, R., 2007. A novel paradigm for patient-cooperative control of upper-limb rehabilitation robots. Adv. Robot. 0169-186421 (8), 843–867. https://doi.org/10.1163/156855307780851975. Available from: http://www.tandfonline.com/doi/abs/10.1163/156855307780851975.

Miller, L.M., Rosen, J., 2010. Comparison of multi-sensor admittance control in joint space and task space for a seven degree of freedom upper limb exoskeleton. In: 2010 3rd IEEE RAS and EMBS International Conference on Biomedical Robotics and Biomechatronics (BioRob). IEEE, pp. 70–75. Available from: http://ieeexplore.ieee.org/lpdocs/epic03/wrapper.htm?arnumber=5628069.

Montagner, A., Frisoli, A., Borelli, L., Procopio, C., Bergamasco, M., Carboncini, M.C., Rossi, B., 2007. A pilot clinical study on robotic assisted rehabilitation in VR with an arm exoskeleton device. In: 2007 International Conference on Virtual Rehabilitation (ICVR). IEEE, pp. 57–64. Available from: http://ieeexplore.ieee.org/document/4362131/.

Morris, S.L., Dodd, K.J., Morris, M.E., 2004. Outcomes of progressive resistance strength training following stroke: a systematic review. Clin. Rehabil. 0269-215518 (1), 27–39. https://doi.org/10.1191/0269215504cr699oa. Available from: http://journals.sagepub.com/doi/10.1191/0269215504cr699oa.

Mostafavi, S.M., Mousavi, P., Dukelow, S.P., Scott, S.H., 2015. Robot-based assessment of motor and proprioceptive function identifies biomarkers for prediction of functional independence measures. J. Neuroeng. Rehabil. 1743-000312 (1), 105. https://doi.org/10.1186/s12984-015-0104-7. Available from: http://www.jneuroengrehab.com/content/12/1/105. http://www.pubmedcentral.nih.gov/articlerender.fcgi?artid=PMC4661950. http://www.ncbi.nlm.nih.gov/pubmed/26611144.

Moubarak, S., Pham, M.T., Moreau, R., Redarce, T., 2010. Gravity compensation of an upper extremity exoskeleton. In: 2010 Annual International Conference of the IEEE Engineering in Medicine and Biology, vol. 2010. IEEE, pp. 4489–4493. Available from: http://ieeexplore.ieee.org/document/5626036/. http://www.ncbi.nlm.nih.gov/pubmed/21095778.

Mrachacz-Kersting, N., Jiang, N., Stevenson, A.J.T., Niazi, I.K., Kostic, V., Pavlovic, A., Radovanovic, S., Djuric-Jovicic, M., Agosta, F., Dremstrup, K., Farina, D., 2016. Efficient neuroplasticity induction in chronic stroke patients by an associative brain-computer interface. J. Neurophysiol. 0022-3077115 (3), 1410–1421. https://doi.org/10.1152/jn.00918.2015. Available from: http://www.physiology.org/doi/10.1152/jn.00918.2015.

Muratori, L.M., Lamberg, E.M., Quinn, L., Duff, S.V., 2013. Applying principles of motor learning and control to upper extremity rehabilitation. J. Hand Ther. 0894113026, 94–103. https://doi.org/10.1016/j.jht.2012.12.007.

Nef, T., Mihelj, M., Riener, R., 2007. ARMin: a robot for patient-cooperative arm therapy. Med. Biol. Eng. Comput. 0140-011845 (9), 887–900. https://doi.org/10.1007/s11517-007-0226-6. Available from: http://link.springer.com/10.1007/s11517-007-0226-6.

Nef, T., Guidali, M., Riener, R., 2009. ARMin III—arm therapy exoskeleton with an ergonomic shoulder actuation. Appl. Bionics Biomech. 6 (2), 127–142. https://doi.org/10.1080/11762320902840179. Available from: http://content.iospress.com/doi/10.1080/11762320902840179.

Novak, D., Riener, R., 2013. Enhancing patient freedom in rehabilitation robotics using gaze-based intention detection. In: 2013 IEEE 13th International Conference on Rehabilitation Robotics (ICORR). IEEE, pp. 1–6. Available from: http://ieeexplore.ieee.org/document/6650507/.

Oblak, J., Cikajlo, I., Matjacic, Z., 2010. Universal haptic drive: a robot for arm and wrist rehabilitation. IEEE Trans. Neural Syst. Rehabil. Eng. 18 (3), 293–302. https://doi.org/10.1109/TNSRE.2009.2034162. Available from: http://ieeexplore.ieee.org/document/5290020/.

O'Connor, R.J., Jackson, A., Makower, S.G., Cozens, A., Levesley, M., 2015. A proof of concept study investigating the feasibility of combining iPAM robot assisted rehabilitation with functional electrical stimulation to deliver whole arm exercise in stroke survivors. J. Med. Eng. Technol. 0309-190239 (7), 411–418. https://doi.org/10.3109/03091902.2015.1088094. Available from: http://www.tandfonline.com/doi/full/10.3109/03091902.2015.1088094. http://www.ncbi.nlm.nih.gov/pubmed/26414146.

Oda, K., Isozumi, S., Ohyama, Y., Tamida, K., Kikuchi, T., Furusho, J., 2009. Development of isokinetic and iso-contractile exercise machine "MEM-MRB" using MR brake. In: IEEE 11th International Conference on Rehabilitation Robotics (ICORR). IEEE, pp. 6–11. Available from: http://ieeexplore.ieee.org/document/5209510/.

Ott, C., Mukherjee, R., Nakamura, Y., 2010. Unified impedance and admittance control. In: 2010 IEEE International Conference on Robotics and Automation (ICRA) pp. 554–561. https://doi.org/10.1109/ROBOT.2010.5509861. Available from: http://ieeexplore.ieee.org/lpdocs/epic03/wrapper.htm?arnumber=5509861.

Oujamaa, L., Relave, I., Froger, J., Mottet, D., Pelissier, J.Y., 2009. Rehabilitation of arm function after stroke. Literature review. Ann. Phys. Rehabil. Med. 1877065752, 269–293. https://doi.org/10.1016/j.rehab.2008.10.003.

Patten, C., Dozono, J., Schmidt, S., Jue, M., Lum, P., 2006. Combined functional task practice and dynamic high intensity resistance training promotes recovery of upper-extremity motor function in post-stroke hemiparesis: a case study. J. Neurol. Phys. Ther. 1557-057630 (3), 99–115. Available from: http://www.ncbi.nlm.nih.gov/pubmed/17029654.

Patton, J.L., Dawe, G., Scharver, C., Mussa-Ivaldi, F.A., Kenyon, R., 2004. Robotics and virtual reality: the development of a life-sized 3-D system for the rehabilitation of motor function. In: The 26th Annual International Conference of the IEEE Engineering in Medicine and Biology Society, vol. 4. IEEE, pp. 4840–4843. Available from: http://ieeexplore.ieee.org/document/1404339/.

Patton, J.L., Kovic, M., Mussa-Ivaldi, F.A., 2006. Custom-designed haptic training for restoring reaching ability to individuals with poststroke hemiparesis. J. Rehabil. Res. Dev. 1938-135243 (5), 643–656. Available from: http://www.ncbi.nlm.nih.gov/pubmed/17123205.

Patton, J.L., Stoykov, M.E., Kovic, M., Mussa-Ivaldi, F.A., 2006. Evaluation of robotic training forces that either enhance or reduce error in chronic hemiparetic stroke survivors. Exp. Brain Res. 0014-4819168 (3), 368–383. https://doi.org/10.1007/s00221-005-0097-8. Available from: http://link.springer.com/10.1007/s00221-005-0097-8. http://www.ncbi.nlm.nih.gov/pubmed/16249912.

Pérez-Rodríguez, R., Rodríguez, C., Costa, Ú., Cáceres, C., Tormos, J.M., Medina, J., Gómez, E.J., 2014. Anticipatory assistance-as-needed control algorithm for a multijoint upper limb robotic orthosis in physical neurorehabilitation. Expert Syst. Appl. 0957417441 (8), 3922–3934. https://doi.org/10.1016/j.eswa.2013.11.047. Available from: http://linkinghub.elsevier.com/retrieve/pii/S0957417413009895.

Perry, J.C., Rosen, J., Burns, S., 2007. Upper-limb powered exoskeleton design. IEEE/ASME Trans. Mechatron. 12 (4), 408–417. https://doi.org/10.1109/TMECH.2007.901934. Available from: http://ieeexplore.ieee.org/document/4291584/.

Pignolo, L., Dolce, G., Basta, G., Lucca, L.F., Serra, S., Sannita, W.G., 2012. Upper limb rehabilitation after stroke: ARAMIS a "robo-mechatronic" innovative approach and prototype. In: 2012 4th IEEE RAS & EMBS International Conference on Biomedical Robotics and Biomechatronics (BioRob). IEEE, pp. 1410–1414. Available from: http://ieeexplore.ieee.org/document/6290868/.

Poli, P., Morone, G., Rosati, G., Masiero, S., 2013. Robotic technologies and rehabilitation: new tools for stroke patients' therapy. Biomed. Res. Int. 23146-1332013, 153872. https://doi.org/10.1155/2013/153872. Available from: http://www.pubmedcentral.nih.gov/articlerender.fcgi?artid=PMC3852950. http://www.ncbi.nlm.nih.gov/pubmed/24350244.

Prisco, G.M., Avizzano, C.A., Calcara, M., Ciancio, S., Pinna, S., Bergamasco, M., 1998. A virtual environment with haptic feedback for the treatment of motor dexterity disabilities. In: Proceedings. 1998 IEEE International Conference on Robotics and Automation (Cat. No. 98CH36146), vol. 4. IEEE, pp. 3721–3726. Available from: http://ieeexplore.ieee.org/document/681418/.

Proietti, T., Jarrasse, N., Roby-Brami, A., Morel, G., 2015. Adaptive control of a robotic exoskeleton for neurorehabilitation. In: 2015 7th International IEEE/EMBS Conference on Neural Engineering (NER). IEEE, pp. 803–806. Available from: http://ieeexplore.ieee.org/document/7146745/.

Proietti, T., Crocher, V., Roby-Brami, A., Jarrasse, N., 2016. Upper-limb robotic exoskeletons for neurorehabilitation: a review on control strategies. IEEE Rev. Biomed. Eng.

1937-33339, 4–14. https://doi.org/10.1109/RBME.2016.2552201. Available from: http://www.ncbi.nlm.nih.gov/pubmed/27071194. http://ieeexplore.ieee.org/document/7450169/.

Rahman, M.H., Ochoa-Luna, C., Saad, M., 2015. EMG based control of a robotic exoskeleton for shoulder and elbow motion assist. J. Autom. Control Eng. 230137023 (4), 270–276. https://doi.org/10.12720/joace.3.4.270-276. Available from: http://www.joace.org/index.php?m=content&c=index&a=show&catid=44&id=242.

Reinkensmeyer, D.J., 2009. Robotic assistance for upper extremity training after stroke. In: Studies in Health Technology and Informatics, vol. 145. pp. 25–39.

Reinkensmeyer, D.J., Kahn, L.E., Averbuch, M., McKenna-Cole, A., Schmit, B.D., Rymer, W.Z., 2000. Understanding and treating arm movement impairment after chronic brain injury: progress with the ARM guide. J. Rehabil. Res. Dev. 0748-771137 (6), 653–662. Available from: http://www.ncbi.nlm.nih.gov/pubmed/11321001.

Reinkensmeyer, D.J., Lum, P., Winters, J.M., 2002. Emerging technologies for improving access to movement therapy following neurologic injury. In: Winters, J., Robinson, C., Simpson, R., Vanderheiden, G. (Eds.), Emerging and Accessible Telecommunications, Information and Healthcare Technologies—Engineering Challenges in Enabling Universal Access. IEEE Press Available from: http://www.eng.uci.edu/dreinken/publications/djrresnachapter.pdf.

Ren, Y., Hoon Kang, S., Park, H.-S., Wu, Y.-N., Zhang, L.-Q., 2013. Developing a multi-joint upper limb exoskeleton robot for diagnosis, therapy, and outcome evaluation in neurorehabilitation. IEEE Trans. Neural Syst. Rehabil. Eng. 1534-432021 (3), 490–499. https://doi.org/10.1109/TNSRE.2012.2225073. Available from: http://ieeexplore.ieee.org/document/6335485/.

Richards, C.L., Malouin, F., 2015. Stroke rehabilitation: clinical picture, assessment, and therapeutic challenge. Prog. Brain Res. 00796123218, 253–280. https://doi.org/10.1016/bs.pbr.2015.01.003.

Riener, R., Lunenburger, L., Jezernik, S., Anderschitz, M., Colombo, G., Dietz, V., 2005. Patient-cooperative strategies for robot-aided treadmill training: first experimental results. IEEE Trans. Neural Syst. Rehabil. Eng. 1534-432013 (3), 380–394. https://doi.org/10.1109/TNSRE.2005.848628. Available from: http://ieeexplore.ieee.org/document/1506824/. http://www.ncbi.nlm.nih.gov/pubmed/16200761.

Rocon, E., Belda-Lois, J.M., Ruiz, A.F., Manto, M., Moreno, J.C., Pons, J.L., 2007. Design and validation of a rehabilitation robotic exoskeleton for tremor assessment and suppression. IEEE Trans. Neural Syst. Rehabil. Eng. 15 (3), 367–378. https://doi.org/10.1109/TNSRE.2007.903917. Available from: http://ieeexplore.ieee.org/document/4303108/.

Rosati, G., Gallina, P., Masiero, S., Rossi, A., 2005. Design of a new 5 D.O.F. wire-based robot for rehabilitation. In: 9th International Conference on Rehabilitation Robotics (ICORR). IEEE, pp. 430–433. Available from: http://ieeexplore.ieee.org/document/1501135/.

Rosati, G., Volpe, G., Biondi, A., 2007. Trajectory planning of a two-link rehabilitation robot arm. In: Proceedings of the 12th IFToMM World Congress, Besancon, France. Available from: http://www.iftomm.org/iftomm/proceedings/proceedings_WorldCongress/WorldCongress07/articles/sessions/papers/A884.pdf.

Rosen, J., Perry, J.C., 2007. Upper-limb powered exoskeleton. Int. J. Humanoid Robot. 4 (3), 529–548. https://doi.org/10.1142/S021984360700114X. Available from: http://www.worldscientific.com/doi/abs/10.1142/S021984360700114X.

Sakurada, T., Kawase, T., Takano, K., Komatsu, T., Kansaku, K., 2013. A BMI-based occupational therapy assist suit: asynchronous control by SSVEP. Front. Neurosci. 1662-453X7, 172. https://doi.org/10.3389/fnins.2013.00172. Available from: http://journal.frontiersin.org/article/10.3389/fnins.2013.00172/abstract. http://www.pubmedcentral.nih.gov/articlerender.fcgi?artid=PMC3779864. http://www.ncbi.nlm.nih.gov/pubmed/24068982.

Sale, P., Lombardi, V., Franceschini, M., 2012. Hand robotics rehabilitation: feasibility and preliminary results of a robotic treatment in patients with hemiparesis. Stroke Res. Treat. 2090-81052012, 1–5. https://doi.org/10.1155/2012/820931. Available from: http://www.hindawi.com/journals/srt/2012/820931/. http://www.pubmedcentral.nih.gov/articlerender.fcgi?artid=PMC3540892. http://www.ncbi.nlm.nih.gov/pubmed/23320252.

Sanchez, R., Reinkensmeyer, D., Shah, P., Liu, J., Rao, S., Smith, R., Cramer, S., Rahman, T., Bobrow, J., 2004. Monitoring functional arm movement for home-based therapy after stroke. In: The 26th Annual International Conference of the IEEE Engineering in Medicine and Biology Society, vol. 4. IEEE, pp. 4787–4790. Available from: http://www.ncbi.nlm.nih.gov/pubmed/17271381.

Sanchez, R.J., Jiayin Liu, J., Rao, S., Shah, P., Smith, R., Rahman, T., Cramer, S.C., Bobrow, J.E., Reinkensmeyer, D.J., 2006. Automating arm movement training following severe stroke: functional exercises with quantitative feedback in a gravity-reduced environment. IEEE Trans. Neural Syst. Rehabil. Eng. 1534-432014 (3), 378–389. https://doi.org/10.1109/TNSRE.2006.881553. Available from: http://ieeexplore.ieee.org/document/1703570/. http://www.ncbi.nlm.nih.gov/pubmed/17009498.

Schabowsky, C.N., Godfrey, S.B., Holley, R.J., Lum, P.S., 2010. Development and pilot testing of HEXORR: hand EXOskeleton rehabilitation robot. J. Neuroeng. Rehabil. 1743-00037, 36. https://doi.org/10.1186/1743-0003-7-36. Available from: http://www.pubmedcentral.nih.gov/articlerender.fcgi?artid=PMC2920290. http://www.ncbi.nlm.nih.gov/pubmed/20667083.

Sharif Razavian, R., Ghannadi, B., Mehrabi, N., Charlet, M., McPhee, J., 2018. Feedback control of functional electrical stimulation for 2D arm reaching movements. In: IEEE Trans. Neural Syst. Rehabil. Eng. https://doi.org/10.1109/TNSRE.2018.2853573. Available from: https://ieeexplore.ieee.org/document/8404077/.

Shaw, S.E., Morris, D.M., Uswatte, G., McKay, S., Meythaler, J.M., Taub, E., 2005. Constraint-induced movement therapy for recovery of upper-limb function following traumatic brain injury. J. Rehabil. Res. Dev. 0748771142 (6), 769–778. https://doi.org/10.1682/JRRD.2005.06.0094. Available from: http://www.ncbi.nlm.nih.gov/pubmed/16680614.

Sheng, B., Zhang, Y., Meng, W., Deng, C., Xie, S., 2016. Bilateral robots for upper-limb stroke rehabilitation: state of the art and future prospects. https://doi.org/10.1016/j.medengphy.2016.04.004. Available from: https://www.sciencedirect.com/science/article/pii/S1350453316300480.

Simkins, M., Hyuchul, K., Abrams, G., Byl, N., Rosen, J., 2013. Robotic unilateral and bilateral upper-limb movement training for stroke survivors afflicted by chronic hemiparesis. In: IEEE 13th International Conference on Rehabilitation Robotics (ICORR). IEEE, pp. 1–6. Available from: http://ieeexplore.ieee.org/document/6650506/.

Skirven, T., Osterman, A., Fedorczyk, J., Amadio, P., 2011. Rehabilitation of the Hand and Upper Extremity, sixth ed. Elsevier, Amsterdam. ISBN 978-0-323-05602-1.

Staubli, P., Nef, T., Klamroth-Marganska, V., Riener, R., 2009. Effects of intensive arm training with the rehabilitation robot ARMin II in chronic stroke patients: four single-cases. J. Neuroeng. Rehabil. 1743-00036 (1), 46. https://doi.org/10.1186/1743-0003-6-46. Available from: http://jneuroengrehab.biomedcentral.com/articles/10.1186/1743-0003-6-46.

Stein, J., Narendran, K., McBean, J., Krebs, K., Hughes, R., 2007. Electromyography-controlled exoskeletal upper-limb-powered orthosis for exercise training after stroke. Am. J. Phys. Med. Rehabil. 0894-911586 (4), 255–261. https://doi.org/10.1097/PHM.0b013e3180383cc5. Available from: http://content.wkhealth.com/linkback/openurl?sid=WKPTLP:landingpage&an=00002060-200704000-00002.

Stienen, A.H.A., Hekman, E.E.G., Van der Helm, F.C.T., Prange, G.B., Jannink, M.J.A., Aalsma, A.M.M., Van der Kooij, H., 2007. Freebal: dedicated gravity compensation for the upper extremities. In: 2007 IEEE 10th International Conference on Rehabilitation Robotics (ICORR). IEEE, pp. 804–808. Available from: http://ieeexplore.ieee.org/document/4428517/.

Stoykov, M.E., Corcos, D.M., 2009. A review of bilateral training for upper extremity hemiparesis. Occup. Ther. Int. 0966790316 (3–4), 190–203. https://doi.org/10.1002/oti.277.

Sugar, T.G., He, J., Koeneman, E.J., Koeneman, J.B., Herman, R., Huang, H., Schultz, R.S., Herring, D.E., Wanberg, J., Balasubramanian, S., Swenson, P., Ward, J.A., 2007. Design and control of RUPERT: a device for robotic upper extremity repetitive therapy. IEEE Trans. Neural Syst. Rehabil. Eng. 15 (3), 336–346. https://doi.org/10.1109/TNSRE.2007.903903. Available from: http://ieeexplore.ieee.org/document/4303112/.

Sukal, T.M., Ellis, M.D., Dewald, J.P.A., 2006. Source of work area reduction following hemiparetic stroke and preliminary intervention using the ACT 3D system. In: 2006 International Conference of the IEEE Engineering in Medicine and Biology Society. IEEE, pp. 177–180. Available from: http://ieeexplore.ieee.org/document/4461714/.

Sveistrup, H., 2004. Motor rehabilitation using virtual reality. J. Neuroeng. Rehabil. 174300031 (1), 10. https://doi.org/10.1186/1743-0003-1-10. Available from: http://jneuroengrehab.biomedcentral.com/articles/10.1186/1743-0003-1-10.

Thielbar, K.O., Triandafilou, K.M., Fischer, H.C., O'Toole, J.M., Corrigan, M.L., Ochoa, J.M., Stoykov, M.E., Kamper, D.G., 2017. Benefits of using a voice and EMG-driven actuated glove to support occupational therapy for stroke survivors. IEEE Trans. Neural Syst. Rehabil. Eng. 25 (3), 297–305. https://doi.org/10.1109/TNSRE.2016.2569070. Available from: http://ieeexplore.ieee.org/document/7470432/.

Thrasher, T.A., Zivanovic, V., McIlroy, W., Popovic, M.R., 2008. Rehabilitation of reaching and grasping function in severe hemiplegic patients using functional electrical stimulation therapy. Neurorehabil. Neural Repair 22 (6), 706–714. https://doi.org/10.1177/1545968308317436. Available from: http://journals.sagepub.com/doi/10.1177/1545968308317436.

Todorov, E., Shadmehr, R., Bizzi, E., 1997. Augmented feedback presented in a virtual environment accelerates learning of a difficult motor task. J. Mot. Behav. 29 (2), 147–158. https://doi.org/10.1080/00222899709600829. Available from: https://www.tandfonline.com/doi/full/10.1080/00222899709600829.

Tsai, B.C., Wang, W.W., Hsu, L.C., Fu, L.C., Lai, J.S., 2010. An articulated rehabilitation robot for upper limb physiotherapy and training. In: IEEE/RSJ 2010 International Conference on Intelligent Robots and Systems, IROS 2010—Conference Proceedings. IEEE, pp. 1470–1475. Available from: http://ieeexplore.ieee.org/document/5649567/.

Tseng, C.M., Lai, C.L., Erdenetsogt, D., Chen, Y.F., 2014. A Microsoft Kinect based virtual rehabilitation system. In: 2014 International Symposium on Computer, Consumer and ControlIEEE, pp. 934–937. Available from: http://ieeexplore.ieee.org/document/6846037/.

Turolla, A., Dam, M., Ventura, L., Tonin, P., Agostini, M., Zucconi, C., Kiper, P., Cagnin, A., Piron, L., 2013. Virtual reality for the rehabilitation of the upper limb motor function after stroke: a prospective controlled trial. J. Neuroeng. Rehabil. 1743-000310 (1), 85. https://doi.org/10.1186/1743-0003-10-85. Available from: http://www.pubmedcentral.nih.gov/articlerender.fcgi?artid=3734026&tool=pmcentrez&rendertype=abstract.

van Delden, A.L.E.Q., Peper, C.L.E., Nienhuys, K.N., Zijp, N.I., Beek, P.J., Kwakkel, G., 2013. Unilateral versus bilateral upper limb training after stroke: the upper limb training after stroke clinical trial. Stroke 1524-462844 (9), 2613–2616. https://doi.org/10.1161/STROKEAHA.113.001969. Available from: http://www.ncbi.nlm.nih.gov/pubmed/23868279.

Van der Linde, R.Q., Lammertse, P., 2003. HapticMaster: a generic force controlled robot for human interaction. Ind. Robot. 30 (6), 515–524. https://doi.org/10.1108/01439910310506783. Available from: http://www.emeraldinsight.com/doi/10.1108/01439910310506783.

Van der Loos, H.F.M., Wagner, J.J., Smaby, N., Chang, K., Madrigal, O., Leifer, L.J., Khatib, O., 1999. ProVAR assistive robot system architecture. In: Proceedings 1999 IEEE International Conference on Robotics and Automation (Cat. No. 99CH36288C), vol. 1. IEEE, pp. 741–746. Available from: http://ieeexplore.ieee.org/document/770063/.

Venkatakrishnan, A., Francisco, G.E., Contreras-Vidal, J.L., 2014. Applications of brain-machine interface systems in stroke recovery and rehabilitation. Curr. Phys. Med. Rehabil. Rep. 2 (2), 93–105. https://doi.org/10.1007/s40141-014-0051-4. Available from: http://www.ncbi.nlm.nih.gov/pubmed/25110624.

Wamsley, C., Rai, R., Johnson, M., 2017. High-force haptic rehabilitation robot and motor outcomes in chronic stroke. Int. J. Clin. Case Stud. 3 (1) https://doi.org/10.15344/2455-2356/2017/115 Available from: https://www.graphyonline.com/archives/IJCCS/2017/IJCCS-115/.

Wei, Y., Bajaj, P., Scheldt, R., Patton, J., 2005. Visual error augmentation for enhancing motor learning and rehabilitative relearning. In: Proceedings of the 2005 IEEE 9th International Conference on Rehabilitation Robotics, vol. 2005. IEEE, pp. 505–510. Available from: http://ieeexplore.ieee.org/document/1501152/.

Wolbrecht, E.T., Chan, V., Le, V., Cramer, S.C., Reinkensmeyer, D.J., Bobrow, J.E., 2007. Real-time computer modeling of weakness following stroke optimizes robotic assistance for movement therapy. In: Proceedings of the 3rd International IEEE EMBS Conference on Neural Engineering. IEEE, pp. 152–158. Available from: http://ieeexplore.ieee.org/document/4227240/.

Wolbrecht, E.T., Chan, V., Reinkensmeyer, D.J., Bobrow, J.E., 2008. Optimizing compliant, model-based robotic assistance to promote neurorehabilitation. IEEE Trans. Neural Syst. Rehabil. Eng. 1534-432016 (3), 286–297. https://doi.org/10.1109/TNSRE.2008.918389. Available from: http://ieeexplore.ieee.org/document/4451797/.

Wu, C.-Y., Yang, C.-L., Chen, M.-D., Lin, K.-C., Wu, L.-L., 2013. Unilateral versus bilateral robot-assisted rehabilitation on arm-trunk control and functions post stroke: a randomized controlled trial. J. Neuroeng. Rehabil. 1743-000310, 35. https://doi.org/10.1186/1743-0003-10-35. Available from: http://www.pubmedcentral.nih.gov/articlerender.fcgi?artid=3640972&tool=pmcentrez&rendertype=abstract.

Xu, R., Jiang, N., Lin, C., Mrachacz-Kersting, N., Farina, K.D.D., 2014. Enhanced low-latency detection of motor intention from EEG for closed-loop brain-computer interface applications. IEEE Trans. Biomed. Eng. 61 (2), 288–296. https://doi.org/10.1109/TBME.2013.2294203. Available from: http://ieeexplore.ieee.org/document/6678728/.

Yamashita, M., 2014. Robotic rehabilitation system for human upper limbs using guide control and manipulability ellipsoid prediction. Proc. Technol. 2212017315, 559–565. https://doi.org/10.1016/j.protcy.2014.09.016. Available from: http://linkinghub.elsevier.com/retrieve/pii/S2212017314001315.

Yao, L., Sheng, X., Zhang, D., Jiang, N., Mrachacz-Kersting, N., Zhu, X., Farina, D., 2017. A stimulus-independent hybrid BCI based on motor imagery and somatosensory attentional orientation. IEEE Trans. Neural Syst. Rehabil. Eng. 25 (9), 1674–1682. https://doi.org/10.1109/TNSRE.2017.2684084. Available from: http://ieeexplore.ieee.org/document/7880549/.

Yeh, S.C., Lee, S.H., Wang, J.C., Chen, S., Chen, Y.T., Yang, Y.Y., Chen, H.R., Hung, Y.P., Rizzo, A., Tsai, T.L., 2013. Stroke rehabilitation via a haptics-enhanced virtual reality system. In: Advances in Intelligent Systems and Applications, vol. 2. Springer, Berlin, Heidelberg, pp. 439–453. Available from: http://link.springer.com/10.1007/978-3-642-35473-1_45.

Yu, W., Rosen, J., Li, X., 2011. PID admittance control for an upper limb exoskeleton. In: Proceedings of the 2011 American Control Conferencepp. 1124–1129. https://doi.org/10.1109/ACC.2011.5991147. ISSN 0743-1619 http://ieeexplore.ieee.org/document/5991147/.

CHAPTER TEN

Current Advances in the Design of Retinal and Cortical Visual Prostheses

Lilach Bareket*, Alejandro Barriga-Rivera*,†, Jeffrey V. Rosenfeld‡,§,¶, Nigel H. Lovell||, Gregg J. Suaning*

*Faculty of Engineering and Information Technologies, University of Sydney, Sydney, NSW, Australia
†Division of Neuroscience, University Pablo de Olavide, Seville, Spain
‡Monash Institute of Medical Engineering and Department of Surgery, Monash University, Clayton, VIC, Australia
§Department of Neurosurgery, Alfred Hospital, Melbourne, VIC, Australia
¶Department of Surgery, F. Edward Hébert School of Medicine, Uniformed Services University, Bethesda, MD, United States
||Graduate School of Biomedical Engineering, University of New South Wales, Sydney, NSW, Australia

Contents

1 INTRODUCTION

Low vision and ultimately profound blindness may be caused due to damage or alterations at different locations along the visual pathway from the eye to the brain including the retina, optic nerve (ON), lateral geniculate nucleus (LGN), and visual cortex. The US Social Security Administration (SSA) defines "Legal blindness" as central visual acuity (VA) of less than

https://doi.org/10.1016/B978-0-12-812539-7.00005-2

6/60 (Snellen VA), or a visual field no greater than 20 degrees, in the better eye with use of correcting lens. In the 10th revision of the World health organization (WHO) International Statistical Classification of Diseases, Injuries, and Causes of Death, "Low vision" function is defined as VA scores (Snellen VA) in the better eye with the best possible correction of less than 6/18 but at least 3/60, or a visual field of less than 20 degrees. VA scores of less than 3/60, or a visual filed of less than 10 degrees are considered "blindness." "Visual impairment" includes both "low vision" and "blindness." VA is a measure of vision that compares the response of a patient to a normative group. The term 6/60 (20/200 in feet) refers to an individual seeing at 6m what the group saw at 60m (Bourne et al., 2013a; Stevens et al., 2013). According to the WHO in 2010, 285 million people were estimated to be visually impaired worldwide, of which 39 million were blind (Bourne et al., 2013b; Mariotti, 2012). The leading causes for blindness are cataracts (33%), uncorrected refractive error (21%), and macular degeneration (7%). Other disorders that cause visual impairment include retinal dystrophies, diabetic retinopathy, brain trauma, and several infectious diseases (Mariotti, 2012; Lehman, 2012).

Visual prostheses operate by applying electrical stimulation to neurons along the visual pathway to induce visual sensations, ultimately toward restoration of visual perception. Visual sensation relates to the pure physiological process of bringing information from the environment into the body and the brain. It includes the detection and reception of incoming light, conversion of the light stimuli into neural impulses, and transmission of the information to the visual center in the brain (occipital lobe). Visual perception is defined as the processing and interpretation of this information by the visual system. For example, classification of the visual stimulus into color, movement, shape, etc., and assembly of these features into patterns. Visual perception and is both physiological and psychological.

Anatomical targets of stimulation that are currently being explored include the retina (da Cruz et al., 2016; Yue et al., 2015; Stingl et al., 2015; Keserü et al., 2012; Fujikado et al., 2016; Ayton et al., 2014), the ON (Sakaguchi et al., 2009; Veraart et al., 1998; Fang et al., 2005), the LGN (Panetsos et al., 2011; Pezaris and Eskandar, 2009), and the visual cortex (Coulombe et al., 2007; Fernandez et al., 2005; Lowery et al., 2015; Troyk et al., 2006) (Fig. 1).

Retinal prostheses replace the phototransduction function of the retina in conditions where the photoreceptor cells in the retina degenerate, targeting conditions such retinitis pigmentosa (RP) and age-related macular

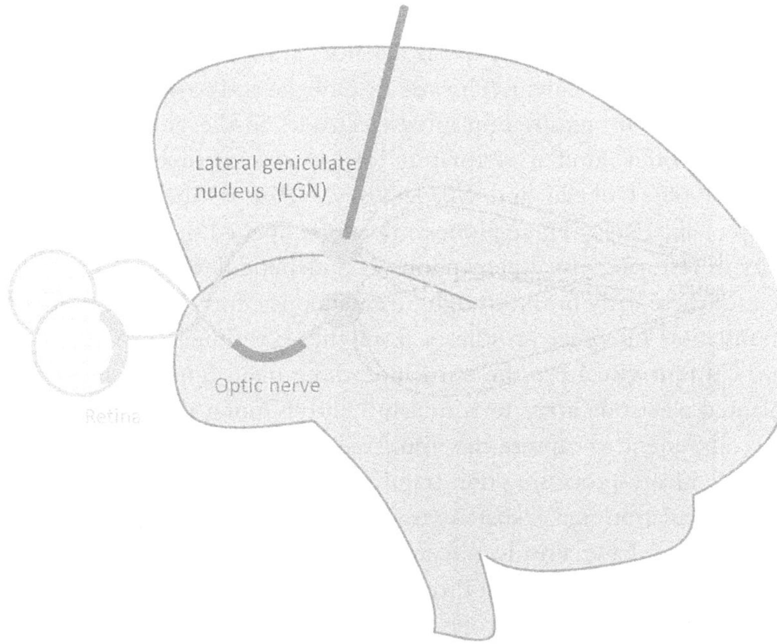

Fig. 1 Schematic representation of the visual pathway from the eyes to the brain, and visual prostheses implantation sites. Flat arrays are placed next to the retina, around the optic nerve or next to the visual cortex, and a penetrating shaft-like array to target the lateral geniculate nucleus (LGN). For cortical and optic nerve prostheses, the electrodes in the array can be either flat or needle-like penetrating electrodes.

degeneration (AMD). While these conditions severely damage the photoreceptors, the remaining neurons in the retina, in particular bipolar cells and the output retinal ganglion cells (RGCs), may still be activated by artificial stimuli, leading to elicitation of visual sensations (Santos et al., 1997; Stone et al., 1992). Cortical prostheses can be applied in cases when the RGCs degenerate or after ON injury, for example, in cases of glaucoma, optic neuropathy, severe retinal disease, diabetic retinopathy, optic neuritis, large pituitary/parasellar tumors, bilateral enucleation, and bilateral retinoblastoma, as well as ON and eye trauma.

The concept of artificial vision was demonstrated by several pioneering studies dating back to the 18th century. In 1755, Charles LeRoy reported that electrical discharge applied to the surface of the eye of a patient blinded from cataract during a surgery, resulted in the sensations of light spots (phosphenes) (LeRoy, 1755). Later, Brindley (1964) showed that visual sensations can be evoked by stimulation of the retina. Probing direct stimulation of the visual cortex to induce phosphenes was indicated by several investigations in the early 1920s (Löwenstein and Borchardt, 1918; Krause, 1924; Foerster, 1929). Brindley and Lewin, followed by Dobelle and coworkers have demonstrated that chronically implanted electrodes in the visual cortex could potentially offer limited restoration of visual sensations (Brindley and Lewin, 1968; Dobelle and Mladejovsky, 1974; Dobelle et al., 1974; Klomp et al., 1977). These pioneering efforts opened the door to the possibility of restoring visual perception via prosthetic devices.

Since these early breakthroughs, tremendous efforts have been invested in translation of visual prostheses from the experimental to the clinical stage. Current visual prostheses include three main functionalities: (1) an implanted electrode array to stimulate neurons along the visual pathway, (2) a component to capture the visual scene, and (3) an image-processing unit. The image-processing unit transduces the captured image into a pattern of stimulation signals that are transferred to the implanted chip. Three retinal devices have already obtained regulatory approvals: the Argus II device (Second Sight Medical Products, United States) received regulatory approval for marketing in Europe (CE mark; 2011), the United States (FDA approval, humanitarian device exemption; 2013) and Canada (2015), and the Alpha IMS prosthesis (Retina Implant AG, Germany) and IRIS II (Pixium Vision SA, France) which gained CE certification in 2014 and 2016, respectively. Patients implanted with retinal devices show improvement in visually guided performance tasks including recognition and discrimination of objects, following a marked trail, grasping objects, and reading. The new generation of visual cortical devices is currently in the experimental or preclinical phase of development, with clinical trials planned within the next several years. The basic device architecture of a camera, vision processing computer, and electrode interface with the brain underpins the design of the new generation cortical visual prosthetics but the difference is the application of computer chips, microelectronic circuit design, new materials, microelectrodes wireless engineering, and advanced manufacturing and neurosurgical techniques.

In this chapter, we review the progress in visual bionics, focusing on retinal and cortical prostheses. We describe state-of-the-art devices undergoing preclinical or clinical trials. Next, we discuss current technological challenges that need to be addressed. Finally, we highlight progress in next-generation technologies, including alternative implantation sites along the visual pathway.

2 CURRENT RETINAL IMPLANT TECHNOLOGIES

Retinal prostheses aim to restore visual capacity lost due to degeneration of the photoreceptor cells in the retina. The photosensitive cells constitute the outer nuclear layer of the retina and include two functional types of cells: rods and cones (Fig. 2A). The rods, positioned primarily in the peripheral areas of the retina, are very sensitive to light and can be triggered by a single photon (Hecht et al., 1942). Therefore, at low light levels, for example, at night, visual signals are primarily initiated by the rod photoreceptors. The cone photoreceptors are located primarily at the macula (center of the retina) and account for high acuity day vision and color experiences. Cones require a significantly higher number of photons in order to produce a signal, and can be distinguished into three different types according to their pattern of response to short, medium, and long wavelengths in the visible light range (Hurvich, 1981; Lennie and D'Zmura, 1987). The average human retina contains about 4.6 million cone cells, with peak foveal density of 199,000 cones/mm^2, 92 million rod cells, and about 1.07 million ganglion

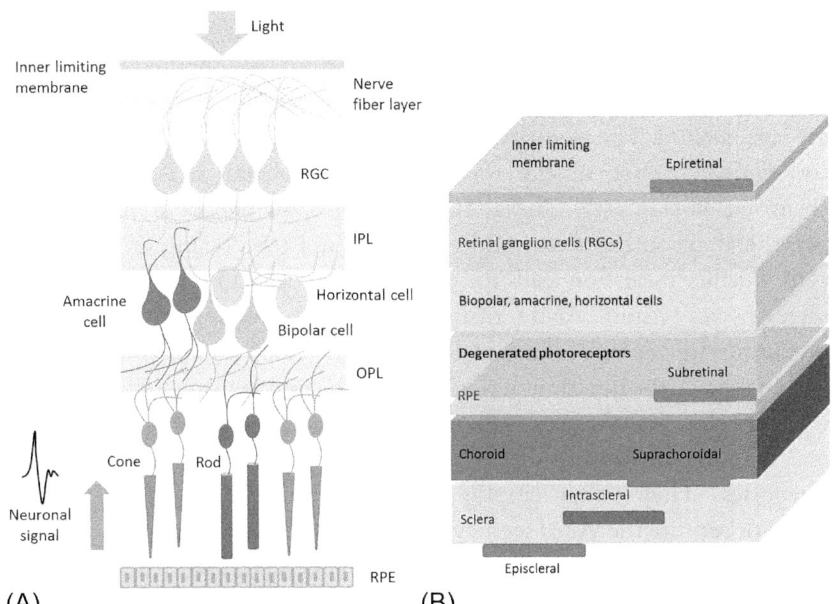

Fig. 2 Schematic representation of (A) organization of cells in the retina and (B) retinal device implantation strategies.

cells (Curcio and Allen, 1990; Curcio et al., 1990). Besides the photoreceptor layer, the retinal tissue is composed of two more layers of nerve cell bodies and two layers of synapses (Fig. 2A). The middle nuclear layer of the retina includes bipolar, horizontal, and amacrine cells and the outer layer consists of RGCs. Between these three structures are two layers where the neurons synapse, the outer and inner plexiform layers (OPL and IPL, respectively) (Fig. 2A). The bottom layer is the retinal pigment epithelium (RPE), which regulates nutrients and waste exchange. In a healthy retina, the photoreceptor cells transduce light into biochemical signals that propagate through the mid-layers up to the output neurons, the ganglion cells. The axons of the RGCs converge into the ON that delivers the visual signals to higher visual centers, the LGN and ultimately the visual cortex.

The most prevalent degenerative disorders of the retina are RP and AMD, both leading to progressive reduction in vision where the typical outcome is legal blindness (Ferris et al., 1984; Hartong et al., 2006; Wright et al., 2010). RP may further decline to profound blindness (no useful vision) (Hartong et al., 2006). AMD primarily affects people aged 60 and older. Currently, approximately 170 million people live with AMD, of which 2 million are blind (Mariotti, 2012). With the aging of the global population, this number is expected to rise to become 196 million by 2020 and 288 million by 2040 (Wong et al., 2014). RP is a group of inherited pathologies, afflicting approximately 1.5 million people worldwide 25% of RP patients are legally blind. The majority of gene defects leading to RP are expressed in rod photoreceptors or the RPE, in the peripheral areas of the retina (Hartong et al., 2006). Initial RP symptoms include poor night vision, peripheral vision loss followed by damage, and remodeling of foveal cone cells. Further escalation leads to "tunnel vision" and in many cases, a complete loss of vision. AMD mainly affects the cones around the fovea, impairing the center of the visual field (Ferris et al., 1984).

To replace the function of the lost photoreceptive tissue, retinal prostheses apply electrical stimulation to the residual neural components of the retina. In order to capture the visual scene, retinal prostheses fall into two categories. They either employ an external camera (Humayun et al., 2012) or capture the visual scene via an array of photodiodes integrated with the electrodes implanted in the vicinity of the retina (Stingl et al., 2015). The camera can be mounted on spectacles worn by the patient. Because the location of the camera is fixed, the perceived image is not coordinated with eye movements (saccades and smooth pursuit) and with the direction of gaze (Sabbah et al., 2014). To compensate for the lack of natural coordination, head movements are used to scan the visual field (Ho et al., 2015). The photodiode-based configuration potentially permits object tracking via eye movements, as the light is transduced to electrical signals at the implanted device. These signals are then processed, amplified, and delivered to the coupled electrodes.

Four sites of implantation for the stimulation array have been clinically investigated: epiretinal, subretinal, intrascleral, and suprachoroidal (Fig. 2B). Epiretinal devices are placed between the vitreous humor and the inner limiting membrane (ILM) to interact with the RGCs (Humayun et al., 2003; Rizzo et al., 2003), while subretinal devices are inserted between the RPE and the outer retina to stimulate the inner retinal neurons (primarily bipolar cells, but also amacrine and horizontal cells) (Chow et al., 2004; Zrenner et al., 2011). Suprachoroidal electrodes are maintained between the choroid and scleral tissues (suprachoroidal) (Ayton et al., 2014; Zhou et al., 2008; Wong et al., 2009) and intrascleral electrodes are embedded in the sclera (Nakauchi et al., 2005; Fujikado et al., 2007) (Fig. 2B). While in the suprachoroidal and intrascleral configurations the electrodes are distant from the target cells, these strategies offer a substantially less invasive surgical procedure. An extraocluar retinal prosthesis (ERP) placed on the surface of the sclera (episcleral) was also proposed by Chowdhury et al. (2005, 2008).

Epiretinal prostheses (Fig. 3A) that have been clinically tested include the devices by Second Sight Medical Products Inc. (Argus I and Argus II; United States) (da Cruz et al., 2016; Yue et al., 2015), EPIRET GmBH (EPIRET3; Germany) (Klauke et al., 2011), and Intelligent Medical Implants GmBH (IMI, Switzerland) (Keserü et al., 2012), which later became part of Pixium Vision SA (France) with the Intelligent Retinal Implant System (IRIS I and IRIS II). Subretinal devices (Fig. 3B) include the artificial silicon retina (ASR) developed by Optobionics (United States) (Chow et al., 2010), and the Alpha IMS by Retina Implant AG (Germany) (Stingl et al., 2015). Finally, the suprachoroidal (Fig. 3C) and intrasceral devices are clinically examined by groups in Australia [Bionic Vision Australia (BVA) consortium] (Ayton et al., 2014), and in Japan (Osaka University) (Fujikado et al., 2016).

Retinal systems that have not yet been tested in humans include the subretinal PRIMA vision restoration system (Pixium Vision) developed by Palanker and coworkers (Stanford University, United States) (Lorach et al., 2015a; Palanker et al., 2005), and the Boston Retinal Implant (BRI) developed by the Boston Retinal Implant Project group (BRIP; Boston, United States) (Rizzo, 2011). The Bioretina epiretinal device is

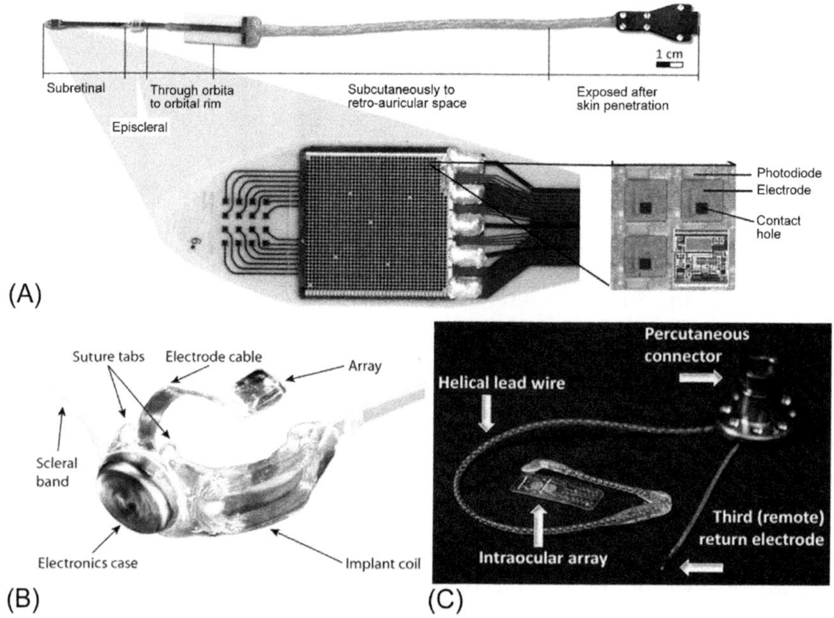

Fig. 3 Retinal devices. (A) The internal part of the Argus II system (epiretinal prosthesis) including the electrode array, electronic case, and implant radio frequency (RF) coil. (B) Prototype of the Alpha-IMS subretinal system. Top bottom panel is a detailed view of the microphotodiode array (MPDA) with an additional 16 TiN electrodes (investigational device). (C) The Bionic Vision Australia (BVA) suprachoroidal implant with 33 stimulating electrodes on the silicone substrate. *(A) Image reproduced with permission from Zrenner, E., et al., 2011. Subretinal electronic chips allow blind patients to read letters and combine them to words. Proc. R. Soc. B Biol. Sci. 278, 1489–1497. (B) Image reproduced with permission from Humayun, M.S., et al., 2012. Interim results from the international trial of second sight's visual prosthesis. Ophthalmology 119, 779–788. (C) Images reproduced with permission from Ayton, L.N., et al., 2014. First-in-human trial of a novel suprachoroidal retinal prosthesis. PLoS ONE 9, e115239.*

being developed by NanoRetina (Israel) (Raz-Prag et al., 2014; Yanovitz et al., 2014). Two suprachoroidal devices are being developed by researchers from the Universities of New South Wales (UNSWs) and Sydney in Australia with the Phoenix99 device (Suaning et al., 2014; Barriga-Rivera et al., 2016b), and by researchers from Seoul National University with a device based on liquid crystal polymer (LCP) technology (Jeong et al., 2016).

2.1 Epiretinal Implants

In epiretinal prostheses, the electrode array is placed on the surface of the retina (Fig. 2B). The first epiretinal device chronically implanted in humans was the Argus I (Second Sight Medical Products) (Humayun et al., 2003, 2012). The intraocular array included 16 platinum (Pt) electrodes in a 4 by 4 arrangement, with an 800 μm pitch which corresponds to a field of view (FOV) of 10 degrees (diagonally) (Caspi et al., 2009; Humayun et al., 2003). The electrodes were fixed in a silicone rubber platform and wired via a trans-scleral cable to the extraocular part of the device. Following implantation with the Argus I, all of the six subjects reported the appearance of phosphenes upon stimulation of the retina (de Balthasar et al., 2008; Yanai et al., 2007; Mahadevappa et al., 2005). The brightness of the phosphenes was directly correlated to the stimulation amplitude and frequency (Caspi et al., 2009; Humayun et al., 2003). A 10-year follow-up study in one subject reported that the tissue-implant interface remained stable (Yue et al., 2015). The second-generation device, Argus II, consists of a higher electrode density with 6 by 10 channels, with a distance of 525 μm between electrodes (FOV 22 degrees diagonally), thus covering a larger area of the FOV than the Argus I. The device was so far implanted in over 100 patients. In a study published in 2013, patients reported an improvement in the ability to localize high contrast objects and to detect motion, with at least two of the recipients able to read letters (da Cruz et al., 2013; Dorn et al., 2013). The best VA achieved with this implant was 20/1262, as tested with the visual grating acuity (VGA) test, which is still considered as legal blindness (<20/200) (Ho et al., 2015; Humayun et al., 2012). Some of the patients experienced ocular damage such as conjunctival erosion (2 out of 30), endophthalmitis (3 out of 30), and hypotony (3 out of 30), which was treated without permanent complications (Humayun et al., 2012). In a 5-year postimplantation study published in 2016, it was reported that the majority of the recipients did not experience severe adverse effects that are related to the device or to the surgical procedure, and that the improvement in visual function was maintained (da Cruz et al., 2016).

The IRIS I and IRIS II epiretinal systems originate from the IMI device. The IMI array with 49 iridium oxide (IrOx) included a polyimide (PI) thin film encapsulation for both the electrodes and the extraocular stimulator (wired to the electrodes). The IMI system employs a radio frequency (RF) wireless transmission route for power and an infrared (IR) link for data (Hornig et al., 2007). Switching on and off of the device by opening and closing of the eyelid is possible through the optical link (Keserü et al., 2009). As evaluated in a chronic implantation trial conducted in three patients for 9 months, the recipients were able to recognize simple light patterns. Later, an acute implantation (up to 45 min) was tested in 20 subjects, confirming the ability to identify and describe phosphenes following stimulation of the retina (Keserü et al., 2012). In a 3-month follow-up test, the majority of the patients showed no sign of damage due to the surgery or to the electrical stimulation, except for one patient with peripheral retinal detachment (which was successfully treated) (Keserü et al., 2012). The company has obtained CE mark for the IRIS II in July 2016 (Hornig et al., 2017).

The EPIRET3 prosthesis was so far evaluated through two clinical trials: a short-term trial (4 weeks) and a long-term trial (7–18 months) conducted in six and three volunteers with blindness due to RP, respectively (Roessler et al., 2009; Schimitzek et al., 2016; Menzel-Severing et al., 2012). The extraocular component of the EPIRET3 includes a computer system, a transmitter unit, and a transmitter coil attached to a holder placed in front of the eye. The intraocular part consists of a receiver coil, all necessary electronics, and an array with 25 electrodes (Roessler et al., 2009). Similar to the IMI device, the EPIRET3 chip is also encapsulated with polymer [polydimethylsiloxane (PDMS)]. The 4-week long trial was designed to evaluate the efficacy and safety of the device and of the surgical procedure (Roessler et al., 2009). The implant remained stable for the duration of the study, and removal was performed successfully in all patients. It was concluded that the tissue reaction was within an acceptable range, with temporary inflammation developed in all of the patients. Two cases of more adverse effects were detected: a sterile hypopyon in one case and a retinal tear that occurred during explantation, in the second case (Roessler et al., 2009). Both of these conditions were successfully treated. All patients reported visual responses triggered by electrical stimulation through the implant (Klauke et al., 2011). Some of the patients were able to discriminate signals applied at different locations of the array as well as pattern orientations. In the second study, some of the patients developed gliosis near the tacks used to attach the implant to the tissue (Menzel-Severing et al., 2012). Although gliosis does not appear to pose a major health risk, it may contribute to deterioration in the functionality of the implant. Strong glial proliferation of glia cells is part of the immune response of the retinal tissue to the presence of the implant (Dyer and Cepko, 2000). Formation of such glial encapsulation can potentially widen the gap between electrodes and their target neurons, thus increasing the amount of charge required for stimulation.

Currently, at preclinical experimental stage is the high-resolution epiretinal system with 600 pixels (Bioretina) that is being developed by Nanoretina. Employing three-dimensional (3D) microelectrodes the high-density array is designed to penetrate from its location at the outer retina into the inner layers. The miniature implantable chip also includes an internal digital imager (camera and processor), and photovoltaic power supply based on IR light. The IR laser light is delivered via a pair of accompanying wireless and rechargeable glasses. The image capture and processing functionalities embedded in the implanted chip permit the use of the natural optical pathway, potentially eliminating the requirement for scanning of the visual field through head movements. Preclinical trials done so far on pigs have validated that the feasibility of the surgical approach (Yanovitz et al., 2014). Histopathological evaluation demonstrates the needle electrodes can be applied to anchor the implant to the retina, and that both the electrodes and the tissue remained intact (Yanovitz et al., 2014; Raz-Prag et al., 2014).

2.2 Subretinal Implants

Subretinal technologies rely on stimulation through either a standard electrode array (similar to the epiretinal scheme) or an array of microphotodiodes integrated with the implanted electrodes. The photodiodes transduce the perceived luminescence into electrical signals. This strategy aims to achieve a wire-free device that is directly stimulated by incident light, and to take advantage of the biological amplification and modulation processing, that occurs in the middle layers of the retina (Chow et al., 2010).

Optobionics Inc. were the first to develop a photovoltaic subretinal device powered by incidence light, and implant it in humans. The ASR consisted of 5000 silicon photodiodes coupled with IrOx electrodes (Chow, 2013; Chow et al., 2004, 2010; Peachey and Chow, 1999). The ASR was implanted in 42 patients with follow-up studies of up to 8 years. The recipients reported improvement in VA, but it was not correlated with the location of implant, suggesting that the neurons were not stimulated through the array (Chow et al., 2004, 2010). This inability to elicit light-mediated neuronal responses was probably because the output currents produced by the photodiodes were below the threshold for neuronal activation (Palanker et al., 2005). The improvement in VA was attributed to a neurotropic effect, slowing the progression of retinal degeneration (Chow, 2013).

To produce enough photocurrent to activate the retina, in the Alpha IMS subretinal photovoltaic device, an amplifying circuit is connected to each microphodiode (MPD), and to matching electrode [titanium nitride (TiN)] (Zrenner et al., 2011; Stingl et al., 2013b, 2015; Schwahn et al., 2001). The amplifier increases the output signal of the photodiode in order to send high enough current to the electrode to stimulate the nearby neurons. There are 1500 stimulating elements in the Alpha-IMS, arranged in a 38 by 40 layout. The distance between the electrodes is 700 μm, corresponding to a FOV 15 degrees (diagonally). The investigational version of the implant includes a percutaneous cable connected to an external battery and additional 16 electrodes used to study pure electrical stimulation in addition to the photovoltaic scheme. In 3 out of the 11 patients implanted with this investigational device direct photostimulation of the retina was achieved (Zrenner et al., 2011). Patients were able to recognize unknown bright objects on a dark background without training, and one of them was even able to read letters (Zrenner et al., 2011). So far 29 people were implanted with the commercial version (Retina Implant GmBH) of this device which is wireless and does not contain the additional 16 electrodes designed to test electrical stimulation (Wilke et al., 2011; Stingl et al., 2013a,b). In a year postimplantation report, the majority of the patients described improvement in perception of light (25 out of 29), in recognition (shape, details) and localization of objects (21 out of 29) (Stingl et al., 2013a, 2015). The best VA demonstrated with the Alpha IMS was 20/200 and 20/546 in VGA and Landolt-C optotypes acuity (LCA) tests, respectively (Stingl et al., 2013a, 2015). These long-term studies also revealed two sources for potential device failure: breaks in the intraorbital cable probably due to mechanical stress induced by eye movement, and corrosion of the electronics seal indicating need to improve hermeticity (Stingl et al., 2015).

Subretinal technologies that are currently at preclinical experimental stage are the PRIMA vision restoration system and the device developed by the BRIP. In the PRIMA vision restoration system, several near infrared (NIR; 880–915 nm) light activated photodetectors are connected to the same stimulating electrode (Palanker et al., 2005; Mathieson et al., 2012). Prototype arrays with 37 stimulating elements each composed of an IrOx electrode (diameter of 70 or 140 μm) connected to two or three NIR photodiodes, and to a return electrode were realized. The purpose for an individual return is to focus the electric field and reduce electrode crosstalk (Mathieson et al., 2012; Lorach et al., 2015b). Incoming light reaching the device is transduced into high-intensity NIR laser pulses that are projected onto the retina, to activate the photodiodes. Optical activation of retinal cells was so far validated in rats with retinal degeneration by recording of visually evoked potentials (VEPs) from the cortex (Lorach et al., 2015a). This scheme allow for direct optical powering of the implant at the expense of increased complexity and power required for production and projection of high-intensity laser pulses. In addition, the illumination intensity for neuronal stimulation is 0.55–10 mW/mm^2, which is more than 100 times higher than bright daylight conditions (Lorach et al., 2015b).

The BRI device does not attempt to drive stimulation through microphotodiode array (MPDA) but rather applies electrical signals (Rizzo, 2011). Prototypes with 15 electrodes that were implanted in 7 minipigs continued to function for up to 1 year (Shire et al., 2009; Kelly et al., 2011, 2013). The group specifically focuses on developing technology for wireless transmission to achieve 256 individually controlled channels, as well as hermetic titanium casing for the extrascleral electronics, and ceramic feedthroughs for the cables (Kelly et al., 2011, 2013). The latest version of the BRIP device (256 channels) is currently undergoing preclinical tests.

2.3 Suprachoroidal Implants

While in the subretinal and epiretinal approaches the electrodes are in direct contact with the neurons in the retina, it has been demonstrated that positioning of the array at a distance to the neurons, in the suprachoroidal space or in the sclera (Fig. 2B), can also be applied to evoke visual sensation. The idea of placing a retinal stimulator between the choroidal and the scleral tissues was first tested by Tassicker (1956). The choroid contains several layers of connective tissue and blood vessels, and the sclera contains mainly collagen and elastic fibers protecting the eye. Following Tassicker's study, the surgical procedure was further developed by several research groups (Sakaguchi et al., 2004a; Ayton et al., 2014; Zhou et al., 2008). The array is inserted through an incision in the sclera with the electrodes facing toward the choroid and the neurons beyond, and then stabilized with sutures (no metal tacks are required). The return electrode can be placed in the front of the eye (vitreous cavity or cornea), integrated into the electrode array, or subcutaneously implanted behind the ear (Villalobos et al., 2012; Ayton et al., 2014; Saunders et al., 2014).

Preclinical investigations of suprachoroidal implantation were conducted by several groups (Sakaguchi et al., 2004a; Zhou et al., 2008; Saunders et al., 2014; Shivdasani et al., 2010, 2012). In 2004, researchers from Osaka University reported implantation of an eight-electrode array in the suprachoroidal space of the eyes of Albino rabbits, and demonstrated that stimulation of the retina evoked neuronal activity measured from the visual cortex (Sakaguchi et al., 2004a). A histological study further revealed the choroidal and scleral tissues became separated. Propagation of electrically induced signals from the retina to the visual cortex was also demonstrated with strip-shaped electrodes (300 μm × 750 μm) implanted between the sclera and the choroid of normally sighted rabbits by Zhou et al. (2008). As expected, the stimulation threshold was higher compared with subretinal or epiretinal stimulation. Optical coherence tomography (OCT) and fundus imaging conducted for a duration of 16 months after the surgery, showed that the implanted arrays maintained their position in the tissue, and there was no sign for inflammation or structural damage (Zhou et al., 2008).

The BVA consortium is so far the only group to report a clinical study with a suprachoroidal retinal prosthesis (Ayton et al., 2013). The surgical procedure was based on the approach demonstrated by the Korean team (Zhou et al., 2008) and was further developed using cadaver studies (feline and human) as well as acute and chronic investigations in a feline model (Shivdasani et al., 2010, 2012; Saunders et al., 2014; Villalobos et al., 2012, 2013; Cicione et al., 2012). The implanted device was fabricated using technology and materials similar to auditory brainstem implants and consisted of 33 Pt electrodes (diameter of 400 or 600 μm) embedded in a silicone support and organized in a hexagonal pattern. Hexagonal layout is designed to focus the electric field and reduce electrode cross talk as the active site is surrounded by six shortened sites, collectively returning the current (Lovell et al., 2005; Matteucci et al., 2013). Hexagonal stimulation with the suprachoroidal prosthesis was clinically tested in a later study by the group (Sinclair et al., 2016) and prior to that in preclinical studies (Wong et al., 2009; Matteucci et al., 2013). In the study by Ayton et al. (2014), the outer ring of the array was shortened to form a return electrode. Therefore, the device had a total of 20 individually addressable stimulating sites, and 4 return electrodes: a shortened "U"-shaped including additional two embedded in the array (diameter of 2000 μm) and one implanted subcutaneously behind the ear. The electrode array was wired to a percutaneous connector behind the ear, to enable a connection to the external stimulator.

Three participants who were blind due to RP, were implanted with the prototype prosthesis (Ayton et al., 2014; Sinclair et al., 2016). While the appearance of phosphenes was markedly different between individuals and between single electrodes in the same patient, all of the participants reported reliable, retinotopically correlated perception of phosphenes, and phosphene size and intensity were correlated with the stimulation charge (Sinclair et al., 2016). Two of the patients were able to identify the location of the active electrode in 57.2% and 23% of the trials by pointing the center of the perceived phosphine (randomized recognition task) (Sinclair et al., 2016). While the calculated limit of acuity with 20 electrodes and a distance of 1 mm (FOV 17 degrees diagonally) is 20/4242, the best VA score reported was 20/8397 (LCA) in one out of the three patients (Ayton et al., 2014). The biocompatibility of the device was investigated using OCT scans during a year following the surgery. It was reported that the implant remained mechanically stable, functional, and without signs for adverse immune response, except for an infection at the percutaneous connector which was treated with antibiotics (Ayton et al., 2014). The researchers noticed that the electrode-retina distance increased with time especially in one of the patients, resulting in an elevation of the stimulation threshold (Ayton et al., 2014).

Two suprachoroidal systems that are currently under investigation at preclinical stage include the Phoenix99 prototype developed by groups at UNSW and the University of Sydney (Schuettler et al., 2005; Suaning et al., 2014; Matteucci et al., 2013; Barriga-Rivera et al., 2016b), and a LCP-based device developed by a group from Seoul National University (South Korea) (Jeong et al., 2012, 2016).

During the fabrication of the Phoenix99, the entire electrode array including electrodes and conducting tracks is patterned through laser micromachining of Pt foil on top of medical grade PDMS film (Schuettler et al., 2005). The high-density Phoenix99 with 98 stimulation electrodes is achieved by folding of the planar array into a multilayer architecture (Suaning et al., 2007, 2014). An additional return electrode is located under the hermetic capsule that contains the electronics. The main advantages of this approach are fewer fabrication steps and the ability to readily implement thick metal layers (tens of micrometers). Thick metallization layers contribute to improved mechanical strength and resistance to electrochemical corrosion. Alternative methods such as photolithography and metal sputtering do not permit deposition of thick metal layer (<hundreds of nanometer), and may not be reliably reproduce (Schuettler et al., 2005). The disadvantages with laser patterning, however, are limited precision (few tens of micrometers) and reduced elasticity due to the use of thick metal foil. The electrodes in this suprachoroidal implant are arranged in a hexagonal pattern where each channel is individually connected to a stimulation circuit and can serve as a stimulating element (Suaning et al., 2014). When hexagonal stimulation is applied, each hexagon cluster of electrodes contains one electrodes serving as stimulating element surrounded by six electrodes

serving as return (Suaning et al., 2004). So far early versions of this device with 14 channels in one layer were examined through acute stimulation of arrays implanted in the suprachoroidal space in the eyes of cats (Wong et al., 2008, 2009; Matteucci et al., 2013; Barriga-Rivera et al., 2016b, 2017) and sheep (Barriga-Rivera et al., 2015). Several stimulation paradigms including monopolar, bipolar and hexapolar stimulation (Wong et al., 2009; Habib et al., 2013; Matteucci et al., 2013, 2016), and the efficacy of new array designs were examined (Barriga-Rivera et al., 2015, 2016a). A nonactive Phoenix99 device has been chronically implanted in sheep (manuscript in preparation).

The suprachoroidal system developed by the Korean group is based on implementation of LCP films as means to achieve a miniaturized, integrated, and entirely intraocular monolithic device (Jeong et al., 2012, 2016). Compared with biopolymers such as PI, parylene, and PDMS, LCP films have superior resistance to penetration of water and ions, potentially contributing longevity of the implant (Jeong et al., 2012; Gwon et al., 2016). In addition, there is no need to use adhesives as the films can be laminated to each other by fusion bonding (combination of heat and pressure) (Lee et al., 2011). Photolithography was used to define the electrode pattern and gold electrodes were deposited via sputtering. Next, lamination of the LCP substrate and precut cover layers was conducted via thermal-press bonding (Lee et al., 2009). The researchers further suggested to overcome the fact that commercially available LCP films are stiffer compared with PI, PDMS, and parylene materials, by adapting a conformable fabrication scheme in which the LCP-based device is molded into a curved shape to fit the eyeball through a thermo-forming process (Jeong et al., 2012; Gwon et al., 2016).

The investigators were able to record cortical potentials in response to electrical stimulation applied to the retina of rabbits via LCP-based arrays with 8–16 channels (acute experiments) (Jeong et al., 2015; Lee et al., 2009). OCT imaging and histological analysis, during 3–4 months post-surgery, further confirm no adverse immune effects for the duration of the study (up to 2.5 years and 4 months, respectively) (Lee et al., 2009; Jeong et al., 2015, 2016). Later, an entire LCP packaged retinal implant including a 16-electrode array (IrOx/Au), telemetry coils, and stimulator circuitry was realized (Jeong et al., 2016). Further, extensive efforts are being invested by the group to investigate the long-term reliability of LCP-based retinal implants, and customize the fabrication process accordingly (Jeong et al., 2011, 2016; Lee et al., 2009, 2011).

2.4 Intrascleral Implants

Similar to the suprachoroidal approach, in intrascleral implantation the electrodes are also spatially separated from the neurons. The suprachoroidal-transretinal (STS) prosthesis, investigated by a group from Osaka University in Japan, is inserted into a surgically formed pocket in the sclera (Fujikado et al., 2011). Preliminary investigations were conducted in blind and normally sighted rats using a single silver electrode and a reference electrode in the vitreous (Kanda et al., 2003, 2004), and in rabbits using an eight-electrode array (Nakauchi et al., 2005). Results from a semichronically clinical study with a 49-electrode prosthesis implanted in two blind patients were reported in 2011 (Fujikado et al., 2011, 2012). During the 5–7-week study period, it was found that nine channels were electrically active, with four to six sites successful in elicitation of phosphenes. Considerably, higher current thresholds were required to evoke this response, as expected due to the physical distance from target cells (Fujikado et al., 2011). A head-mounted video camera was used to control the stimulation strength based on visual scene, applying head movements to scan the FOV in front of them. Patients were able to better than chance level with the device turned on in several visual tasks including: discrimination of shape thickness and object localization. In object localization task patient performance improved with training (Fujikado et al., 2011). Recently, the team reported a 1-year clinical trial with a similar device implanted in three patients with RP (Fujikado et al., 2016). All of the recipients were able to describe the appearance of phosphenes in 24–28 out of the 49 electrodes. Two out of three patients were able to perform significantly better with the device turned on as compared with the device turned off in localization, discrimination, and mobility tasks (Fujikado et al., 2016).

The recent progress with retinal devices reveals their potential in restoration of visual perception in individuals blind due to retinal degeneration. There are many advantages to placing electrode grids on or near to the retina including a large area of the visual field which can be covered, avoidance of intracranial surgery, as well as the possibility to place an optical sensor in the eye to avoid the need for tracking of eye position to compensate for gaze shifts. However, retinal implants target a relatively small proportion of overall blindness conditions with the rare inherited disease RP being the main indication. Cortical prostheses can potentially target multiple other cases of blindness where retinal or ON prostheses cannot be applied.

3 CURRENT VISUAL CORTEX IMPLANT TECHNOLOGIES

Cortical visual prostheses employ electrodes that are placed in contact with the primary visual cortex (V1 or striate cortex). The visual sensory input that arrives from the eyes goes first through the LGN (located in the thalamus) and then reaches V1. Each hemisphere in the visual cortex receives information from the ipsilateral LGN. The left visual cortex receives signals from the right visual field and vice versa. V1 is part of the occipital lobe (in the back of the skull), that encompasses buried portions of cortex in the calcarine sulcus and its upper and lower banks, extending posterolaterally to the occipital pole.

A major advantage of the visual cortex as the stimulation site is the possibility to treat indications where the ganglion cells in the retina or the ON's are severely damaged, such as: severe retinal disease such as diabetic retinopathy and glaucoma, optic atrophy, optic neuritis, large pituitary or parasellar tumors, trauma to the eyes or the ON's, or bilateral retinoblastoma following enucleation of both eyes. Other advantages are the relatively large area to place electrodes compared with the retina and the LGN, which means that a large number of electrodes can potentially be implanted. The total available surface area of V1 varies between 1400 and 6300 mm^2 (depending on the method of estimation) with approximately 67% of that area buried inside the calcarine fissure (Andrews et al., 1997; Stensaas et al., 1974). Electrodes can be placed on the surface of the visual cortex (subdural) or inserted into the brain tissue (intracortical).

Two key pioneers in the development of cortical bionic vision prostheses were Giles Brindley and William Dobelle (Brindley and Lewin, 1968; Dobelle and Mladejovsky, 1974; Dobelle et al., 1974). By 1972, Brindley and Lewin had implanted two devices. In the second device, enhancements to the electronic design by Donaldson (1973) produced a "row-column" transmission strategy which dramatically reduced the number of receivers necessary on the subpericranial portion of the implant while maintaining a similar quantity of electrodes (75). This reduction in receiver quantity (and associated increase in separation between receivers) had a corresponding reduction in "cross talk" between neighboring receivers. This was indicated as a significant problem in the original design as transmitter misalignment of as little as 5 mm had caused stimulation or partial stimulation of electrodes attached to neighboring receivers. The second patient could read braille characters presented to him using the device alone at a rate of seven letters per minute and with an accuracy of up to 90% (Donaldson, 1973). Dobelle also implanted subdural electrodes that were connected via a plug in the skull, and via an electrical cable to a computer processing images from a video camera. Dobelle's patients were able to perceive phosphenes, identify shapes, navigate the environment, and in one patient, to read Braille. The currents required to generate the phosphenes with this device were in the low-mA range (Dobelle and Mladejovsky, 1974; Dobelle, 2000; Dobelle et al., 1974).

Penetrating cortical microelectrodes [intracortical microstimulation (ICMS)] produce more localized phosphenes at currents in the μA range. This is a >100-fold reduction in current output that significantly lowers the risk for seizures (Lewis et al., 2016a; Davis et al., 2012). A preliminary human study examining ICMS of visual cortex was published in 1990, the results of which added significant impetus to the effort to develop a cortical visual prosthesis (Bak et al., 1990). The three sighted volunteers were able to perceive phosphenes from ICMS at currents up to 100 times lower than those required by surface stimulation. Moreover, the phosphenes were discriminable when stimulated by electrodes 700 μm apart (Bak et al., 1990). Further work identifying thresholds of total charge delivered and charge density, beyond which neuronal damage could be expected to occur (McCreery et al., 2010), supported the progression to a more systematic evaluation of ICMS of visual cortex in a blind volunteer (Schmidt et al., 1996). A key finding from this study was that the chronically blind subject, who was unable to perceive phosphenes from surface stimulation, perceived phosphenes from ICMS in a similar manner to sighted volunteers in the previous report (Schmidt et al., 1996).

Currently, there are five groups around the world that are developing brain implants for bionic vision. Intracortical electrodes are investigated by the "CORTIVIS" group at the Universidad Miguel Hernandez (Spain) (Fernandez et al., 2005, 2007), Sawan and coworkers at the Polytechnique Montreal (Canada) (Coulombe et al., 2007; Mohammadi et al., 2012), Troyk and coworkers at the Illinois Institute of Technology (United States) (Troyk et al., 2006; Srivastava and Troyk, 2006a,b), and the Monash Vision Group (MVG) at Monash University (Australia) with the commercial entity Gennaris Neural Systems Inc. (Lowery et al., 2015). Subdural electrodes were recently implanted in humans by Second Sight Medical Products Inc. (United States). The subdural array (Orion) was placed on the medial occipital cortex of a blind subject and phosphenes have been generated through cortical stimulation. The results of this study have not yet been reported in the scientific literature.

Ferrandez et al. (2007) developed a system that includes a transistor-based current source, a microprocessor circuit with programmable waveforms, and an array of penetrating electrodes (Utah array). The main advantage of using a transistor-based stimulator is the very low power consumption. The current level could be adapted to the different working electrodes and tissue impedances. The penetrating electrodes were tested in rabbit cortex and the system was found to be safe, and able to deliver continuous stimuli during 240 h using a 9 V battery (Ferrandez et al., 2007). The disadvantage of this array is its limited geographic coverage of the visual cortex, unless multiple arrays can be deployed.

Troyk and coworkers at Illinois Institute of Technology developed a wireless floating microelectrode array (WFMA), and an image processing system. In-vivo testing of a 16-channel WFMA was conducted in a rat sciatic nerve model over a period of 143 days (Bredeson et al., 2015; Romero-Ortega et al., 2015). Different fascicles of the rat sciatic nerve have been selectively stimulated and motor evoked potentials remain stable overtime and nerve stimulation charges were within tissue safety limits. The testing of these arrays in cerebral cortex has not yet been reported (Bredeson et al., 2015; Romero-Ortega et al., 2015). An instrument for cortical implantation procedures in nonhuman primates was also developed (Tawakol et al., 2016).

The MVG is developing the "Gennaris" device which includes a camera, a vision processing computer, the electrode interface with the brain, the interconnecting links (wired and wireless), and the power system. The electrode carrier is a ceramic box containing an ASIC, wireless coil, and 43 Pt-Ir microelectrodes (2.5 mm height and 125 μm in diameter), spaced 1 mm apart on a hexagonal grid (Fig. 4A). The electrodes are insulated with Parylene C and have an annular exposed area on the shaft (Fig. 4B).

Potentially, more than 300 electrodes could be placed in the visual cortex. A pocket-sized vision processor has been developed to run software algorithms that convert photographic images from a small digital camera mounted on a spectacle frame into pixelated patterns which represent relevant shapes and contours in the environment. The "transformative reality" algorithms are similar to those used in robotic vision applications, and are suitable for such as object recognition and indoor navigation. The system could potentially enable the recipient to recognize objects, avoid obstacles, and recognize the outlines of people and their gestures, but the anticipated

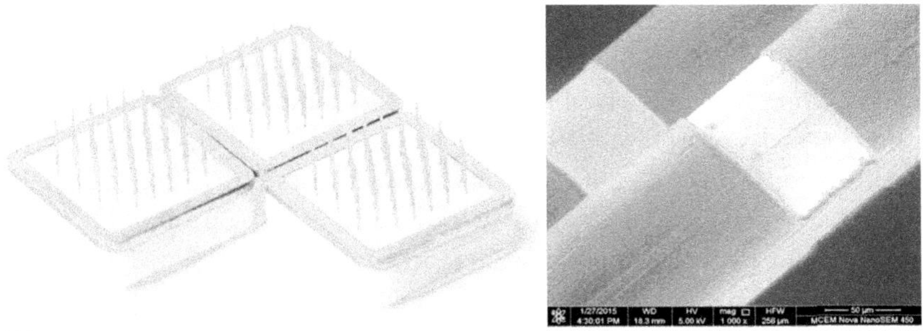

(A) (B)

Fig. 4 Monash Vision Group (MVG) cortical prosthesis: (A) electrode carrier and (B) electrodes under scanning electron microscopy showing the laser etched annular exposed area of Pt-Ir with the parylene-C insulation surrounding it.

resolution is not sufficient to provide facial recognition. Stimulation parameters using the Gennaris device were examined in normally sighted rodents. Biocompatibility and the effects of chronic electrical stimulation have been studied in sheep with normal vision. Chronic cortical stimulation was conducted in 18 sheep for up to 9 months, confirming that the wireless connectivity and electrode activity of the device remain functional. Further, over 2400 h of stimulation achieved in 6 of these sheep. Detailed histological examination of the animal's brains is underway. Sheep studies were also used to confirm the temperature rise of dummy tiles fitted with heaters and thermistors. The temperature rise of the implants was less than 1.5°C. Computer simulations of heat generation rise produced a similar temperature rise (unpublished data). The artificial construct of discrete phosphenes produced by electrical stimulation of the cortex is very different from the normal physiological processing of visual signals through the complex hierarchies of visual cortex. Therefore, it cannot be assumed that the production of hundreds of phosphenes would be synthesized by the brain to produce an accurate representation of the external world. A further challenge is the visual scanning of the scene through a camera interface which rapidly updates the phosphene patterns. The rapid updating of the camera images requires an advanced vision processing algorithms. This challenge also applies to vision prostheses implanted in other sites in the visual system which also do not compensate for eye movement.

Potential health risks associated with cortical prostheses are the insertion of the electrodes including the requirement to remove a part of the bone from the skull to expose the brain (craniotomy), and brain injury caused by the trauma of insertion, as well as epileptic seizures. Craniotomy carries risks of: infection (1%–2%) which could necessitate re-exploration and removal of the cranial bone flap or its replacement, hemorrhage requiring reoperation (1%), and possible injury to the sagittal or transverse venous sinuses which are at the edge of an occipital craniotomy exposing the visual cortex (Bjornsson et al., 2006). A craniotomy carries a small risk of infection (1%–2%) which could necessitate re-exploration and removal of the cranial bone flap or its replacement, and hemorrhage requiring reoperation (1%). There is a small risk of injury to the sagittal or transverse venous sinuses which are at the edge of an occipital craniotomy exposing the visual cortex. Local brain injury due to the insertion of the electrodes may cause micro-hemorrhage, local scarring, and loss of neurons. This would further impair the transmission of current into the brain (Bjornsson et al., 2006).

Chronic focal stimulation of cerebral cortex may also cause development of epilepsy through a process called kindling, where this focal area of cortex becomes an independent generator of seizure activity (Goddard, 1967; Morimoto et al., 2004). The seizure activity may spread to other parts of the brain depending on the magnitude of electric currents passing through the cortex, the duration of the stimulation, and the individual's threshold for seizure activation (Goddard, 1967; Morimoto et al., 2004). The potential for kindling of epileptic seizures may be increased if the patient has a history of epilepsy. The risk of seizures can be mitigated by low-stimulation currents (10s of μA range), limiting temperature rise, intermittent stimulation patterns rather than the same electrode firing constantly, and prophylactic anti-epileptic drugs (AEDs). These drugs, however, may also render the neurons less responsive (refractory) to the stimulation by the electrodes so a careful balance should be achieved (Bezard et al., 1999).

4 ON AND LGN PROSTHESES

Besides the great progress in restoration of visual sensation through retinal and cortical prostheses, electrical stimulation of the ON (Veraart et al., 2003; Sakaguchi et al., 2012) or the LGN (Panetsos et al., 2011; Pezaris and Eskandar, 2009) is also being investigated.

The first to attempt a visual prosthesis based on stimulation of the ON were Veraart et al. (1998). The patient (blind due to RP) received a four-contact silicon cuff array implanted around the intracranial section of the ON. The four electrodes were located at 90-degree intervals, to enable stimulation to the entire visual field. Colored phosphenes were reproducibly elicited from all four quarters of the visual field (Veraart et al., 1998, 2003). The indications for ON prosthesis are the same as retinal implants. The surgical approaches to the ON are via a craniotomy or an intraorbital approach (which is less invasive than craniotomy) (Brelén et al., 2006). The axons of the ON must be intact for this site to be applicable for electrode implantation. Electrodes can either be placed as a cuff on the surface of the ON or as penetrating electrodes. The ON is surrounded by a dural sheath within which is contained with cerebrospinal fluid. Thus, cuff electrodes may require higher currents and their spatial resolution would be inferior to penetrating electrodes, because of the larger distance from the target ON axons. The main advantages of this approach are the relatively easier surgical access to the ON (compared with accessing the brain or the retina), and the possibility of targeting the entire visual field (Nishida et al., 2015; Veraart et al., 2003). However, with more than 1 million axons passing through a 2 mm diameter nerve fiber, achieving focal stimulation is a challenge.

In the following studies by Veraat and coworkers, the influence of stimulation parameters including amplitude, duration, frequency, and number of pulses per phase on phosphene appearance (location and size) (Delbeke et al., 2003). Despite the poor resolution of this prosthesis (four pixels), following training the patient was able to achieve some pattern recognition, shape orientation, object localization (Veraart et al., 2003; Brelén et al., 2005), as well as perform grasping tasks (Duret et al., 2006). The second volunteer received an eight-contact array implanted around the intraorbital section of the ON (Brelén et al., 2006). Cortical evoked potentials and electroretinograms (ERG) generated during electrical stimulation of the ON (in both of the recipients) were recorded and compared with light and electrical surface eye stimulation conducted in the normally sighted subjects (control group) (Brelén et al., 2010).

Researchers from the Department of Ophthalmology in Osaka University (Japan) are also developing an ON prosthesis. Their approach is named the artificial vision by direct optic nerve electrode (AV-DONE) (Sakaguchi et al., 2009). Here, metal wires are designed to penetrate into the optic disc (the ON head) (Fang et al., 2006; Sakaguchi et al., 2004b, 2012). This is the point of exit for ganglion cell axons leaving the eye and converging into the ON. Two blind patients with RP were so far implanted with three Pt wire electrodes penetrating into the optic disc. Round, oval, or linear phosphenes which were primarily yellow, and focally distributed, were experienced by the patients in response to electrical activation (Sakaguchi et al., 2009; Nishida et al., 2015).

Finally, the "C-sight" project group has demonstrated stimulation of the ON using a two-channel penetrating electrode array implanted into the ON in rabbits and in cats, and a 16-channel array implanted in rabbits (Chai et al., 2008; Li et al., 2009; Yan et al., 2016; Sun et al., 2011). They observed differences in cortical recordings in response to modification of the stimulation parameters. For example, it was found that monopolar stimulation with three penetrating electrodes induced more localized cortical responses than the bipolar stimulation (Cao et al., 2015), and that a charge-balanced biphasic pulse with an interphase gap of 0.2 ms was the most efficacious (Sun et al., 2013). In addition, it was demonstrated that a five-electrode array with electrodes placed at different depths in the ON elicited visuotopic electrical stimulation of the visual cortex with a spatial resolution of about 2–3 degrees (Lu et al., 2013). Current steering was also used to enhance the spatial resolution of the stimulation (Yan et al., 2016). The investigators suggested that these observations imply that the resolution of an ON visual prosthesis may not be limited by the number of electrodes only (Lu et al., 2013). Recent investigations also demonstrate the potential of targeting the LGN to evoke visual response (Panetsos et al., 2011; Pezaris and Reid, 2007). The LGN is a small nucleus (about 250 mm^3 volume and up to 10 mm long) situated in the posterior/inferior part of the thalamus. Most of the fibers of the optic tract terminate on neurons in the LGN, and it has six distinctive layers of neurons. The visual information from the retina to passes through the LGN before reaching the primary visual cortex. Key features of the LGN are physical segregation, in form of different cellular layers, for the magnocellular and parvocellular visual pathways that are specific to motion/localization and color/detail respectively and projection of retinal structure (Mullen et al., 2008; Wiesel and Hubel, 1966). Also, it has a geographic representation of the retina (retinotopic projection). The retinotopic projection spans a larger area of tissue (compared with the retina) and therefore it is potentially possible to achieve higher spatial resolution for similar size electrodes. Beyond these unique features, the relatively small dimensions and simple structure of the LGN compared with the visual cortex, as well as the possibility to apply surgical techniques similar deep brain stimulation procedures are additional advantages (Pezaris and Eskandar, 2009). The LGN could potentially be used as a stimulation target in cases where the retina and ON have lost function if the visual cortices have been injured or lost (cortical blindness). The disadvantages of the LGN are that only a hemifield passes through each side so that bilateral placement would be required to cover the entire visual field. Similar to cortical and ON prostheses, an image compensation for gaze direction would be required. The effect of corticothalamic activity on modifying LGN output is also little understood (Pezaris and Eskandar, 2009).

Stimulation of the LGN to evoke visual sensation was so far examined in nonhuman primates by analysis of eye movements in response to electrical stimulation (Pezaris and Eskandar, 2009; Pezaris and Reid, 2007, 2009), as well as in rats and rabbits, comparing visually and electrically generated signals recorded from V1 (Panetsos et al., 2011). Retinotopic models of macaque and human LGN were used to simulate phosphene patterns to investigate the electrode spacing (Pezaris and Reid, 2009). It was concluded that a spacing of 600 μm between the microelectrodes in the LGN (in three dimensions) would theoretically allow for over 250 phosphenes per visual hemifield in macaques, and over 800 in humans (Pezaris and Reid, 2009). To simultaneously place so many electrodes in such a small deep structure, the investigators described a concept of a brush-like electrode bundle splaying out from the end of the electrode sheath into the LGN. With electrodes spaced 1 mm apart in three dimensions, 250 electrodes for each hemisphere could be placed on each side giving a total of 500 electrodes.

Pezaris and coworkers have further created simulated prosthetic vision (SPV) for letter recognition (Bourkiza et al., 2013) and reading (Vurro et al., 2014) with a thalamic prosthesis. In SPV studies virtual reality models of prosthetic vision are administrated to normally sighted subjects (Chen et al., 2009). The effects of electrode count on VA, learning rate, and response time were examined, providing the first reports for thalamic designs. The group recently reported a nonhuman primate model for visual prostheses where animals are capable of performing similarly to humans on the letter recognition task (Killian et al., 2016).

5 ENGINEERING CONSIDERATIONS FOR CORTICAL AND RETINAL STIMULATION

In past two decades, tremendous efforts have led to substantial progress in the ability to evoke visual sensation through prosthetic devices. In particular, patients implanted with retinal prostheses have demonstrated ability to distinguish recognize objects, detect motion and orientation, perform simple navigation tasks, and even read large letters (Zrenner et al., 2011; da Cruz et al., 2013; Ayton et al., 2014; Fujikado et al., 2016). Regained VA was improved to 20/1590 or better in 8 out of 21 people as tested 5 years after impanation with the Argus II epiretinal device, and even up to 20/200 when this device was coupled with an external optical magnification system (Sahel et al., 2013), while VA of 20/546 was reported in one patient with a sub-retinal implant (Stingl et al., 2015). While these results represent a great promise for rehabilitation of impaired vision, none of these devices have yet to achieve VA above 20/200, which is considered as legal blindness. Further, more than 75% of patients who received retinal implants could not achieve measurable VA (Stingl et al., 2015; Humayun et al., 2012; Yue et al., 2015; Ha et al., 2016). Although several retinal devices have been approved as safe for commercial use, avoiding health complications and damage to the retina are still a major concern. Complications such as conjunctival erosion, endophthalmitis, hypotony, retinal detachment, retinal microaneurysms, gliosis, and retinal atrophy have been reported, and removal of the device was required in several cases (Humayun et al., 2012; Keserü et al., 2009, 2012; Sailer et al., 2007; Menzel-Severing et al., 2012; Butterwick et al., 2009). Improvements in the VA performance achievable by visual implants and in their biocompatibility and long-term reliability are highly desirable. In particular, high-resolution stimulation through small electrodes with safe and focused electrical fields, minimizing tissue damage, and facilitating long-term operation. In this section, we will discuss important engineering considerations that need to be taken into account in the development of retinal and cortical prostheses, including: the placement and fixation of the electrode array in the proximity of the tissue, electrode size, and materials and parameters for stimulation.

5.1 Placement and Fixation of Retinal and Cortical Electrode Arrays

The anatomic location of the retinal array has implications on the complexity of the surgery, risk for injury and retinal detachment, long-term mechanical stability, threshold for electrical stimulation, as well as spatial resolution and visual perception.

In epiretinal prostheses, the electrode array is placed on the surface of the retina (Fig. 2B). The close positioning of the electrodes to the target cells, the RGCs, potentially improves the efficiency of electrical stimulation. However, in this configuration, there is a potential of stimulating the bundle of neural extensions passing through the center of the retina introduces nonspecific stimulation pathways that may disrupt the accuracy of electrical activation (Humayun et al., 2003; Rizzo et al., 2003). While insertion of the implant via the vitreous chamber (between the lens and the retina), makes the surgical procedure simpler and safer compared with the subretinal approach, there is a difficulty in attaching the array to the retina, and sharp metal tacks are required to hold it in place (Fernandes et al., 2012). Finally, another potential benefit is that the possibility for heat (generated from electrical stimulation) dissipation through the gel mass of the vitreous cavity (Opie et al., 2012).

On the contrary, in subretinal implantation strategy, it is potentially possible to gain from the biological signal processing that occurs in the middle layers of the retina (Chow et al., 2010). The constrained space between the RPE and the neurons (Fig. 2B) contributes to fixation of the implant and no tacks are required (Stingl et al., 2013a). It was further suggested that improved electrode-tissue attachment accounts for lower stimulation thresholds observed with subretinal compared with epiretinal stimulation (Jensen and Rizzo, 2006). Avoiding the activation of RGC axon bundle, often distorting visual perception is another advantage of this approach (Humayun et al., 2003; Rizzo et al., 2003). A potential disadvantage is the risk of retinal atrophy in vicinity of the implant due to blocking of the retina-choroid fluid link, preventing heat dissipation and nutrient transport (Peachey and Chow, 1999; Sailer et al., 2007).

While in the subretinal and epiretinal approaches, the electrodes are in direct contact with the neurons in the retina, in the suprachoroidal or intrascleral approaches, the array is placed at a distance and there is no direct contact with neurons (Fig. 2B). The main advantage of these approaches is that the surgical procedure is significantly less invasive and less technically challenging. There is no need for removal of the vitreous humor (vitrectomy) or for direct manipulation to the retina (Gekeler et al., 2010; Zrenner et al., 2011). These surgical manipulations may contribute to complications such as conjunctival erosion, endophthalmitis, hypotony, and retinal detachment detected with epiretinal prostheses (Keserü et al., 2009, 2012; Menzel-Severing et al., 2012; Humayun et al., 2012), or gliosis (Butterwick et al., 2009), retinal microaneurysms (Stingl et al., 2013c), and retinal atrophy (Peachey and Chow, 1999) reported with subretinal prostheses. Further, fixation of the prosthesis requires only sutures to stabilize the implant (no metal tacks), and a larger array can be implanted (Fujikado et al., 2007, 2011; Saunders et al., 2014). This means that potentially a large FOV can be addressed (Villalobos et al., 2012, 2013). The close proximity to blood vessels in the choroid also contributes to reducing the risk for heat induced damage (Parver et al., 1980). However a major disadvantage is the distance from the neurons, with approximately 250–400 μm vs 180 μm reported for a suprachoroidal and subretinal configuration (Argus II), respectively (Ahuja et al., 2013; Ayton et al., 2013). This means higher stimulation currents are required to activate the neurons (relative to other stimulation sites). Another issue is that more current is spreading to neighboring cells. Focal activation of the RGCs is highly desired to achieve spatially distinct phosphenes perception, and therefore current spreading means lower resolution.

The visual cortex is a relatively large area to place electrodes compared with the retina, and thus a large number of electrodes can be employed without requiring further reduction in electrode dimensions. The electrodes are placed to interface the surface of the cortex or penetrate into the tissue cortical arrays. The penetration of the electrode into the tissue contributes to an increase in the precision of stimulation and to a decrease in the power consumption of the device. There is a dramatic reduction in the threshold to stimulation of phosphenes as the electrodes penetrate deeper into the tissue (Bak et al., 1990). The ideal depth of electrode penetration for human intracortical stimulation is approximately 2–2.25 mm where stimulation thresholds are within the 1.9–77 μA range (Bak et al., 1990; Schmidt et al., 1996). Therefore, due to the lower stimulation currents, the risk for epilepsy is lower with ICMS compared with surface stimulation (Bezard et al., 1999; Goddard, 1967).

A significant proportion of the visual cortex is submerged in the calcarine fissure. If this part of the cortex is not included in the stimulation, an hourglass-shaped visual field results. Head scanning which places the camera at different angles will help the individual "fill in the gaps" and create a more complete picture of the visual scene. This has been studied using simulation in normally sighted individuals wearing a heads up mounted display in goggles (Lowery et al., 2015). Placing electrodes within the calcarine cortex may also require varying length electrodes and different carrier designs. Alternatively, additional electrodes can be positioned in secondary visual cortex to cover some of the missing visual field (if the electrodes do not penetrate this hidden calcarine cortex) (Lewis et al., 2015). Placement of electrodes on the medial surface of the hemisphere potentially covers the peripheral vision and central vision, but the orientation of these electrodes would be orthogonal to a transmitting coil on the scalp and wireless data transfer would be restricted.

Insertion of penetrating cortical arrays cannot be achieved with pressure from the surgeon's finger and requires a pneumatic-actuated inserter (Utah array insertion tool) (Normann et al., 1999). The Utah array insertion tool operates through a pneumatically controlled piston which pushes the electrodes through the pia, a relatively tough fibrous layer covering the brain surface. The bone flap or synthetic material (acrylic) that is inserted instead of the scalp area removed during exposure of the cortex should not apply any pressure on the electrode array. Direct pressure on these grids may gradually force the array to sink into the brain causing deformation and damage to the cortex. If the active tip or shaft of the electrodes passes beyond the neurons of the gray matter and enters the white matter, the electrodes will be non-functional. In addition there is possible interference in cross-modal sensory adaptation such as the reading of Braille which the blind person depends on for activities of daily living (Lewis et al., 2015).

5.2 Parameters for Retinal and Cortical Stimulation

In 1968, Brindley and Lewin demonstrated that restoration of visual perception is possible through electrical stimulation of the visual cortex (Brindley and Lewin, 1968). Nowadays, cortical and retinal neurostimulators are emerging as a therapy (Lewis et al., 2016a). Despite the enormous progress over the last decades, integration between the electrodes and the neural tissue is still poor. While electrode locations in the visual cortex can directly map visual perception (Lewis et al., 2015), at the retina, information is codified by the retinal neural network and transmitted to the brain via RGCs (Koch, 2013). These anatomical and functional differences involve a need for different stimulation strategies that target excitable cells more specifically.

The parameters of electrical stimulation are critical in the nervous system. These include: monophasic or biphasic stimulation; pulse durations; anodic or cathodic stimulation (first or last); anodic scaling; pulse repetition frequencies; whether the interpulse intervals have high or low impedance, and the placement and size of the active area of the electrode, and the current return path (or whether bipolar stimulation is used). Examples of stimulation waveforms are illustrated in Fig. 5. Electrical stimulation itself (aside from the electrodes' physical presence) may also injure the surrounding neurons and cause neurodegeneration. Electrochemical Faradic reactions can occur at the

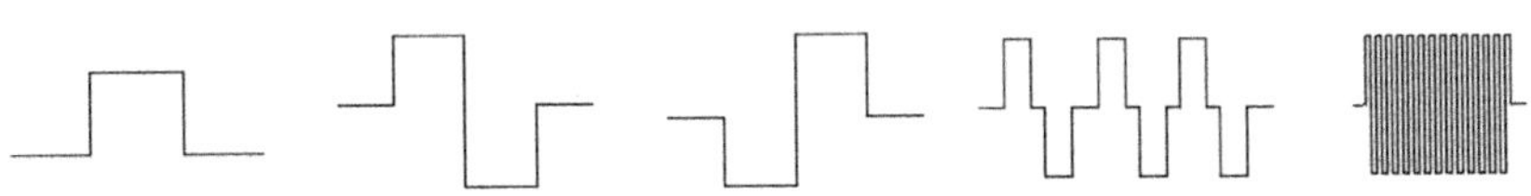

(A) (B) (C) (D) (E)

Fig. 5 Common stimulation waveforms used in electrical stimulation. (A) shows a monophasic pulse. (B) and (C) are biphasic pulses, anodic first and last respectively. (D) is a train of pulses and (E) is an example of a high-frequency stimulus.

electrode/tissue interface (Merrill et al., 2005), leading to formation of gas bubbles and pH shifts that damage the tissue (McCreery et al., 1994, 2010). To prevent these processes, the net charge used in stimulation should be neutralized in some way, such as shorting poststimulus, series capacitors in line with each electrode, and carful circuit design (Bartlett et al., 1977; Brummer et al., 1983). The electrodes can be intermittently stimulated in various patterns of anode/cathode geometry. This may enhance the focus of the stimulation (Berenstein et al., 2008). Some electrodes may fail with adjacent electrodes maintaining function. Reconfiguration of stimulation parameters and patterns may maintain device efficacy overtime.

In the case of ICMS, the goal is to activate small areas of the visual cortex around each individual electrode to produce single visual percepts, thus reducing crosstalk. According to a recent study in rats, the optimal stimulus is an extended pulse train of low amplitude and low frequency (Watson et al., 2016). In the retina, challenges of stimulation strategies include not only the containment of neural activation as in the former case, but also others derived from the neural architecture of the retina. For electric field containment, different return configurations are commonly used, as this technique allows for modifying the shape of the electric field, and therefore the extent of the neural excitation. In a monopolar configuration, the return electrode is placed far from the active electrode thus producing wide neural activation with low activation thresholds (Matteucci et al., 2013). In a bipolar strategy, the return electrode is placed in the vicinity of the active. This configuration produces more contained activation, but when biphasic pulses are used, it may also activate the area around the return electrode during charge recovery (Dokos et al., 2005). In a multipolar disposition (Matteucci et al., 2013; Spencer et al., 2016), current returns through a group of electrodes (see Fig. 6). Containment of the electric field is also being targeted by using field overlapping techniques, a strategy that combines different electric fields, for example, to reduce activation thresholds while containing neural activation (Matteucci et al., 2013, 2016). Concomitant stimulation has to be managed carefully as it can also produce neural inhibition (Barriga-Rivera et al., 2017).

Another important challenge in retinal electrostimulation is to become able to elicit neural responses that encode appropriately visual stimuli. With more than 32 functional RCGs (Baden et al., 2016), researchers are using high-frequency stimulation to selectively target different information streams (Twyford et al., 2014). Although these approaches are promising

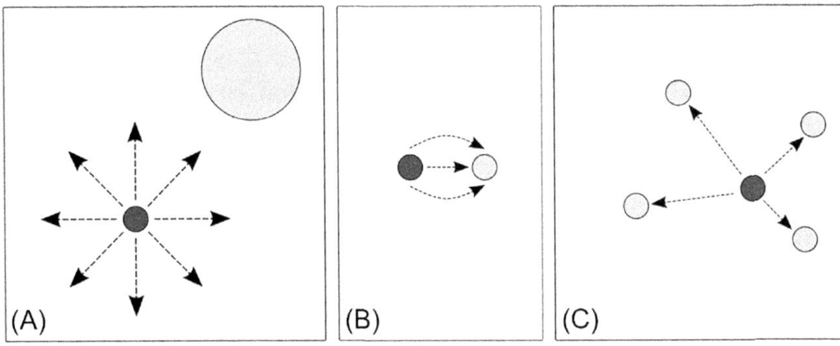

Fig. 6 Return configuration in retinal electrostimulation. *Black circles* represent active electrodes, whereas *gray circles* are the electrodes used as a return. (A) Monopolar configuration, in which current returns via a distant electrode. (B) Bipolar configuration where current returns via a neighbor electrode. (C) Multipolar approach where electric current returns via a group of electrodes.

more meaningful percepts with increased visual resolution, there is a clear need for new technologies that provide better integration between biology and the device.

6 CONCLUSIONS AND PERSPECTIVES

In this chapter, we reviewed the progress achieved with the application of bioelectronics devices for restoration of visual perception, describing the different system architectures applied along the visual path, with emphasis on retinal and cortical prostheses. We further discuss key challenges facing current devices focusing on the direct interface between the device and the neuronal tissue, the electrode.

Recent developments in materials and nanoengineering open new routes in interfacing with neurons, that diverges from the traditional approach based on electrical stimulation through metal electrodes. The next generation of visual prostheses is designed to be minimally invasive, biocompatible, with high spatial resolution, cell specificity, and improved safety. A new generation of materials such as carbon nanotubes (CNTs) (Gabriel et al., 2009; David-Pur et al., 2014; Eleftheriou et al., 2017; Shoval et al., 2009), nanocrystalline diamond (NCD) (Ganesan et al., 2014; Hadjinicolaou et al., 2012; Ahnood et al., 2016), conjugated polymers (CPs) (Samba et al., 2015), and silicon nanowires (Si NWs) (Khraiche et al., 2011, 2013; Ha et al., 2016) has attracted attention as promising candidates for improved electrical activation of neurons, and were demonstrated for activation of retinal neurons. These materials offer enhanced electrochemical properties and superior neuron-electrode mechanical attachment through unique surface topography and charge injection mechanism.

A different approach employs optical activation of retinal neurons through photoactive interfaces, offering a new rout for wire-free, self-powering autonomous retinal prostheses (Antognazza et al., 2015; Bareket-Keren and Hanein, 2014). Several investigations have proposed photovoltaic polymers (Antognazza et al., 2012, 2016; Ghezzi et al., 2013; Gautam et al., 2014; Feyen et al., 2016), quantum dot (QD) films (Pappas et al., 2007; Molokanova et al., 2008; Bareket et al., 2014), and QDs directly interfacing the cell membrane (Winter et al., 2001). Bareket et al. (2014) demonstrated photostimulation of light insensitive retina explants with ambient light intensity using a nanomaterial film. In the composite film of semiconductor nanorods and CNTs (SCNR-CNT), absorbance of light was followed by charge separation at the NR-CNT interface, that activated the RGCs. Maya-Vetencourt et al. (2017) showed a fully organic retinal prosthesis restoring vision to a rat model of RP. The device is made of poly(3-hexylthiophene) (P3HT) and poly(3,4-ethylenedioxythiophene)-poly(styrenesulfonate) (PEDOT:PSS) deposited on a silk fibroin substrate. When light is absorbed by the organic polymer, it results in a mobile excited state (excitons) that enable stimulation of nearby neurons. The researchers demonstrated recovery of cortical responses and visually driven behavior (Maya-Vetencourt et al., 2017).

For cortical stimulation, a novel method of magnetic cortical stimulation was developed by Fried and coworkers (Lee et al., 2016). This approach is based on the concept that magnetic fields are not affected by encapsulating glial scarring, which impairs the functionality of conventional electrodes. The absence of an electrode-tissue interface opens the possibility of improved device longevity and patient safety. Micro-coils ($50 \times 100\,\mu m$) were able to activate neurons in the fifth layer of the visual cortex of mouse brain (Lee et al., 2016). Alternatively placing miniature field generators on the surface of the brain can be applied as a method of delivering focused magnetic fields. This would be less invasive than intracortical electrodes but also less precise (Lewis et al., 2016b).

These investigations are exciting breakthroughs describing powerful enabling tools in neuroprostheses in general and in artificial vision in particular. However, these systems are in early stages of development, and additional research is required before they can be considered for clinical use. Primarily understanding chronic effects related to biocompatibility and long-term stability of these materials in the biological environment should be explored (Polikov et al., 2005; Schwahn et al., 2001; Cogan et al., 2016). For example, CPs have low stability under continuous electrical stimulation, exposure to ultraviolet (UV) light or heat which may gradually deplete their conductive elements AND degrade their mechanical properties. In some approaches, toxic residues arising from the synthesis process may exist. While CNT, Si NWs, and NCDs are highly stable, owing their nano-dimensionality, they may penetrate into the living cells. This may be an advantage achieving intracellular connection, but may also lead to damage to the integrity and operation of the cell. In particular, in case the nanomaterials are not properly anchored to the substrate and may disintegrate and freely float in the tissue and promote cellular uptake. On the other hand, compared with CPs, these materials are still relatively stiff relative to the tissue. With micro-magnetic stimulation, it should be noted that the region of activation is confined to a near-field region around the implanted coil (Lee et al., 2016). Beyond the materials and design of the electrodes, implementation of high-density prostheses with more than hundreds of stimulating channels requires additional challenges to be overcome, including high-resolution connections between the stimulation circuitry and the electrode array, design, and integration of the stimulator chip to individually control each electrode, and wireless transfer of data and power to the electrodes. For example, the connecting leads can tether the device and increase its chance of migration and also the chance of device failure by lead fracture or disconnection. Designing wireless transmission should consider that coupling of the transmitter and receiver coils is most efficient with the coils apposing each other in parallel (Rasouli and Phee, 2010). The internal coil may be integral to the electrode array housing or may be a separate unit receiving wired connections from the electrode arrays. With cortical prostheses, the distance between the coils and the absorption of electromagnetic energy by the scalp and skull are also important design considerations (RamRakhyani et al., 2011; Schwarz et al., 2014).

When trying to artificially replace the biological function, additional considerations, beyond engineering of materials and stimulation paradigms should be taken into account. For example, the postoperative rehabilitation process and evaluation of visual performance in individuals implanted with prosthesis. As part of the rehabilitation process, the recipients will need to be trained to use the bionic device and then engage in daily practice sessions to optimize their performance. Psychophysics tests to evaluate visual function of the recipients provide an objective assessment of visual function, and should be provided throughout the rehabilitation phase and beyond. These tests involves visuotopic mapping of the phosphenes and assessments of light perception, direction and motion perception, object and shape recognition, navigational tasks, and various activities of daily living (Dagnelie, 2008). However, current approaches to assess visual function in sighted people may not be suitable for blind individuals learning to use a bionic device and may not provide accurate information to achieve successful rehabilitation. It has been demonstrated with retinal stimulation that experimental results may be different to this theoretical relationship between VA and electrode size (Lorach et al., 2013). The highest VA demonstrated with the Alpha IMS device was 20/546, lower than the expected acuity of 20/250, possibly a result of crosstalk between neighboring electrodes, indirect stimulation of passing axons and/or local inhibition from concomitant stimulation. Local return electrodes surrounding each stimulation element could provide more focal stimulation (Lovell et al., 2005; Matteucci et al., 2013; Joucla and Yvert, 2009). In fact, VA as a quantitative measure may not necessarily corresponds to the spatial resolution produced by the retinal prosthesis. The traditional tests for estimating VA may need to be modified when artificial vision is evaluated, taking additional factors into account such as visual field scanning movements with the head and the time to complete the visual tests. Head scanning was demonstrated to improve VA score in studies applying SPV (Cha et al., 1992; Chen et al., 2006), and it seems to be a natural mechanism that the subjects in the experimental SPV conditions employed (Caspi and Zivotofsky, 2015). In addition, patients with low vision may need longer duration to complete the visual task (Caspi and Zivotofsky, 2015). This additional time may need to be added into the assessment of visual performance with prosthesis. SPV in normally sighted individuals is also used as means to test image processing algorithms which transform the visual scene into pixelated forms that can be elicited electrically into the visual pathways (Zapf et al., 2015; Bourkiza et al., 2013; Lui et al., 2012; Chen et al., 2009). For example, Dagnelie et al. (2006) developed a simulation of pixelated vision with cortical prosthesis and concluded that a $3 \times 3\,\text{mm}^2$ prosthesis with 16×16 electrodes should allow paragraph reading.

A further consideration is the need for consistent and uniform testing regimes that groups doing visual psychophysics assessment can commonly adopt. One positive step in this direction is the formation of an international task force to develop substantive recommendations for the assessment baseline (preimplant) visual status of potential patients (including specification of the disease diagnosis and impact on visual functioning, as well as postoperative visual function) (Rizzo III and Ayton, 2014).

The complexity of the phosphene percepts delivered via neuromodulation has thus far been a key limitation to the advancement of the visual prosthesis field that significantly influences the efficacy of these early-stage devices. While one would hope that a grid of electrodes would produce a corresponding perception of a grid of similarly spaced, punctate phosphenes, the reported experiences of implant recipients have been significantly different to this aspirational goal. While the electric fields that emanate from the stimulating electrodes may indeed be shaped by way of charge containment, altering the return path of the stimulation circuit, or manipulation of the stimulus pulse profile, it remains that without an electrode-neuron interface that approaches one-to-one, the phosphenes elicited will remain more complex than in natural vision. Phosphene shapes are likely influenced significantly by a combination of concomitant activation of multiple neurons, varying degrees of recruitment of retinal network elements vs direct activation of RGCs, or extraneous stimulation of passing axons from distal RGCs. Stimulation of columns of neurons in the primary visual cortex may produce geographic maps of phosphenes but this does not necessarily translate into synthesis of images. Greater points of stimulation in the retina, geniculate nuclei, or cortex do not necessarily mean there will be a higher resolution of visual perception as a result. This may well be an intractable problem to do with the fundamental approach of electrical stimulation and until efficacious steps are taken to address this issue, this field of research may stagnate at a point close to where we are now. However, as has been demonstrated in the cochlear implant experience, very meaningful sensory input can be delivered in spite of the fact that single electrodes elicit responses from multiple neurons. In the example of the cochlear implant, the lack of a one-to-one correspondence between electrodes and neurons has been managed rather than solved by way of accepting that it occurs, and implementing novel stimulation strategies that deliver the meaningful elements that describe sound. Perhaps, this is the most plausible approach to be taken with visual prosthesis—that is, to accept that phosphene complexity exists and focus on the delivery of the most meaningful elements that describe the visual scene.

To conclude, the future of new visual bionic devices is directly linked to expanding our understanding of the mechanisms and key barriers underlying the generation of and transduction of electrical current across the biohybrid interface. As we await the arrival of genetic and stem cell therapies where individual neural elements may become independent of their neighbors, neuromodulation remains a most powerful tool that has yet to be harnessed to its fullest potential. The development of new biomaterials is reducing the integration gap between biology and technology thus providing better connections between the electrodes and the excitable tissue. While most research efforts are being directed toward reducing the size of the electrodes, there is a key factor that impacts significantly the quality of the visual percepts elicited by bionic technologies: the ability to encode neural activity elicited by electrical stimulation as occurs in normal vision. The combination of more effective electrode-tissue interfaces with new neuromodulation techniques will soon allow for a more selective activation of the neural tissue with controlled timing of the elicited neural responses, taking vision prostheses further toward the goal of mimicking the lost sensory functionalities.

REFERENCES

Ahnood, A., Meffin, H., Garrett, D.J., Fox, K., Ganesan, K., Stacey, A., Apollo, N.V., Wong, Y.T., Lichter, S.G., Kentler, W., 2016. Diamond devices for high acuity prosthetic vision. Adv. Biosyst. 2 (3), 1–10.

Ahuja, A.K., Yeoh, J., Dorn, J.D., Caspi, A., Wuyyuru, V., Mcmahon, M.J., Humayun, M.S., Greenberg, R.J., Dacruz, L., Argus II, S.G., 2013. Factors affecting perceptual threshold in Argus II retinal prosthesis subjects. Transl. Vision Sci. Technol. 2, 1.

Andrews, T.J., Halpern, S.D., Purves, D., 1997. Correlated size variations in human visual cortex, lateral geniculate nucleus, and optic tract. J. Neurosci. 17, 2859–2868.

Antognazza, M.R., Ghezzi, D., Maschio, M.D., Lanzarini, E., Benfenati, F., Lanzani, G., 2012. A hybrid bio-organic interface for neuronal photo-activation. arXiv preprint arXiv:1202.1189.

Antognazza, M.R., Martino, N., Ghezzi, D., Feyen, P., Colombo, E., Endeman, D., Benfenati, F., Lanzani, G., 2015. Shedding light on living cells. Adv. Mater. 27, 7662–7669.

Antognazza, M.R., Di Paolo, M., Ghezzi, D., Mete, M., Di Marco, S., Maya-Vetencourt, J.F., Maccarone, R., Desii, A., Di Fonzo, F., Bramini, M., 2016. Characterization of a polymer-based, fully organic prosthesis for implantation into the subretinal space of the rat. Adv. Healthcare Mater. 5, 2271–2282.

Ayton, L.N., Guymer, R.H., Luu, C.D., 2013. Choroidal thickness profiles in retinitis pigmentosa. Clin. Exp. Ophthalmol. 41, 396–403.

Ayton, L.N., Blamey, P.J., Guymer, R.H., Luu, C.D., Nayagam, D.A., Sinclair, N.C., Shivdasani, M.N., Yeoh, J., Mccombe, M.F., Briggs, R.J., 2014. First-in-human trial of a novel suprachoroidal retinal prosthesis. PLoS ONE. 9, e115239.

Baden, T., Berens, P., Franke, K., Rosón, M.R., Bethge, M., Euler, T., 2016. The functional diversity of retinal ganglion cells in the mouse. Nature 529, 345–350.

Bak, M., Girvin, J., Hambrecht, F., Kufta, C., Loeb, G., Schmidt, E., 1990. Visual sensations produced by intracortical microstimulation of the human occipital cortex. Med. Biol. Eng. Comput. 28, 257–259.

Bareket, L., Waiskopf, N., Rand, D., Lubin, G., David-Pur, M., Ben-Dov, J., Roy, S., Eleftheriou, C., Sernagor, E., Cheshnovsky, O., 2014. Semiconductor nanorod–carbon nanotube biomimetic films for wire-free photostimulation of blind retinas. Nano Lett. 14, 6685–6692.

Bareket-Keren, L., Hanein, Y., 2014. Novel interfaces for light directed neuronal stimulation: advances and challenges. Int. J. Nanomed. 9, 65–83.

Barriga-rivera, A., Eiber, C.D., Dodds, C.W., Fung, A.T., Tatarinoff, V., Lovell, N.H., Suaning, G.J., 2015. In: Electrically evoked potentials in an ovine model for the evaluation of visual prosthesis efficacy. Proceedings of the 37th Annual International Conference of the Engineering in Medicine and Biology Society (EMBC). IEEE, pp. 3359–3362.

Barriga-Rivera, A., Eiber, C.D., Matteucci, P.B., Spencer, C.C., Morley, J.W., Lovell, N.H., Suaning, G.J., 2016a. A 4 + 1 architecture for in vivo electrophysiology visual prosthesis. J. Access. Des. All 6, 81–101.

Barriga-Rivera, A., Morley, J.W., Lovell, N.H., Suaning, G.J., 2016b. In: Cortical responses following simultaneous and sequential retinal neurostimulation with different return configurations. Proceedings of the 38th Annual International Conference of the Engineering in Medicine and Biology Society (EMBC). IEEE, pp. 5435–5438.

Barriga-Rivera, A., Guo, T., Yang, C.-Y., Al-Abed, A., Lovell, N.H., Morley, J.W., Suaning, G.J., 2017. High-amplitude electrical stimulation can reduce elicited neuronal activity in visual prosthesis. Sci. Rep. 7, 42682.

Bartlett, J.R., Doty, R.W., Lee, B.B., Negrao, N., Overman JR., W.H., 1977. Deleterious effects of prolonged electrical excitation of striate cortex in macaques. Brain Behav. Evol. 14, 46–66.

Berenstein, C.K., Mens, L.H., Mulder, J.J., Vanpoucke, F.J., 2008. Current steering and current focusing in cochlear implants: comparison of monopolar, tripolar, and virtual channel electrode configurations. Ear Hear. 29, 250–260.

Bezard, E., Boraud, T., Nguyen, J.-P., Velasco, F., Keravel, Y., Gross, C., 1999. Cortical stimulation and epileptic seizure: a study of the potential risk in primates. Neurosurgery 45, 346.

Bjornsson, C., OH, S.J., Al-Kofahi, Y., Lim, Y., Smith, K., Turner, J., DE, S., Roysam, B., Shain, W., Kim, S.J., 2006. Effects of insertion conditions on tissue strain and vascular damage during neuroprosthetic device insertion. J. Neural Eng. 3, 196.

Bourkiza, B., Vurro, M., Jeffries, A., Pezaris, J.S., 2013. Visual acuity of simulated thalamic visual prostheses in normally sighted humans. PLoS ONE. 8, e73592.

Bourne, R., Price, H., Taylor, H., Leasher, J., Keeffe, J., Glanville, J., Sieving, P.C., Khairallah, M., Wong, T.Y., Zheng, Y., 2013a. New systematic review methodology for visual impairment and blindness for the 2010 Global Burden of Disease study. Ophthalmic Epidemiol. 20, 33–39.

Bourne, R.R., Stevens, G.A., White, R.A., Smith, J.L., Flaxman, S.R., Price, H., Jonas, J.B., Keeffe, J., Leasher, J., Naidoo, K., 2013b. Causes of vision loss worldwide, 1990–2010: a systematic analysis. Lancet Glob. Health 1, e339–e349.

Bredeson, S., Kanneganti, A., Deku, F., Cogan, S., Romero-Ortega, M., Troyk, P., 2015. In: Chronic in-vivo testing of a 16-channel implantable wireless neural stimulator. 2015 37th Annual International Conference of the IEEE Engineering in Medicine and Biology Society (EMBC). IEEE, pp. 1017–1020.

Brelén, M.E., Duret, F., Gerard, B., Delbeke, J., Veraart, C., 2005. Creating a meaningful visual perception in blind volunteers by optic nerve stimulation. J. Neural Eng. 2, S22.

Brelén, M.E., de Potter, P., Gersdorff, M., Cosnard, G., Veraart, C., Delbeke, J., 2006. Intraorbital implantation of a stimulating electrode for an optic nerve visual prosthesis: case report. J. Neurosurg. 104, 593–597.

Brelén, M.E., Vince, V., Gérard, B., Veraart, C., Delbeke, J., 2010. Measurement of evoked potentials after electrical stimulation of the human optic nerve. Investig. Ophthalmol. Vis. Sci. 51, 5351–5355.

Brindley, G.S., 1964. A new interaction of light and electricity in stimulating the human retina. J. Physiol. 171.

Brindley, G.S., Lewin, W.S., 1968. The sensations produced by electrical stimulation of the visual cortex. J. Physiol. 196, 18.

Brummer, S.B., Robblee, L.S., Hambrecht, F.T., 1983. Criteria for selecting electrodes for electrical stimulation: theoretical and practical considerations. Ann. N. Y. Acad. Sci. 405, 159–171.

Butterwick, A., Huie, P., Jones, B.W., Marc, R.E., Marmor, M., Palanker, D., 2009. Effect of shape and coating of a subretinal prosthesis on its integration with the retina. Exp. Eye Res. 88, 22–29.

Cao, P., Sun, J., Yan, Y., Chen, Y., Chai, X., Sun, X., Ren, Q., Li, L., 2015. Properties of electrically evoked potentials activated by optic nerve stimulation with penetrating electrodes of different modes in rabbits. Graefes Arch. Clin. Exp. Ophthalmol. 253, 2171–2180.

Caspi, A., Zivotofsky, A.Z., 2015. Assessing the utility of visual acuity measures in visual prostheses. Vis. Res. 108, 77–84.

Caspi, A., Dorn, J.D., Mcclure, K.H., Humayun, M.S., Greenberg, R.J., Mcmahon, M.J., 2009. Feasibility study of a retinal prosthesis. Arch. Ophthalmol. 127, 398–401.

Cha, K., Boman, D.K., Horch, K.W., Normann, R.A., 1992. Reading speed with a pixelized vision system. J. Opt. Soc. Am. A 9, 673–677.

Chai, X., Li, L., Wu, K., Zhou, C., Cao, P., Ren, Q., 2008. C-sight visual prostheses for the blind. IEEE Eng. Med. Biol. Mag. 27 (5), 20–28.

Chen, S., Hallum, L., Suaning, G., Lovell, N., 2006. Psychophysics of prosthetic vision: I. Visual scanning and visual acuity. Proceedings of the 28th Annual International Conference of the IEEE Engineering in Medicine and Biology Society. IEEE, pp. 4400–4403.

Chen, S.C., Suaning, G.J., Morley, J.W., Lovell, N.H., 2009. Simulating prosthetic vision: I. Visual models of phosphenes. Vis. Res. 49, 1493–1506.

Chow, A.Y., 2013. Retinal prostheses development in retinitis pigmentosa patients—progress and comparison. Asia Pac. J. Ophthalmol. 2, 253–268.

Chow, A.Y., Chow, V.Y., Packo, K.H., Pollack, J.S., Peyman, G.A., Schuchard, R., 2004. The artificial silicon retina microchip for the treatment of vision loss from retinitis pigmentosa. Arch. Ophthalmol. 122, 460–469.

Chow, A.Y., Bittner, A.K., Pardue, M.T., 2010. The artificial silicon retina in retinitis pigmentosa patients (an American Ophthalmological Association thesis). Trans. Am. Ophthalmol. Soc. 108, 120–154.

Chowdhury, V., Morley, J.W., Coroneo, M.T., 2005. Stimulation of the retina with a multielectrode extraocular visual prosthesis. ANZ J. Surg. 75, 697–704.

Chowdhury, V., Morley, J.W., Coroneo, M.T., 2008. Development of an extraocular retinal prosthesis: evaluation of stimulation parameters in the cat. J. Clin. Neurosci. 15, 900–906.

Cicione, R., Shivdasani, M.N., Fallon, J.B., Luu, C.D., Allen, P.J., Rathbone, G.D., Shepherd, R.K., Williams, C.E., 2012. Visual cortex responses to suprachoroidal electrical stimulation of the retina: effects of electrode return configuration. J. Neural Eng. 9, 036009.

Cogan, S.F., Ludwig, K.A., Welle, C.G., Takmakov, P., 2016. Tissue damage thresholds during therapeutic electrical stimulation. J. Neural Eng. 13, 021001.

Coulombe, J., Sawan, M., Gervais, J.-F., 2007. A highly flexible system for microstimulation of the visual cortex: design and implementation. IEEE Trans. Biomed. Circuits Syst 1, 258–269.

Curcio, C.A., Allen, K.A., 1990. Topography of ganglion cells in human retina. J. Comp. Neurol. 300, 5–25.

Curcio, C.A., Sloan, K.R., Kalina, R.E., Hendrickson, A.E., 1990. Human photoreceptor topography. J. Comp. Neurol. 292, 497–523.

da Cruz, L., Dorn, J.D., Humayun, M.S., Dagnelie, G., Handa, J., Barale, P.-O., Sahel, J.-A., Stanga, P.E., Hafezi, F., Safran, A.B., 2016. Five-year safety and performance results from the Argus II retinal prosthesis system clinical trial. Ophthalmology 123, 2248–2254.

da Cruz, L., Coley, B.F., Dorn, J., Merlini, F., Filley, E., Christopher, P., Chen, F.K., Wuyyuru, V., Sahel, J., Stanga, P., Humayun, M., Greenberg, R.J., Dagnelie, G., Argus II, S.G., 2013. The Argus II epiretinal prosthesis system allows letter and word reading and long-term function in patients with profound vision loss. Br. J. Ophthalmol. 97, 632–636.

Dagnelie, G., 2008. Psychophysical evaluation for visual prosthesis. Annu. Rev. Biomed. Eng. 10, 339–368.

Dagnelie, G., Barnett, D., Humayun, M.S., Thompson, R.W., 2006. Paragraph text reading using a pixelized prosthetic vision simulator: parameter dependence and task learning in free-viewing conditions. Investig. Ophthalmol. Vis. Sci. 47, 1241–1250.

David-Pur, M., Bareket-Keren, L., Beit-Yaakov, G., Raz-Prag, D., Hanein, Y., 2014. All-carbon-nanotube flexible multi-electrode array for neuronal recording and stimulation. Biomed. Microdevices 16, 43–53.

Davis, T., Parker, R., House, P., Bagley, E., Wendelken, S., Normann, R., Greger, B., 2012. Spatial and temporal characteristics of V1 microstimulation during chronic implantation of a microelectrode array in a behaving macaque. J. Neural Eng.. 9065003.

de Balthasar, C., Patel, S., Roy, A., Freda, R., Greenwald, S., Horsager, A., Mahadevappa, M., Yanai, D., Mcmahon, M.J., Humayun, M.S., Greenberg, R.J., Weiland, J.D., Fine, I., 2008. Factors affecting perceptual thresholds in epiretinal prostheses. Investig. Ophthalmol. Vis. Sci. 49, 2303–2314.

Delbeke, J., Oozeer, M., Veraart, C., 2003. Position, size and luminosity of phosphenes generated by direct optic nerve stimulation. Vis. Res. 43, 1091–1102.

Dobelle, W.H., 2000. Artificial vision for the blind by connecting a television camera to the visual cortex. ASAIO J. 46, 3–9.

Dobelle, W., Mladejovsky, M., 1974. Phosphenes produced by electrical stimulation of human occipital cortex, and their application to the development of a prosthesis for the blind. J. Physiol. 243, 553.

Dobelle, W.H., Mladejovsky, M., Girvin, J., 1974. Artificial vision for the blind: electrical stimulation of visual cortex offers hope for a functional prosthesis. Science 183, 440–444.

Dokos, S., Suaning, G.J., Lovell, N.H., 2005. A bidomain model of epiretinal stimulation. IEEE Trans. Neural Syst. Rehabil. Eng 13, 137–146.

Donaldson, P., 1973. In: Experimental visual prosthesis. Proceedings of the Institution of Electrical Engineers. IET, pp. 281–298.

Dorn, J.D., Ahuja, A.K., Caspi, A., da Cruz, L., Dagnelie, G., Sahel, J.A., Greenberg, R.J., Mcmahon, M.J., Argus II, S.G., 2013. The detection of motion by blind subjects with the epiretinal 60-electrode (Argus II) retinal prosthesis. JAMA Ophthalmol. 131, 183–189.

Duret, F., Brelén, M.E., Lambert, V., Gérard, B., Delbeke, J., Veraart, C., 2006. Object localization, discrimination, and grasping with the optic nerve visual prosthesis. Restor. Neurol. Neurosci. 24, 31–40.

Dyer, M.A., Cepko, C.L., 2000. Control of Müller glial cell proliferation and activation following retinal injury. Nat. Neurosci. 3, 873–880.

Eleftheriou, C.G., Zimmermann, J.B., Kjeldsen, H.D., David-Pur, M., Hanein, Y., Sernagor, E., 2017. Carbon nanotube electrodes for retinal implants: a study of structural and functional integration over time. Biomaterials 112, 108–121.

Fang, X., Sakaguchi, H., Fujikado, T., Osanai, M., Kanda, H., Ikuno, Y., Kamei, M., Ohji, M., Gan, D., Choi, J., 2005. Direct stimulation of optic nerve by electrodes implanted in optic disc of rabbit eyes. Graefes Arch. Clin. Exp. Ophthalmol. 243, 49–56.

Fang, X., Sakaguchi, H., Fujikado, T., Osanai, M., Ikuno, Y., Kamei, M., Ohji, M., Yagi, T., Tano, Y., 2006. Electrophysiological and histological studies of chronically implanted intrapapillary microelectrodes in rabbit eyes. Graefes Arch. Clin. Exp. Ophthalmol. 244, 364–375.

Fernandes, R.A.B., Diniz, B., Ribeiro, R., Humayun, M., 2012. Artificial vision through neuronal stimulation. Neurosci. Lett. 519, 122–128.

Fernandez, E., Pelayo, F., Romero, S., Bongard, M., Marin, C., Alfaro, A., Merabet, L., 2005. Development of a cortical visual neuroprosthesis for the blind: the relevance of neuroplasticity. J. Neural Eng. 2, R1.

Ferrandez, J.M., Liaño, E., Bonomini, P., Martínez, J., Toledo, J., Fernández, E., 2007. In: A customizable multi-channel stimulator for cortical neuroprosthesis. 29th Annual International Conference of the IEEE Engineering in Medicine and Biology Society. EMBS 2007. IEEE, pp. 4707–4710.

Ferris, F.L., Fine, S.L., Hyman, L., 1984. Age-related macular degeneration and blindness due to neovascular maculopathy. Arch. Ophthalmol. 102, 1640–1642.

Feyen, P., Colombo, E., Endeman, D., Nova, M., Laudato, L., Martino, N., Antognazza, M.R., Lanzani, G., Benfenati, F., Ghezzi, D., 2016. Light-evoked hyperpolarization and silencing of neurons by conjugated polymers. Sci. Rep. 6, 1–14.

Foerster, O., 1929. Beitrage zur pathophysiologie der sehbahn und der spehsphare. J. Psychol. Neurol. 39, 435–463.

Fujikado, T., Morimoto, T., Kanda, H., Kusaka, S., Nakauchi, K., Ozawa, M., Matsushita, K., Sakaguchi, H., Ikuno, Y., Kamei, M., Tano, Y., 2007. Evaluation of phosphenes elicited by extraocular stimulation in normals and by suprachoroidal-transretinal stimulation in patients with retinitis pigmentosa. Graefes Arch. Clin. Exp. Ophthalmol. 245, 1411–1419.

Fujikado, T., Kamei, M., Sakaguchi, H., Kanda, H., Morimoto, T., Ikuno, Y., Nishida, K., Kishima, H., Maruo, T., Konoma, K., Ozawa, M., Nishida, K., 2011. Testing of semichronically implanted retinal prosthesis by suprachoroidal-transretinal stimulation in patients with retinitis pigmentosa. Investig. Ophthalmol. Vis. Sci. 52, 4726–4733.

Fujikado, T., Kamei, M., Sakaguchi, H., Kanda, H., Morimoto, T., Ikuno, Y., Nishida, K., Kishima, H., Maruo, T., Sawai, H., Miyoshi, T., 2012. Clinical trial of chronic implantation of suprachoroidal-transretinal stimulation system for retinal prosthesis. Sens. Mater. 24, 181–187.

Fujikado, T., Kamei, M., Sakaguchi, H., Kanda, H., Endo, T., Hirota, M., Morimoto, T., Nishida, K., Kishima, H., Terasawa, Y., Oosawa, K., Ozawa, M., 2016. One-year outcome of 49-channel suprachoroidal–transretinal stimulation prosthesis in patients with advanced retinitis pigmentosa. Investig. Ophthalmol. Vis. Sci. 57, 6147–6157.

Gabriel, G., Gómez, R., Bongard, M., Benito, N., Fernández, E., Villa, R., 2009. Easily made single-walled carbon nanotube surface microelectrodes for neuronal applications. Biosens. Bioelectron. 24, 1942–1948.

Ganesan, K., Garrett, D.J., Ahnood, A., Shivdasani, M.N., Tong, W., Turnley, A.M., Fox, K., Meffin, H., Prawer, S., 2014. An all-diamond, hermetic electrical feedthrough array for a retinal prosthesis. Biomaterials 35, 908–915.

Gautam, V., Rand, D., Hanein, Y., Narayan, K., 2014. A polymer optoelectronic interface provides visual cues to a blind retina. Adv. Mater. 26, 1751–1756.

Gekeler, F., Szurman, P., Besch, D., Rohrbach, M., 2010. Implantation and explantation of active subretinal visual prostheses using a combined transcutaneous and transchoroidal approach. Nova Acta Leopoldina NF 111, 205–216.

Ghezzi, D., Antognazza, M.R., Maccarone, R., Bellani, S., Lanzarini, E., Martino, N., Mete, M., Pertile, G., Bisti, S., Lanzani, G., 2013. A polymer optoelectronic interface restores light sensitivity in blind rat retinas. Nat. Photonics 7, 400–406.

Goddard, G.V., 1967. Development of epileptic seizures through brain stimulation at low intensity. Nature 214, 1020–1021.

Gwon, T.M., Kim, J.H., Choi, G.J., Kim, S.J., 2016. Mechanical interlocking to improve metal–polymer adhesion in polymer-based neural electrodes and its impact on device reliability. J. Mater. Sci. 51, 6897–6912.

Ha, S., Khraiche, M.L., Akinin, A., Jing, Y., Damle, S., Kuang, Y., Bauchner, S., Lo, Y.-H., Freeman, W.R., Silva, G.A., 2016. Towards high-resolution retinal prostheses with direct optical addressing and inductive telemetry. J. Neural Eng.. 13056008.

Habib, A.G., Cameron, M.A., Suaning, G.J., Lovell, N.H., Morley, J.W., 2013. Spatially restricted electrical activation of retinal ganglion cells in the rabbit retina by hexapolar electrode return configuration. J. Neural Eng.. 10036013.

Hadjinicolaou, A.E., Leung, R.T., Garrett, D.J., Ganesan, K., Fox, K., Nayagam, D.A., Shivdasani, M.N., Meffin, H., Ibbotson, M.R., Prawer, S., 2012. Electrical stimulation of retinal ganglion cells with diamond and the development of an all diamond retinal prosthesis. Biomaterials 33, 5812–5820.

Hartong, D.T., Berson, E.L., Dryja, T.P., 2006. Retinitis pigmentosa. Lancet 368, 1795–1809.

Hecht, S., Shlaer, S., Pirenne, M.H., 1942. Energy, quanta, and vision. J. Gen. Physiol. 25, 819–840.

Ho, A.C., Humayun, M.S., Dorn, J.D., da Cruz, L., Dagnelie, G., Handa, J., Barale, P.-O., Sahel, J.-A., Stanga, P.E., Hafezi, F., Safran, A.B., Salzmann, J., 2015. Long-term results from an epiretinal prosthesis to restore sight to the blind. Ophthalmology 122, 1547–1554.

Hornig, R., Zehnder, T., Velikay-Parel, M., Laube, T., Feucht, M., Richard, G., 2007. The IMI retinal implant system. In: Artificial Sight. Springer, New York, NY.

Hornig, R., Dapper, M., Le Joliff, E., Hill, R., Ishaque, K., Posch, C., Benosman, R., Lemer, Y., Sahel, J.-A., Picaud, S., 2017. Pixium vision: first clinical results and innovative developments. In: Artificial Vision. Springer, Cham, Switzerland.

Humayun, M.S., Weiland, J.D., Fujii, G.Y., Greenberg, R., Williamson, R., Little, J., Mech, B., Cimmarusti, V., van Boemel, G., Dagnelie, G., de Juan, E., 2003. Visual perception in a blind subject with a chronic microelectronic retinal prosthesis. Vis. Res. 43, 2573–2581.

Humayun, M.S., Dorn, J.D., da Cruz, L., Dagnelie, G., Sahel, J.A., Stanga, P.E., Cideciyan, A.V., Duncan, J.L., Eliott, D., Filley, E., Ho, A.C., Santos, A., Safran, A.B., Arditi, A., Del Priore, L.V., Greenberg, R.J., Argus II, S.G., 2012. Interim results from the international trial of second sight's visual prosthesis. Ophthalmology 119, 779–788.

Hurvich, L.M., 1981. Color Vision. Sinauer Associates, Sunderland, MA.

Jensen, R.J., Rizzo, J.F., 2006. Thresholds for activation of rabbit retinal ganglion cells with a subretinal electrode. Exp. Eye Res. 83, 367–373.

Jeong, J., Lee, S.W., Min, K., Eom, K., Bae, S.H., Kim, S.J., 2011. Eye-surface conformable telemetric structure for polymer-based retinal prosthesis. Proceedings for the 33rd Annual International Conference of the IEEE Engineering in Medicine and Biology Society. IEEE, pp. 1097–1100.

Jeong, J., Lee, S.W., Min, K.S., Shin, S., Jun, S.B., Kim, S.J., 2012. Liquid crystal polymer (LCP), an attractive substrate for retinal implant. Sens. Mater 24, 189–203.

Jeong, J., Bae, S.H., Min, K.S., Seo, J.-M., Chung, H., Kim, S.J., 2015. A miniaturized, eye-conformable, and long-term reliable retinal prosthesis using monolithic fabrication of liquid crystal polymer (LCP). IEEE Trans. Biomed. Eng. 62, 982–989.

Jeong, J., Bae, S.H., Seo, J.M., Chung, H., Kim, S.J., 2016. Long-term evaluation of a liquid crystal polymer (LCP)-based retinal prosthesis. J. Neural Eng. 13, 1–12.

Joucla, S., Yvert, B., 2009. Improved focalization of electrical microstimulation using microelectrode arrays: a modeling study. PLoS ONE. 4, e4828.

Kanda, H., Sawai, H., Morimoto, T., Fujikado, T., Tano, Y., Fukuda, Y., 2003. Suprachoroidal-transretinal stimulation (STS) can elicit localized evoked responses from the superior colliculus in normal and RCS rats. Investig. Ophthalmol. Vis. Sci. 44, 5053.
Kanda, H., Morimoto, T., Fujikado, T., Tano, Y., Fukuda, Y., Sawai, H., 2004. Electrophysiological studies of the feasibility of suprachoroidal-transretinal stimulation for artificial vision in normal and RCS rats. Investig. Ophthalmol. Vis. Sci. 45, 560–566.
Kelly, S.K., Shire, D.B., Chen, J., Doyle, P., Gingerich, M.D., Cogan, S.F., Drohan, W.A., Behan, S., Theogarajan, L., Wyatt, J.L., Rizzo, J.F., 2011. A hermetic wireless subretinal neurostimulator for vision prostheses. IEEE Trans. Biomed. Eng. 58, 3197–3205.
Kelly, S.K., Shire, D.B., Chen, J., Gingerich, M.D., Cogan, S.F., Drohan, W.A., Ellersick, W., Krishnan, A., Behan, S., Wyatt, J.L., Rizzo, J.F., 2013. In: Developments on the Boston 256-channel retinal implant.Electronic Proceedings of the 2013 IEEE International Conference on Multimedia and Expo Workshops.
Keserü, M., Post, N., Hornig, R., Zeitz, O., Richard, G., 2009. Long term tolerability of the first wireless implant for electrical epiretinal stimulation. Investig. Ophthalmol. Vis. Sci. 50, 4226.
Keserü, M., Feucht, M., Bornfeld, N., Laube, T., Walter, P., Rössler, G., Velikay-Parel, M., Hornig, R., Richard, G., 2012. Acute electrical stimulation of the human retina with an epiretinal electrode array. Acta Ophthalmol. 90, e1–e8.
Khraiche, M.L., Lo, Y., Wang, D., Cauwenberghs, G., Freeman, W., Silva, G.A., 2011. In: Ultra-high photosensitivity silicon nanophotonics for retinal prosthesis: electrical characteristics.2011 Annual International Conference of the IEEE Engineering in Medicine and Biology Society (EMBC), pp. 2933–2936.
Khraiche, M.L., El Emam, S., Akinin, A., Cauwenberghs, G., Freeman, W., Silva, G.A., 2013. In: Visual evoked potential characterization of rabbit animal model for retinal prosthesis research.2013 35th Annual International Conference of the IEEE Engineering in Medicine and Biology Society (EMBC), pp. 3539–3542.
Killian, N.J., Vurro, M., Keith, S.B., Kyada, M.J., Pezaris, J.S., 2016. Perceptual learning in a non-human primate model of artificial vision. Sci. Rep. 6, 1–16.
Klauke, S., Goertz, M., Rein, S., Hoehl, D., Thomas, U., Eckhorn, R., Bremmer, F., Wachtler, T., 2011. Stimulation with a wireless intraocular epiretinal implant elicits visual percepts in blind humans. Investig. Ophthalmol. Vis. Sci. 52, 449–455.
Klomp, G., Womack, M., Dobelle, W., 1977. Fabrication of large arrays of cortical electrodes for use in man. J. Biomed. Mater. Res. A 11, 347–364.
Koch, C., 2013. Cracking the retinal code. Sci. Am. Mind 23, 20–21.
Krause, F., 1924. Die Sehbahn in chirurgischer Beziehung und die faradische Reizung des Sehzentrums. J. Mol. Med. 3, 1260–1265.
Lee, S.W., Seo, J.-M., Ha, S., Kim, E.T., Chung, H., Kim, S.J., 2009. Development of microelectrode arrays for artificial retinal implants using liquid crystal polymers. Investig. Ophthalmol. Vis. Sci. 50, 5859–5866.
Lee, S.W., Min, K.S., Jeong, J., Kim, J., Kim, S.J., 2011. Monolithic encapsulation of implantable neuroprosthetic devices using liquid crystal polymers. IEEE Trans. Biomed. Eng. 58, 2255–2263.
Lee, S.W., Fallegger, F., Casse, B.D., Fried, S.I., 2016. Implantable microcoils for intracortical magnetic stimulation. Sci. Adv. 2, e1600889.
Lehman, S.S., 2012. Cortical visual impairment in children: identification, evaluation and diagnosis. Curr. Opin. Ophthalmol. 23, 384–387.
Lennie, P., D'Zmura, M., 1987. Mechanisms of color vision. Crit. Rev. Neurobiol. 3, 333–400.
LeRoy, C., 1755. Où l'on rend compte de quelques tentatives que l'on a faites pour guérir plusieurs maladies par l'électricité. Hist. Acad. R. Sci. (Paris) Mémoire Math. Phys. 60, 87–95.

Lewis, P.M., Ackland, H.M., Lowery, A.J., Rosenfeld, J.V., 2015. Restoration of vision in blind individuals using bionic devices: a review with a focus on cortical visual prostheses. Brain Res. 1595, 51–73.

Lewis, P.M., Ayton, L.N., Guymer, R.H., Lowery, A.J., Blamey, P.J., Allen, P.J., Luu, C.D., Rosenfeld, J.V., 2016a. Advances in implantable bionic devices for blindness: a review. ANZ J. Surg. 86, 654–659.

Lewis, P.M., Thomson, R.H., Rosenfeld, J.V., Fitzgerald, P.B., 2016b. Brain neuromodulation techniques: a review. Neuroscientist 22, 406–421.

Li, l., Cao, P., Sun, M., Chai, X., Wu, K., Xu, X., Li, X., Ren, Q., 2009. Intraorbital optic nerve stimulation with penetrating electrodes: in vivo electrophysiology study in rabbits. Graefes Arch. Clin. Exp. Ophthalmol. 247, 349–361.

Lorach, H., Marre, O., Sahel, J.-A., Benosman, R., Picaud, S., 2013. Neural stimulation for visual rehabilitation: advances and challenges. J. Physiol.-Paris 107, 421–431.

Lorach, H., Goetz, G., Mandel, Y., Lei, X., Kamins, T.I., Mathieson, K., Huie, P., Dalal, R., Harris, J.S., Palanker, D., 2015a. Performance of photovoltaic arrays in-vivo and characteristics of prosthetic vision in animals with retinal degeneration. Vis. Res. 111, 142–148.

Lorach, H., Goetz, G., Smith, R., Lei, X., Mandel, Y., Kamins, T., Mathieson, K., Huie, P., Harris, J., Sher, A., 2015b. Photovoltaic restoration of sight with high visual acuity. Nat. Med. 21, 476–482.

Lovell, N.H., Dokos, S., Cloherty, S.L., Preston, P.J., Suaning, G.J., 2005. In: Current distribution during parallel stimulation: implications for an epiretinal neuroprosthesis.2005 27th Annual International Conference of the IEEE Engineering in Medicine and Biology Society, vols. 1–7.

Löwenstein, K., Borchardt, M., 1918. Symptomatologie und elektrische Reizung bei einer Schußverletzung des Hinterhauptlappens. J. Neurol. 58, 264–292.

Lowery, A.J., Rosenfeld, J.V., Lewis, P.M., Browne, D., Mohan, A., Brunton, E., Yan, E., Maller, J., Mann, C., Rajan, R., 2015. In: Restoration of vision using wireless cortical implants: the Monash Vision Group project. 2015 37th Annual International Conference of the IEEE Engineering in Medicine and Biology Society (EMBC), pp. 1041–1044.

Lu, Y., Yan, Y., Chai, X., Ren, Q., Chen, Y., Li, L., 2013. Electrical stimulation with a penetrating optic nerve electrode array elicits visuotopic cortical responses in cats. J. Neural Eng. 10, 036022.

Lui, W.L.D., Browne, D., Kleeman, L., Drummond, T., Li, W.H., 2012. In: Transformative reality: improving bionic vision with robotic sensing. 2012 Annual International Conference of the IEEE Engineering in Medicine and Biology Society (EMBC). IEEE, pp. 304–307.

Mahadevappa, M., Weiland, J.D., Greenberg, R.J., Humayun, M.S., 2005. Perceptual thresholds and electrode impedance in three retinal prosthesis subjects. IEEE Trans. Neural Syst. Rehabil. Eng 13, 201–206.

Mariotti, S.P., 2012. Global Data on Visual Impairments 2010. World Health Organization, p. 20.

Mathieson, K., Loudin, J., Goetz, G., Huie, P., Wang, L.L., Kamins, T.I., Galambos, L., Smith, R., Harris, J.S., Sher, A., Palanker, D., 2012. Photovoltaic retinal prosthesis with high pixel density. Nat. Photonics 6, 391–397.

Matteucci, P.B., Chen, S.C., Tsai, D., Dodds, C.W., Dokos, S., Morley, J.W., Lovell, N.H., Suaning, G.J., 2013. Current steering in retinal stimulation via a quasimonopolar stimulation paradigm. Investig. Ophthalmol. Vis. Sci. 54, 4307–4320.

Matteucci, P.B., Barriga-Rivera, A., Eiber, C.D., Morley, J.W., Lovell, N.H., Suaning, G.J., 2016. The effect of electric cross-talk in retinal neurostimulation. Investig. Ophthalmol. Vis. Sci. 57, 1031–1037.

Maya-Vetencourt, J.F., Ghezzi, D., Antognazza, M.R., Colombo, E., Mete, M., Feyen, P., Desii, A., Buschiazzo, A., Di Paolo, M., Di Marco, S., 2017. A fully organic retinal prosthesis restores vision in a rat model of degenerative blindness. Nat. Mater. 681–689.

Mccreery, D.B., Yuen, T.G., Agnew, W.F., Bullara, L.A., 1994. Stimulus parameters affecting tissue injury during microstimulation in the cochlear nucleus of the cat. Hear. Res. 77, 105–115.

Mccreery, D., Pikov, V., Troyk, P.R., 2010. Neuronal loss due to prolonged controlled-current stimulation with chronically implanted microelectrodes in the cat cerebral cortex. J. Neural Eng. 7, 036005.

Menzel-Severing, J., Laube, T., Brockmann, C., Bornfeld, N., Mokwa, W., Mazinani, B., Walter, P., Roessler, G., 2012. Implantation and explantation of an active epiretinal visual prosthesis: 2-year follow-up data from the EPIRET3 prospective clinical trial. Eye 26, 502–509.

Merrill, D.R., Bikson, M., Jefferys, J.G.R., 2005. Electrical stimulation of excitable tissue: design of efficacious and safe protocols. J. Neurosci. Methods 141, 171–198.

Mohammadi, H.M., Ghafar-Zadeh, E., Sawan, M., 2012. An image processing approach for blind mobility facilitated through visual intracortical stimulation. Artif. Organs 36, 616–628.

Molokanova, E., Bartel, J., Zhao, W., Naasani, I., Ignatius, M., Treadway, J., Savtchenko, A., 2008. Quantum dots move beyond fluorescence imaging the unique properties of quantum dots allow them to be optimized for voltage sensing and for light-controlled electrical activation of cells. Biophoton. Int. 15, 26.

Morimoto, K., Fahnestock, M., Racine, R.J., 2004. Kindling and status epilepticus models of epilepsy: rewiring the brain. Prog. Neurobiol. 73, 1–60.

Mullen, K.T., Dumoulin, S.O., Hess, R.F., 2008. Color responses of the human lateral geniculate nucleus: selective amplification of S-cone signals between the lateral geniculate nucleno and primary visual cortex measured with high-field fMRI. Eur. J. Neurosci. 28, 1911–1923.

Nakauchi, K., Fujikado, T., Kanda, H., Morimoto, T., Choi, J.S., Ikuno, Y., Sakaguchi, H., Kamei, M., Ohji, M., Yagi, T., Nishimura, S., Sawai, H., fukuda, Y., Tano, Y., 2005. Transretinal electrical stimulation by an intrascleral multichannel electrode array in rabbit eyes. Graefes Arch. Clin. Exp. Ophthalmol. 243, 169–174.

Nishida, K., Sakaguchi, H., Kamei, M., Cecilia-Gonzalez, C., Terasawa, Y., Velez-Montoya, R., Fujikado, T., Sanchez-Fontan, R., Ozawa, M., Quiroz-Mercado, H., 2015. Visual sensation by electrical stimulation using a new direct optic nerve electrode device. Brain Stimul 8, 678.

Normann, R.A., Maynard, E.M., rousche, P.J., Warren, D.J., 1999. A neural interface for a cortical vision prosthesis. Vis. Res. 39, 2577–2587.

Opie, N.L., Burkitt, A.N., Meffin, H., Grayden, D.B., 2012. Heating of the eye by a retinal prosthesis: modeling, cadaver and in vivo study. IEEE Trans. Biomed. Eng. 59, 339–345.

Palanker, D., Vankov, A., Huie, P., Baccus, S., 2005. Design of a high-resolution optoelectronic retinal prosthesis. J. Neural Eng. 2, 16.

Panetsos, F., Sanchez-Jimenez, A., Diaz-de Cerio, E.R., Diaz-Guemes, I., Sanchez, F.M., 2011. Consistent phosphenes generated by electrical microstimulation of the visual thalamus. An experimental approach for thalamic visual neuroprostheses. Front. Neurosci. 5, 84.

Pappas, T.C., Wickramanyake, W.S., Jan, E., Motamedi, M., Brodwick, M., Kotov, N.A., 2007. Nanoscale engineering of a cellular interface with semiconductor nanoparticle films for photoelectric stimulation of neurons. Nano Lett. 7, 513–519.

Parver, L.M., Auker, C., Carpenter, D.O., 1980. Choroidal blood-flow as a heat dissipating mechanism in the macula. Am J. Ophthalmol. 89, 641–646.

Peachey, N.S., Chow, A.Y., 1999. Subretinal implantation of semiconductor-based photodiodes: progress and challenges. J. Rehabil. Res. Dev. 36, 371–376.

Pezaris, J.S., Eskandar, E.N., 2009. Getting signals into the brain: visual prosthetics through thalamic microstimulation. Neurosurg. Focus. 27, E6.

Pezaris, J.S., Reid, R.C., 2007. Demonstration of artificial visual percepts generated through thalamic microstimulation. Proc. Natl. Acad. Sci. 104, 7670–7675.

Pezaris, J.S., Reid, R.C., 2009. Simulations of electrode placement for a thalamic visual prosthesis. IEEE Trans. Biomed. Eng. 56, 172–178.

Polikov, V.S., Tresco, P.A., Reichert, W.M., 2005. Response of brain tissue to chronically implanted neural electrodes. J. Neurosci. Methods 148, 1–18.

RamRakhyani, A.K., Mirabbasi, S., Chiao, M., 2011. Design and optimization of resonance-based efficient wireless power delivery systems for biomedical implants. IEEE Trans. Biomed. Circ. Syst. 5, 48–63.

Rasouli, M., Phee, L.S.J., 2010. Energy sources and their development for application in medical devices. Expert Rev. Med. Device 7, 693–709.

Raz-Prag, D., Weinberger, D., Gefen, R.A., 2014. Implantation procedure for retinal prosthesis: adaptation of extracapsular cataract procedure. Investig. Ophthalmol. Vis. Sci. 55, 1809.

Rizzo, J.F., 2011. Update on retinal prosthetic research: the Boston retinal implant project. J. Neuroophthalmol. 31, 160–168.

Rizzo III, J.F., Ayton, L.N., 2014. Psychophysical testing of visual prosthetic devices: a call to establish a multi-national joint task force. J. Neural Eng. 11, 020301.

Rizzo, J.F., Wyatt, J., Loewenstein, J., Kelly, S., Shire, D., 2003. Perceptual efficacy of electrical stimulation of human retina with a microelectrode array during short-term surgical trials. Investig. Ophthalmol. Vis. Sci. 44, 5362–5369.

Roessler, G., Laube, T., Brockmann, C., Kirschkamp, T., Mazinani, B., Goertz, M., Koch, C., Krisch, I., Sellhaus, B., Trieu, H.K., Weis, J., Bornfeld, N., Rothgen, H., Messner, A., Mokwa, W., Walter, P., 2009. Implantation and explantation of a wireless epiretinal retina implant device: observations during the EPIRET3 prospective clinical trial. Investig. Ophthalmol. Vis. Sci. 50, 3003–3008.

Romero-Ortega, M., Kanneganti, A., Bendale, G., Seifert, J., Bredeson, S., Troyk, P., Deku, F., Cogan, S., 2015. In: Chronic and low charge injection wireless intraneural stimulation in vivo.2015 37th Annual International Conference of the IEEE Engineering in Medicine and Biology Society (EMBC), pp. 1013–1016.

Sabbah, N., Authié, C.N., Sanda, N., Mohand-Said, S., Sahel, J.-A., Safran, A.B., 2014. Importance of eye position on spatial localization in blind subjects wearing an argus ii retinal prosthesiseye position, localization, and retinal prosthesis. Investig. Ophthalmol. Vis. Sci. 55, 8259–8266.

Sahel, J., Mohand-Said, S., Stanga, P., Caspi, A., Greenberg, R., 2013. Acuboost™: enhancing the maximum acuity of the Argus II retinal prosthesis system. Investig. Ophthalmol. Vis. Sci. 54, 1389.

Sailer, H., Shinoda, K., Blatsios, G., Kohler, K., Bondzio, L., Zrenner, E., Gekeler, F., 2007. Investigation of thermal effects of infrared lasers on the rabbit retina: a study in the course of development of an active subretinal prosthesis. Graefes Arch. Clin. Exp. Ophthalmol. 245, 1169–1178.

Sakaguchi, H., Fujikado, T., Fang, X., Kanda, H., Osanai, M., Nakauchi, K., Ikuno, Y., Kamei, M., Yagi, T., Nishimura, S., 2004a. Transretinal electrical stimulation with a suprachoroidal multichannel electrode in rabbit eyes. Jpn. J. Ophthalmol. 48, 256–261.

Sakaguchi, H., Fujikado, T., Kanda, H., Osanai, M., Fang, X., Nakauchi, K., Ikuno, Y., Kamei, M., Ohji, M., Yagi, T., 2004b. Electrical stimulation with a needle-type electrode inserted into the optic nerve in rabbit eyes. Jpn. J. Ophthalmol. 48, 552–557.

Sakaguchi, H., Kamei, M., Fujikado, T., Yonezawa, E., Ozawa, M., Cecilia-Gonzalez, C., Ustariz-Gonzalez, O., Quiroz-Mercado, H., Tano, Y., 2009. Artificial vision by direct optic nerve electrode (AV-DONE) implantation in a blind patient with retinitis pigmentosa. J. Artif. Organs 12, 206–209.

Sakaguchi, H., Kamei, M., Nishida, K., Terasawa, Y., Fujikado, T., Ozawa, M., Nishida, K., 2012. Implantation of a newly developed direct optic nerve electrode device for artificial vision in rabbits. J. Artif. Organs 15, 295–300.

Samba, R., Herrmann, T., Zeck, G., 2015. PEDOT–CNT coated electrodes stimulate retinal neurons at low voltage amplitudes and low charge densities. J. Neural Eng.. 12, 016014.

Santos, A., Humayun, M.S., Dejuan, E., Greenburg, R.J., Marsh, M.J., Klock, I.B., Milam, A.H., 1997. Preservation of the inner retina in retinitis pigmentosa—a morphometric analysis. Arch. Ophthalmol. 115, 511–515.

Saunders, A.L., Williams, C.E., Heriot, W., Briggs, R., Yeoh, J., Nayagam, D.A., Mccombe, M., Villalobos, J., Burns, O., Luu, C.D., Ayton, L.N., Mcphedran, M., 2014. Development of a surgical procedure for implantation of a prototype suprachoroidal retinal prosthesis. Clin. Exp. Ophthalmol. 42, 665–674.

Schimitzek, H., Roessler, G., Walter, P., 2016. Clinical results after implantation of epiretinal visual prostheses. Klin. Monbl. Augenheilkd. 233, 1227.

Schmidt, E., Bak, M., Hambrecht, F., Kufta, C., O'Rourke, D., Vallabhanath, P., 1996. Feasibility of a visual prosthesis for the blind based on intracortical micro stimulation of the visual cortex. Brain 119, 507–522.

Schuettler, M., Stiess, S., King, B.V., Suaning, G.J., 2005. Fabrication of implantable microelectrode arrays by laser cutting of silicone rubber and platinum foil. J. Neural Eng. 2, S121–S128.

Schwahn, H.N., Gekeler, F., Kohler, K., Kobuch, K., Sachs, H.G., Schulmeyer, F., Jakob, W., Gabel, V.P., Zrenner, E., 2001. Studies on the feasibility of a subretinal visual prosthesis: data from Yucatan micropig and rabbit. Graefes Arch. Clin. Exp. Ophthalmol. 239, 961–967.

Schwarz, D.A., Lebedev, M.A., Hanson, T.L., Dimitrov, D.F., Lehew, G., Meloy, J., Rajangam, S., Subramanian, V., Ifft, P.J., Li, Z., 2014. Chronic, wireless recordings of large scale brain activity in freely moving rhesus monkeys. Nat. Methods 11, 670.

Shire, D.B., Kelly, S.K., Chen, J., Doyle, P., Gingerich, M.D., Cogan, S.F., Drohan, W.A., Mendoza, O., Theogarajan, L., Wyatt, J.L., Rizzo, J.F., 2009. Development and implantation of a minimally invasive wireless subretinal neurostimulator. IEEE Trans. Biomed. Eng. 56, 2502–2511.

Shivdasani, M.N., Luu, C.D., Cicione, R., Fallon, J.B., Allen, P.J., Leuenberger, J., Suaning, G.J., Lovell, N.H., Shepherd, R.K., Williams, C.E., 2010. Evaluation of stimulus parameters and electrode geometry for an effective suprachoroidal retinal prosthesis. J. Neural Eng. 7, 036008.

Shivdasani, M.N., Fallon, J.B., Luu, C.D., Cicione, R., Allen, P.J., Morley, J.W., Williams, C.E., 2012. Visual cortex responses to single- and simultaneous multiple-electrode stimulation of the retina: implications for retinal prostheses. Investig. Ophthalmol. Vis. Sci. 53, 6291–6300.

Shoval, A., ADAMS, C., David-Pur, M., Shein, M., Hanein, Y., Sernagor, E., 2009. Carbon nanotube electrodes for effective interfacing with retinal tissue. Front. Neuroeng. 2, 4.

Sinclair, N.C., Shivdasani, M.N., Perera, T., Gillespie, L.N., Mcdermott, H.J., Ayton, L.N., Blamey, P.J., 2016. The appearance of phosphenes elicited using a suprachoroidal retinal prosthesisphosphenes of a suprachoroidal retinal prosthesis. Investig. Ophthalmol. Vis. Sci. 57, 4948–4961.

Spencer, T.C., Fallon, J.B., Thien, P.C., Shivdasani, M.N., 2016. Spatial restriction of neural activation using focused multipolar stimulation with a retinal prosthesisspatial restriction of neural activation. Investig. Ophthalmol. Vis. Sci. 57, 3181–3191.

Srivastava, N., Troyk, P., 2006a. In: A proposed intracortical visual prosthesis image processing system.27th Annual International Conference of the IEEE Engineering in Medicine and Biology Society, 2005. IEEE-EMBS 2005, pp. 5264–5267.

Srivastava, N., Troyk, P., 2006b. In: Some solutions to technical hurdles for developing a practical intracortical visual prosthesis device.28th Annual International Conference of the IEEE Engineering in Medicine and Biology Society, 2006. EMBS'06, pp. 2936–2939.

Stensaas, S.S., Eddington, D.K., Dobelle, W.H., 1974. The topography and variability of the primary visual cortex in man. J. Neurosurg. 40, 747–755.

Stevens, G.A., White, R.A., Flaxman, S.R., Price, H., Jonas, J.B., Keeffe, J., Leasher, J., Naidoo, K., Pesudovs, K., Resnikoff, S., 2013. Global prevalence of vision impairment and blindness: magnitude and temporal trends, 1990–2010. Ophthalmology 120, 2377–2384.

Stingl, K., Bach, M., Bartz-Schmidt, K.U., Braun, A., Bruckmann, A., Gekeler, F., Greppmaier, U., Hörtdörfer, G., Kusnyerik, A., Peters, T., Wilhelm, B., Wilke, R., Zrenner, E., 2013a. Safety and efficacy of subretinal visual implants in humans: methodological aspects. Clin. Exp. Optom. 96, 4–13.

Stingl, K., Bartz-Schmidt, K.U., Besch, D., Braun, A., Bruckmann, A., Gekeler, F., Greppmaier, U., Hipp, S., Hortdorfer, G., Kernstock, C., Koitschev, A., Kusnyerik, A., Sachs, H., Schatz, A., Stingl, K.T., Peters, T., Wilhelm, B., Zrenner, E., 2013b. Artificial vision with wirelessly powered subretinal electronic implant alpha-IMS. Proc. R. Soc. B Biol. Sci. 280, 1–8.

Stingl, K., Gekeler, F., Bartz-Schmidt, K.U., Kögel, A., Zrenner, E., Gelisken, F., 2013c. Fluorescein angiographic findings in eyes of patients with a subretinal electronic implant. Curr. Eye Res. 38, 588–596.

Stingl, K., Bartz-Schmidt, K.U., Besch, D., Chee, C.K., Cottriall, C.L., Gekeler, F., Groppe, M., Jackson, T.L., Maclaren, R.E., Koitschev, A., Kusnyerik, A., Neffendorf, J., Nemeth, J., Naeem, M.A.N., Peters, T., Ramsden, J.D., Sachs, H., Simpson, A., Singh, M.S., Wilhelm, B., Wong, D., Zrenner, E., 2015. Subretinal visual implant alpha IMS—clinical trial interim report. Vis. Res. 111, 149–160.

Stone, J.L., Barlow, W.E., Humayun, M.S., Dejuan, E., Milam, A.H., 1992. Morphometric analysis of macular photoreceprors and ganglion-cells in retinas with retinis-pigmentosa. Arch. Ophthalmol. 110, 1634–1639.

Suaning, G.J., Hallum, L.E., Preston, P.J., Lovell, N.H., 2004. In: Efficient multiplexing method for addressing large numbers of electrodes in a visual neuroprosthesis. Proceedings of the 26th Annual International Conference of the IEEE Engineering in Medicine and Biology Society. IEEE, pp. 4174–4177.

Suaning, G.J., Schuettler, M., Ordonez, J.S., Lovell, N.H., 2007. In: Fabrication of multilayer, high-density micro-electrode arrays for neural stimulation and bio-signal recording. Proceedings of the 3rd International IEEE/EMBS Conference on Neural Engineering. IEEE, pp. 5–8.

Suaning, G.J., Lovell, N.H., Lehmann, T., 2014. In: Neuromodulation of the retina from the suprachoroidal space: the Phoenix 99 implant. Biomedical Circuits and Systems Conference (BioCAS). IEEE, pp. 256–259.

Sun, J., Lu, Y., Cao, P., Li, X., Cai, C., Chai, X., Ren, Q., Li, L., 2011. Spatiotemporal properties of multipeaked electrically evoked potentials elicited by penetrative optic nerve stimulation in rabbits. Investig. Ophthalmol. Vis. Sci. 52, 146–154.

Sun, J., Chen, Y., Chai, X., Ren, Q., Li, L., 2013. Penetrating electrode stimulation of the rabbit optic nerve: parameters and effects on evoked cortical potentials. Graefes Arch. Clin. Exp. Ophthalmol. 251, 2545–2554.

Tassicker, G., 1956. Preliminary report on a retinal stimulator. Br. J. Physiol. Optics 13, 102–105.

Tawakol, O., Bredeson, S.D., Troyk, P.R., 2016. In: Preparation of a neural electrode implantation device for in-vivo surgical use.2016 IEEE 38th Annual International Conference of the IEEE Engineering in Medicine and Biology Society (EMBC), pp. 4507–4510.

Troyk, P., Bradley, D., Bak, M., Cogan, S., Erickson, R., Hu, Z., Kufta, C., Mccreery, D., Schmidt, E., Sung, S., 2006. Intracortical visual prosthesis research-approach and progress. 27th Annual International Conference of the Engineering in Medicine and Biology Society. IEEE-EMBS 2005. IEEE, pp. 7376–7379.

Twyford, P., Cai, C., Fried, S., 2014. Differential responses to high-frequency electrical stimulation in ON and OFF retinal ganglion cells. J. Neural Eng. 11, 025001.

Veraart, C., Raftopoulos, C., Mortimer, J.T., Delbeke, J., Pins, D., Michaux, G., Vanlierde, A., Parrini, S., Wanet-Defalque, M.-C., 1998. Visual sensations produced by optic nerve stimulation using an implanted self-sizing spiral cuff electrode. Brain Res. 813, 181–186.

Veraart, C., Wanet-Defalque, M.C., Gerard, B., Vanlierde, A., Delbeke, J., 2003. Pattern recognition with the optic nerve visual prosthesis. Artif. Organs 27, 996–1004.

Villalobos, J., Allen, P.J., Mccombe, M.F., Ulaganathan, M., Zamir, E., Ng, D.C., Shepherd, R.K., Williams, C.E., 2012. Development of a surgical approach for a wide-view suprachoroidal retinal prosthesis: evaluation of implantation trauma. Graefes Arch. Clin. Exp. Ophthalmol. 250, 399–407.

Villalobos, J., Nayagam, D.A.X., Allen, P.J., Mckelvie, P., Luu, C.D., Ayton, L.N., Freemantle, A.L., Mcphedran, M., Basa, M., Mcgowan, C.C., Shepherd, R.K., Williams, C.E., 2013. A wide-field suprachoroidal retinal prosthesis is stable and well tolerated following chronic implantation. Investig. Ophthalmol. Vis. Sci. 54, 3751–3762.

Vurro, M., Crowell, A.M., Pezaris, J.S., 2014. Simulation of thalamic prosthetic vision: reading accuracy, speed, and acuity in sighted humans. Front. Hum. Neurosci. 8, 1–14.

Watson, M., Dancause, N., Sawan, M., 2016. Intracortical microstimulation parameters dictate the amplitude and latency of evoked responses. Brain Stimul. 9, 276–284.

Wiesel, T.N., Hubel, D.H., 1966. Spatial and chromatic interactions in the lateral geniculate body of the Rhesus Monkey. Notes 27, 47.

Wilke, R.G., Gabel, V.P., Sachs, H., Schmidt, K.U.B., Gekeler, F., Besch, D., Szurman, P., stett, A., Wilhelm, B., Peters, T., Harscher, A., Greppmaier, U., Kibbel, S., Benav, H., Bruckmann, A., Stingl, K., Kusnyerik, A., Zrenner, E., 2011. Spatial resolution and perception of patterns mediated by a subretinal 16-electrode array in patients blinded by hereditary retinal dystrophies. Investig. Ophthalmol. Vis. Sci. 52, 5995–6003.

Winter, J.O., Liu, T.Y., Korgel, B.A., Schmidt, C.E., 2001. Recognition molecule directed interfacing between semiconductor quantum dots and nerve cells. Adv. Mater. 13, 1673–1677.

Wong, Y.T., Chen, S.C., Kerdraon, Y.A., Allen, P.J., Mccombe, M.F., Morley, J.W., Lovell, N.H., Suaning, G.J., 2008. Efficacy of supra-choroidal, bipolar, electrical stimulation in a vision prosthesis. Proceedings of the 30th Annual International Conference of the IEEE Engineering in Medicine and Biology Society. IEEE, pp. 1789–1792.

Wong, Y.T., Chen, S.C., Seo, J.M., Morley, J.W., Lovell, N.H., Suaning, G.J., 2009. Focal activation of the feline retina via a suprachoroidal electrode array. Vis. Res. 49, 825–833.

Wong, W.L., Su, X., Li, X., Cheung, C.M.G., Klein, R., Cheng, C.-Y., Wong, T.Y., 2014. Global prevalence of age-related macular degeneration and disease burden projection for 2020 and 2040: a systematic review and meta-analysis. Lancet Glob. Health 2, e106–e116.

Wright, A.F., Chakarova, C.F., El-Aziz, M.M.A., Bhattacharya, S.S., 2010. Photoreceptor degeneration: genetic and mechanistic dissection of a complex trait. Nat. Rev. Genet. 11, 273–284.

Yan, Y., Lu, Y., Li, M., Ma, Z., Cao, P., Chen, Y., Sun, X., Chai, X., Ren, Q., Li, L., 2016. Electrically evoked responses in the rabbit cortex induced by current steering with penetrating optic nerve electrodescurrent steering with penetrating optic nerve electrodes. Investig. Ophthalmol. Vis. Sci. 57, 6327–6338.

Yanai, D., Weiland, J.D., Mahadevappa, M., Greenberg, R.J., Fine, I., Humayun, M.S., 2007. Visual performance using a retinal prosthesis in three subjects with retinitis pigmentosa. Am J. Ophthalmol. 143, 820–827.

Yanovitz, L., Raz-Prag, D., Eden, K.-O., Saini, R., Hanein, Y., Gefen, R.A., 2014. Retinal responses evoked and recorded with 3D electrodes designated for a novel prosthetic device. Investig. Ophthalmol. Vis. Sci. 55, 1808.

Yue, L., Falabella, P., Christopher, P., Wuyyuru, V., Dorn, J., Schor, P., Greenberg, R.J., Weiland, J.D., Humayun, M.S., 2015. Ten-year follow-up of a blind patient chronically implanted with epiretinal prosthesis Argus I. Ophthalmology 122, 2545–2552.

Zapf, M.P.H., Boon, M.-Y., Matteucci, P.B., Lovell, N.H., Suaning, G.J., 2015. Towards an assistive peripheral visual prosthesis for long-term treatment of retinitis pigmentosa: evaluating mobility performance in immersive simulations. J. Neural Eng.. 12036001.

Zhou, J.A., Woo, S.J., Park, S.I., Kim, E.T., Seo, J.M., Chung, H., Kim, S.J., 2008. A suprachoroidal electrical retinal stimulator design for long-term animal experiments and in vivo assessment of its feasibility and biocompatibility in rabbits. J. Biomed. Biotechnol. 2008, 1–10. Article ID 547428.

Zrenner, E., Bartz-Schmidt, K.U., Benav, H., Besch, D., Bruckmann, A., Gabel, V.P., Gekeler, F., Greppmaier, U., Harscher, A., Kibbel, S., Koch, J., Kusnyerik, A., Peters, T., Stingl, K., Sachs, H., Stett, A., Szurman, P., Wilhelm, B., Wilke, R., 2011. Subretinal electronic chips allow blind patients to read letters and combine them to words. Proc. R. Soc. B Biol. Sci. 278, 1489–1497.

CHAPTER ELEVEN

The Artificial Pancreas

Graham Brooker
Australian Centre for Field Robotics, University of Sydney, Sydney, NSW, Australia

Contents

1 INTRODUCTION

Before embarking on a review of the history and the technology that has led to the advent of an artificial pancreas (AP), it is important to understand the natural processes involved in glucose balance, and how it goes wrong in people suffering from diabetes.

Glucose is mainly produced by the liver after which it is distributed and utilized in both insulin-independent (e.g., central nervous system and red blood cells) and insulin-dependent (muscle and adipose) tissues. Insulin is secreted by pancreatic beta-cells and enters the blood stream after liver

Handbook of Biomechatronics
https://doi.org/10.1016/B978-0-12-812539-7.00015-5

degradation following which it is cleared from the blood stream, primarily by the kidneys. The glucose and insulin systems interact using feedback control signals. For example, if a glucose perturbation occurs after eating, beta-cells secrete more insulin in response to increased glucose concentration in the blood. In turn insulin signaling promotes glucose utilization and inhibits glucose production so as to return blood glucose to normal levels. These control interactions are usually referred to as insulin sensitivity and beta-cell responsivity, respectively.

In type 2 diabetes, this degradation starts as prediabetes which is characterized by a slow deterioration of both insulin sensitivity and beta-cell responsivity. In contrast, in type 1 diabetes, beta-cells rapidly become virtually silent and insulin must be administered by the patient in an attempt to avoid hyperglycemia. However, insulin treatment risks potentially fatal hypoglycemia and thus people with type 1 diabetes face a life-long optimization problem: to maintain strict glycemic control and reduce hyperglycemia, without increasing their risk of hypoglycemia.

Blood glucose level is both the measurable result of this optimization and the principal feedback signal to the patient or an AP for the control of diabetes (Cobelli et al., 2009).

The ultimate objective of an AP is to replace that function of the natural pancreas, which is lost in diabetics. It combines the accurate measurement of blood glucose levels with a smart insulin dispenser to maintain the correct glucose balance irrespective of external disturbances.

2 HISTORICAL BACKGROUND

The diabetes epidemic of the late 20th and early 21st century would presuppose that diseases related to pancreatic function were a modern phenomenon. However that is not the case, and since the dawn of civilization documents have referred to people with diabetic-like symptoms and medicinal remedies for the problem. The earliest known record is found in the Ebers Papyrus, a collection of medical texts from Egypt, dated to about 1550 BCE now kept in the library of the University of Leipzig. This papyrus preserved an extensive record of Egyptian medical history including references to anatomy, physiology, and toxicology along with treatments and prescriptions for myriad common ailments (Crystalinks). A single reference to the symptoms of diabetes stated "...to eliminate urine which is too plentiful." Unfortunately this reference is ambiguous with the crucial word, *asha* meaning either plentiful or often, and so could relate to cystitis rather than diabetes (Loriaux, 2006).

At around the same time, ancient Hindu writings noted that ants were attracted to the urine of people suffering from a mysterious emaciating disease, but it was only in about 500 BCE that the first references to sugar in the urine of obese people were made (Swidorski, 2014).

Both Galen and Hippocrates refer to periods of study at the temple of Imhotep in Memphis but neither mentions any symptoms consistent with diabetes in their subsequent writings. In fact, the first-known clear reference to the disease was by Aretaeus of Cappadocia (AD 129–199), a Greek physician who introduced the term "diabetes" from the Greek word for "syphon" as he noted that it caused a constant flow of urine. The reason for this increased flow is due to the kidneys trying to flush excess glucose from the blood. It is worth quoting Aretaeus in full as his observations accurately reflect the full horror of the disease if left untreated.

> *Diabetes is a wonderful affliction, not very frequent among men being a melting down of the flesh and limbs into urine. The patients never stop making water, but the flow is incessant, as if the opening of aqueducts. Life is too short, disgusting, and painful, thirst unquenchable, excessive drinking, which, however, is disproportionate to the large quantity of urine, for more urine is passed; and one cannot stop them either from drinking or making water, or, if for a time they abstain from drinking, their mouth becomes parched and their body dry, the viscera seems as if scorched up; they are affected with nausea, restlessness and burning thirst, and in no distant term they expire*
>
> ***(Adams, 1856).***

For the next two millennia, diagnosis of diabetes was most often made by "water tasters" who drank the urine of those suspected of having diabetes to determine whether it tasted sweet. Mellitus, the Latin word for honey, was later added to the term "diabetes" for this reason. This diagnostic method was formalized by Matthew Dobson in 1776 when he found a substance like sugar in appearance and taste when diabetic urine evaporated. He also noted a sweetish taste of sugar in the blood of diabetics and that diabetes could be fatal in <5 weeks in some cases while in others it was a chronic condition. This was the first time that a distinction is made between type 1 and type 2 diabetes (Macfarlane, 2014).

In 1848, Claude Bernard discovered that glycogen was formed in the liver by identifying different concentrations of the substance in the portal and hepatic veins. He speculated that it was the same sugar that was found in the urine of diabetics. This was the first time that glycogen metabolism was identified and linked to diabetes (Tattersall, 2009). Bernard also explored the significance of the pancreas in diabetes and noted that ligature of the pancreatic ducts was not associated with diabetes (Leiva-Hidalgo et al., 2011).

During the second half of the 19th century a number of important discoveries relating to diabetes occurred. These included observations by Paul Langerhans of two different cell types in the pancreas of which one produced normal pancreatic fluid but the function of the other was unknown. These cells were later discovered to produce insulin and named the "Islets of Langerhans" in his honor. A few years later, two researchers from the University of Strasbourg, Joseph von Mering and Oskar Minkowski, removed the pancreas from a dog to identify its effect on digestion, but instead discovered that the dog developed diabetes. A year later Minkowski administered dried pancreas and later injected pancreatic extracts subcutaneously without obtaining an effective response from the dogs (Leiva-Hidalgo et al., 2011).

During the first two decades of the 20th century, the relation between diet and diabetes was thoroughly investigated. A number of researchers including Rennie and Frazer from the Aberdeen Royal Infirmary and Georg Zuelzer in Berlin investigated the effects of both oral and injected pancreatic extracts with Zuelzer having some success in his treatment of patients. However, during these trials many animals and some people suffered from serious side effects including fever, convulsions, as well as shivering and sweating caused by hypoglycemic shock. Consequently, human tests were discontinued for a time. Further research conducted by Israel Kleiner and Nicolae Paulescu confirmed the effects of pancreatic extracts on the glucose levels in dogs that had had their pancreases surgically removed. Paulsecu's research describing the isolation of "pancreatine" was only published in 1920 at about the same time that Frederick Banting apparently conceived of the idea of insulin after reading Moses Barron's paper relating the Islets of Langerhans to diabetes. He and a number of other researchers also experimented with different pancreatic extracts on the health of de-pancreatized dogs. Banting and John Macleod were awarded the 1922 Nobel Prize for Physiology and Medicine for creating usable insulin. Paulescu wrote to the Nobel Prize committee claiming that he had discovered and used insulin first, but his claims were rejected.

Banting and Macleod were not successful in producing useful pancreatic extracts for their experiments until they were joined by James Collip who developed a successful process for extracting insulin using a double precipitation process in ethanol. His extract was tested on a 14-year-old child and appeared to be successful over the following months. This and successful tests on more diabetic patients lead to Eli Lilly and the University of Toronto collaborating on the production of large quantities of insulin for the

North American market. The Toronto group called their substance insulin and Eli Lilly called their product iletin (Leiva-Hidalgo et al., 2011).

3 BLOOD SUGAR MONITORING

In 1841, Karl Trommer developed a qualitative test for sugar, which involved treating a urine sample with a strong acid that resulted in the conversion of disaccharides into monosaccharides. This solution was then neutralized and a solution of copper sulfate and further alkali added. After boiling, a brick-red cuprous oxide precipitate formed if glucose was present. A decade later Hermann von Fehling developed a qualitative test based on this work (Kirchhof et al., 2008).

The first semiquantitative urine glucose test was devised in 1907 by Stanley Benedict and this remained the only method to monitor glucose levels in diabetic patients for many years. By 1925 home testing for glucose in urine was possible using Benedict's reagent generally provided by the family doctor. The principle of Benedict's test is that when reducing sugars are heated in the presence of an alkali they are converted to powerful reducing agents called enediols. In turn, these reduce the cupric compounds ($Cu^{2+)}$ present in the reagent to cuprous compounds (Cu^{+}) which precipitate as insoluble copper oxide (Cu_2O). The color of the precipitate provides a reasonably quantitative indication of the glucose concentration.

In the home tests, eight drops of urine were mixed with 6 cm^3 of Benedict's reagent in a test tube. The tube was placed in boiling water for 5 min after which a color change occurred and a color chart could be used to determine the amount of glucose present. A greenish color indicated a small amount of sugar, while increasing sugar concentrations were indicated by yellow (moderate) and finally red/orange precipitates indicating a high concentration. This test process was later simplified by providing an effervescent tablet that could be dropped into a test tube containing urine. By 1956 dry test strips using the glucose oxidase (GOx) method had become available (Bronzino, 2006).

The measurement of glucose in urine is not a satisfactory method for the control of diabetes as only after the blood glucose level has been high for a number of hours does it spill over into the urine.

Initially, testing for glucose in blood required large volumes of the liquid. However, micromethods initially developed by Ivar Bang and soon improved by Lewis and Benedict in 1915 and in 1918 by Hagedorn and Jensen became available. Bang's test functioned by fixing blood proteins to filter paper before measuring glucose in the precipitate using copper sulfate and potassium chloride. A description of the more complex but extremely sensitive Hagedorn-Jensen method that used only $0.1\,cm^3$ of blood is described by Miller and Van-Slyke (1936). These tests remain extremely complicated in their execution and were not amenable to at-home testing.

In 1941, the Ames company introduced the first colorimetric strip test (Clinitest) using the old copper sulfate reduction method. Shortly thereafter, the company introduced the more accurate Clinistix test based on the enzymatic reaction of GOx. In this test, the enzyme generates hydrogen peroxide (H_2O_2) when it reacts with the glucose, which in turn reacts with horseradish peroxidase to produce oxygen that then oxidizes orthotoluidine to produce a blue or purple color (Kirchhof et al., 2008). Whole blood contains hemoglobin that will interfere with the measurement color of a test strip, so to avoid staining the strip with red blood cells, an ethyl cellulose layer was applied over the enzyme and dye-impregnated paper substrate. To maintain stiffness, this was in turn attached to a plastic support (Bronzino, 2006).

Clinic-based colorimetric strips that doctors could use to measure their patient's blood sugar levels became available in the late 1960s. Plastic Dextrostix use enzymatic reaction process developed by the Ames company, as described by Rennie et al. (1964) and could provide measurements in 1 min. Like earlier urine-based tests, the strip changed color based on the amount of sugar in a drop of blood and its color was compared to a color chart. These strips needed to be washed to remove the blood cell residue to allow this comparison to be made. However, through various generations of products the formulation of the strips was slowly improved to eliminate the washing and wiping steps.

3.1 Automatic Glucose Concentration Measurement Using Colorimetric Strips and Optical Reflectance Meters

Visually comparing the color of a strip to a chart did not provide a very accurate measure of the glucose levels, particularly at the extremes of the blood-glucose spectrum, so methods to provide this function automatically were devised.

The first attempts at automation of the color identification process involved the use of a lightweight battery operated reflectance meter in conjunction with the Dextrostix. Using a stabilized light source, the meter measures the light reflected from the surface of the reacted reagent strip and converts this to a reading on a calibrated scale.

A trial consisting of 34 volunteers compared the performance of this technique with a reference Technicon AutoAnalyzer. Blood was drawn from the antecubital vein, allowed to clot and the serum removed for processing in the AutoAnalyzer while simultaneously finger-prick blood was applied directly to the strip, allowed to react for 60 s before the strip was washed and the color measured using the reflectance meter. These tests provided the first confirmation that the technique was sound, as can be seen from a comparison graph shown in Fig. 1, though slight differences between the values measured by the two techniques were noted at low and high glucose concentrations (Mazaferri et al., 1970).

Optically based strips are generally made up of various layers, each of which provides a specific function: support, reflective, analytical, and sample spreading, as illustrated in Fig. 2. The support function provides a foundation for the dry reagent and may also include the reflective material. Otherwise scattering or reflective components including TiO_2, $BaSO_4$, MgO, or ZnO, are added to the dry reagent. The analytical layer contains the active enzyme, while the spreading function must rapidly disperse the blood to cover the active layer as uniformly as possible. These have been made from

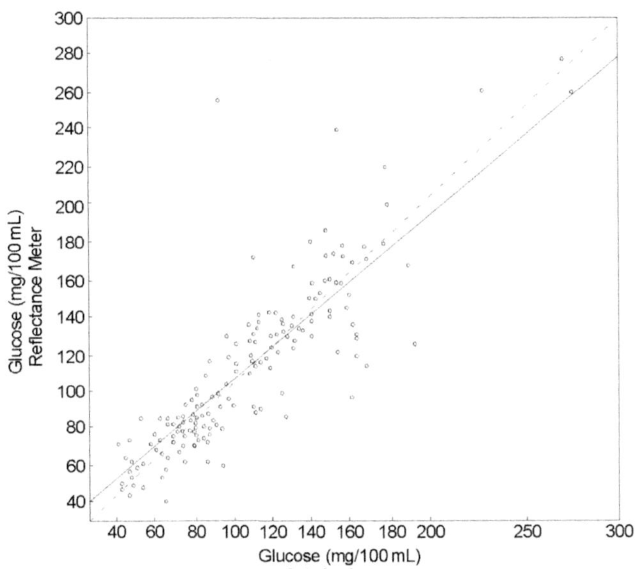

Fig. 1 Comparison between the blood glucose levels measured by an AutoAnalyzer and a Color Reflectance Meter. *(Data from Mazaferri, E., Skillman, T., Lanese, R., Keller, M., 1970. Use of test strips with colour meter to measure blood glucose. Lancet 265(7642): 331–333.)*

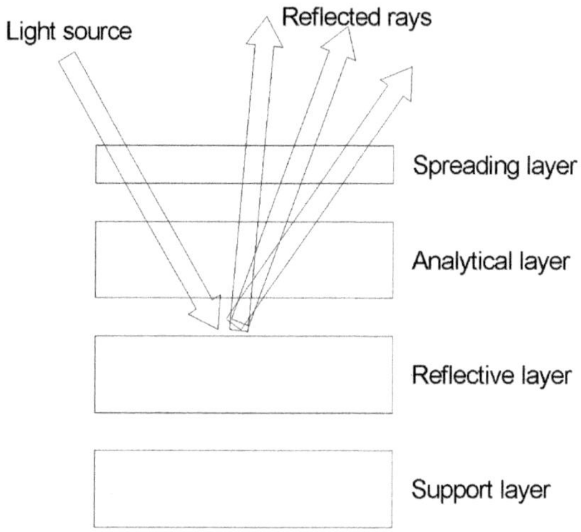

Fig. 2 Basic functions of the individual layers of a reflectance-based test strip. *(Based on Bronzino, J. (Ed.), 2006. Medical Devices and Systems, CRC Press.)*

swellable films or semipermeable membranes and commonly use glass-fiber wool to separate the plasma from the whole blood.

Once the reaction product is formed, the percentage of diffuse light reflected from the analytical layer of the strip, $R_{\%}$, decreases in accordance with the following equations.

$$R_{\%} = \frac{I_u}{I_s} R_s \tag{1}$$

where I_u is the reflected light intensity from the sample, I_s is the reflected light intensity from a standard reference, and R_s is the percentage reflectivity of the standard.

A more usual way of representing this relationship is to use the Kubelka-Munk (K-M) equation.

$$C \propto \frac{K}{S} = \frac{(1-R^2)}{2R} \tag{2}$$

where C is the concentration, K is the absorption coefficient, S is the scattering coefficient, and $R = R_{\%}/100$ (the percentage reflectance from the strip divided by 100).

The K-M transform of the measured reflectance is approximately proportional to the absorption coefficient and hence is approximately proportional to the concentration, assuming that the scattering coefficient is constant. In general the scattering coefficient is dominated by particle size and the refractive index of the sample so is not a strong function of the wavelength or the absorption coefficient. This allows it to be treated as a constant by the K-M model. Thus, to determine the concentration, the measured reflectance is just transformed by the K-M model, and scaled appropriately.

An example of the reaction used by the Lifescan ONETOUCH and the SureStep provides an indication of the complexity of the chemical process involved in these strips. GOx catalyzes the oxidation of glucose to form gluconic acid and H_2O_2. Blood oxygen concentration is much lower than that of the air, so oxygen from the air must diffuse into the strip to bring the reaction to completion. Peroxidase then catalyzes the reaction of the H_2O_2 with 3-methyl-2-benzothiazolinone hydrazone (MBTH) and 3-dimethyllaminobenzoic acid (DMAB) to form MBTH-DMAB that has an absorption peak at 635 nm. This, and similar reactions used by other manufacturers requires between 3 and 15 μL of blood, can be read using a reflectance meter within 10–30 s.

One of the confounding factors is the range of red blood cell concentrations (% hematocrit) found in whole blood. To overcome this problem, the ONETOUCH meter includes two LEDs operating at 635 and 700 nm, respectively. The absorbance peak at 635 nm is fairly sharp, so the absorbance by the dye at 700 nm is negligible. Therefore, the second reflectance measurement at 700 nm provides a correction factor in a band where oxyhemoglobin provides some absorbance. The two signals are multiplexed to allow independent measurements using the same photometer.

All photometers measure the reflectance as a function of time to ensure that the curve follows a prescribed path as the reaction progresses. Some correct for temperature that affects the reaction speed, as well.

Optical systems have slowly lost market share to biosensor (electrochemical)-based technologies as the latter require less blood, are faster and are easier to calibrate (Bronzino, 2006).

3.2 Biosensor-Based Glucose Monitoring

A biosensor is a device that uses a biologically derived sensitive element in association with a physiochemical transducer to recognize an organic molecule. The process to achieve this involves firstly differentiating target molecules from other chemicals and then implementing the physiochemical transducer and signal processor to convert the signal to a readable form. The molecular differentiation element can include receptors, enzymes, antibodies, nucleic acids, microorganisms, and lectins. There are five major transducer classes including electrochemical, optical, thermometric, piezoelectric, and magnetic, with most of the current glucose biosensors of the electrochemical type because of their good sensitivity, reproducibility, and easy maintenance (Yoo and Lee, 2010).

Electrochemical sensors can use current, voltage, or conductivity to measure concentration, with devices that measure current generated when electrons are exchanged between the biological system and the electrode being the most common for glucose monitoring. Transducers used for glucose measurements are based on interactions with one of three enzymes: hexokinase, GOx, or glucose-1-dehydrogenase (GDH), with GOx being the most common as it has a reasonably high affinity for glucose and can withstand more extreme conditions than the other transducer types. The process starts with the immobilized GOx catalyzing the oxidation of β-D-glucose by molecular oxygen to produce gluconic acid and H_2O_2. The H_2O_2 is in turn oxidized at a catalytic, platinum anode where electron flow (current) is proportional to the number of glucose molecules present in the blood sample (Yoo and Lee, 2010).

The basic principles for this technology were introduced in 1956 by Leland Clark Jr. in a paper on the oxygen (later Clarke) electrode. In 1962, he and Ann Lyons from the Cincinnati Children's Hospital developed the first glucose enzyme electrode. This glucose biosensor consisted of an oxygen electrode, an inner oxygen semipermeable membrane, a thin layer of GOx, and an outer dialysis membrane. Enzymes were immobilized on an electrochemical detector to form an enzyme electrode. A decrease in the measured oxygen concentration was proportional to the glucose concentration. This process was greatly simplified by Updike and Hicks who bound the GOx in a polyacrylamide gel on an oxygen electrode and showed for the first time that it was possible to measure the glucose concentration of biological fluids (Updike and Hicks, 1967).

This technology was first used commercially in 1975 in the Model 23A YSI analyzer developed by the Yellow Springs Instrument Company. It measured current flow in a platinum electrode. The 1980s saw significant efforts to reduce interference by other reactive molecules including uric and ascorbic acid as well as certain drugs. A further problem was the limited solubility of oxygen in biological fluids that resulted in fluctuations in the oxygen tension, known as the "oxygen deficit" (Yoo and Lee, 2010).

Mediator-based glucose biosensors overcame the oxygen deficit problem and allowed for the introduction of commercial screen-printed strips in the

1980s. Sensor performance was further improved with the use of modified electrodes and tailored membranes. The first electrochemical blood glucose monitor for self-monitoring of diabetic patients was pen-sized and was launched in 1987 as ExacTech by Medisense Inc. Its success led to a revolution in the health care of diabetic patients (Yoo and Lee, 2010).

In the process, the mediator is oxidized at a solid electrode with an applied positive potential. The rate of the electron transfer reaction is given by the Butler-Volmer equation. However, when the potential is sufficiently large, the mediator reaching the electrode reacts rapidly and the reaction becomes diffusion controlled. The current flow then follows the Cottrell equation (Bronzino, 2006).

$$i = \frac{nFAD^{1/2}C}{\sqrt{\pi t}} \tag{3}$$

where i (A) is the current, n is the number of electrons to oxidize one molecule of the mediator, F is Faraday's constant (96,485 C/mol), A (cm^2) is the electrode area, D (cm^2/s) is the diffusion coefficient, C (mol/cm^3) is the initial concentration, and t (s) is the time. The current will therefore decay with the square root of the time, which means that the strongest chemical signal occurs at the start, in contrast to the color reactions in which the color becomes more intense as the time increases. However, when a voltage is first applied to the electrodes, the dipole moments of the solvent molecules align with the electric field on the electrode and this provides an interfering current for a short period. The measurement process needs to wait until this has decayed before readings are taken.

The first commercial test strip from MediSense was based on the oxidation of ferrocene at 0.6 V. Unfortunately at this voltage it also oxidizes other molecules present in the blood such as ascorbic acid. This is corrected for by incorporating an additional electrode on the strip that does not contain any GOx. The newer generation strips function slightly differently with the glucose dehydrogenase (GDH) enzyme not reacting with oxygen, allowing the phenanthroline quinine (PQ) mediator to be oxidized at 0.2 V that is lower than the oxidation potential of most interfering substances. This reaction is described by the following equations (Bronzino, 2006).

$$\text{Glucose} + \text{GDH/NAD}^+ \rightarrow \text{GDH/NADH} + \text{gluconolactone}$$
$$\text{GDH/NADH} + \text{PQ} \rightarrow \text{GDH/NAD}^+ + \text{PQH}_2$$
$$\text{PQH}_2 \rightarrow \text{PQ} + \text{electrons (reaction at solid electrode surface)}$$

The package insert describes the test principle and composition for these strips, an example of which is the ACCU-CHEK Performa shown in Table 1. Calibration codes are often included with each batch of strips, and this must be entered manually in at-home testers. The more expensive meters used in hospitals include a barcode that can be read by the meter to facilitate automatic calibration.

User interaction has been simplified over the years to minimize the potential for errors in the sampling and testing process. One of the main problems was smearing a drop of blood on top of the strip, so modern strips use capillary action to draw blood from the end of the strip onto the active surface, as shown in Fig. 3. This small capillary space only requires about 300 nL of blood. Lancets have become thinner and sharper and are available in carrousels that are loaded into a pen-like device that automatically lances and then retracts the point. This controls the insertion depth and minimizes the pain from each finger prick while still providing sufficient blood for a measurement.

Alternate sites provide capillary blood with slightly different glucose and hematocrit values to the fingertip and need separate calibration. Because of the larger volume available to provide a blood sample, some alternate site testers extract blood by drawing a vacuum over a lanced site to increase the width of the lancet wound opening as the skin is stretched.

3.3 Glucose Meter Hardware

The hardware and software functionality for a typical glucose meter is discussed in relation to Maxim and Microchip components (DiCristina, 2017; Dalvi, 2013).

Table 1 Test Principle and Composition of the ACCU-CHEK Performa Strips

Results for blood glucose concentrations <75 mg/dL (<4.2 mmol/L)		
Within ±5 mg/dL (within ±0.28 mmol/L)	**Within ±10 mg/dL (within ±0.56 mmol/L)**	**Within ±15 mg/dL (within ±0.83 mmol/L)**
32/36 (88.9%)	34/36 (94.4%)	36/36 (100%)

Results for blood glucose concentrations ≥75 mg/dL (≥4.2 mmol/L)			
Within ±5%	**Within ±10%**	**Within ±15%**	**Within ±20%**
98/164 (59.8%)	150/164 (91.5%)	162/164 (98.8%)	163/164 (99.4%)

Repeatability (within-series imprecision): The mean imprecision is <3.5%. In a typical series of tests, a coefficient of variation of 3.3% was obtained.

Reproducibility (day-to-day imprecision): The mean imprecision is <1.7%. In a typical series of tests, a coefficient of variation of 1.6% was obtained.

Test principle: The enzyme on the test strip, mutant variant of quinoprotein glucose dehydrogenase (Mut. Q-GDH) from *Acinetobacter calcoaceticus*, recombinant in *E. coli*, converts the glucose in the blood sample to gluconolactone. This reaction creates a harmless DC electrical current that the meter interprets for the blood glucose result. The sample and environmental conditions are evaluated using AC and DC signals.

Reagent composition¤

Mediator	6.72%
Quinoprotein glucose dehydrogenase¤¤	15.27%
Pyrroloquinoline quinone	0.14%
Buffer	34.66%
Stabiliser	0.54%
Nonreactive ingredients	42.66%

¤Minimum at time of manufacture.

¤¤From *A. calcoaceticus*, recombinant in *E. coli*, detailed description in patent application WO 2007/118647.

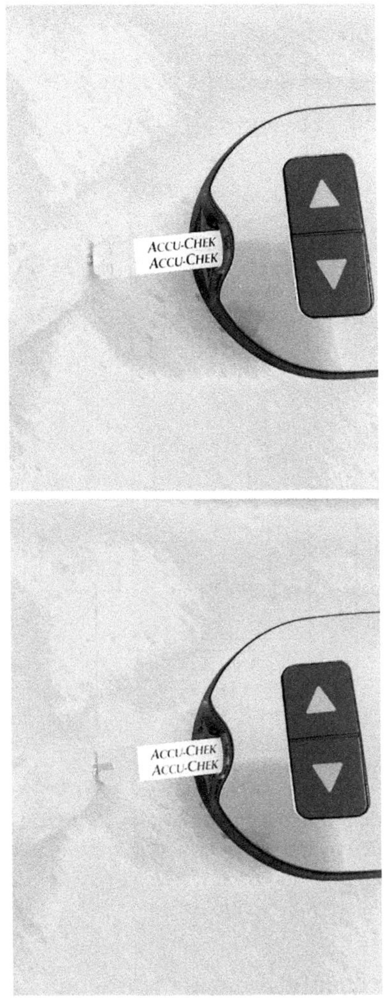

Fig. 3 Process of introducing blood into the capillary space of an ACCU-CHEK strip.

The electrochemical strips used by most glucose meters include an electrode that is provided with a bias voltage usually generated using a digital-to-analogue converter (DAC) as shown in Fig. 4. The current flow through the strip to a second electrode is proportional to the glucose concentration, and that current is generally fed into a transimpedance amplifier (current-to-voltage converter) followed by an analogue-to-digital converter. The amount of current is small, with the full-scale range between 10 and 50 μA, so sensitive, low noise circuitry is required. In addition, the

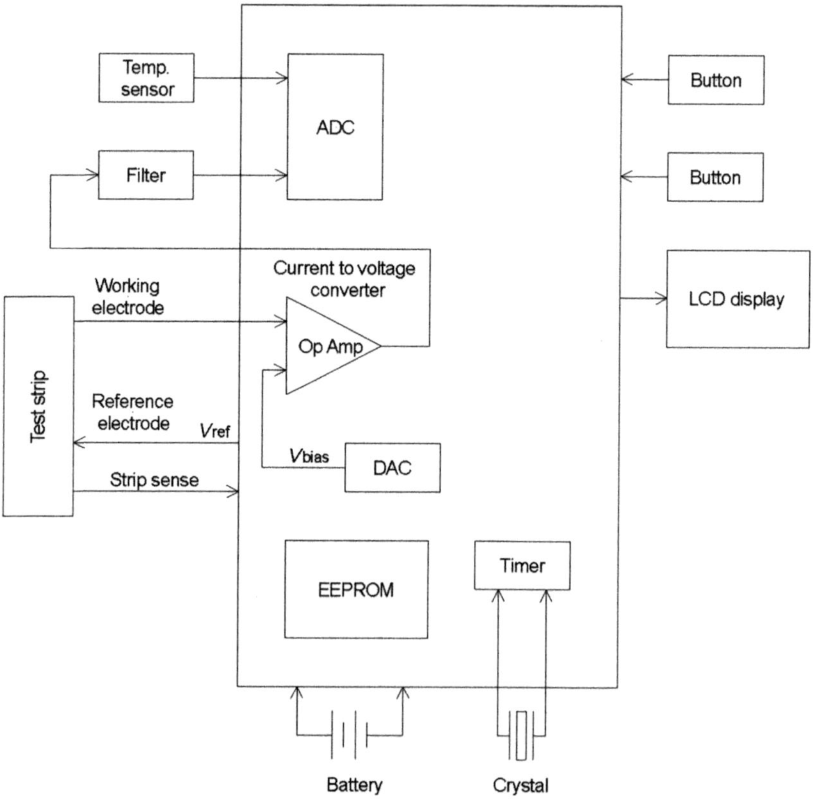

Fig. 4 Functional block diagram of a glucose meter. *(Based on DiCristina, J., 2017. Maxim Tutorial 4659: Blood Glucose Meters. https://www.maximintegrated.com/en/app-notes/index.mvp/id/4659 (Retrieved 10 August 2017); Dalvi, N., 2013. Microchip AN1560: Glucose Meter Reference Design, Microchip Technology.)*

temperature needs to be monitored as the reaction on the test strip is temperature dependent.

Peripheral functions including timing, sensing the strip presence, reading push buttons, and driving the LCD display are integral to the ease of use and the reliability of the hardware.

As discussed earlier, optical reflectometry test strips use color to determine the glucose concentration in blood. For this method, a pair of LEDs with different color characteristics are powered by a constant current source and energized alternately to flash onto the test strip. A photodiode generates a current proportional to the reflected light intensity that is fed into a transimpedance amplifier, as with the electrochemical sensors, to provide a voltage that is then digitized. The full-scale current from the photodiode ranges

Fig. 5 Photos of colorimetric and electrochemical glucose monitors.

from 1 to 5 μA with a resolution of <5 nA. The temperature of the strip is again required.

Examples of colorimetric and electrochemical glucose monitors are shown in Fig. 5.

Calibration is achieved by entering a code manually or by inserting a memory device from the package of test strips. An EPROM enables additional information to be transferred to the monitor, which can be useful. Some meters are "self-calibrating" either by applying tight manufacturing control, built-in calibration on each strip, or built-in calibration on a pack of test strips loaded into the meter.

3.3.1 Accuracy and Precision

Both reflective and electrochemical meters need to accurately measure currents in the nA range. This requires components with low leakage and minimal drift with changes in supply voltage, temperature, and time. For these applications, an opamp needs an ultralow bias current (typically <1 nA) as well as maintaining good linearity and stability when connected to a capacitive test strip (for the electrochemical option). The reference voltage needs a temperature coefficient of <50 ppm/°C, low drift with time and excellent line and load regulation. A 10- or 12-bit DAC is required to set the bias voltage for electrochemical test strips and to set the LED current in reflectance meters. A 14- or a 12-bit ADC with a programmable gain stage is required to provide the required dynamic range for reliable operation.

3.3.2 Temperature Measurement

The temperature of the blood on the test strip should be measured, but that is impractical so the ambient temperature near the strip is measured instead. This should be accurate to ±1°C and can be achieved using a thermometer IC or a thermistor and one of the ADC channels.

3.3.3 Electrochemical Test-Strip Configurations

Test strips are proprietary and can vary in the number of electrodes and bias method used. The simplest electronic configuration is for a self-biased strip and is shown in Fig. 6. It consists of two electrodes with the current measured at the working electrode (WE) and the common electrode (CE), grounded.

In this configuration, the opamp provides a bias reference voltage, V_B, to the test strip because the positive and negative inputs of the opamp will remain at the same potential due to its high gain. The configuration as an inverting transimpedance amplifier produces an output V_{OUT} given by the following equation, when the current flow, I_S, is as defined in the figure.

$$V_{OUT} = V_B + I_S \times R_S \tag{4}$$

The feedback capacitor C_S in conjunction with the feedback resistor R_S forms a low-pass filter to reduce the signal bandwidth, and hence the noise at the output of the amplifier.

A more advanced configuration is shown in Fig. 7. In this alternative configuration the current is measured at the WE using the standard transimpedance configuration described above, while a force-sense circuit based on a second opamp drives the CE and reference electrodes (REs). In this configuration, the bias voltage on the test strip is maintained more accurately irrespective of the current flow. The disadvantages of this design include the added complexity and the larger headroom needed to allow the force-sense amplifier to swing negative to maintain the bias voltage while current is flowing through the strip.

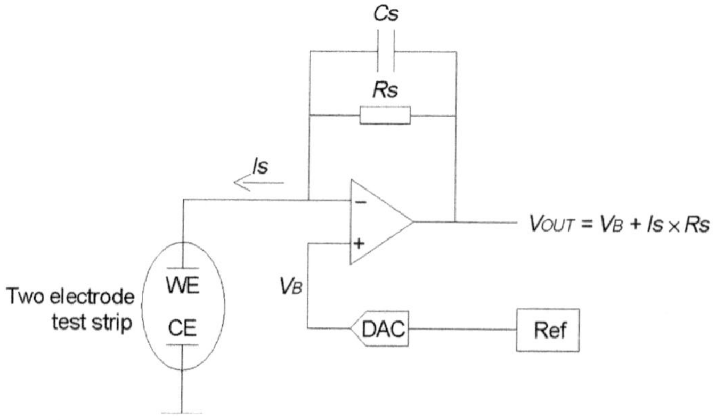

Fig. 6 Schematic diagram for a self-biased test-strip monitor. *(Based on DiCristina, J., 2017. Maxim Tutorial 4659: Blood Glucose Meters. https://www.maximintegrated.com/en/app-notes/index.mvp/id/4659 (Retrieved 10 August 2017).)*

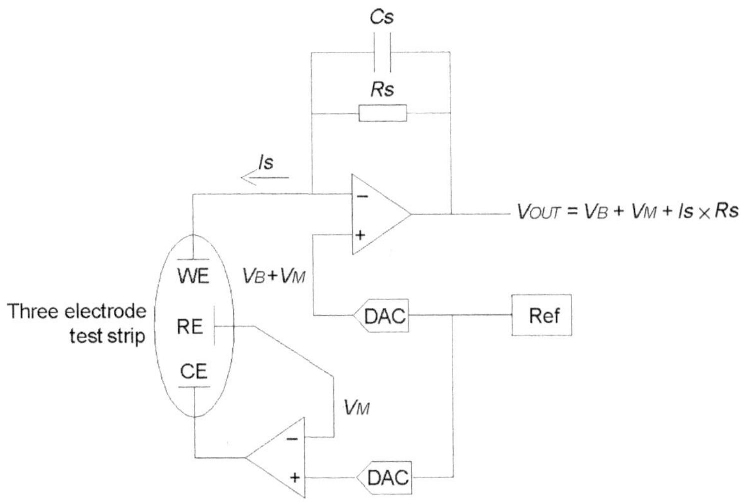

Fig. 7 Schematic diagram of a three electrode force-sense test-strip monitor. *(Based on DiCristina, J., 2017. Maxim Tutorial 4659: Blood Glucose Meters. https://www.maximintegrated.com/en/app-notes/index.mvp/id/4659 (Retrieved 10 August 2017).)*

Analogue front ends from Maxim and Microchip, among many, integrate all of the functionality described above into a single package with the specification and performance required by blood glucose meters. This integration drives the low cost of these devices in an extremely competitive market.

3.3.4 Peripheral Functions

Most blood glucose meters use a custom liquid-crystal display (LCD) that can be driven with an LCD driver integrated in the microcontroller. Others feature a more complicated dot-matrix LCD that requires using a module with the glass, bias voltages, and drivers assembled together into a single integrated module. A dot-matrix display also requires memory to store the messages to be displayed. High-end instruments provide color displays that require additional and higher voltages than both the monochrome types. Backlighting can be added using one or two white LEDs or an electroluminescent source.

Most meters include the ability to upload test results to a computer. These are typically based on industry standard USB or Bluetooth interfaces and are generally used to upload patient data to the user's health-care provider.

Low-end meters with simple displays can run directly off of a single lithium coin cell or two alkaline AAA batteries. To maximize battery life, electronics capable of running from 3.6 V down to 2.2 V for the lithium coin cell or 1.8 V for the alkaline AAAs is needed. A boost switching regulator is required in meters with more complex displays that require higher voltages

Finally, all meters must meet electrostatic discharge (ESD) specifications in their country of use. Electronics with built-in ESD protection or the addition of ESD line protectors to exposed traces can help meet this requirement (DiCristina, 2017).

3.4 Continuous Glucose Monitoring Systems

Continuous glucose monitoring systems (CGMSs) are reagentless and based on the direct transfer between the enzyme and the electrode without a mediator stage. This is advantageous as, by removing potentially toxic mediators, the electrode performs direct electron transfers using the charge transfer complexes of conductive organic materials. This has facilitated the development of implantable needle-based devices for continuous in vivo glucose monitoring. Conducting organic salts, such as tetrathiafulvalene-tetracyanoquinodimethane mediate the electrochemistry of pyrrole-quinolinequinone enzymes as well as of flavoproteins including GOx. As a bonus, the absence of mediators also provides the biosensors with superior selectivity. However, only a few enzymes including peroxidases have been proved to exhibit direct electron transfer at normal electrode surfaces.

CGMSs can be subcutaneous or blood based. However because of problems with surface contamination of electrodes by blood proteins and the possibility of embolisms due to blood coagulation, few measure blood glucose directly, but rather sample glucose levels in interstitial fluid. These are known as subcutaneous glucose monitoring systems (SCGMSs), a schematic example of which is shown in Fig. 8. The first commercial SCGMS was marketed by MiniMed. It did not provide a local readout but 3 days of logged data could be downloaded by the health-care professional for analysis.

At present SCGMS devices including the MiniMed Guardian REAL-Time system by Medtronic, the SEVEN by Dexcom and the Freestyle Navigator by Abbott are most widely used monitors on the market. These devices display updated real-time glucose concentrations every few minutes using a disposable sensor with a lifespan of between 3 and 7 days. They all consist of flexible strips implanted in the hypodermis, but are neither non-invasive, nor painless, as their insertion requires a large bore hypodermic needle. The smallest sensor strips on the market are 3 mm long and about

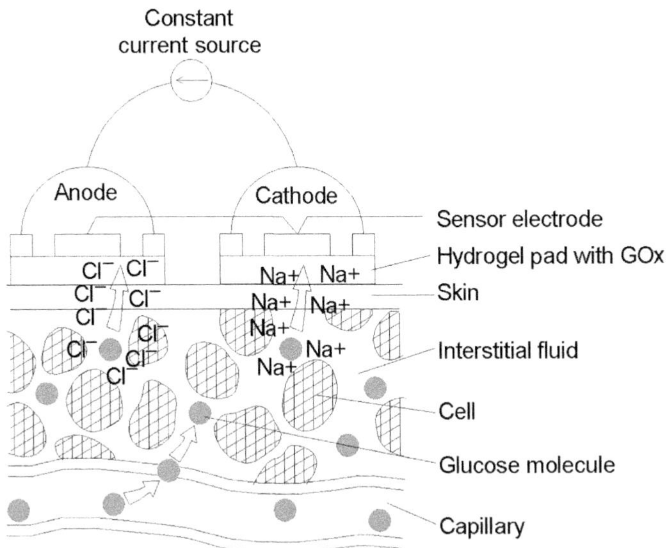

Fig. 8 Cross section showing the SCGM sensor attached to the skin.

0.5 mm wide. However, research is underway to develop a sensor that is less than 1 mm long and embedded in the epidermis (Ribet et al., 2017).

The accuracy of subcutaneous devices is generally lower than that provided by traditional glucose monitors. They also have much more stringent design specifications to cater for biocompatibility, stability, specificity, linearity, and miniaturization, so they are more expensive to develop and manufacture.

Microneedle-based arrays can be utilized for a variety of therapeutic and diagnostic systems. They have a potential for painless sampling in the intradermal space, so as a result there have been attempts to use them as a means of drawing blood or other fluids for electrochemical glucose sensing. Initial systems made use of capillary action to draw the analyte out into a separate sensor where analysis could take place. An alternative is to use the needle itself as the functional electrode array. This is a more elegant and simple design that uses a platinum-coated stainless steel in-line two-dimensional (2D) microneedle array coated with a film of poly(3,4-ethylenedioxythiophene) PEDOT in which GOx has been immobilized. GOx converts glucose that contains oxygen into gluconic acid and produces H_2O_2. Flavin adenosine dinucleotide (FAD) cofactor, associated with GOx, undergoes reversible oxidation and reduction reactions during this process. As the glucose is converted, a current is generated proportional to the glucose concentration that can be sensed by applying a low voltage. An example of a microneedle array is shown in Fig. 9 (Invernale et al., 2014).

An alternative mechanism SCGM uses microdialysis that avoids direct contact of the interstitial fluid and the sensor. This method provides superior precision and accuracy and lower signal drift than needle-based sensors and is used by GlucoDay from Menarini, and SCGM from Roche. This is a closed fluid containing system as can be seen in Fig. 10. The microdialysis probe consists of a tubular dialysis membrane with a 20,000 molecular weight cut-off glued to the end of a double lumen catheter. The outer in-going lumen of the catheter is continuously perfused with Ringer's solution at a flow rate of 0.3 mL/min by a miniature peristaltic pump. The dialysate leaves the probe through the inner out-going lumen. The peristaltic pump operates as both a pumping and suction unit to generate identical flow rates into and out of the catheter. The dialysate is mixed with the GOx containing reagent solution via the Y mixer inside the cassette. The GOx then catalyzes the oxidation of the glucose in the dialysate to gluconolactone and H_2O_2 in the subsequent tube segment. To prevent an incomplete reaction caused by an oxygen deficit when the glucose concentration is high, this segment is

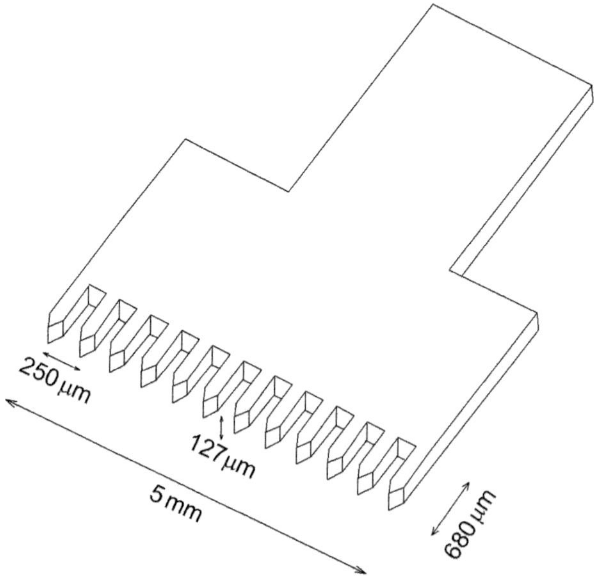

Fig. 9 Image of the microneedle array. *(Based on Invernale, M., Tang, B., York, R., Le, L., Hou, D., 2014. Microneedle electrodes toward an amperometric glucose-sensing smart patch. Adv. Healthc. Mater. 3: 338–342.)*

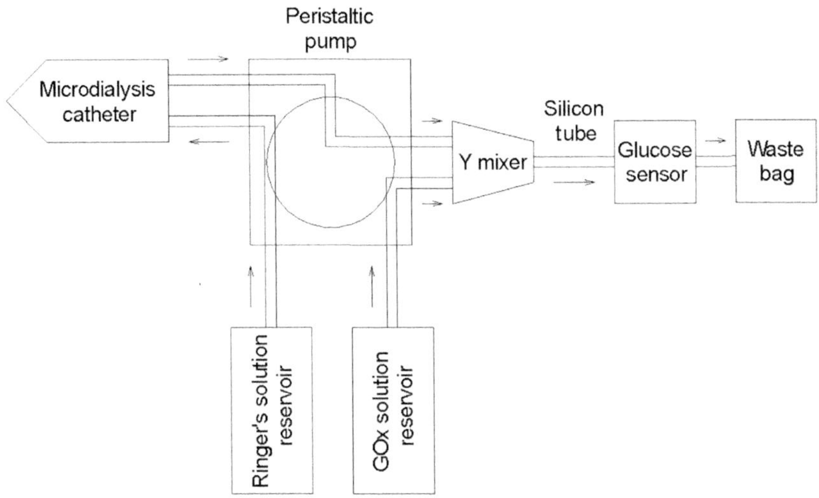

Fig. 10 Schematic representation of a microdialysis SCGM system. *(Based on Schoemaker, M., Andreis, E., Röper, J. Kotulla, R., Lodwig, V., Obermaier, K., Stephan, P., Reuschling, W., Rutschmann, M., Schwaninger, R., Wittmann, U., Rinne, H., Kontschieder, H., Strohmeier, W., 2003. The SCGM1 system: subcutaneous continuous glucose monitoring based on microdialysis technique. Diabetes Technol. Ther. 5: 599–608.)*

made of a highly oxygen-permeable silicon tube. The fluid then passes through the sensor where the H_2O_2 is oxidized on an electrode in the normal fashion to produce a small current proportional to the glucose concentration in the dialysate. The electrode consists of a screen-printed three-electrode system with an Ag/AgCl RE, a WE, and a carbon counter electrode. Calibration ensures that the measured current is converted to the correct blood glucose level. Finally, the fluid is collected in the waste bag (Schoemaker et al., 2003).

3.5 Noninvasive Glucose Monitoring

The glucose monitoring techniques discussed so far leave a lot to be desired. The standard lancet and strip methods are expensive in terms of consumables and boring to administer resulting in repeated measurements being skipped. Subcutaneous sensors that measure glucose concentration in interstitial fluid are limited by patient discomfort, onerous calibration procedures, and biofouling. These limitations have led to a strong research drive to find a completely noninvasive glucose monitoring method. The methods investigated so far are mostly based on measuring the optical properties of interstitial fluid, the aqueous humor in the anterior chamber of the eye, or the transdermal extraction of interstitial fluid using vacuum methods or reverse iontophoresis. These techniques are reviewed below, as described by Vashist (2012).

3.5.1 Reverse Iontophoresis

This process draws glucose out through the skin using by the potential difference between two electrodes attached to the skin. It works by drawing Na^+ cations and Cl^- anions through the skin to the cathodic and anodic electrodes, respectively. Some of the uncharged glucose molecules present in the interstitial fluid are carried along with the ions and collected at the electrodes. Glucose concentration is then measured using one of the conventional methods described previously (Vashist, 2012). This technique was used by the now discontinued GlucoWatch shown in Fig. 11.

3.5.2 Ultrasound

This method provides an alternative to reverse iontophoresis, as vibrations at a frequency of 20 kHz (beyond the range of human hearing) increases the permeability of skin to interstitial fluids, and so glucose is transported to the epidermis where it can be measured using conventional glucose-sensing technology (Vashist, 2012).

3.5.3 Bioimpedance Spectroscopy

This method examines the tissue impedance at different frequencies by introducing an alternating current in the frequency range from 100 to 100 MHz. Changes in the glucose concentration in blood plasma alter the membrane potential of red blood cells that can be inferred from changes in the bioimpedance. However, it is susceptible to water content and diseases that affect the cell membrane. This technique is used in the Pendra glucose monitor (Vashist, 2012).

3.5.4 Infrared Spectroscopy

These techniques range from the thermal in which glucose absorption at 9.8 and 10.9 μm is monitored against the background thermal signature of the body to the near and mid infrared bands where glucose affects the absorption and scattering effects at specific frequencies. Unfortunately, the signal is either weak compared to that of water or the penetration is poor. The principle used is attenuated total reflection and relies on the light beam guided through a crystal and a thin film of squalane oil in contact with the skin. SugarTrack, Sensys, and Orsense use variations of this technique for glucose monitoring (Vashist, 2012).

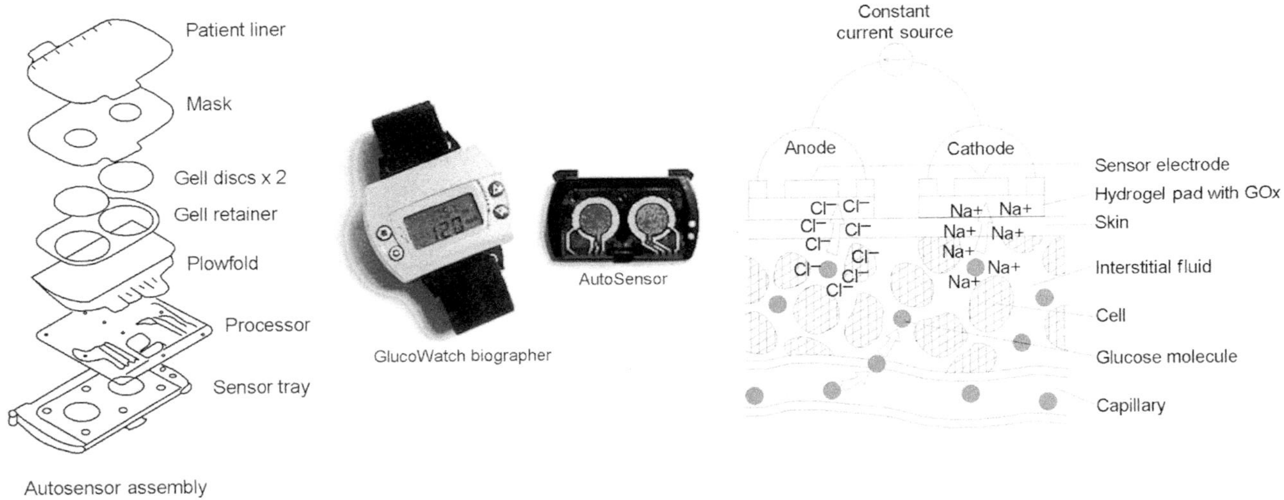

Fig. 11 Photo for the now discontinued GlucoWatch and a diagram showing the principle of operation using reverse iontophoresis.

3.5.5 Photoacoustic Spectroscopy

This is a method based on pressure variations generated in tissue when heated by a laser pulse. Selective detection of blood glucose uses specific laser frequencies. It is not affected by water because of its poor photoacoustic response of the liquid and a wide range of frequencies from the IR through to the UV that can be used. It is used by the Aprise system for glucose monitoring (Vashist, 2012).

3.5.6 Raman Spectroscopy

This form of spectroscopy measures the characteristics of scattered light at a higher frequency and lower intensity than the incident source. Water has weak scattering indices so does not interfere with the measurements. It is generally used to measure for glucose in the aqueous humor of the eye (Vashist, 2012).

3.5.7 Ocular Spectroscopy

Another spectroscopy method that measures glucose concentration in tears using a boric acid derivative hydrogel wafer bound to a contact lens. The boric acid derivatives form reversible covalent bonds with the glucose in tears resulting in a change in frequency of light reflected from the lens that is identified using a spectrometer. Issues with this technique include significant lag and poor correlation between glucose in tears and that in blood (Vashist, 2012).

3.5.8 Fluorescence

This uses UV excitation of primed tissues. Currently, polymerized crystalline colloidal arrays that respond to glucose concentration by altering their refractive index are used. The latest research focuses on the development of contact lenses that change color in response to changes in glucose concentration (Vashist, 2012).

3.5.9 Polarimetry

This technique has been used in industry for decades and so is well researched as a method of measuring glucose concentration. The technique relies on the rotation of linearly polarized light as it passes through a glucose solution. The aqueous humor of the eye, being clear, is an ideal measurement site. A path length of 10 mm gives 10 mdeg of rotation for a glucose concentration 10 mmol/L at a wavelength of 670 nm. Unfortunately its specificity in body fluids is compromised by other optically active compounds in solution (Vashist, 2012).

3.5.10 Electromagnetic Sensing

This method can be used to measure changes in the dielectric characteristics of blood that vary with glucose concentration. Eddy current or conductivity measurements using a radiofrequency (RF) probe can be used to obtain this information (Vashist, 2012).

3.6 The Future of Noninvasive Glucose Monitoring

In 2014 Google through a medical spin-off, Verily Life Sciences and the Novartis eye care division, Alcon, joined forces to develop and commercialize contact lens-based glucose monitors. This is a hard call as contact lenses must be compliant to fit the shape of the eye and with a diameter of only about 14 mm and between 0.1 and 0.2 mm thick, they must house the sensor as well as the electronics to control the sensor, handle power consumption, and provide telemetry to an external device such as a smart phone.

Batteries are not practical for a number of reasons including their power-to-weight ratio and the possibility of leaking toxic chemicals into the eye. This leaves two major alternatives, the first being wireless energy transfer and the second, the incorporation of a photocell onto the contact lens. Wireless transfer generally uses near-field inductive coupling using transmitter coils mounted within the frames of eye glasses. Unfortunately, both of these techniques are only good during the day or while the user is awake and wearing glasses. Possible alternatives are to embed a piezoelectric element into the contact lens to extract energy from eye movement or even tiny fuel cells that metabolize tears to provide electric power (Barrettino, 2017).

Contact lenses are designed to be disposable, so the sensors and electronics need to be low cost and reliable over the life of the device. However, standard glucose sensing techniques are considered to be too short lived for this application, so research is focusing on the incorporation of hundreds or thousands of nanoscale biosensors into the lens. The ultimate objective of the sensor development is the ability to detect a single molecule of glucose.

There are, of course, problems with monitoring glucose levels in tears. These include the low concentration, typically only 0.1–0.5 mmol/L in tears compared with 4–6 mmol/L in blood, as well as a poor correlation and significant time lag linking the concentration of glucose in tears and that in blood. Finally there are three different sorts of tears: basal tears that lubricate the eye, tears caused by irritation, and finally those caused by emotion, each of which could have different glucose levels relative to the blood concentration (Barrettino, 2017).

The alternative according to Lucisano et al. (2017), for a continuous glucose monitor would be one that is not percutaneous, unobtrusive, not attached to the skin, able to retain long-term calibration and requires no maintenance by the user. To that end an implantable device within a hermetically sealed titanium puck housing the battery-powered signal conditioning and telemetry circuitry has been developed. A ceramic insert with a planar antenna on the inside and eight 300-μm diameter platinum electrodes and counter electrodes on the outside forms the core of the device. The WE are negatively polarized for oxygen reduction and covered with a thin electrolyte layer, a protective layer of polydimethylsiloxane (PDMS), and finally a further layer of PDMS with hollows over the active electrodes filled with gel-containing immobilized GOx and a catalyst.

In trials, these pucks were implanted in the abdominal tissue below any subcutaneous fat with electrodes facing inward in contact with muscle tissue but not tethered. The telemetry antenna faced outward. Tests were conducted over a 6-month period, with glucose clamp studies conducted at monthly intervals. These clamp studies involved the infusion of glucose and insulin to achieve specified blood glucose levels in the normal, hypoglycemic, and hyperglycemic ranges. Venous blood was extracted and analyzed on a regular basis and was used as a retrospective calibration reference.

Between studies, the diabetic patients managed their condition in the normal way.

Results of this study showed excellent correlation between the calibrated outputs of the sensor and measured glucose levels both during the clamp studies as shown in Fig. 12, and intervening finger-prick measurements. This allows for the construction of an accurate glucose/insulin model for individual patients (discussed later).

Another developer of this technology now under FDA review is the Eversense from Senseonics designed to provide continuous monitoring for between 3 and 6 months. It is small enough to be replaceable with a simple surgical procedure doable by any trained health-care provider (Castle et al., 2017).

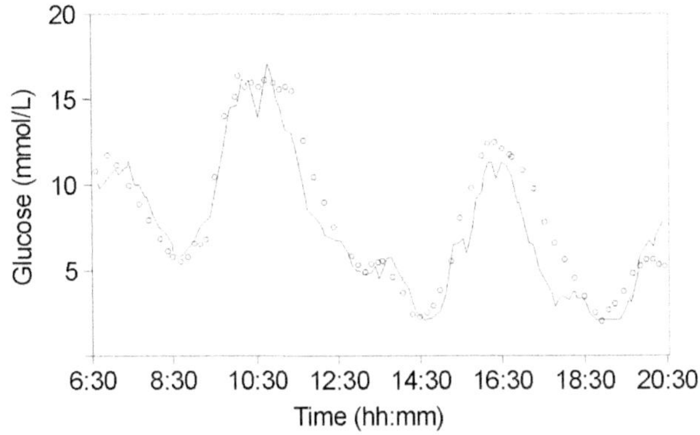

Fig. 12 Clamp test results showing the relationship between the glucose sensor output and measured glucose samples in venous blood. *(Data from Lucisano, J., Routh, T., Lin, J., Gough, A., 2017. Glucose monitoring in individuals with diabetes using a long term implanted sensor/telemetry system and model. IEEE Trans. Biomed. Eng. 64(9).)*

4 INSULIN DISPENSING

4.1 Insulin Pumps—Historical Perspective

In essence, a modern insulin pump tries to mimic insulin secretion from pancreatic β-cells by delivering rapid-acting insulin both at preset continuous basal rates and in extra bolus doses at mealtimes, on demand. Typical pumps provide for hourly basal rate settings over a 24-h period. For bolus doses, pump users input their current blood glucose level and the number of carbohydrates they expect to consume during the meal. The pump then calculates their required dose based on insulin currently "on board" (i.e., the remaining active insulin from the previous dose), their unique insulin-to-carbohydrate ratio, and their insulin sensitivity factor—the expected drop in blood glucose from a single unit of insulin. This results in the pump delivering insulin in a more physiological manner than manually injected insulin regimens.

Insulin pumps consist of a reservoir, a pump, and an infusion set. The reservoir, which is generally similar to a syringe, holds a 2–3 days supply of insulin and is inserted into the battery-powered pump module. The infusion set consists of tubing that connects the reservoir to a cannula and transports the insulin from pump to where it is inserted into the patient. The needle of the infusion set can be inserted into the abdomen, upper thigh, or upper arm. Typically the infusion set and reservoir are both replaced every 2–3 days (Rubin and Peyrot, 2010; Valla, 2010).

The development of insulin pumps started in the early 1960s, when Dr. Arnold Kadish developed the first insulin pump. It delivered both glucagon and insulin, and was the size of a rucksack, as can be seen from Fig. 13 (Fernandez and Marcus, 1996).

Over the years, insulin pumps have become much more refined and have decreased to the size and weight of a mobile phone. Insulin pump therapy is no longer seen as experimental, but is viewed as an acceptable alternative to multiple daily injection therapy in the management of diabetes (American_Diabetes_Association, 2013). However, it was a long road from the Kadish backpack to this point. In 1976 a battery operated infusion pump was developed at the National Institute for Medical Research, based in Mill Hill, London. This "Mill Hill Infuser" shown in Fig. 14 is claimed to be first portable insulin infusion pump made to treat diabetes. It provided a constant "basal rate" dose of insulin, and the speed could be altered manually to produce an increased "bolus dose" as required.

At about the same time, Dean Kamen, better known as the driving force behind the Segway and later the DEKA prosthetic arm, also developed a

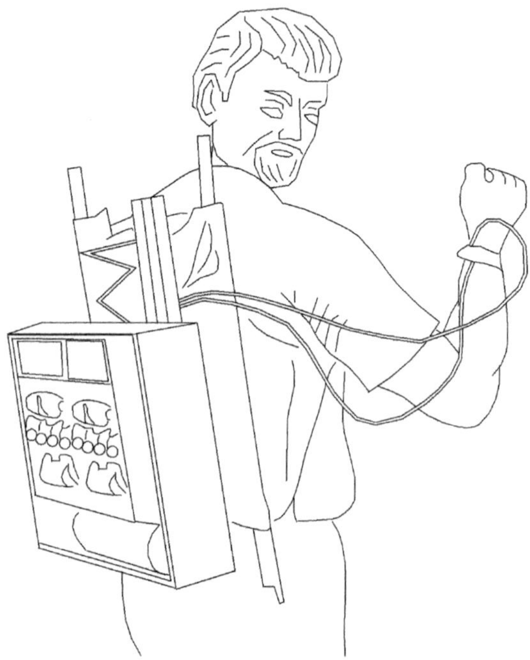

Fig. 13 Drawing of the first rucksack sized insulin pump developed by Dr. Arnold Kadish. *(Adapted from Fernandez, M., Marcus, A., 1996. Insulin pump therapy: acceptable alternative to injection therapy. Postgrad. Med. 99, 125–132, 141–144.)*

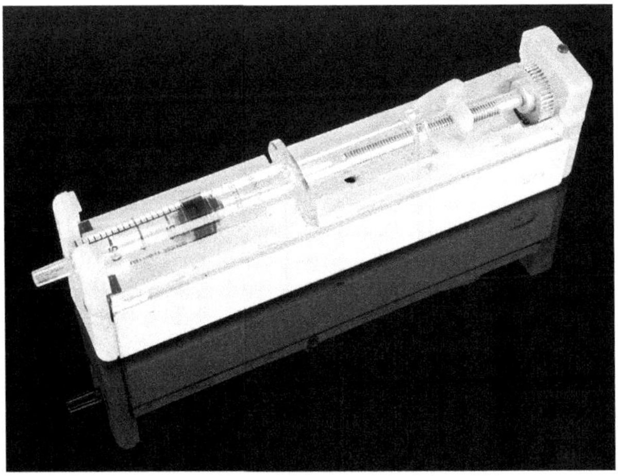

Fig. 14 Mill Hill portable insulin infusion pump developed in 1976. *(Science and Society Picture Library.)*

syringe-based wearable infusion pump. The pump rapidly gained acceptance from a wide range of medical specialties including chemotherapy, neonatology, and endocrinology, so in 1976 Kamen founded AutoSyringe, Inc., to manufacture and market the pumps. By the time this was sold to Baxter, it included the first insulin pump. This was followed in the early 1980s with the Auto-Syringe AS6C pump with a variable delivering speed. Typical pumps from that time are shown in Fig. 15.

As the number of pump users and companies manufacturing pumps increased, disadvantages of the technology became apparent. Batteries did not last for more than a few days and some pumps had expensive insulin reservoirs. In addition having a needle in place continuously was an irritation and often led to infection of the site. It was also noted that because of the low levels of insulin available in the blood, users were more prone to diabetic ketoacidosis if the flow was interrupted. Given their problems and the fact that they did not provide much better control than multiple injections, pumps fell out of favor and by the end of the 1980s pump manufacturers where battling to maintain sales. That notwithstanding, two companies, MiniMed (now part of Medtronic) and Disetronic Medical Systems of Switzerland, were continued to provide innovative new pumps during the 1990s.

4.2 Modern Insulin Pumps

Pump technology has matured in the past 25 years, with pumps becoming smaller and more reliable. Most pumps still use syringe mechanisms driven

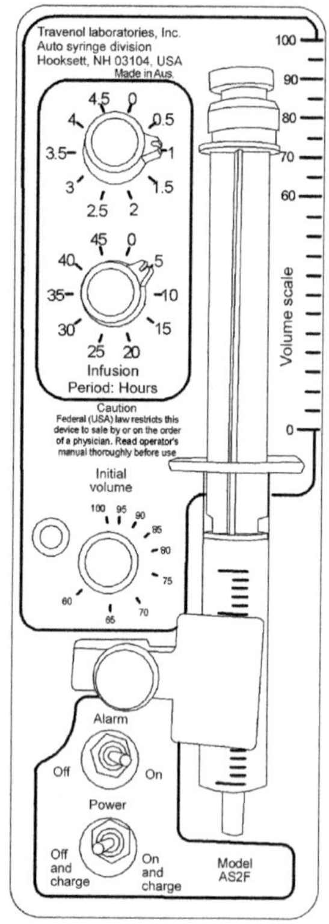

Fig. 15 A syringe insulin pump with adjustable dose rate.

by geared motors and ball screw mechanisms that depresses the plunger, much as the original Mill Hill device did. They generally still deliver insulin subcutaneously through a cannula. A drawing of a generic modern insulin pump is shown in Fig. 16.

Insulin is measured in "units" where there are 100 units per mL, making each unit 10 μL. Basal rates are of the order of one unit per hour administered every 3–10 min, while bolus doses are typically several units. Cartridge volumes are generally 200–300 units (2–3 mL) and can last 4–5 days.

Because pumps are FDA regulated in the United States, their design and construction must be precisely documented and their performance must meet strict guidelines. In addition, the hardware must provide comprehensive self-test and fault reporting capabilities. This requires significant amounts of additional circuitry not directly involved in the pump process, as can be seen in Fig. 17 (Mossman, 2010).

To ensure they are truly portable, insulin pumps should not weigh more than 100 g and be no larger than a mobile phone, yet should only require

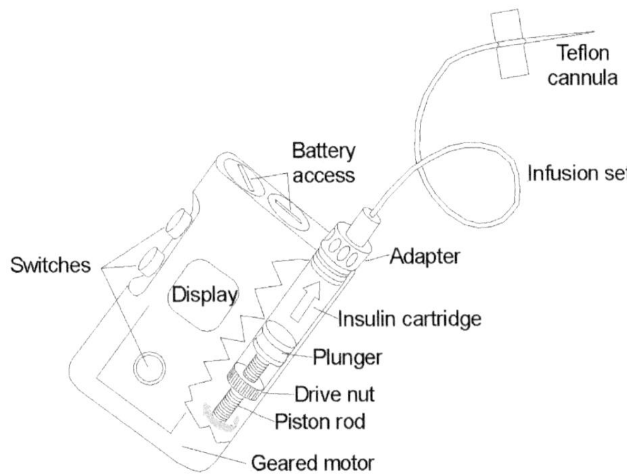

Fig. 16 A generic syringe-based insulin pump. *(Based on from Valla, V., 2010. Therapeutics of diabetes mellitus: focus on insulin analogueues and insulin pumps. Exp. Diab. Res. (Hindawi) 2010: 1–14, (ID 148372).)*

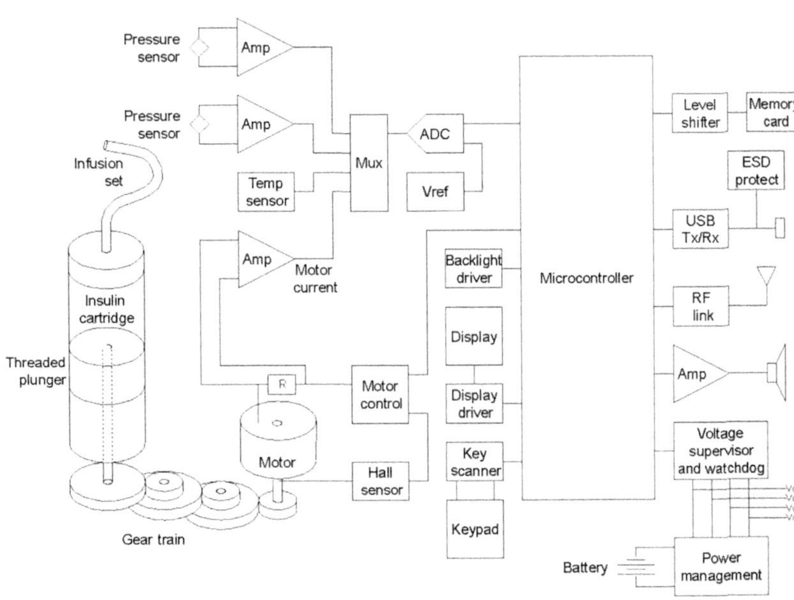

Fig. 17 Detailed schematic of an insulin pump. *(Based on Mossman, J., 2010. Insulin Pumps: Design Basics and Tradeoffs. EE Times. http://www.eetimes.com/document.asp?doc_id=1278073 (Retrieved 24 August 2017).)*

recharging (or battery replacement) once a week. These form factor and power consumption requirements drive component selection.

4.2.1 Portability

The size and shape of insulin pumps is determined largely by the reservoir. However, that presupposes that the electronics can be highly integrated and low power, leading to a small battery pack. To achieve tight integration, chip scale, or wafer level packaging is used and any circuit not required is turned off. A good example of this device is the low standby power 16-bit MAXQ2010 microcontroller that only draws 1 mA when operating at 1 MHz and 2.7 V, and just 370 nA in stop mode. As with the MCUs from other manufacturers, this device includes USARTs, timers, 64 kbytes of Flash-based program storage, 2 kbytes of RAM, some general-purpose I/O pins, a 312.5 ksample/s 12-bit successive-approximation ADC with reference, and a 160-segment LCD controller (Mossman, 2010; Ganesan, 2013).

4.2.2 Pumping and Sensing

To achieve flow rates measured in fractions of a μL, the motor driving the pump is geared down and coupled to a threaded screwdriver that advances the reservoir piston very slowly with many revolutions of the motor. This concept is illustrated in Fig. 17 above. As a consequence motor angle need to only be coarsely measured, typically with optical or Hall-effect sensors counting individual revolutions. Conventional DC micro-motors are generally used, though some designers have used stepper motors or MEMS-based pumps (Mossman, 2010; Ganesan, 2013).

Some innovative new pump technologies include an shape memory alloy (SMA) motor used in the Omnipod pump and the V-Go, a disposable spring-driven pump from Valeritas shown in Fig. 18.

Similar to a typical patch pump, the V-Go sticks to the body and delivers a continuous stream of rapid-acting insulin at a rate of 20, 30, or 40 units of insulin per 24 h, depending on the model. At mealtimes, the user can press a button to deliver an additional 2 units per push, for a maximum bolus volume of 36 units.

The Tandem Diabetes Care t:slim is another clever new pump design. It consists of a small cylinder containing a reciprocating shuttle that transfers the one-third of a unit from the reservoir to the infusion set with each cycle. Because the whole reservoir is not pressurized to drive the infusion process, the motor size can be reduced, power requirements decreased, and overall safety improved.

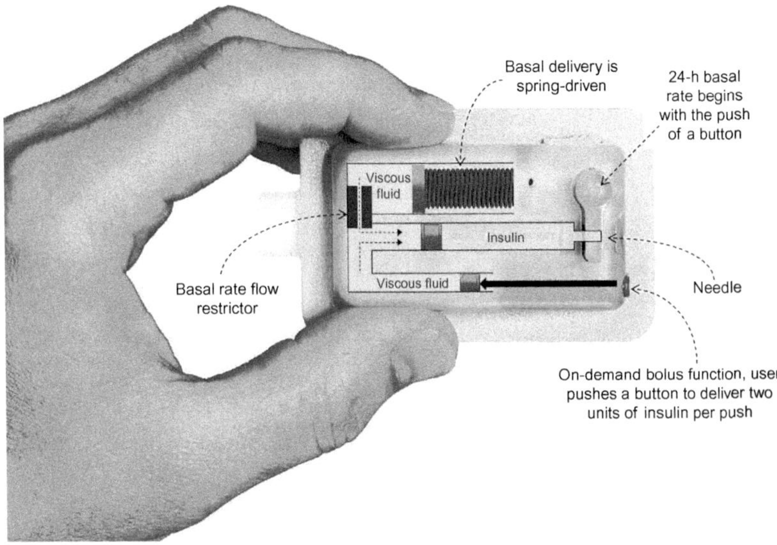

Fig. 18 V-Go spring-driven disposable insulin pump. *(Courtesy Valeritas Inc.)*

Pressure sensors are incorporated into most insulin pumps to ensure normal operation and detect occlusions. These are usually based on silicon strain gauges that provide signals in the millivolt range, rather than the microvolt level provided by bonded-wire strain gauges. The strain gauges are configured in a bridge configuration, which provides a differential signal at a common-mode voltage that is about half of the supply voltage. Differential instrumentation amplifiers followed by programmable gain amplifiers and analogue to digital converters interface pressure measurements to the microcontroller. Precision pressure measurements are not required as they are only used to verify normal operation and are not concerned with actual drug delivery (Mossman, 2010; Ganesan, 2013).

4.2.3 Power Management

Power is generally provided using a single 1.5 V alkaline or rechargeable LiPo cell that is boosted to a nominal operating voltage of 2 V to obtain the most life from a cell. Typical boost regulators used in insulin pumps can run from voltages as low as 0.6 V. An example of the genre is the MAX1947 has a 0.7–3.6 V input-voltage range. Its 2-MHz switching frequency and current-mode control reduces component sizes, achieves over 94% efficiency while integrating all the required switches (power switch, synchronous rectifier, and reverse-current blocker) to minimize solution size. For precision voltages required by the various electronic components in the device, low dropout (LDO) linear regulators are typically used as they provide excellent efficiency and small size if the minimum headroom is maintained.

Insulin pump manufacturers have made good progress in reducing power consumption to maximize battery life, with modern pumps able to operate between 3 and 10 weeks before battery replacement is required. Many pumps on the market use single AA or AAA alkaline or lithium batteries. Primary cells are common, but rechargeable cells can be used for the long-term cost savings. However, because of their lower capacity, the latter need to be recharged more often.

Because of the variety of batteries that can be used in pumps, they typically rely on simple battery-voltage and, sometimes, temperature measurement. These readings of voltage and temperature are digitized and processed by the microcontroller to determine the remaining capacity of the cell. This is indicated by the number of charge bars on the LCD display in the usual manner, with an audible warning when the battery needs to be replaced (Mossman, 2010; Ganesan, 2013).

4.2.4 External Interface

Simple external interfaces consisting of a custom alphanumeric monochrome LCD display and a few push button switches allow the user to tailor basal and bolus dosages to their requirements. The display typically includes information about insulin dose and rate, battery capacity as well as date and time and various alarms (Mossman, 2010).

4.2.5 Self-Test and Alarms

All insulin pumps must perform a self-test on switch on to meet FDA requirements. This includes tests of all critical processors, critical circuitry, indicators, and displays. Some models even incorporate a separate processor to monitor the main processor for anomalous operation. The most common form of self-test is a separate watchdog timer that monitors the time taken by the processor to execute various functions and flags an error if this is outside normal limits.

Power supplies are monitored for undervoltage and overvoltage conditions and motor loading is monitored with motor-stall detection being a priority as it is a critical failure.

Insulin pumps require audible and visible alarms. These use an audio beeper and the LCD display as well as separate LEDs. Typical systems use a flashing green LED during normal operation and a red LED and an audio alarm indicating a fault condition. Even the beeper is provided with a self-test capability based either on its impedance or feedback through a separate microphone. Some pumps also incorporate eccentric rotating mass vibrators to provide an additional warning capability.

All pumps must be ESD protected. This can be built into the electronics or provided across any exposed lines using separate electronics (Mossman, 2010).

4.2.6 Data Exchange

Data ports (typically USB) are included on most pumps to allow logged data to be transferred to a PC and to download firmware upgrades. In addition, RF interfaces are common in modern pumps, allowing them to be linked to continuous glucose monitors for closed-loop control. Wireless interfaces are generally implemented using Bluetooth or other ISM band links (Mossman, 2010; Ganesan, 2013).

4.3 Implantable Insulin Pumps

Given the requirements for continuous insulin provision, a totally implantable insulin pump would appear to be the ideal way to minimize regular cannulization and potential infection resulting from the use of external pumps. The process started in 1970 with a paper describing a multipurpose permanently implantable infusion pump that could be used for heparin or insulin delivery (Blackshear et al., 1970). This was followed by a paper that described an ingenious method to use the vapor pressure of an isolated fluid as an energy source to deliver the drug at a single constant rate as shown in Figs. 19 and 20 (Blackshear et al., 1972).

For a liquid infusate with viscosity, η (pa.s), and a pressure differential along the capillary is ΔP (Pa), the volume flow rate Q (m^3/s) described by Poiseuille's law is

$$Q = \frac{dV}{dt} = \frac{\Delta P \pi R^4}{8 \eta L} \tag{5}$$

where R (m) is the radius of the cannula and L (m) is its length.

Later revisions of the pump used a magnet to alter the settings of an internal valve that could change the flow rate by actuating a valve to lengthen or shorten the capillary length and hence change the flow rate according to the Poiseuille relationship.

Human trials of intravenous (IV) infusion of insulin started in 1980 using the Infusaid Model 400, Constant Rate infusion pump (Buchwald et al., 1981).

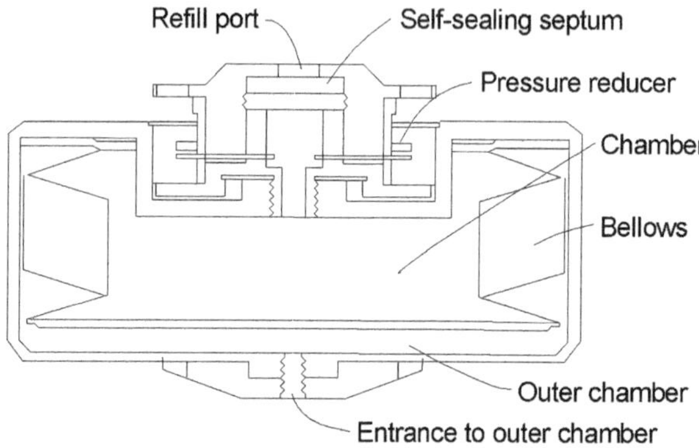

Fig. 19 Cross section of an implantable infusion pump. *Reprinted with permission from the Journal of the American College of Surgeons, formerly Surgery Gynecology & Obstetrics.*

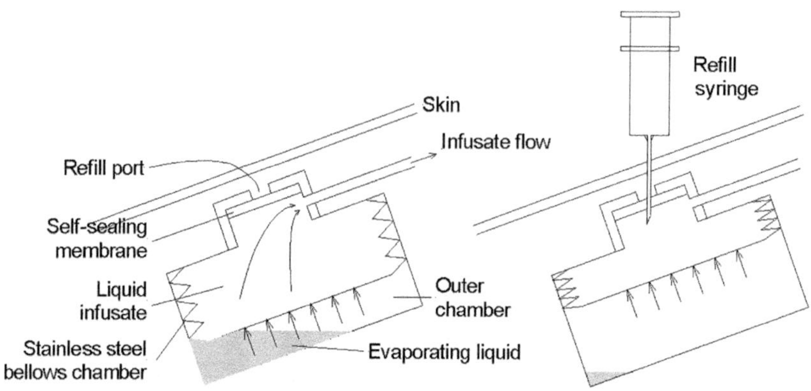

Fig. 20 Schematic representation of the operation of the implantable infusion pump. *(Reprinted with permission from the Journal of the American College of Surgeons, formerly Surgery Gynecology & Obstetrics.)*

However, IV insulin delivery was associated with a high incidence of catheter obstructions and thus the intraperitoneal (IP) option was promoted as a better option.

A completely different design for an implantable pump using a peristaltic pump driven by rotary solenoid motor was developed in 1980 in a cooperative effort between the researchers at the University of New Mexico (UNM) and Sandia Laboratories (Spencer et al., 1980).

Details of the UNM/Sandia pump design are shown in Fig. 21. The pump was not limited to a single basal flow rate and could be programmed following implantation to deliver any of 15 different basal

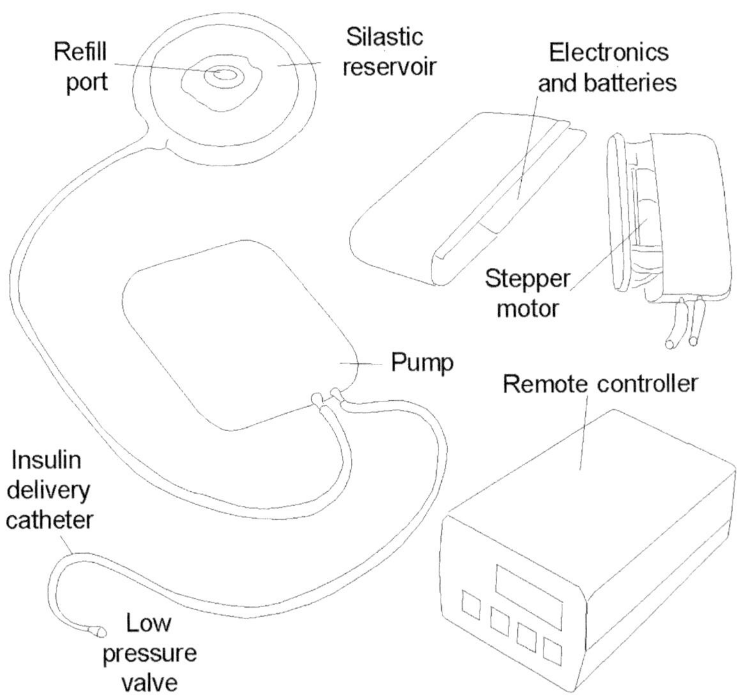

Fig. 21 Schematic diagram of an early implantable rate-controlled insulin delivery system. *(Based on Schade, D., Eaton, R., Edwards, W., Doberneck, R., Spencer, W., Carlson, G., Blair, R., Love, J., Urenda, R., Gaona, J., 1982. A remotely programmable insulin delivery system. JAMA 247(13): 1848–1853.)*

rates (from 0.78 to 11.7 units/h) and any of 15 different bolus deliveries (from 1.66 to 24.9 units). In this case, insulin was delivered intraperitoneally (Schade et al., 1982).

The pump, batteries, and electronics is housed in a welded titanium that is connected to a reservoir with a refill port as well as a silicone rubber intraperitoneal insulin delivery catheter surrounding an embedded stainless steel spring and a terminal low pressure valve. The implantable components communicate with an external controller.

Other manufacturers were quick to develop similar pumps, with the Siemens PFA 01 insulin pump (shown in Fig. 22) being typical of the genre (Selam et al., 1982).

The clinical experience with implantable pumps remained limited for a period until the development of a stable insulin, U400 Insulin by Hoechst (semisynthetic human insulin at neutral pH, stabilized by adding Genapol). This led the way for several manufacturers to develop implantable insulin

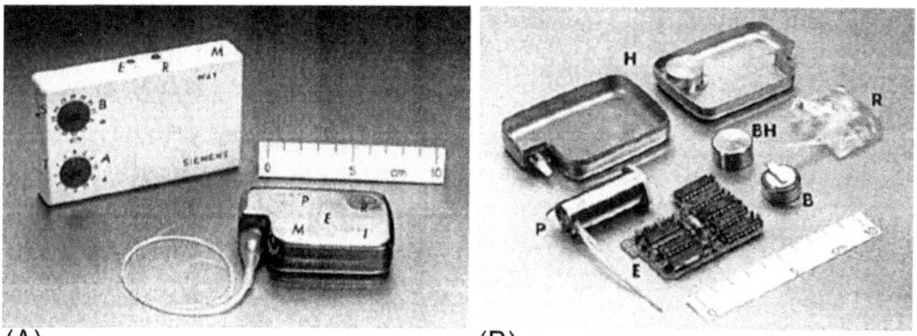

(A) (B)

Fig. 22 Photographs showing (A) the pump with its controller and (B) the pump opened showing H, housing; P, pump; E, electronics; B, battery; BH, battery housing; R, reservoir (Selam et al., 1982).

pumps, such as the Infusaid M1000, Siemens ID1/ID3, and MiniMed PIMS/MIP (later Medtronic).

Following the development of the early versions of their implantable insulin pump, MiniMed proceeded to develop their Programmable Implantable Medication System (the PIMS). The first MiniMed implantable insulin pump came in 1986, but it was not until nearly a decade later that the device received regulatory approval in Europe. MiniMed improved its technology with more memory and improved battery life, releasing its final model shown in Fig. 23, in 2000.

Medtronic bought MiniMed in 2001, and only minimal improvements were made in the years following that. Finally in 2007 the company announced that it would be discontinuing its clinical R&D for the implantable insulin pumps altogether because they had shifted their focus to closed-loop AP technology (Hoskins, 2017).

5 THE ARTIFICIAL PANCREAS

Recent advances in the technology for continuous glucose monitoring in conjunction with miniature automated insulin dispensing pumps have at last made the prospect of a true closed-loop AP more than just a dream. However, to achieve this in a complex, ever changing environment involves more than just the technology of monitoring and dispensing, it requires sophisticated algorithms that can accommodate significant response lags, the complex interaction between glucose and insulin levels, exercise as well as spikes in glucose levels caused by eating.

Early attempts to close the loop date back to the 1980s with the development of the Biostator shown in Fig. 24. It required a continuous flow of

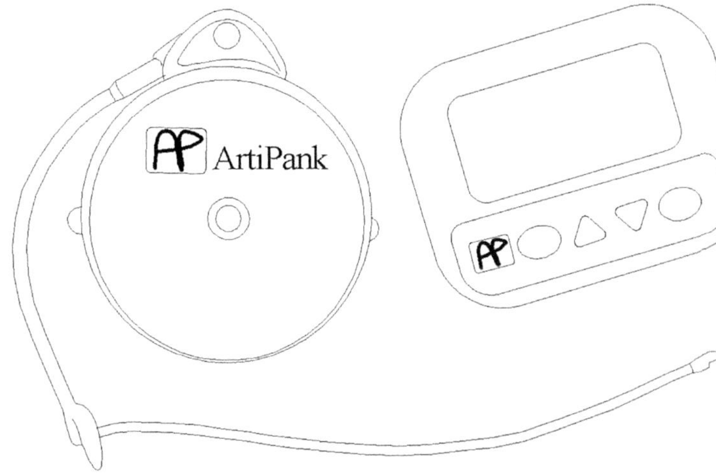

Fig. 23 MiniMed implantable insulin pump and controller. *(Redrawn from Hoskins, M., 2017. Implantable Insulin Pumps are Near Extinction, But Still Alive.... Diabetes Mine, http://www.healthline.com/diabetesmine/implantable-insulin-pumps. (Retrieved 24 August 2017).)*

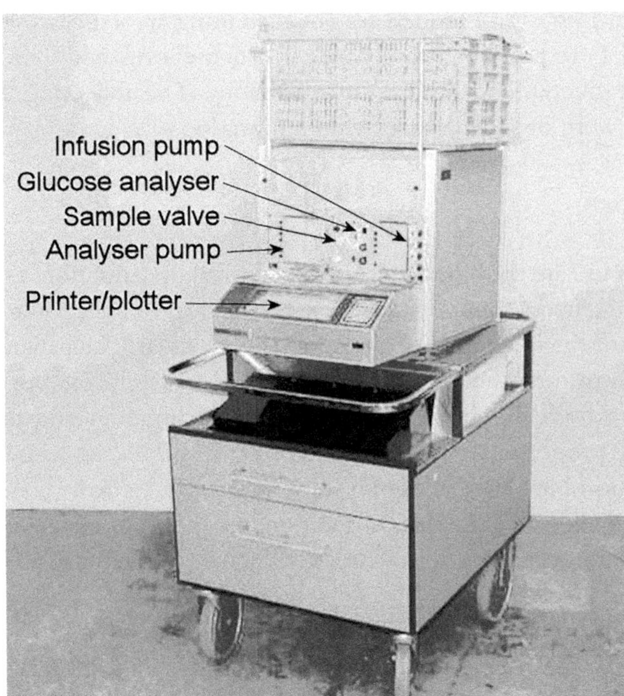

Fig. 24 Biostator—glucose controlled insulin infusion device (Young and Herf, 1984). https://www.recycledgoods.com/brands/Life-Science-Instruments-Miles-Lab.html.

venous blood for glucose level analysis and could provide a continuous measure of insulin requirements. However, although the IV route is ideal it is not practical and despite delayed absorption of insulin and in reading glucose levels provided by subcutaneous operation, it has become the method of choice for closed-loop operation (Leiva-Hidalgo, 2011).

5.1 Modeling

The development of adequate closed-loop control systems relies on comprehensive in vitro models of the interaction between insulin and glucose in diverse environments. According to Cobelli, two mechanistic physiologically based models have been developed. Minimal (coarse) models describe key components of system functionality and provide insight into glucose metabolism and insulin control in both healthy and diabetic patients. Maximal (fine grain) models include all available knowledge about system functionality and are thus capable of accurately simulating glucose-insulin systems in diabetics, making it possible to create accurate simulation scenarios to evaluate closed-loop treatments (Cobelli et al., 2009).

The simplest coarse models are obtained using an oral glucose tolerance test (OGTT) to provide a step change in glucose levels and then observing the blood glucose and insulin concentrations. The following differential equations were obtained using IV administration of glucose.

$$\begin{aligned} \dot{G}(t) &= a_1 G(t) - a_2 I(t) + J(t) \\ \dot{I}(t) &= a_3 G(t) - a_4 I(t) \end{aligned} \tag{6}$$

where G and I are the blood glucose and insulin concentrations respectively and J is the glucose input that can be either a glucose injection providing a step input or the more gradual absorption after a meal. The parameters a_1 to a_4 are used to tune the response It assumes that the glucose use is a linear function of both the glucose and insulin concentrations and that insulin secretion is proportional to glucose concentration. This model is too simple as the relationship between insulin secretion rate and glucose concentration is nonlinear and very complex. In addition, the model does not consider the complex interactive control of hepatic glucose production and uptake of glucose and insulin.

Later models isolated the various contributions by opening the loop and decomposing the closed-loop system into two independent subsystems that are linked together by measured variables. In this case, an insulin subsystem represents all tissues that secrete, distribute, or degrade insulin running in parallel with a glucose subsystem that represents all tissues producing, distributing, or metabolizing glucose. When the system is perturbed the concentrations of glucose, $G(t)$ and insulin $I(t)$ can each be considered in terms of a known input and a noisy output and modeled independently. Cobelli describes seven coarse models of increasing complexity to explain blood glucose concentration using blood insulin as a known input. The differential equations describing the simplest of these glucose models is

$$\begin{aligned} \dot{Q}_1(t) &= NHGB(Q(t), I'(t)) - R_d(Q(t), I'(t)) + D \times \delta(t) \\ \dot{I}'(t) &= -k_3 I'(t) + k_2[I(t) - I_b] \\ G(t) &= \frac{Q(t)}{V} \end{aligned} \tag{7}$$

For initial conditions

$$\begin{aligned} Q(0) &= Q_b \\ I'(0) &= 0 \end{aligned}$$

where Q is the blood glucose mass with Q_b being the basal value, I is the blood insulin concentration with I_b being the basal value. $I'(t)$ is the above basal remote insulin, D is the glucose dose, and V is the glucose distribution volume. k_2 and k_3 are rate parameters and $NHGB$ is the net hepatic glucose balance that depends on the blood glucose and the remote insulin I'.

$$NHGB(Q(t), I'(t)) = NHGB_0 - [k_5 + k_6 I'(t)]Q(t)$$

and R_d is the rate of glucose disappearance from the peripheral tissues, which is also a function of blood glucose and remote insulin I'.

$$R_d(Q(t), I'(t)) = R_{d0} + [k_1 + k_4 I'(t)]Q(t)$$

Various nonlinear insulin models have been described. However, their parameterization is complicated because it is only possible to measure the posthepatic insulin concentration that is smaller than the pancreatic secretion levels. To overcome this, the C-peptide secretion that is equivalent in concentration to the insulin secretion and is not extracted by the liver, is measured as a proxy.

The insulin secretion model can be described by fast and slow components. The first phase secretion has a 2-min turnover and is probably from previously primed insulin secretory granules. It exerts derivative control as it is proportional to the rate of increase of glucose from the basal up to the maximum. The slow secretion is generated by new insulin secretory granules and is proportional to the glucose concentration.

$$SR(t) = mF(t) \tag{8}$$

where SR is the secretion rate and F is the ready realizable insulin described by

$$\dot{F}(t) = -mF(t) + Y(G, t)$$

with initial conditions $F(0) = F_0$, where F_0 is the amount of insulin released immediately after the glucose stimulus. $Y(G,t)$ is the provision of new insulin that depends on the glucose level

$$\dot{Y}(G, t) = -\frac{1}{T}[Y(G, t) - Y(G, \infty)]$$

where $Y(0) = 0$ and

$$Y(G, \infty) = \left\{ \begin{array}{ll} 0 & \text{if } G(t) < h \\ \beta[G(t) - h] & \text{if } G(t) \geq h \end{array} \right\}$$

These rather complex subsystem models can be integrated into a complete model of the glucose-insulin control system as shown in Fig. 25.

For conventional physical systems, equations can be developed from first principles to accurately describe their behavior, whereas maximal models of physiological processes rely on the interpretation of measurements. Those issues notwithstanding, complex models provide answers to "what if" questions in a teaching environment and more importantly to assess control algorithms and different insulin diffusion techniques. This method is sufficiently advanced that the FDA has accepted in silico trials using maximal models as a substitute to preclinical animal studies (Cobelli et al., 2009).

5.2 Closed-Loop Control

Control schemes can be open or closed loop, both of which aim to keep blood glucose within a desired range by compensating for disturbances using insulin. Typically, open-loop methods do not use real-time data to make their decisions whereas the closed-loop system exploits real-time measurements correlated with the control variable to react to disturbances.

A completely open-loop system would rely on fixed basal insulin administration throughout the day with additional boluses at meal times, based on patient characteristics without glucose measurements. Such a control system

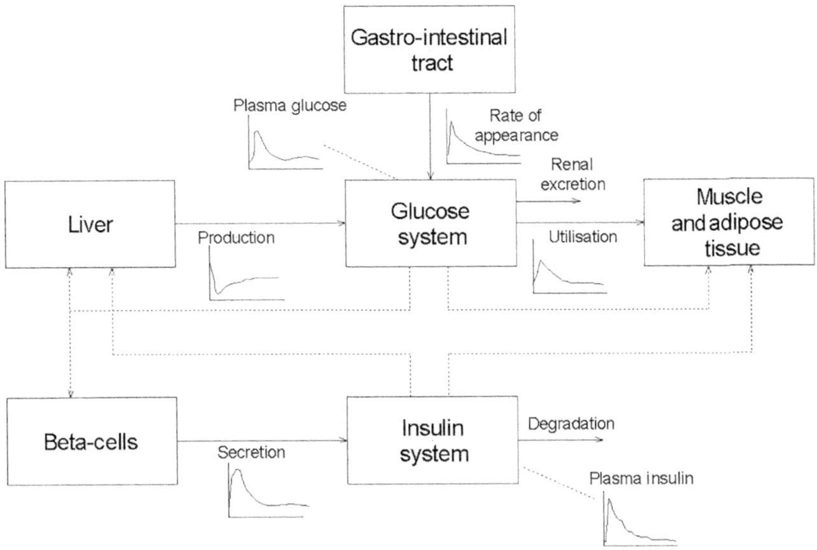

Fig. 25 Glucose-insulin control system that relates the measured blood concentrations of glucose and insulin to glucose fluxes (the rate of appearance, production, utilization and extraction) and insulin fluxes (secretion and degradation). *(Based on Cobelli, C., DallaMan, C., Sparacino, G., Magni, L., DeNicolao, G., Kovatchev, B., 2009. Diabetes: models, signals and control. IEEE Rev. Biomed. Eng. 2: 54–96.)*

could exploit knowledge of expected disturbances by altering the bolus size to account for different meal types. In theory, if patient dynamics and disturbances were perfectly known, it would be possible to design an open-loop insulin profile that ensured the desired glycemic control. However, in reality patient dynamics and the nature of disturbances are not well known, hence the need for corrections based on the actual patient state. Conventional therapy including the occasional finger-prick-based measurement is considered partially closed loop, but provides insufficient data for effective feed-forward compensation. It is only since the availability of continuous glucose monitoring (CGM) that it has been possible to design minimally invasive closed-loop control, as illustrated in Fig. 26. In general, only small corrections are required by the control system, but it can accommodate unpredicted events or disturbances as well as changes in the patient's dynamics (Patek et al., 2012).

Unfortunately developing suitable control algorithms is not that simple, primarily because of inherent delays in both glucose measurements and response times, as shown if Fig. 27. These delays can result in responses that

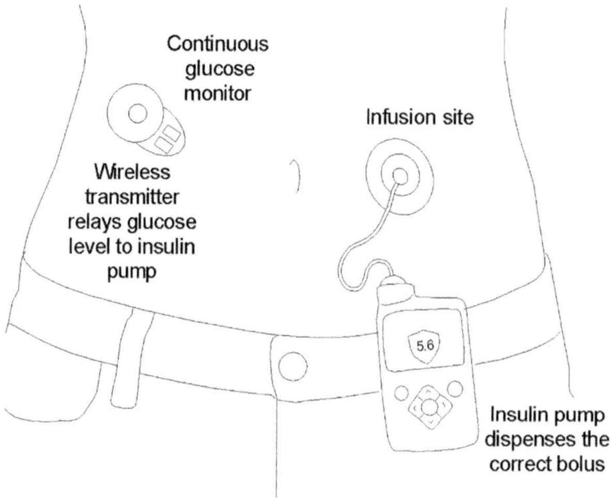

Fig. 26 Drawing showing the operation of a closed-loop insulin delivery system.

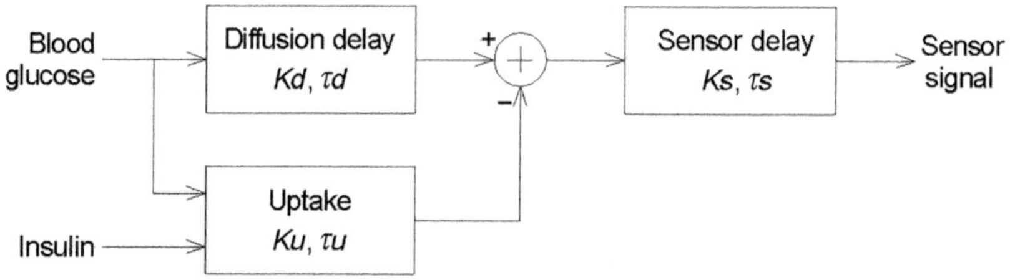

Fig. 27 Model to predict sensor signals from blood glucose and insulin. *(Based on Lucisano, J., Routh, T., Lin, J., Gough, A., 2017. Glucose monitoring in individuals with diabetes using a long term implanted sensor/telemetry system and model. IEEE Trans. Biomed. Eng. 64(9).)*

come too late to correct for the disturbance and often lead to overshoot. For example, an excessive administration of insulin following postprandial hyperglycemia can lead to a hypoglycemic episode a short while later.

The deployment of a successful control system relies heavily on the mathematical models used to describe the physiological functions over timescales relevant to the user's short-term wellbeing. The example described by Lucisano models the delays using a simple quasi-steady-state model with parameters determined from initial guesses and adjusted iteratively using the system identification toolbox in MATLAB until a good match between the predicted and observed values was obtained. The sensor delay, τ_s, was

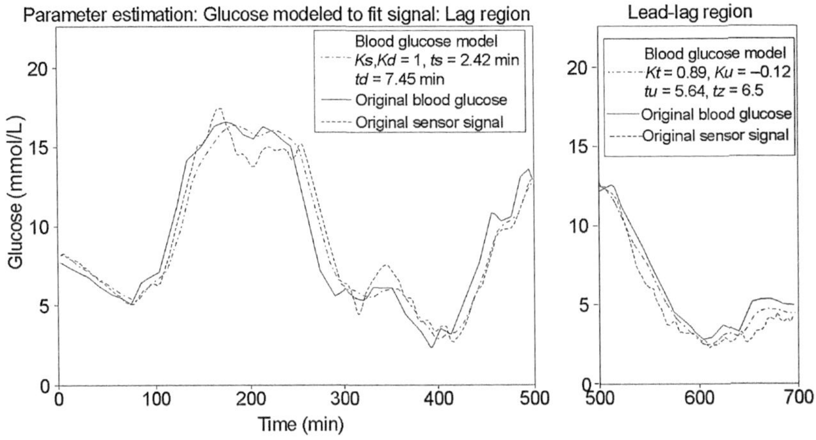

Fig. 28 Example of parameter estimation showing the ability of the model to identify glucose concentration in both lag and lead-lag situations. *(Data from Lucisano, J., Routh, T., Lin, J., Gough, A., 2017. Glucose monitoring in individuals with diabetes using a long term implanted sensor/telemetry system and model. IEEE Trans. Biomed. Eng. 64(9).)*

first determined in vitro for the response to a glucose step in well-stirred solutions. The diffusion delay, τ_d, was identified from the signal in the lag region with the gains set to unity, and finally the tissue uptake delay, τ_u, and gain, K_u, were then calculated from model terms (Lucisano et al., 2017). The results of the fitting process for one glucose clamp test are shown in Fig. 28.

However, in the wider scheme, there is still a problem with reconciling nocturnal control that is well suited to mild control actions and postprandial regulation that calls for prompt and energetic response. Compact models are mostly used with linear time-invariant models obtained either by linearization of the insulin-glucose response from a maximal model or by using a minimal model. Alternatively, the black box approach using system identification methods to obtain the relevant parameters, as described above, can be used.

Proportional, integral, derivative (PID) control is used widely in industry because it is simple to implement, flexible and easy to tune. The primary problem with this algorithm for glucose control relates to the integral term, particularly in the light of the transient response to a disturbance. For example, after a meal where blood glucose exceeds the set point, the use of a PID controller will result in an undershoot with a blood glucose level below the set point. It can be shown that the area of the undershoot will be comparable to the area of the overshoot. For that reason, it is convenient to dispense with the integral component and use a PD controller in which the proportional term is determined by the current glucose level, while the derivative term is determined by its variation in the previous half hour or so. These control algorithms are not very effective at providing a slow nocturnal response as well as a more aggressive response after eating, and so also require the addition a feed forward component to accommodate regulation after meals (Cobelli et al., 2009).

Model predictive control (MPC) provides the most effective approach to glucose control to date. The main components of MPC are the model, the cost function, and the constraints. The model is required to predict current and future states as well as system outputs and variables. The actual algorithm is not that important and can be linear, nonlinear, continuous, or discrete time. The cost function, usually quadratic, measures the quality of the closed-loop control and provides a penalty on future deviations from the required glucose concentration out to a prediction horizon. Finally, there may be constraints on the manipulated variables. For example, insulin flow will always be greater than zero and less than some maximum. The principal merit of MPC is that it reduces the control design problem to a sequence of finite-horizon optimization problems that allow it to deal easily with nonlinear dynamics. Fig. 29 shows schematically the operation of a MPC control system (Cobelli et al., 2009).

In addition to the basic PID, PD, and MPC methods, a number of alternative control strategies have been investigated in the recent years. These include nonlinear MPC and various neural network-based approaches (Semizer et al., 2012) as well as fuzzy logic controllers (Mauseth et al., 2013; DoyleIII et al., 2014).

Because the biggest risk of a closed-loop system is hypoglycemia caused by an over delivery of insulin, intense exercise, or the consumption of alcohol, modern devices generally include safety systems. These can include a low glucose prediction module or an insulin-on-board (IOB) calculator (DoyleIII et al., 2014). A good example is the modular closed-loop control structure described by Patek et al. (2012). The structure of the controller is shown in Fig. 30. Three primary modules are defined of which the lowest level one is the interface module (IM) that communicates with the glucose monitor and the insulin pump as well as providing an external interface and a data logging function. The real-time control module (RTCM) implements one of the control algorithms discussed above. Finally, the continuous safety module (CSM) monitors the patient's state and authorizes insulin recommendations that come to it form the RTCM. It can either use IOB estimates or can involve the patient in the approval process.

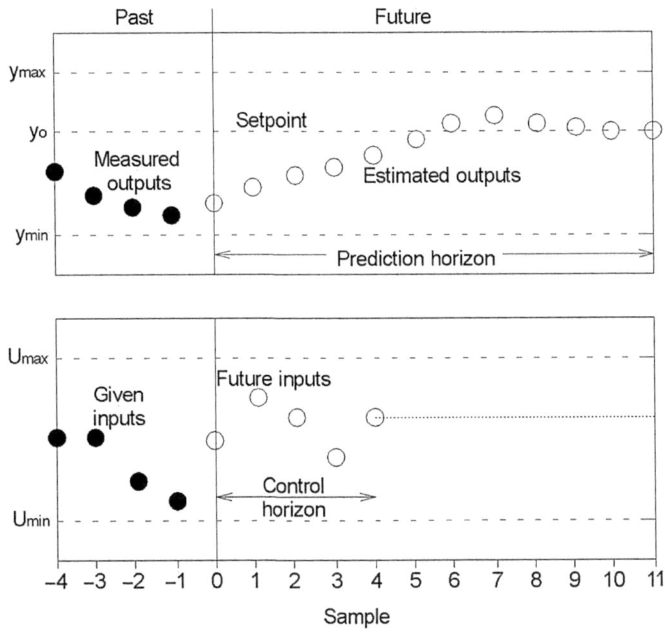

Fig. 29 The MPC prediction scheme relies on past inputs and outputs with future outputs predicted as a function of future inputs within the control horizon. *(Based on Cobelli, C., DallaMan, C., Sparacino, G., Magni, L., DeNicolao, G., Kovatchev, B., 2009. Diabetes: models, signals and control. IEEE Rev. Biomed. Eng. 2: 54–96.)*

5.3 The Future of Automated Insulin Delivery

As discussed earlier, managing diabetes without hypoglycemia is complicated by the wide fluctuation of insulin requirements between people and even for the same person from day to day and under different circumstances. These fluctuations are driven by a host of factors including meal size, activity level, illness, sleep deprivation, emotional stress, and menstrual cycle. Any success in effectively controlling levels is driven primarily by the accuracy and reliability of CGM systems and effective modeling of individual responses.

CGM accuracy has improved over the years due to improved filtering and denoising, a reduction in biofouling and enzyme degradation as well as numerous proprietary techniques used by various manufacturers. A consensus guideline relating to the best way to assess CGM accuracy is available but it has not been agreed upon by researchers. However, mean absolute relative difference (MARD) is quite common, and in conjunction with the percentage of readings exceeding 20% error is a good measure of CGM performance. A threshold below a 10% MARD and below a 12% probability of readings exceeding 20% error has is considered to be the sweet spot beyond

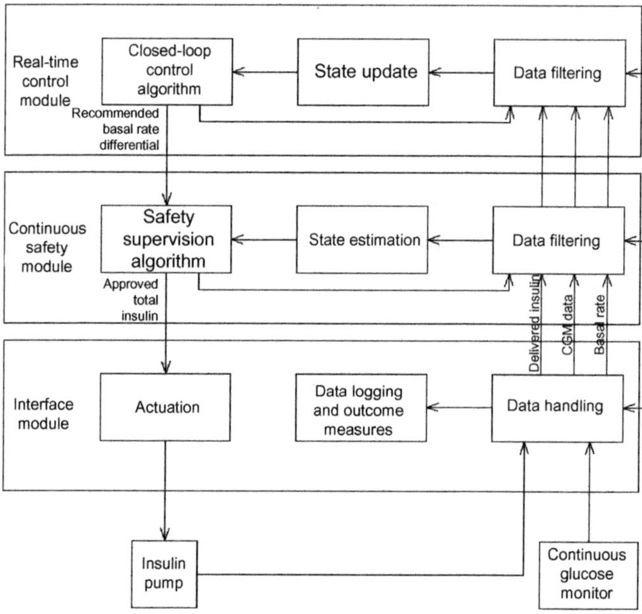

Fig. 30 The MPC prediction scheme relies on past inputs and outputs with future outputs predicted as a function of future inputs within the control horizon. *(Based on Cobelli, C., DallaMan, C., Sparacino, G., Magni, L., DeNicolao, G., Kovatchev, B., 2009. Diabetes: models, signals and control. IEEE Rev. Biomed. Eng. 2: 54–96.)*

which improvements in accuracy have little effect on clinical outcomes. To date only one system CGM system meets this criterion (Castle et al., 2017).

There is a drive toward producing CGM systems that are accurate enough without finger-prick calibrations for both control and insulin shut off. Commercial systems developed by Dexcom and Medtronic obtained FDA approval for both of these functions in 2016. Novel CGM systems include implantable types discussed earlier, and flash glucose monitoring (CGM on demand). The latter can replace finger-prick measurements completely in some cases.

One of the issues surrounding the incorporation of sophisticated control, safety, and multihormonal delivery into the insulin pump is that it complicates what would otherwise be a comparatively simple device into a complex power-hungry computer with a pump attachment. An alternative is to keep both the CGM and the pump as simple as possible and provide wireless communication to a smart phone as shown in Fig. 31. This approach leverages a powerful, yet cheap computing platform with performance driven by a massive user base beyond diabetics. This allows for new and improved

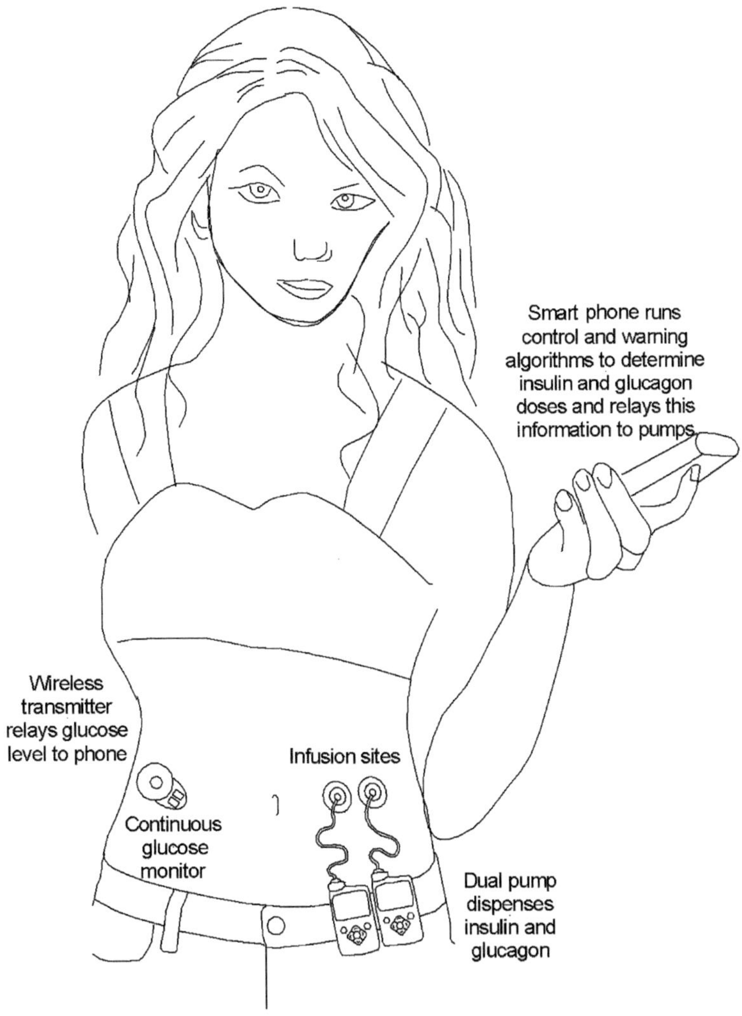

Fig. 31 Modular architecture for a closed-loop artificial pancreas. *(Based on Patek, S., Magni, L., Hughes-Karvetski, C., Toffanin, C., DeNicolao, G., DelFavero, S., Breton, M., DallaMan, C., Renard, E., Zisser, H., DoyleIII, F., Cobelli, C., Kovatchev, B., 2012. Modular closed-loop control of diabetes. IEEE Trans. Biomed. Eng. 59(11): 2986–2999.)*

algorithms to be easily implemented without requiring changes to the core CGM or pump hardware.

Future automation of insulin delivery will be further supported by the introduction of algorithms that are capable of using multiple signals to better determine the physiological state of the person. Studies of multihormonal systems will gauge the viability of insulin plus glucagon control and of alternatives like insulin plus pramlintide. Finally, all of these systems will benefit from the emerging understanding that there is no regime that fits everyone and that customized algorithms tailored to each individual diabetic are necessary to achieve the tight glucose range seen in the healthy people (Castle et al., 2017).

REFERENCES

Adams, F. (Ed.), 1856. Aretaeus the Kappadokia: The Extant Works. Francis Adams LLD, London.

American_Diabetes_Association, 2013. Standards of medical care for patients with diabetes mellitus—2013. Diabetes Care 36, 11–66.

Barrettino, D., 2017. Smart Contact Lenses and Eye Implants Will Give Doctors Medical Insights. http://spectrum.ieee.org/biomedical/devices/smart-contact-lenses-and-eye-implants-will-give-doctors-medical-insights. (Accessed 20 August 2017).

Blackshear, P., Dorman, F., Blackshear Jr., P.L., Varco, R.L., Buchwald, H., 1970. A permanently implantable self-recycling low flow constant rate multipurpose infusion pump of simple design. Surg. Forum 21, 136–137.

Blackshear, P., Dorman, F., Blackshear Jr., P.L., Varco, R.L., Buchwald, H., 1972. The design and initial testing of an implantable infusion pump. Surg. Gynecol. Obstet. 134, 51–56.

Bronzino, J. (Ed.), 2006. Medical Devices and Systems. CRC Press, Boca Raton, FL.

Buchwald, H., Barbosa, J., Varco, R., Rhode, T., Rupp, W., Schwartz, R.A., Goldenberg, F., Rubelin, T.G., Blackshear, P., 1981. Treatment of a type II diabetic by a totally implantable insulin infusion device. Lancet 1233–1234.

Castle, J., DeVries, J., Kovatchev, B., 2017. Future of automated insulin delivery systems. Diabetes Technol. Ther. 19 (S3), S67–S72.

Cobelli, C., DallaMan, C., Sparacino, G., Magni, L., DeNicolao, G., Kovatchev, B., 2009. Diabetes: models, signals and control. IEEE Rev. Biomed. Eng. 2, 54–96.

Dalvi, N., 2013. Microchip AN1560: Glucose Meter Reference Design. Microchip Technology, Chandler, AZ.

DiCristina, J., 2017. Maxim Tutorial 4659: blood glucose meters. https://www.maximintegrated.com/en/app-notes/index.mvp/id/4659. (Accessed 10 August 2017).

DoyleIII, F., Huyett, L., BokLee, J., Zisser, H., Dassau, E., 2014. Closed-loop artificial pancreas systems: engineering the algorithms. Diabetes Care 37 (5), 1191–1197.

Fernandez, M., Marcus, A., 1996. Insulin pump therapy: acceptable alternative to injection therapy. Postgrad. Med. 99, 125–132, 141–144.

Ganesan, A., 2013. Electronics in Medicine: Insulin Pump Design. http://www.eetimes.com/document.asp?doc_id=1280826. (Accessed 24 August 2017).

Hoskins, M., 2017. Implantable Insulin Pumps are Near Extinction, But Still Alive.... Diabetes Mine, http://www.healthline.com/diabetesmine/implantable-insulin-pumps. (Retrieved 24 August 2017).

Invernale, M., Tang, B., York, R., Le, L., Hou, D., 2014. Microneedle electrodes toward an amperometric glucose-sensing smart patch. Adv. Healthc. Mater. 3, 338–342.

Kirchhof, M., Popat, N., Malowany, J., 2008. A historical perspective of the diagnosis of diabetes. UWOMJ 78 (1), 7–11.

Leiva-Hidalgo, A., 2011. From pancreatic extracts to artiicial pancreas: history, science and controversies about the discovery of the pancreatic antidiabetic hormone VIII: step by step towards the artiicial pancreas. Adv. Diabetologia 27 (1), 27–38.

Leiva-Hidalgo, A., Brugues-Brugues, E., Leiva-Perez, A., 2011. From pancreatic extracts to artificial pancreas: history, science and controversies about the discovery of the pancreatic antidiabetic hormone. VIII Step by step towards the artificial pancreas. Adv. Diabetologia 27 (1), 27–38.

Loriaux, L., 2006. Diabetes and the Ebers Papyrus: 1552 B.C. Endocrinologist 16 (2), 55–56.

Lucisano, J., Routh, T., Lin, J., Gough, A., 2017. Glucose monitoring in individuals with diabetes using a long term implanted sensor/telemetry system and model. IEEE Trans. Biomed. Eng. 64 (9), 1982–1993.

Macfarlane, I., 2014. 1776: revolution in liverpool: Matthew Dobson discovers hyperglycaemia. Endocr. Abstr. 34, (SE1.2).

Mauseth, R., Hirsch, I., Bollyky, J., Kircher, R., Matheson, D., Sanda, S., Greenbaum, C., 2013. Use of a "fuzzy logic" controller in a clised-loop artificial pancreas. Diabetes Technol. Ther. 15 (8), 628–633.

Mazaferri, E., Skillman, T., Lanese, R., Keller, M., 1970. Use of test strips with colour meter to measure blood glucose. Lancet 265 (7642), 331–333.

Miller, B., Van-Slyke, D., 1936. A direct microtitration method for blood sugar. J. Biol. Chem. 114, 583–595.

Mossman, J., 2010. Insulin Pumps: Design Basics and Tradeoffs. EE Times, http://www.eetimes.com/document.asp?doc_id=1278073. (Accessed 24 August 2017).

Patek, S., Magni, L., Hughes-Karvetski, C., Toffanin, C., DeNicolao, G., DelFavero, S., Breton, M., DallaMan, C., Renard, E., Zisser, H., DoyleIII, F., Cobelli, C., Kovatchev, B., 2012. Modular closed-loop control of diabetes. IEEE Trans. Biomed. Eng. 59 (11), 2986–2999.

Rennie, I., Keen, H., Southon, A., 1964. A rapid enzyme-strip method for estimating blood sugar. Lancet 2, 884–886.

Ribet, F., Stemme, G., Roxhed, N., 2017. Ultra-miniaturisation of a planar amperometric sensor targeting continuous intradermal glucose monitoring. Biosens. Bioelectron. 90 (2017), 577–583.

Rubin, R., Peyrot, M., 2010. Patient-reported outcomes and diabetes technology: a systemic review of literature. Pediatr. Endocrinol. Rev. 7 (3), 405–412.

Schade, D., Eaton, R., Edwards, W., Doberneck, R., Spencer, W., Carlson, G., Blair, R., Love, J., Urenda, R., Gaona, J., 1982. A remotely programmable insulin delivery system. JAMA 247 (13), 1848–1853.

Schoemaker, M., Andreis, E., Röper, J., Kotulla, R., Lodwig, V., Obermaier, K., Stephan, P., Reuschling, W., Rutschmann, M., Schwaninger, R., Wittmann, U., Rinne, H., Kontschieder, H., Strohmeier, W., 2003. The SCGM1 system: subcutaneous continuous glucose monitoring based on microdialysis technique. Diabetes Technol. Ther. 5, 599–608.

Selam, J., Slingeneyer, A., Chaptal, P., Franetzki, M., Prestele, K., Mirouze, J., 1982. Total implantation of a remotely controlled insulin minipump in a human insulin-dependent diabetic. Artif. Organs 6 (3), 315–319.

Semizer, E., Yuceer, M., Atasoy, J., Berber, R., 2012. Comparison of control algorithms for the blood glucose concentration in a virtual patient with an artificial pancreas. Chem. Eng. Res. Design 90 (2012), 926–937.

Spencer, W., Bair, R., Carlson, G., Love, J., Urenda, R., Eaton, R., Schade, D., 1980. Some engineering aspects of insulin delivery systems. Diabetes Care 3, 345–350.

Swidorski, D., 2014. Diabetes History. https://www.defeatdiabetes.org/diabetes-history/. (Accessed 5 August 2017).

Tattersall, R., 2009. Diabetes:The Biography. Oxford Univeristy Press, Oxford.

Updike, S., Hicks, G., 1967. The enzyme electrode. Nature 216, 986–988.

Valla, V., 2010. Therapeutics of diabetes mellitus: focus on insulin analogueues and insulin pumps. Exp. Diab. Res. (Hindawi) 2010, 1–14 148372.

Vashist, S., 2012. Non-invasive glucose monitoring technology in diabetes management: a review. Anal. Chim. Acta 750, 16–27.

Yoo, E., Lee, S., 2010. Glucose biosensors: an overview of use in clinical practise. Sensors (Basel) 10 (5), 4558–4576.

Young, A., Herf, S., 1984. Biostator glucose controller: a building block of the future. Diabetes Educ. 10 (2), 11–12.

FURTHER READING

Crystalinks, 2007. Ancient Egyptian Medicine. http://www.crystalinks.com/egyptmedicine.html. (Accessed 5 August 2007).

CHAPTER TWELVE

Bioinspired and Biomimetic Micro-Robotics for Therapeutic Applications

Ahmet Fatih Tabak
Max-Planck Institute for Intelligent Systems, Stuttgart, Germany

Contents

1 MOTIVATION

The notion of tiny robots performing inside biological systems is one of the most science-fiction ideas subtly explored in the pop culture with different versions envisioned in movies such as Fantastic Voyage (20th Century Fox, 1966), Innerspace (Amblin Entertainment & Warner Bros., 1987), and The Island (Parkes/MacDonald Productions & DreamWorks & Warner Bros., 2005), and in TV series such as Star Trek Voyager (Paramount Network Television & CBS Television Distribution, 1995–2001). If one looks for the pattern of evolution of such systems portrayed in the silver screen, it will be clear that the perception changed from manned generic vehicles shrunk in

Handbook of Biomechatronics
https://doi.org/10.1016/B978-0-12-812539-7.00010-6

size to the specialized molecular machines that are programmed to conduct predefined tasks. However, after more than half a decade since Feynman's inspirational lecture "*There's Plenty of Room at the Bottom*" (Feynman, 1992), we did end up trying to recruit the single-celled organisms that fabled to eventually save the humanity from the Martian invasion as brilliantly narrated by Orson Welles (October 1938). The one thing remaining relatively common to this date is the idea that the microscale robotic agent should be able to exploit *swimming* as the method of traveling inside the living tissue.

At the time Feynman talked about the idea, he humbly conveyed as "*Hibbs' swallowable surgeon*" that was supposed to carry out therapeutic applications inside living organisms eliminating the need for the invasive medicine, he mainly discussed the possible ways to transmit energy to such systems and how to control them doing so. His suggestions were mainly focused on exploiting electromagnetic (EM) fields in comparison to the use of adenosine triphosphate (ATP) molecules by the single-celled organisms (Feynman, 1993). Today, we are still going around the same delicate issue but with more intense arguments as the problem proved to be far more complex and interdisciplinary due to the nature of the desired working environment for such robotic devices.

The aforementioned systems can also be classified as biomechatronic systems which are a combination of engineering methods in diverse areas with organic and inorganic materials actively interacting with each other and with their surrounding. If these systems are autonomous or semiautonomous, we would like to call them as biomedical robots. The general definition of robotics evolved within this micro-robotics context and what is known as the "*end-effector*" of the entire robotic system came to be referred as the robot itself. Furthermore, when biological components, such as living cells, are introduced as an integral part, these micro-robots can also be referred as cybernetic microsystems. The terms micro-swimmer and end-effector will be used interchangeably in this chapter, to emphasize the swimming robot as a separate entity and to emphasize it as an integral part of a much larger robotic system, respectively, whenever the context allows doing so.

Robotics and cybernetics will inevitably dominate certain parts of the global market, industry, military, and medicine as a workforce. Tentative medical applications are still considered to be highly challenging and arduous research topics. Conventionally, it is safe to say that we are interfering with the useful bacteria indirectly and fighting an active war against harmful ones. Another perception has emerged in the last couple of decades. Scientists are constantly inspired by natural micro-swimmers, and we are trying to employ the micro-swimmers in a more direct way with an increasing enthusiasm. This evolution of robotics research efforts that will be humbly outlined here has a long past and an even longer journey ahead. Therefore, the reader will be initiated to the two main conundra waiting to be solved, that is, energy efficiency and biocompatibility, for the particular field of bioinspired micro-robotic swimmers for medicine as focused on the entirety of this chapter.

It is safe to say that the bulk of the effort spent to this date has been indeed to address these two main problems and their implications for the design, actuation, and control of such robotic systems. The success of these systems will revolutionize the modern medicine in terms of elevated patience complacency and reduced drug use as the need of invasive procedures may be greatly reduced in some cases and the total amount of consumption of potent chemicals will significantly drop. Cancer treatment, the remedy to a specific group of terminal diseases, is a very vivid example of bombarding the entire body with drugs to kill only a certain group of cells. Due to special conditions in the cancerous tissue, such as high pressure, a significant amount of the administrated drug does not concentrate on the targeted area. In that regard, the current efforts are orchestrated to achieve drug delivery in small doses with high accuracy, thus sparing the patient of a rather overwhelming treatment procedure (Felfoul and Martel, 2013).

The subject matter hereafter is mainly focused on swimming micro-robots as some of the major constituents such as circulatory, lymphatic, digestive, and central nervous systems contain ducts and channels filled with liquids that can be used as conduits for a planned journey. In this text, untethered, cm- (10^{-2}) to nm-scale (10^{-9}), and robotic experiments and demonstrations have been considered as suitable examples of application. However, it should be noted that there lies a massive bulk of scientific and engineering effort behind. Although the focus will be on bioinspired robots, the reader will find that the manufacturing techniques and the actuation methods are far agiler, and therefore could be useful in a wider range of robotic systems (Kummer et al., 2010). It is also important to note that nano-robotics and nano-science are not included in great detail in this context; however, the reader will find the flavor whenever it is deemed important to mention.

All variables are assumed to be in standard international (SI) units as presented following the designated symbol of the respective parameter throughout the chapter; however, coefficients and dimensionless number groups are exceptions to this practice.

2 BACKGROUND

The concept of swimming inside the human body is more promising for a micro-robot than other methods of commuting as human beings are basically electrochemical machines with an intense and interconnected web of ducts and channels filled with liquids of different properties. Purcell put the conditions of micro-swimming in perspective with his famous "*Scallop theorem*" providing calculations and demonstrating how conventional methods of propulsion should endure accomplishing very little in terms of controllable displacement. He also argued that the efficiency of swimming in micro-realm is expected to be very low as the entire process is rather dependent on continuous time-irreversible action to harness propulsive effect mainly from the shear resistance of the surrounding medium instead of inertia of the swimmer (Purcell, 1977). This is indeed the condition on which bacteria and spermatozoa swim as an inspirational answer, to high viscous forces dominating inertia, provided by nature. However, this also means that one of the biggest challenges is to supply the energy in order to sustain the continuous swimming as long as it is needed. Hence, that requirement should be minimized with increased efficiency.

The natural micro-swimmers, such as bacteria and spermatozoa that inspire us, are fully functional, "*nonholonomic*," that is, velocities are not independent of each other and the position is dependent on the total trajectory history of the moving object (Kane and Levinson, 1985), and autonomous systems with onboard energy generation and steering capabilities. They possess rudimentary survival instincts based on chemical reactions and spatial gradients. They continuously search for nutrients with a series of motions generally classified with the suffix-taxis; in a sense helping them to decide their next move based on their interaction with the surroundings as they produce the energy required to commute (Lighthill, 1976).

What makes them successful is that their ability to move in an environment not suitable to move with conventional propulsion methods mainly relying on inertial forces. A helpful analogy would be thinking of a human being trying to swim in a pool filled with honey. The repeated swimming stroke that person will exploit in that viscous liquid will prove to be useless since an action tracing its own steps back as the time progress will generate only a minute amount of displacement with respect to the body length. Indeed, the overall performance can be envisioned as a hysteresis behavior with which almost all power is spent in terms of heat loss due to enormous viscous drag that reigns over the momentum transfer from arms to the liquid at each cycle as articulated with the "*Scallop theorem*" (Purcell, 1977).

It has been observed that, in order to overcome this energy-loss obstacle, a majority of natural micro-swimmers employ microstructures and chemical motors to generate wave propagation, either planar or helical, along with slender structures hanging from their cell membrane. Thus, they break spatial and temporal symmetry to harness forward thrust from viscous friction. That friction is present as long as the wave is propagating, meaning that once the bacterium or sperm cell ceases propagation of waves the swimming motion is supposed to stop instantaneously (Purcell, 1977; Lighthill, 1976).

The very first comprehensive theoretical work exploring the physics to understand this phenomenon was presented by Sir Taylor in 1951 (Taylor, 1951). He introduced the particular stream function, Ψ (m^2/s); a mathematical function representing the flow field around a moving object in viscous domains, as depicted in Fig. 1, for simple two-dimensional (2D) wave propagation and combined it with force- and torque-free swimming concept to demonstrate the asymptotic relationship between wave propagation and swimming efficiency. In a sense, the stream function represents the diffusion of flow field within the liquid by representing it via a cross-sectional view of the bulk. From that point onward, in over 60 years, the efforts mounted toward the very first computer-controlled artificial cybernetic micro-swimmer, towing a red blood cell, presented by Dreyfus et al. (2005).

The body of work presented in this subsection can be grouped into the following categories: observations of bacteria and spermatozoa, mathematical models explaining the swimming behavior based on observations,

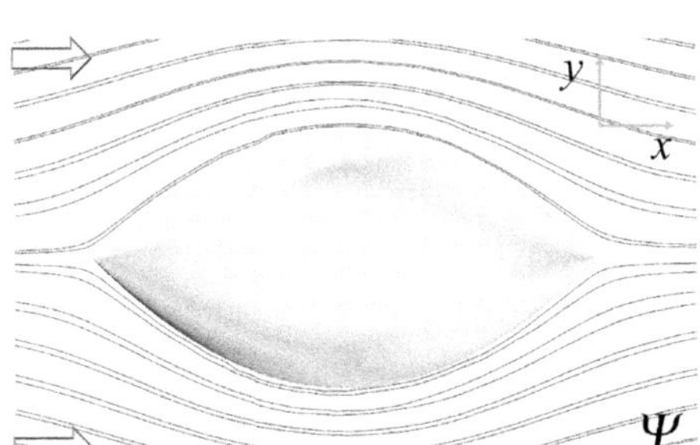

Fig. 1 The flow field, that is, *black lines*, around a stream-lined object immersed in viscous flow. The *black lines* can be represented by a mathematical function, Ψ (m^2/s), from which the local velocity could be deduced, for example, the *x*-component is obtained by $u_x = d\Psi/dy$ and the *y*-component is given by $u_y = -d\Psi/dx$, in Cartesian coordinates. The *arrows* show the flow direction of the far field.

numerical investigations (especially after powerful numerical tools have emerged), conceptual work on swimming micro-robots, experiments on artificial micro-swimmers, experiments with cybernetic micro-robots, control studies, robotic swarm studies, and conceptual demonstrations.

2.1 Observations of Bacteria and Spermatozoa

At this juncture, it should be sufficient to briefly present key elements to this context gathered from observation. The first example is the sea-urchin spermatozoa which possess a single slender tail known as a flagellum. The flagellum is actuated locally by sliding protein chains embedded within. Thus, planar wave propagation is induced from the base of the flagellum, where the tail and head are connected with a flexible joint, toward the end trailing freely (Gray and Hancock, 1955; Brennen and Winet, 1977). On the other hand, bacterium species utilize helical-wave propagation along their flagellum or bundle of flagella. The flagellum, in this case, is attached to the cell membrane with nanoscale machines called bacterial motor. Surprisingly, this structure has a bearing feature similar to an electric motor that allows it to rotate at the base canceling motion on the other axes, thus rotating the flagellum. The elastic flagellum takes a helical shape due to this rotation. As for the bacterium species with bundles, such as *Salmonella abortus equi*, the wave propagation is more of a combination of rotation and active deformation along the intertwined flagella group (Brennen and Winet, 1977; Flores et al., 2005; Berg and Anderson, 1973).

It is also important to note that geometry of waves is not constant under all conditions. Given that the structural deformation occurs as a result of force balance between internal and external stress, the distinctive wave parameters such as wavelength, λ, wave amplitude, A, and angular frequency of the wave, ω, are subject to change under different viscosities (Brokaw, 1966). A very interesting observation made over the years is that some sperm cells can induce helical-wave deformation along their flagellum. It is now understood that certain spermatozoa cells, such as of the *Echinus esculentus*, are capable of switching to helical-wave propagation when the viscosity of the liquid they are immersed in is lower than certain values (Woolley and Vernon, 2001). It is safe to say that this behavior is to optimize the efficiency of swimming performance thanks to the overall survival strategy. The reader will later be introduced to the possible mathematical definitions efficiency, η, followed by informative discussions on its importance and implications within the contexts of geometric and structural design, motion control, and maneuvering of swimming micro-robots, as a key issue.

The wave patterns are also found to be of different external and internal stimuli. The wave parameters are not necessarily homogeneous throughout the entire swimming action. This is caused by either viscosity or sporadic changes in direction of swimming (Lighthill, 1976; Brennen and Winet, 1977; Brokaw, 1965; Gibbons and Gibbons, 1980). The reader is encouraged to study the references for the details of changes in wave patterns. However, one particular stimulus important for all intents and purposes is the wall effects. Flow fields with the presence of a solid boundary nearby the micro-swimmer will induce attenuated or amplified shear stress on the head or the tail. This will inevitably affect the wave pattern and the swimming behavior (Brennen and Winet, 1977; Fauci and McDonald, 1995; Lauga et al., 2006). Unfortunately, it is not straightforward to determine the overall effect; the local viscous stress is either inversely proportional to the local proximity to the solid boundary, h (m), or linearly proportional to the logarithm of $1/h$. Only after taking the integral over the entire flagellum and summing up with the effect on the head, one can predict the instantaneous influence (Brennen and Winet, 1977; Lauga et al., 2006) which leads us to the mathematical modeling to the physics of swimming in micro-realm via wave propagation.

2.2 Physics of Swimming in Micro-Realm

Normally the most powerful approach, to solve the hydrodynamics of a complex geometry going through six degrees of freedom (6 DOF) rigid-body motion while it is carrying out wave propagation on only a certain portion of its surface area, is to summon Navier-Stokes equations with proper boundary conditions and appropriate meshing. These set of equations articulate the conservation of momentum and mass, respectively, but for flowing matter instead of for objects experiencing rigid-body motion:

$$\rho\left(\frac{\partial \mathbf{U}}{\partial t}+(\mathbf{U}\cdot\nabla)\mathbf{U}\right)-\rho\mathbf{g}=-\nabla p+\mu\nabla^2\mathbf{U} \tag{1}$$

$$\rho\nabla\cdot\mathbf{U}=0 \tag{2}$$

where t (s) denotes time, $\mathbf{g}$ (m/s^2) is the gravity vector, ρ (kg/m^3) is the density, μ (Pa s) denotes the dynamic viscosity, p (Pa) is the static pressure of the fluid, and $\mathbf{U}$ (m/s) signifies the flow field vector. It should be noted that these equations are valid for incompressible liquids. The left-hand side of the first equation is responsible for the acceleration and inertial effects of the fluid. It is important to acknowledge that the much-dreaded nonlinearity of the equation happens to be due to the $(\mathbf{U}\cdot\nabla)\mathbf{U}$ term which also dictates that the acceleration is not necessarily a time-dependent phenomenon in the explicit form.

However, taking into account the fact that computers and the numerical methods being used in abundance today were not simply available at the time when the problem was first studied, mathematicians had to avoid tackling this highly nonlinear equation as a whole, and came up with ingenious simplified models to represent the problem without higher-order effects which are expected to be insignificant due to dimensions and the *effective viscosity* in micro-realm. In other words, one aspect that engineers try to eliminate from macroscale systems whenever convenient, that is, friction, gave rise to a completely new and much simpler modeling approach for micro-swimmers. Effective viscosity can be explained best by the analogy of "*real feel*" of the temperature due to humidity levels, and it is directly related to the famous Reynolds number (Purcell, 1977), which will be discussed at length later. Although this simplification will be of great help, it will be evident later that including all possible physical stimuli acting on the micro-swimmer renders the modeling laborious and delicate, nevertheless.

2.2.1 Mathematical Models for Hydrodynamics of Flagellar Motion

From the mathematical and, arguably, the chronological points of view, the first approach to explore is the slender-body theory (SBT). In 1953, Hancock talked about singularities representing the flow fields around slender objects and relatively simple geometries in order to explain the fluid dynamics of the swimmers (Hancock, 1953). Although it was in 1955 that the resistive force theory (RFT) approach was fully articulated with the use of local force coefficients introduced by Gray and Hancock, as a mathematical conclusion of the previous analysis followed by a series of simplifications and assumptions (Gray and Hancock, 1955), the core of the SBT method lingered in the literature for 22 years before Sir Lighthill brilliantly took it to the next step (Lighthill, 1976) while referring to the paper published by Hancock (1953) on the distribution of Stokeslets along a beating flagellum. SBT employs "*Stokeslet functions*" that represent an approximation for the flow field invoked by a singularity, that is, dipole force, $\mathbf{f}$, which articulates the physical stimuli driving the resultant local flow field, $\mathbf{U}$. Hence, the flow field can be modeled in terms of that particular singularity associated with the local structural deformation and the dynamic viscosity, μ, of the surrounding liquid. The solution could be obtained after $\mathbf{f}$ is introduced as an external force term in the simplified conservation of momentum equation governing the viscous domain, also known as the Stokes equations for incompressible creeping flow conditions (Lighthill, 1976):

$$0 = -\nabla p + \mu \nabla^2 \mathbf{U} + \mathbf{f} \tag{3}$$

The approximated local flow field is usually represented by stream functions, Ψ, such as the one Sir Taylor made use of (Taylor, 1951). The stream function is a geometric representation of the path of fluid packets around a stationary or moving object. The reason the term "*fluid packet*" is preferred is to look at the flow field at a safe distance, with a zoomed enough view, while staying clear of the quantum limit where it is possible to observe the motion of individual molecules. In such a case, usual conservation equations do not apply as the continuity assumption is violated. Hence, when a singularity or stream function is modeled, the volume of interest is much larger than a single molecule and all intermolecular interactions are lumped as bulk properties of flowing matter at an arbitrary location. For instance, Sir Lighthill (Lighthill, 1976) articulated the "*Stokeslet velocity distribution*" associated with the dipole force, **f**, as follows:

$$\mathbf{U} = \nabla \Psi = \frac{\mathbf{f}}{6\pi\mu r_f} + \frac{1}{4} r_f^2 \nabla \left[\nabla \cdot \left(\frac{\mathbf{f}}{6\pi\mu r_f} \right) \right] \tag{4}$$

where r_f (m) denotes the radial position from the singularity.

Similarly, it is possible to come up with a "*rotlet function*" that represents the flow field based on the local rotation of a specific point. The magnitude of the Stokeslet function is inversely proportional to the distance to the selected point of analysis whereas the magnitude of the rotlet function is inversely proportional to the cube of the same distance. The overall hydrodynamic force acting on the flagellum is found by the integration of forces, $\mathbf{f}(s,t)$, represented by the resultant singularity distribution throughout the flagellum, s, at any given instance, t, (Higdon, 1979; Johnson and Brokaw, 1979). The intricacies of modeling the dipoles and obtaining the solution of the Stokes equation are left to the reader.

Once the local force distribution is obtained, it is possible to reevaluate the local force-velocity relation in terms of local resistance force coefficients, which brings us to the next approach known as the RFT. RFT is a "*suboptimal solution*" to the SBT and allows us to model the local force in the form of $\mathbf{f}(s,t) = \mathbf{C}\mathbf{u}$, where $\mathbf{C}$ (N s/m^2) denotes the diagonal matrix of local resistive force coefficients written in Frenet-Serret coordinates, that is, c_t, c_n, and c_b, and $\mathbf{u}$ (m/s) represent the local velocity vector of the flagellum

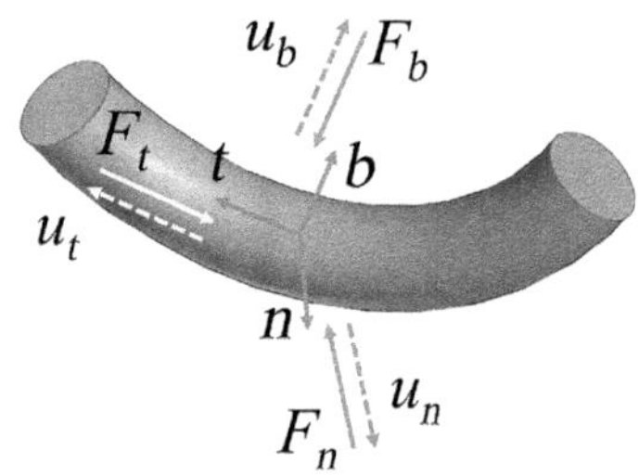

Fig. 2 Local Frenet-Serret coordinates on a curved slender body in micro-realm (Higdon, 1979). The local velocity of the slender surface and the related viscous drag acting on it are found to be in opposite directions. The subscripts *t*, *n*, and *b* denote the local tangent, normal, and binormal directions as illustrated by red arrows. *F* (m/s) signifies the local force per unit length and *u* (m/s) signifies velocity at an arbitrary point on the surface of the slender body.

with respect to the surrounding liquid also written in the same coordinate frame, that is, u_t, u_n, and u_b (Lighthill, 1976; Gray and Hancock, 1955; Hancock, 1953; Keller and Rubinow, 1976), as depicted in Fig. 2. It is important to acknowledge that with RFT approach, the far field around the flagellum is omitted, thus SBT is inherently superior to RFT in terms of hydrodynamic analysis (Johnson and Brokaw, 1979). There are several RFT coefficient sets in the literature; however, two of them are worth mentioning the most: the suboptimal set presented by Sir Lighthill for swimmers in fluid bulk (Lighthill, 1976), and the coefficient set, provided as a function of proximity to rigid walls, compiled by Brennen and Winet (1977). Once the suitable local coefficients are selected, they are used to construct the overall resistance matrix of the entire micro-swimmer, which is incorporated into the overall robotic model. This procedure will be discussed at length in the following subsections.

2.2.2 Numerical Investigations on Hydrodynamics of Micro-Swimmers

When we talk about numerical methods, we enter the world of computational analysis. The nature of any profound physical phenomenon is represented by a specific governing equation written for an infinitesimal volume of the bulk of the domain in question. The governing equation and following boundary conditions are valid down to the limiting dimensions where the continuity assumption holds as discussed before. The domain, regardless of its dimensions, should be discretized with some sort of "*node*" at which the solution will be obtained. The rest is predicted with another method, such as interpolation with shape function. Because of the demanded accuracy of the numerical problem, the solution could be quite costly in terms of time and computational resources. Unfortunately, the nonlinear nature of the fluid dynamics problem at hand does not allow most of the numerical tools to be used for fast and cheap calculations let alone real-time control. On top of that, a powerful computational tool, either a commercial package or a custom code, is inevitably required to investigate the hydrodynamics of micro-swimmers with higher accuracy for micro-robotic research.

It is important to concisely acknowledge the numerical methods, namely the computational fluid dynamics (CFD), used to further understand the hydrodynamics of the problem that calls for considerable computational power to resolve higher-order effects taking place during the momentum transfer from a micro-swimmer to the surrounding liquid. CFD methods can be primarily categorized into three main branches: (i) finite element method (FEM), with which either full Navier-Stokes equations or only Stokes equations, subject to continuity, are solved with discretizing the entire domain occupied by the liquid (Tabak and Yesilyurt, 2014a); (ii) boundary element method (BEM), with which only boundaries are discretized and the flow field is predicted by mathematical functions such as Stokeslet functions (Phan-Thien et al., 1987); and (iii) immersed boundary (IB) method, with which the internal structural stresses are coupled with the fluid stresses at the boundary that is discretized as presented by Maniyeri et al. (2012), and the authors utilized the finite volume method (FVM) instead of FEM to discretize the domain. In the FEM analysis, boundary conditions are set as "*no-slip*" for stationary solid boundaries, "*velocity vector*" for moving boundaries, and the main constituents of the solution are the viscous force on boundaries, velocity in the bulk of the liquid, and pressure profile in the entire domain (Tabak and Yesilyurt, 2014a) (see Fig. 3). The details of and differences between FEM and FVM are left to the reader to investigate.

Indeed, many more modeling studies are present in the literature. At this juncture, the reader should focus the attention on the fact that numerical analysis is invaluable to examine the hydrodynamic interactions. Ascertaining the hydrodynamic interactions between solid boundaries and the micro-swimmer (Fauci and McDonald, 1995), within flagella bundles (Flores et al., 2005), or between body and tail of a micro-swimmer (Tabak and Yesilyurt, 2014a) is crucial for gait planning and robust robot control. It should be also noted that mathematicians demonstrated the implication of these secondary effects with delicate methods (Lighthill, 1996); however, the analysis gets rather laborious, and sometimes to the point of

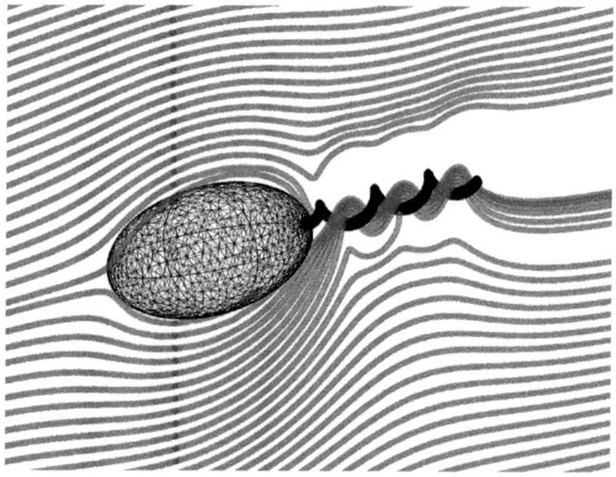

Fig. 3 Computational fluid dynamics (CFD) simulation example: a bacteria-like micro-swimmer swimming in the liquid bulk, depicted with the plane-confined streamlines of surrounding flow field. Tetrahedral elements are used to mesh the surface of the micro-swimmer and the entire fluid-domain. Streamlines are found by finite element method (FEM) solution of Navier-Stokes equations subject to conservation of mass. The solution is obtained by COMSOL Multiphysics (COMSOL AB, Stockholm, Sweden).

impractical, as the combined geometry of the micro-swimmer and the surrounding boundaries forces induced flow fields to be asymmetric and intricate.

2.2.3 Importance of Dimensionless Analysis for Robotic Design and Applications

Before going further deep into design and analysis of micro-swimmers, the reader should be familiarized with a list of dimensionless numbers. Envisioned micro-robotic systems, either experimental or numerical, vary in size and vary in the method of actuation of the end-effector. Furthermore, the manufacturing methods could present distinct physical properties that are not common in general. In short, dimensionless numbers are required to examine, characterize, and compare the micro-robotic systems discussed in the literature, in terms of hydrodynamics, elasto-hydrodynamics, magnetohydrodynamics, and magneto-elastic characteristics, whenever they are pertinent to the application.

The dimensionless numbers presented here are referred to help researchers to find the optimum conditions of the micro-swimmers that are fully submerged in liquids and of the swimming conditions these micro-swimmers are subject to. Moreover, dimensionless numbers are helpful determining the design space in which the optimum will be searched for. Finally, they allow researchers to study the behavior of such systems with large-scale mock-ups and with CFD simulations as building and tracking a self-sustaining and autonomous swimmer in actual micro dimensions are not always trivial.

The most important dimensionless number is the Reynolds number (Re) (Purcell, 1977). It provides information on the ratio of inertial forces to shear forces. If Re number is smaller than unity, Navier-Stokes equations could be simplified to Stokes equations. This is known as the Stokesian-flow assumption where RFT and SBT methods are also subject to this condition for validity. It is important to note that the smaller the Re number is, the more accurate the Stokesian-flow approach will be. The Re number is given as

$$\mathrm{Re}=\frac{\rho UL}{\mu}=\frac{\rho L^2 f}{\mu} \tag{5}$$

where L (m) is the characteristic length of the micro-swimmer, such as the diameter of the body, and U (m/s) is the swimming velocity of the center of mass of the micro-swimmer. The letter f signifies the frequency, in Hz, of the wave propagation should the characteristic length is selected to be the wavelength, λ (m). So, the characteristic length and the characteristic velocity remains subjective depending on the analysis. Consequently, different Re numbers can be defined and used. Re number is virtually the most preferred dimensionless quantity to prove the validity of the experiments and analysis, numerical or theoretical, in the literature to this date. Moreover, the reciprocal of Re number, that is, 1/Re, gives the effective viscosity of the flow field; as Re number attains smaller values, the effective viscosity of the flow field increases. Another number that is dependent on velocities is the Strouhal number (St) (Lovalenti and Brady, 1993; Ido et al., 2016): Although the St number is associated with high Re number flows, it can be used as an indication of efficiency. Because, while investigating within a design space, relatively smaller St number means loss of useful momentum in terms of "*parasitic circulations*" within the bulk of the flow. The St number is given as

$$\mathrm{St}=\frac{fL}{U} \tag{6}$$

Next, we study the Sperm number (Sp) (Dreyfus et al., 2005; Wiggins and Goldstein, 1998). Sp represents the ratio of viscous forces acting on the flagellum to the balancing structural stresses. It is also the ratio of characteristic length to the "*viscous penetration length*," L_ν (m), when the flagellum is undulating. It defines the shape of the flagellum that will give the optimum momentum transfer to the fluid. If the Sp number is smaller than unity, reciprocation of the elastic flagellum is converging to the limit of a rigid slender object. As the Sp number becomes very large, the undulation is not effective as the flagellum is virtually indistinguishable from the surrounding liquid at such conditions. Therefore, there exists a distinct Sp number with maximum hydrodynamic thrust for the selected natural micro-swimmer or the tailored artificial flagellum. The Sp number is given as

$$\mathrm{Sp} = \frac{L}{L_v} = \frac{L}{\left(\kappa / c_n \omega\right)^{0.25}} \tag{7}$$

where κ (J/m) is the bending rigidity of the flagellum and ω (rad/s) is the angular frequency of the actuation. Next comes the magnetoelastic number, $\mathrm{Mn_e}$ (Dreyfus et al., 2005; Roper et al., 2006). This number is also known as the "*dimensionless magnetic strength*" or square of the ratio of characteristic length to the "*magnetoelastic persistence length*," and it becomes important if the elastic flagellum has significant magnetic properties and reacts to the external magnetic field: as $\mathrm{Mn_e}$ increases, the forward velocity is also expected to increase; and *vice versa*. However, distinct regions with structural stability are observed, and it has been reported that buckling is expected to occur with $\mathrm{Mn_e} = \pi^2/4$ (Roper et al., 2006). The $\mathrm{Mn_e}$ number is given as

$$\mathrm{Mn_e} = \frac{2\pi (rBL)^2}{3\mu_0 \kappa} \left(\frac{\chi_t - \chi_n + \chi_t \chi_n / 4}{(1 - \chi_t/6)(1 + \chi_n/12)} \right) \tag{8}$$

where χ denotes the direction-dependent susceptibility given by respective subscripts, r (m) is the radius of the tail given that it has a somewhat homogeneous magnetic quality, B (T) is the magnitude of the applied magnetic field-density, and μ_0 (H/m) is the permeability.

In order to quantify the magneto-hydrodynamic behavior of the micro-swimmer, we use the Mason number, $\mathrm{Mn_h}$, which is the ratio of the viscous torque to magnetic torque, and akin to magnetoelastic number, structural stability is only observed under certain values of $\mathrm{Mn_h}$ after exceeding which the artificial flagellum is expected to rupture (Ido et al., 2016). The $\mathrm{Mn_h}$ number is given as

$$\mathrm{Mn_h} = \frac{32 \mu \omega}{\mu_0 \chi^2 |\mathbf{H}|^2} \tag{9}$$

where $\mathbf{H}$ (A/m) is the magnetic field-strength vector acting on the magnetic flagellum or at the joint between the magnetic head and the flagellum.

Another useful phenomenon to characterize is the ratio of convective displacement to diffusive displacement which is given by Peclet number, Pe (Kasyap et al., 2014). This number indicates the effectiveness of a micro-swimmer towing a particle in the viscous medium, and the larger it is the more beneficial to use a micro-swimmer to deliver a cargo to a predefined destination. It is important to note that, indeed, there is a theoretical limit to Pe number because Re number cannot exceed unity if Scallop theorem and Stokesian-flow assumption are expected to be valid. The Pe number is given as

$$\mathrm{Pe} = \frac{UL}{D} \tag{10}$$

where D (m^2/s) stands for the diffusivity coefficient of the cargo in question. Diffusivity is a term related to the mass transfer of species and very small particles, and it is dependent on temperature and the dynamic viscosity of the liquid medium (Probstein, 2003) where the micro-swimmer is submerged in.

The last dimensionless number we will review is the Deborah number, De (Teran et al., 2010; Liu et al., 2011). De provides qualitative information on the physical condition of the flow field: De number being smaller than unity indicates that the relaxation time of the liquid is short enough so the applied stress is conducted without spatial or temporal attenuation in the liquid bulk, thus the shear rate is not time dependent. On the other hand, De number larger than unity signifies rheological behavior where the stress is not propagating in the liquid with isotropic fashion (Guyon et al., 2015) leading to spatially and temporally inhomogenous stress tensor, $\boldsymbol{\sigma}$ (N/m^2), and time-dependent viscous shear stress. Also, there exists an optimum De for planar wave propagation: undulating flagellum is supposed to be more efficient in non-Newtonian fluids up to De $\sim$3 after which propulsive thrust gradually decreases for a micro-swimmer in the non-Newtonian environment (Teran et al., 2010). The flagellum can be elastic or rigid, and still, propagate helical waves under which condition again an optimum expected to present itself (Liu et al., 2011). The De number is given as

$$\mathrm{De} = \delta_t f \tag{11}$$

where δ_t (s) is the relaxation time of the liquid which can be a polymer solution or a Boger fluid which has a shear rate-dependent dynamic viscosity (Teran et al., 2010; Liu et al., 2011). This also means that, if the other physical constraints allow, a Newtonian liquid will start exhibiting viscoelastic behavior with the influence increasing actuation frequency, although this is not expected under normal conditions.

As it should be obvious to the reader by now, the nondimensional analysis is meaningful if only the physical limits are observed properly. For instance, increasing the actuation frequency while studying Re, Sp, Mn_h, or De numbers of a magnetic micro-swimmer with an elastic flagellum does not necessarily mean something impressive will happen as there are two possibilities at the extremum: either the elastic tail will lose structural integrity at some point, or the deformation will go to zero as viscous friction will simply overwhelm structural deformations. Furthermore, even before structural failure occurs, the step-out problem comes to the attention. Step-out happens when the magnetic body cannot properly follow the magnetic field (Tabak et al., 2011). The magnetic field might rotate sweeping all quadrants or just oscillate, as depicted in Fig. 4, with a smaller reciprocating angle. However, if that rotation or oscillation is very rapid than the magnetic body will not be able to align itself properly, regardless of the angle swept by the external magnetic field. In other words, if the viscous torque to overcome in order to instantaneously align with the magnetic field is higher than the magnetic torque, desired wave propagation will not be achieved. Should this condition arise, step-out phenomenon will be observed with elastic and rigid tails alike.

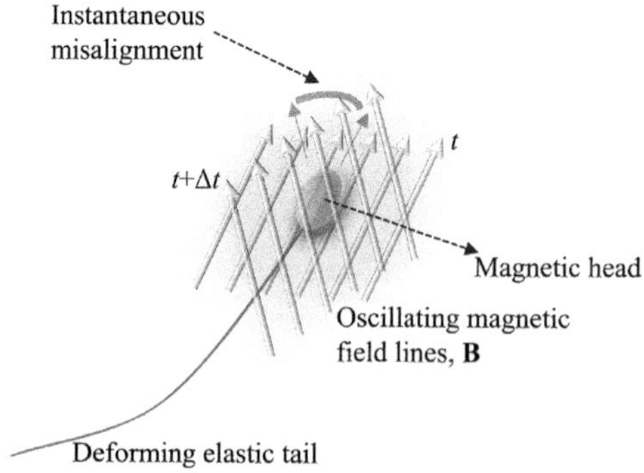

Fig. 4 Depiction of an autonomous micro-swimmer with passively deforming elastic tail, reacting to the oscillation of external magnetic field vector, **B** (T). The consecutive realignment to the field-oscillation produces reciprocating elastic motion along the tail. As a result, the time-irreversible planar wave propagation is induced pushing the head in the opposite direction.

2.2.4 Conceptual Optimization Studies on Swimming Microrobots

The hydrodynamic efficiency for micro-swimmers is conventionally defined for the relation between forward thrust and forward motion of the robot, but we know by observation that this is purely based on the assumption of perfect symmetry for geometry and for induced flow field. It is also known that any aberration of surface morphology, geometric imperfections, the presence of nearby boundaries, or a nonhomogeneous wave pattern is enough to break this symmetry, and even sperm cells may exhibit three-dimensional (3D) trajectories (Corkidi et al., 2008). Thus, to abide by the conservation of energy, the efficiency definition should be revisited to include full 6-DOF motion. This is imperative as, in robotics, researchers are also striving to supply the power, not in abundance or shortage but in the right amount and control the motion with high accuracy. The total rigid-body motion may not be a concern for some applications; however, if the robot has to be steered toward a predefined target within a maze of ducts and intersections than, from a robotics and medical standpoint, one needs to have an idea of what will happen with the power supplied to the system. The ultimate efficiency could be defined as the ratio of power drain on the entire micro-swimmer, $P_{\text{viscous-loss}}$ (W), to the total power supplied, P_{input} (W), and this will deliver yet another dimensionless number:

$$\eta = \frac{P_{\text{viscous-loss}}}{P_{\text{input}}} \tag{12}$$

The micro-swimmer being sought after is the fastest and the most robust one with the minimum energy requirement. Earliest studies solely focused on the fluid resistance as it dominates the overall swimming performance; however, some recent ones do take the overall system performance into account. However the micro-swimmer is actuated, the goal is to tow a cargo on which the largest portion of the input power is dissipated by means of viscous friction. Thus, the first question is to ask what the optimum head geometry should be to give the minimum hydrodynamic drag. The earliest answer was given by Bourot (1974), and it is not a perfect sphere as one might first think but more like the shape of a rugby ball with a numerically tuned aspect ratio; very close to the shape presented in Fig. 1.

Next question should be what the overall geometry of the swimmer should be. There are several optimization studies on the most efficient geometric combination of the head (cargo) and the tail (flagellum) from the point of view of hydrodynamics. These studies mostly carried out in nondimensional form, that is, dimensions being scaled with a selected characteristic length, so results will be applicable to a wide design space as long as Re number is smaller than unity.

For instance, Phan-Thien et al. (1987) presented the optimization of a bacteria-like micro-swimmer, with a single flagellum, based on BEM analysis. Authors, first, swept the design space for the wave propagation parameters, that is, wavelength and wave amplitude, of the tail while attached to a spherical head. Then, they studied the effect of the shape of the head while the wave geometry is kept constant. The efficiency definition they used is:

$$\eta = \frac{6\pi\mu R U^2}{-\Sigma(\boldsymbol{\tau}\cdot\mathbf{U})_{\text{tail}}} \tag{13}$$

where the denominator stands for the total power required to sustain the wave propagation along the tail, which is in the definitive form for all pure hydrodynamic intents and purposes. Moreover, R (m) stands for the diameter of the head. Here, $\mathbf{U}$ (m/s) denotes the total velocity vector at each node along the boundary of the tail as those nodes also coincide with the spatial limits of the fluid domain, and $\boldsymbol{\tau}$ (N/m^2) signifies the hydrodynamic force vector associated with each respective node calculated as $\boldsymbol{\tau}=\boldsymbol{\sigma}\mathbf{n}$ with the surface normal of each individual node, $\mathbf{n}$. Thus, the total power is calculated by summing the hydrodynamic power spent on the nodes at the surface in contact with the liquid.

A very similar study is carried out by Shum et al. (2010), however, extending the analysis to the effect of distance of a nearby solid boundary, again using BEM analysis. The design space analyzed by the authors is the same as the previous example; however, surface plots showing the effect of simultaneous variation of tail and head geometries are presented for the efficiency instead. Furthermore, the stability of swimming motion, that is, steadiness of the swimming direction under the influence of hydrodynamic interactions with the nearby boundary, is investigated by the representation of stable and unstable geometric combinations. In addition, authors discussed that efficiency could also be defined in terms of motor torque since the helical flagellum is being rotated with the bacterial motor present in natural micro-swimmers. The new definition is given in the form of:

$$\eta = \frac{8\pi\mu R^2 \Omega_{\text{body}}}{-\int_{\text{tail}} (\mathbf{r}\times\boldsymbol{\tau})\,\mathrm{d}a} \tag{14}$$

where **r** (m) is the position vector of any arbitrary node along the tail with respect to the axis of rotation, Ω_{body} (rad/s) signifies the rotational velocity of the head, and a (m^2) denotes the surface of the rotating tail that is in contact with the liquid, thus, the denominator yields the motor torque to overcome the viscous resistance acting against helical-wave propagation throughout the tail.

Efficiency and optimization studies, of course, cannot be limited to pure hydrodynamic analysis. The previous two examples consider the tail as rigid helices. As we go one step further and investigate elastic tails, it becomes obvious that internal stresses play an important role. Spagnolie and Lauga (2010) introduced the internal power spent by an undulating flagellum of single-celled organisms on total bending, $P_{bending}$ (W), and sliding, $P_{sliding}$ (W), of the internal filaments along with the hydrodynamic power to the efficiency equation. As for artificial flagellum, there exists an earlier example provided for piezoelectric actuation specifically (Kósa et al., 2007): if the undulation is invoked by the piezoelectric effect, the power input on the system could be modeled as the power consumed by the laminates of lead zirconate titanate (PZT), which behave like the resistance-capacitance circuit, as

$$P_{input} = \frac{\omega \varepsilon S V^2}{t_{PZT}} \tag{15}$$

where ε (F/m) is the permittivity, S (m^2) is the cross-sectional area of the PZT that is perpendicular to the applied electric field, V (V) is the amplitude of the applied voltage to generate that field, and t_{PZT} (m) is the thickness of the PZT laminate. Although the authors focused on swimming speed as the indication for optimized wave pattern and shape (Kósa et al., 2007), this electrical power calculation could be cast to predict the overall efficiency of the entire swimmer if plugged in the denominator of Eq. (12). The composite material PZT employed by Kósa et al. (2007) is a member of piezoelectric materials and presents an appropriate example of mechatronic systems as it transforms electric potential to mechanical strain and *vice versa* (Preumont, 2006), thus it is very suitable for several actuation and sensing applications such as the one mentioned here.

2.3 Experiments

In this section, the reader will find some examples of the most inspirational experimental studies presented in the literature on micro-swimmers, some of which are completely manufactured out of inorganic materials and components whereas the others have the merit to be called as cybernetic micro-swimmers. The size scale varies between a few centimeters to a few tens of micrometers owing to different manufacturing and actuation techniques. However, the flow conditions are always limited to Re number being smaller than unity. Reynolds numbers much smaller than unity signifies that momentum transfer from the micro-swimmer occurs owing to the viscous friction within the fluid rather than mass flow, and this signifies the typical case experienced by bacteria and spermatozoa. This condition must be simulated in the physical experiments or the results are simply not comparable with the natural micro-swimmers as a very different physical process is responsible for the induced fluid resistance. Selected applications are of a broad scope such as motion control studies, object manipulation, and visualization with different techniques. Furthermore, some applications are focused on demonstrating the controllability and observability of such micro-swimmers while the others focused on exhibiting the abilities of micro-swimmers as biomedical micro-robotic tools.

2.3.1 *Conceptual and Proof-of-Concept Applications With Artificial Micro-Swimmers*

The entire micro-robotic system, which is composed of actuation, power supply, sensing, control, and the swimming end-effector, is supposed to be deterministic. This means the Brownian noise in the immediate vicinity should not affect the swimming characteristics, as it is a stochastic phenomenon due to random collisions of molecules associated with temperature. Brownian noise is considered a random effect leading to chaotic jumps of molecules and 6-DOF oscillation of relatively larger objects within the bulk owing to the temperature, that is, average kinetic energy, of the molecules (Einstein, 1956; Berg, 1993). When we say chaos in this context, we mean a series of rapid and coupled events completely bound by causality and the laws of physics, but overwhelming in numbers to predict with acceptable temporal resolution in a deterministic fashion before they take place. The condition to overcome the Brownian noise is theoretically simple: the power input should be larger than Brownian noise and for helical-wave propagation actuated via external magnetic field this condition could be expressed in terms of $\mu_0 B > k_B T$, where k_B (J/K) is the Boltzmann constant and T (K) is the temperature (Ghosh and Fischer, 2009). A considerable portion of the examples provided in this chapter is based on actuation by rotating magnetic fields.

The rotating magnetic field, **B** (T), which the micro-robot should align with, can be applied by permanent magnets or electromagnets, the details of which will be partially given in the coming subsections. The magnetic torque is given by

$$\mathbf{T} = \mathbf{M} \times \mathbf{B} \tag{16}$$

where **M** (N m/T) denotes the effective magnetization vector of the magnetic head. Because at least one of the vectors **M** (N m/T) and **B** (T) is of alternating direction, the magnetic torque, **T** (N m), is expected to be time dependent rather than a constant. One of the earliest examples of such actuation is given by Honda et al. (1996). Authors experimented with a cm-scale, rigid, and helical robot with a magnetic head, and the robot was submerged in silicone oil to keep Re number smaller than unity. Rotation of the rigid helix results in helical-wave propagation, which in return generates the forward thrust to propel the robot.

Now that we know the basics for helical micro-swimmers, we can move on to more profound examples. Mahoney et al. (2011) experimented on the visually assisted velocity control scheme on a magnetically driven helical robot with compensation for gravitational attraction. The helical robot used in this study consists of a magnetic body with a rigid helical tail glued on one side, and its total length is approximately 6 mm. Authors observed that the gravitational pull disturbs the alignment between magnetic head of the helical swimmer reducing the torque and pulling the swimmer downwards. To be able to overcome this problem, they modified the external magnetic field with real-time visual feedback, also known as "*visual-servoing*" that can depend on various imaging approaches and used for medical robotics (Zhang et al., 2017), and "*gravity compensation*." Gravity compensation means generating the necessary force component in the direction opposite to the gravity, and this force is generated by tilting the axis of the robot with a pitch angle as it rotates. Hence, a portion of the propulsive thrust cancels the gravitational pull. This is a very good example of position and velocity control, one being along the direction of gravity and the other being perpendicular to that, respectively. An illustration of this method, with the visual servoing and controlled motion of the tilted micro-swimmer, is given in Fig. 5.

Visualization, for example, and detecting the location and orientation with respect to surroundings, is one of the key elements in the control studies as demonstrated in Mahoney et al. (2011). But in that particular example,

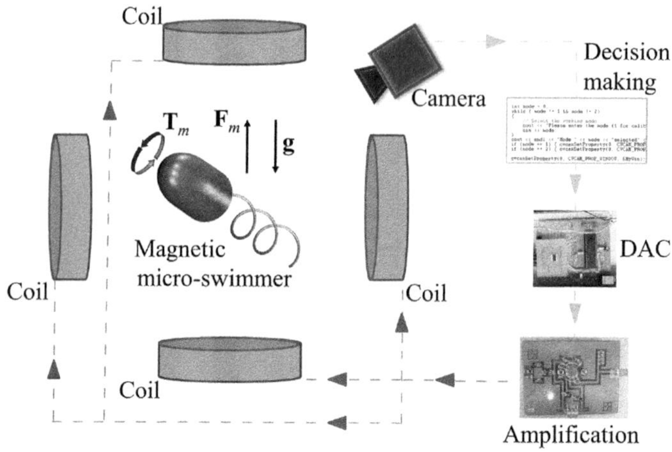

Fig. 5 Depiction of motion control incorporated with gravity compensation. The electromagnetic coils, which are separately driven by a decision making and controller unit. The controller unit accepts visual servoing data relayed by the camera. Depending on the position of the micro-swimmer, the digital, that is, discrete, a control signal is generated by the decision-making code. Then, the signal passes through a digital to analog converter (DAC) so a continuous signal is obtained which will be amplified to adequate power before entering the coils. Coils effectively generate the magnetic field and field gradient superimposed to induce magnetic torque, $\mathbf{T}_m$ (N m), and magnetic force, $\mathbf{F}_m$ (N), respectively. The magnetic force counteracts the gravitational attraction, $\mathbf{g}$ (m/s^2), hence keeping the position along the vertical lab coordinate fixed. In the meantime, as the magnetic field rotates, the helical micro-swimmer follows that rotation harnessing the forward thrust (Mahoney et al., 2011).

there was an unquestionable advantage of in vitro, performing in a physical simulation rather than in an actual living tissue, experimentation where a camera was able to capture images in real time. The micro-robotic system should be able to sense their environment, as all robotic systems are expected to. However, being very small in dimensions and being immersed in living tissue, that is, in vivo, present certain problems in that regard. For instance, because of size constraints and energy consumption, the artificial end-effector itself will probably not be able to accommodate the necessary hardware onboard to relay all the relevant information on the basic physical and geometric conditions of the ever-changing surroundings. Therefore, the computer, that is, the autonomous controller, or the user, that is, the pilot, is expected to infer to a certain degree based on what is available at any instant. In such cases, one type of most valuable data comes from various versions of visual-servoing methods.

Fluorescence imaging is one of those methods (Servant et al., 2015): a number of magnetically actuated helical swimmers are coated with near-infrared fluorescent dye molecules (NIR-797) then released into "*the intraperitoneal cavity of a 4-week-old BALB/c mouse*"; an organism bred in laboratory conditions and commercially available for research purposes. The fluorescent signal of the swarm, which has the intensity that can penetrate the tissue of certain thickness, is collected by a fluorescence imaging system. Location and velocity of the swarm were calculated by the displacement of its common center of mass. This particular example is important because the study combines a swarm of cybernetic micro-swimmers with visual servoing-based data acquisition in an in vivo demonstration.

Once the robot perceives and knows about its environment, special tasks could be achieved. A recent example is object manipulation. Zhang et al. (2009) demonstrated the magnetically actuated rigid helical swimmers, only around 50 μm in length, with a magnetic patch on one end are suitable to push polystyrene microspheres with a radius of 6 μm. Authors demonstrated the approach trajectory and manipulation route to be a controlled with the help of an open-loop control scheme.

A very recent application example is removing blood clots by abrasive effect of the rotating helical tail of a magnetic robot (Khalil et al., 2017a), as illustrated in Fig. 6. Authors developed an in vitro scenario where the tip of the rigid helical tail will remove material by rubbing on the surface of an actual blood clot obtained from a patient and placed inside a transparent channel of cylindrical cross section. They were able to demonstrate that the material removal process was advantageous to pure chemical lysis and predictable by means of robotic modeling, although visually not easy to access for the naked eye which, again, represents the problem of accurate visual investigation during the process.

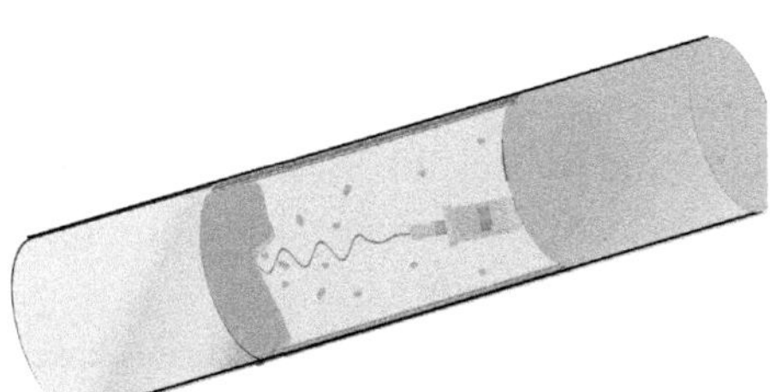

Fig. 6 Depiction of an autonomous micro-swimmer with rotating rigid-helical tail removing a blood clot, from a zoomed-in section of a blood vessel, by machining the surface (Khalil et al., 2017a). The fretting depends on material properties of the blood clot and the energy transfer to the abrasive contact by the tip of the rigid helical tail.

2.3.2 *Conceptual and Proof-of-Concept Applications With Cybernetic Micro-Swimmers*

The general idea being conveyed in this section is to exploit the cells as the end-effector or part of the end-effector of the micro-robotic system. Here, the reader will be presented with the examples in the literature with which actual living cells are involved in manufacturing, actuation, maneuvering and control studies as well as being exploited as a cargo container for drug delivery purposes (Dreyfus et al., 2005). Active use of living cells of different sorts such as natural micro-swimmers (Martel et al., 2009), or cardiac cells (Williams et al., 2014), in fact, solves one of the biggest conundrums as how to store or deliver energy to micro-swimmers for the desired flight time for the intended noninvasive task in a computer-controlled manner. For instance, the magnetotactic bacteria (MTB), which can align themselves with the field lines of a magnetic or EM field owing to magnetic crystals embedded in their structure (Martel et al., 2009; Blakemore, 1982) propose a promising candidate for such applications. Although, employing an active microorganism will probably introduce a stochastic behavior as Brownian noise may be intermittently in effect.

Magnetic resonance imaging (MRI) systems can be used to control and visualize single-celled organisms. These systems are very expensive and delicate magnetic sensors registering small deviations and ripples caused by tracer objects in a homogeneous and a powerful magnetic field. A magnetic head of a micro-swimmer constitutes a tracer object. Martel et al. (2009) discussed the possibility of visualizing and controlling the position of MTBs with possible software and hardware improvements to an existing MRI system. Software improvements are mostly related to real-time data processing, whereas hardware modification is based on adding extra coils to the MRI system to introduce the necessary degrees of freedom to be able to maneuver the MTB without interfering the sensing part of the software. The MTBs, in fact, could be used to deliver potent chemicals to tumor cells (Felfoul and Martel, 2013). If achieved, this therapy approach is believed to be more effective in certain conditions; for instance, before the cancer cells metastasize in the body.

In addition, velocity and position control studies based on direct physical interference with the environment could be found in the literature. Although the propulsion mechanism is autonomous, the velocity of single-celled organisms can also be controlled by adding certain ions, such as sodium ion (Na^+), in the environment to stimulate the bacterial motor to rotate faster (Nogawa et al., 2009). Furthermore, position control of single-celled microorganisms by means of standing acoustic waves is possible (Saito et al., 2002). In theory, when the micro-swimmer reaches its target it should hold its position for some time interval to carry out the task it is given to complete. Now, assuming that the actuation system is also able to visualize the instantaneous position of the swimmer and the distance to the destination, one can summon standing waves at specific locations in the tissue via acoustic transducers. By controlling the frequency, it will be possible to further fine-tune the position of the micro-swimmer around the target area by controlling the nodes of acoustic pressure waves (Saito et al., 2002). Indeed, this effort may require mapping the entire volume of the cavity in which the micro-swimmer is expected to perform.

Similar to artificial micro-swimmers, single-celled organisms are also being used for proof of concept object manipulation studies. It has been shown that a group of bacteria species such as *Serratia marcescens* could be exploited in order to generate directional thrust by asymmetric self-attachment as to propel solid objects and cargo filled with drugs by means of in vitro and in vivo experiments (Park et al., 2010; Hong et al., 2010), as illustrated in Fig. 7.

As an alternative to the use of single-celled organisms, there are some cells suitable for self-actuation such as cardiac cells, which undergo continuous contractions. Actuation with cardiac cells, that is, cardiomyocytes, is a rather novel method in terms of generating undulation along an artificial flagellum (Williams et al., 2014). In this recent study by Williams et al. (2014),

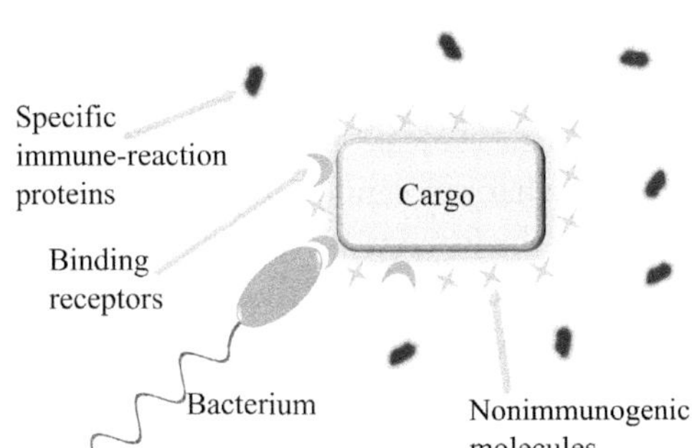

Fig. 7 The cargo towed by a group of bacteria attached to the surface in the asymmetrical order. The cargo could be a capsule containing enzymes and potent chemicals as to carry out a desired therapeutic task (Park et al., 2010; Hong et al., 2010).

attached cardiac cells are used to apply local moment, approximately 38 μN μm (10^{-6}N · 10^{-6}m), on a sperm-like thin and 2-mm-long poly(dimethylsiloxane) (PDMS) filament of optimized cross-sectional aspect ratio to restrict the structural bending on the swimming plane. Therefore, propagation of plane waves with nonsinusoidal profile was achieved followed by meaningful hydrodynamic thrust for Re numbers smaller than unity (Williams et al., 2014).

3 BUILDING A MICRO-SWIMMER ROBOT

Now the reader is introduced to the basic concept of micro-swimmers and related micro-robotic systems with respective possible applications, it is time to go deeper into the subject of how to build an artificial or cybernetic micro-swimmer as a micro-robotic system that will perform inside the living tissue. Here, the reader will be presented with the three most important aspects of swimming micro-robots: manufacturing, biocompatibility, and energy supply. Although some materials and methods do not seem as feasible as the others at the moment, for sake of completeness, a brief discussion will be provided on a broad variety of possible approaches for manufacturing and energy supply of the micro-robotic systems in general.

Before moving forward with the subject, it must be acknowledged that such robotic devices will be subject to harsh physiological responses which are addressed as "*the immune system*" in totality. Briefly, when a foreign substance or object is introduced in the living tissue, especially after it comes into contact with the specific types of cells dwelling in the bloodstream, an almost insurmountable defense mechanism is triggered within seconds creating a positive feedback loop till the perceived risk is properly eliminated. However, this mechanism cannot control itself to limit the response as it is entirely focused on the foreign invader thus probably resulting in serious tissue damage should certain materials and specific surface morphologies are present (Remes and Williams, 1992). Hence, the micro-swimmer should blend into the in vivo environment, that is, living tissue, precisely like a covert operative without triggering any biological alarms till the predefined task at hand is successfully completed and it is extracted or neutralized without a trace. It will be evident that conventional manufacturing techniques are not entirely feasible to that effect. Table 1 summarizes the techniques addressing the cybernetics and biocompatibility aspects as they are briefly

Table 1 Examples of Contemporary Proof-of-Concept Solutions for Discussed Biocompatibility Issues

Targeted Application	Main Chemical Agent/Material	Main Biological Agent	References
Biocompatible surfaces	Phospholipids (e.g., phosphoryl choline)		Bruck (1977) and Durrani et al. (1986)
Biocompatible surfaces	Hydrogels		Peppas et al. (2000)
Biocompatible surfaces	Polymers [e.g., poly(ethylene glycol), poly(vinyl chloride), poly(e-caprolactone)]		Merrill and Salzman (1983) and Rusu et al. (2006)
Biocompatible surfaces	Metals [e.g., titanium (Ti)]		Servant et al. (2015)
Biocompatible surfaces	Anticoagulants (e.g., heparin) and thrombin inhibitor compounds		Larm et al. (1983) and Salvagnini et al. (2005)
Biocompatible cybernetic robot		Magnetotactic bacteria (e.g., MC-1 MTB)	Martel et al. (2009)
Nontoxic cybernetic (ghost) robot		*Mycoplasma mobile* and its bacterial motor	Uenoyama and Miyata (2005)
Biocompatible biohybrid cybernetic robot	PDMS	Cardiac cells (e.g., cardiomyocytes)	Williams et al. (2014)
Biocompatible biohybrid cybernetic robot	Streptavidin, superpara magnetic iron nanoparticles	Red blood cells, DNA strands	Dreyfus et al. (2005) and Alapan et al. (2018)
Biocompatible biohybrid passive cargo		Red blood cells, targeting antibodies	Shi et al. (2014)
Biodegradable cargo with cybernetic robot	PLGA	*Serratia marcescens*	Park et al. (2010) and Hong et al. (2010)
Biodegradable cybernetic robot	Magnetite (Fe_3O_4)	*S. platensis*	Yan et al. (2017)

discussed here. Furthermore, it must be stressed out that genetic manipulation is not discussed within this context to steer clear of the exhaustive debates on the ethical implications of the field.

3.1 Manufacturing Techniques

Some micro-swimmers have no distinct head to be towed; they are rather on the side of proof of concept studies to demonstrate the manufacturing technique or application value. Some micro-swimmers are, although are not micro in dimensions so better to call them mock-ups to be correct, are complete with onboard power packs and control circuits sealed within their head. Similarly, some devices listed in this section are manufactured in a series of subsequent steps borrowed from microfabrication and are assembled together with different techniques. It seems possible to use "*top-to-down*" or "*down-to-top*" approaches depending on the application.

Micro-robotic devices are fundamentally a member of a much larger family known as microelectromechanical systems (MEMS) and, therefore, MEMS manufacturing techniques with semiconductors and soft metals are often cast, especially when the dimensions are smaller than 1 mm. Further, it is safe to say that MEMS manufacturing techniques are improved hand in hand with the research on micromachines as new methods and approaches are realized based on sheer need. A perfect recent example is given by Zhang et al. (2009): The authors presented a helical tail manufacturing method based on the self-folding ribbon of InGaAs/GaAs/Cr with a total thickness of 42 nm, deposited and etched on a GaAs (001) wafer with molecular beam epitaxy and electron-beam (E-beam) evaporation with the help of a sacrificial layer of AlGaAs. This very ingenious layer combination is what forces the ribbon to fold into a helix after it is freed from the substrate surface. The helical structure itself is not magnetic and thus to that effect, before the self-folding step, authors deposited a Cr/Ni/Au layer arrangement of a total thickness of 200 nm, at one end of the ribbon as a square patch of $4.5\,\mu m \times 4.5\,\mu m$ in dimensions. Finally, authors freed the ribbon from the van der Waals forces on the substrate surface with a micromanipulator. The total length of the ribbon is around $50\,\mu m$ and is actuated with a set of three Helmholtz coil pairs, that is, one pair for each axis.

At this juncture, it is important to emphasize that the numbers (001) emphasized in parenthesis are known as Miller indices (Hsu, 2002) denoting the orientation of the atoms in the wafer after it is sliced to a thin and smooth cross-sectional layer of single crystal. This direction is important for wafer-based micro-manufacturing techniques such as the epitaxy, for acoustic applications to conduct mechanical waves in certain directions, and for the realization of stress preloading again as demonstrated in the previously discussed study. The reader is encouraged to refer to Microsystems Engineering and Materials Science textbooks for further details of lattice directions and MEMS manufacturing.

Another interesting MEMS manufacturing technique is demonstrated by Ghosh and Fischer (2009). Authors used "*glancing-angle deposition*," that is, a method relying on changing the incident angle of the targetted substrate, which is depicted in Fig. 8, during material deposition of silicone dioxide (SiO_2) hence generating 3D helices perpendicular to the Si-wafer. The resultant tails are only a few micrometers in length, thus much smaller than an average blood cell. After, the tails are freed from the surface by the method of sonication followed by cobalt deposition by thermal evaporation. Hence, they have ferromagnetic property required to be actuated by EM coil pairs generating rotating homogeneous magnetic field.

MEMS-based manufacturing is also used for generating elastic filaments as demonstrated by Williams et al. (2014). Authors used capillary forces, that is, "*capillary draw*," to fill the high aspect-ratio mold, which is etched by photolithography on Si-wafer, with liquid PDMS. The dried filament can be

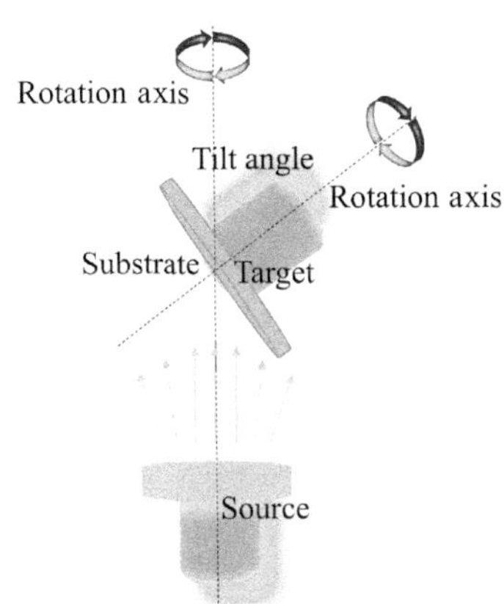

Fig. 8 Glancing angle deposition of source material on the target substrate. Ions are removed from the source under vacuum or by excitation by inert gas molecules. The required energy to remove ions from the surface of the source could be obtained from a direct-current (DC) electric field or radio frequency (RF) power transmission. The material will eventually deposit on the substrate. This method is generally known as sputtering. Furthermore, if the target is rotated in one or more axes than the incident ions will deposit on the substrate forming high aspect ratio geometries overtime, rather than forming a homogenous layer of coating (Ghosh and Fischer, 2009).

freed from the surface by etching the mold material. Furthermore, authors demonstrated that it is possible to manufacture spermatozoa-like micro-swimmers via this method in one molding step. Finally, authors attached live cardiac cells, that is, cardiomyocytes, on the selected locations of the filament generating 1–10 μN force leading to local deformations hence inducing plane wave propagation.

A more recent MEMS manufacturing strategy to build a complex structure, such as a helix, is known as two-photon polymerization technique. It is based on polymerization of a resin material by heating while diluted in a liquid solution; a very well-known method to manufacture 3D structures with detailed features in small scales. The process is perfected for micro-manufacturing by piezo-motors to control the traverse and with directed short-wavelength laser beams, thus generating tiny heated zones in 3D with high spatial precision. The helical shapes later can be coated with magnetic materials or the resin solution may contain nanoparticles with magnetic properties to attain the essential magnetic property, to begin with. There exist studies using this technique to manufacture μm-scale rigid helical tails in the literature (Servant et al., 2015; Nelson and Peyer, 2014).

Another recent novel technique is manufacturing elastic micro-swimmers via electrospinning technique in one step (Khalil et al., 2016, 2017b). This approach is based on the fact that polymer solution can be forced to fly from source to target under the influence of the high electrical field, just like a sputtering operation, by very thin streams. But as opposed to sputtering, the stream is under atmosphere pressure letting the water content to dry out. When the stream hits the target, it forms a spermatozoa-like elastic filament with a head and slender long tail. If the polymer solution contains paramagnetic nanoparticles, then it is the head where they will accumulate due to the applied electric field, thus giving the required magnetic property for actuation. The length of the filaments could vary from a few tens of micrometers to a few millimeters.

In these examples, the tails can mostly considered to be passive and the generation of motion or deformation is invoked via outside stimuli alone, that is, there is only internal resistance at best. There are two distinct examples on the contrary to this approach. In 2005, Dreyfus et al. presented the very first artificial cybernetic micro-swimmer, which was propelling itself via propagation of traveling plane waves (Dreyfus et al., 2005). In this particular example, the flagellum is made out of streptavidin-coated superparamagnetic, that is, able to be magnetized rapidly in room temperature with minimal magnetic hysteresis, particles and double-stranded deoxyribonucleic acid (DNA) chains with biotin: these two main components form slender filaments by self-assembly under a homogeneous magnetic field for an optimized period of time. The diameter of the superparamagnetic particles is 1 μm, and the filaments are around 30 μm in length. The tail is attached to a red blood cell, representing a cargo which will be towed by the magnetically actuated filament acting like the flagellum of a sperm cell. Authors demonstrated the swimming characteristics with respect to Sp and Mn_e numbers.

A very similar method to generate active tails is to cast polymer solutions mixed with magnetic particles in a mold to form a sheet. Once the polymer solution is dried, the sheet can be rolled into a hollow ring and placed in a powerful homogeneous magnetic field to form a predefined magnetization profile, that is, remnant magnetization. This magnetization profile will later help to generate plane wave propagation under oscillating EM fields (Diller et al., 2014). The dimensions of the polymer sheets could vary from mm-scale to cm-scale.

The aforementioned studies are examples of the down-to-top manufacturing approach, that is, adding small amounts of material to the micro-swimmer in one or several steps and giving them shape. Now, we will review some of the top-to-down methods by which removing material and large-scale assembly are common practices, although the sizes are mostly in mm-scales. A suitable example is to manufacture the helical rigid tail via removing material from a cylindrical solid using the electrical discharge method (Mahoney et al., 2011): authors used a wire electrical discharge machining (WEDM) system modified in-house with extra degrees of freedom to obtain a certain attack angle while cutting through a nitinol tubing with an outer diameter of 1 mm and length of 5 mm to form the helical profile with three full turns. The neodymium-iron-boron (NdFeB) grade N50 permanent magnet was then glued to the tail using epoxy. Grade Nxx is the measure of energy potential that could be generated by a magnet. A much earlier and simpler proof-of-concept as to demonstrate how such end-effectors would be actuated presented in the literature is to coil copper wire of 0.15 mm into a rigid helical tail and gluing it to a cubic samarium-cobalt (SmCo) magnet with a volume of 1 mm^3 (Honda et al., 1996), although the overall dimensions are much larger than what is aimed for.

Although cm-scale robots do not really fit into the overall scheme of this review of micro-manufacturing efforts, still there are three large-scale assembly examples worth mentioning very briefly as in the future it might be possible to manufacture artificial systems of much smaller versions. Kim et al.

(2004) and Chen et al. (2010) implemented onboard-powered swimming robots in cm-scale. In the former study, authors manufactured an ionic polymer-metal composite (IPMC) tail. IPMC systems can undergo structural deformations by letting water molecules passing through the internal polymer matrix resulting in local deformations. If this is done under in-phase electric potentials applied to different segments, planar wave propagation will be generated thus resulting in useful forward thrust as demonstrated in the study. The batteries were placed in the body of the untethered tadpole-type swimmer. In the latter study, authors experimented on a bacteria-like robot with four rigid helical tails, placed symmetrically with respect to the long axis of the robot and driven separately by dedicated direct-current (DC) motors. All DC motors and the power pack were sealed inside the head of the swimming mock-up. Authors studied the 6-DOF motion of the robot including maneuvering capability by independent control of motor speeds.

Finally, Kósa et al. (2007) proposed an elastic tail actuated by piezoelectric laminates to propagate plane waves to harness meaningful thrust. Piezomaterials constitute a good example of mechatronic systems as they transfer electrical fields to structural deformation and *vice versa* owing to their asymmetric molecular arrangement. They can reach up to high frequencies; however, it is not easy to achieve large deformations with short piezo-laminates. In Kósa et al. (2007), there exist several cm-scale sections with separate piezo-laminates driven by in-phase electric potentials with different amplitudes to generate the desired wave shape. The system was tethered and powered by conductive cables as piezo-ceramic laminates require high voltage but low current to generate structural deformation. Consequently, one needs elaborate circuits to regulate the current drawn from a high-density power source in accordance with the desired wave dimensions. It is important to note that in these last three examples, the constituents were attached together by proper water proofing.

3.2 Biocompatibility and Functionalization

In this section, the reader will first be introduced to the problem of biocompatibility with the help of three excellent reviews on the subject (Remes and Williams, 1992; Liu et al., 2014; Gorbet and Sefton, 2004). After the possible pitfalls are identified, the reader will be briefly presented with a variety of efforts to overcome the problem presented and made available so far. The problem is intricate as both surface morphology and chemical composition of the active molecules and ions may lead to unwanted physiological reactions.

Biocompatibility, or nonimmunogenicity, is a serious concern with medical micro-robotics defining the feasibility of the proposed material and tool for in vivo use. The immune system is the relentless defense of the living organism to protect itself against all possible dangers. The immune system is twofold in reaction; one being the specific and the other being nonspecific. Nonspecific defense does not recognize the risk but reacts against everything the same way with the same intensity. However, with the specific defense mechanism, scientists, researchers, and medical doctors have the biggest problem. Specific defense works by means of activation of certain enzymes and cells creating a positive feedback loop with growing intensity till the risk is completely eliminated. In fact, this particular mechanism is responsible for graft rejection, or pulmonary dysfunction during heart surgery. Moreover, sensitivity to metals, triggering of T-cells (T-lymphocyte) against polymer surfaces, and systematic attack on ceramic surfaces are also the work of specific immune reaction. Furthermore, the local response differs with respect to tissue conditions within the living organism. The reaction could even cause digestion of the tissue if the enzymes activated due to inflammation, owing to the accumulation of certain reactants based on chemotaxis caused by metals if they exceed a certain concentration (Remes and Williams, 1992). Some of the metals used in the studies presented here are obviously not biocompatible although cobalt (Co) and chromium (Cr) alloys and titanium (Ti) are widely used in implants (Abramson et al., 2004). Besides, it has been known that even biomaterials can trigger cell-mediated immune reactions, also known as "*hypersensitivity*." The reaction may happen very fast or develop in time as the small doses of the foreign substance is released accumulate within the tissue in time due to mechanical or chemical wear (Remes and Williams, 1992).

When an implant is surgically introduced into the tissue, coming into contact with blood is an inescapable eventuality that requires regular use of drugs to avoid blood-compatibility issues. There are attempts to modify surfaces into inert "*pseudo-endothelium*." When the material surface comes into contact with blood, the first stage of the immune mechanism is triggered within seconds attracting fibrinogens, an already available protein, to the surface only to be replaced with kininogens, a protein becomes available afterward, which is known as the Vroman effect. Basically, proteins of certain characteristics cover the surface as the foundation of the entire mechanism. If the mechanism is not disturbed somehow, thrombins may get attracted to the surface leading to platelets resulting in coagulation. One way or the other, blood vessels could be damaged. Thus, modification of the surface is very important. There are three main strategies being used:

(i) generating bioactive surfaces by functionalization of the surface by bioactive components inhibiting the interaction, such as heparin thus preventing the initial triggering; (ii) coating surfaces with chemicals to destroy the fibrin molecules and other molecules to extinguish the process; and (iii) surface endothelialization by modifying the surface exploiting live endothelial cells. The biggest obstacle emerging with this method is the cell recovery and surface coverage rate (Liu et al., 2014).

The problem with these methods comes from the nature of the immune system. Unfortunately, the totality of the elements and critical steps playing role in the immune reactions are not fully understood. Thus, one simply cannot give a definitive guarantee of a universal non-immunogenicity for materials and devices in every medical scenario. It should also be emphasized that each patient presents a uniquely hostile environment. It would not be too ambitious to state that each solution discussed so far addresses only one piece of a much bigger puzzle. Blood is exceptionally important in this whole scheme because it simply gets almost everywhere, and everything may end up in blood by diffusion. It is very important to acknowledge that there are different reactants taking a role in different stages of immune reactions that cause incompatibility. Furthermore, reactions are dependent on local flow conditions (Gorbet and Sefton, 2004).

Although the picture painted here looks pretty grim, there is a vast pile of applications, addressing the issue of biocompatibility, in the form of artificial tissue and organs. In a sense, we are learning as we go, and usually, the solution to an individual problem can be figured out at a time if an additional problem does not present itself along the way. The struggle might be never over as every patient presents a new and different hostile environment that the medical tools and chemicals, from a simple catheter to the most complicated robotic device, will be subject to and expected to stand the test of time or get improved accordingly. Here, briefly given are some of the outstanding solutions currently being used or studied.

Masking the surface in question by coating it with chemicals or functionalizing it with active particles to address the problem at hand, that is, avoiding possible allergic reactions and coagulation as discussed, are common practices. For instance, polymer surfaces can be chemically built and used such that they mimic the inner endothelial surfaces of blood vessels, resulting in less reaction with biological elements and live cells in the bloodstream (Bruck, 1977). A similar surface imitation procedure is the replication of the "*phospholipid ingredient*" of the living membranes using phosphorylcholine (Durrani et al., 1986). The intention is to simply hide the surface, with an "*engineered natural look*" as the authors put it, using synthetic polymers.

Another solution is to use hydrogels, which have a cross-linked matrix in between long polymer chains, as biocompatible coatings (Peppas et al., 2000). Hydrogels are soft materials having porous polymer matrices containing more aqueous content in volume than the polymer molecule itself. There are several areas hydrogels are used and they can be engineered in a way that they can imitate the elasticity of living tissue. A different coating approach is immobilization of polymers such as poly(ethylene glycol), poly(vinyl chloride), and poly(e-caprolactone) in order to make surfaces more invisible to blood cells and blood proteins (Merrill and Salzman, 1983; Rusu et al., 2006). Also, Ti coating can be used as a biocompatible material concealing other toxic elements in some cases, since it has a rather inert nature (Servant et al., 2015).

Furthermore, there are special chemicals to that effect known as "*surface additives*." They are used to engineer biocompatible surfaces while preserving the physical properties of the material beneath the surface, such as heparin (Larm et al., 1983), and "*thrombin inhibitor compounds*" (Salvagnini et al., 2005). Heparin can also be used for surface functionalization during manufacturing process (Dreyfus et al., 2005) thus serving two purposes at once. Surface functionalization is depicted in Fig. 9.

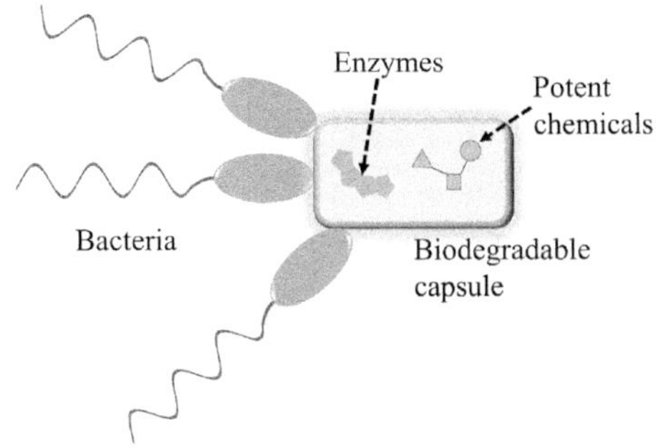

Fig. 9 Surface functionalization of a cargo: nonimmunogenic molecules of the organic or inorganic source could be used to mask the surface as to keep the specific immune-response proteins away. Similarly, molecules (proteins) of various sorts could be used to functionalize the surface, for example, bacteria could use these proteins as an umbilical to the surface and they can tow the cargo. Either way, the expected end result is a clandestine micro-robotic agent.

One of the best options is to employ the actual cell as cargo hall, such as the red blood cell in the manner Dreyfus et al. (2005) and Alapan et al. (2018) did, or rendering a bacterium cell inert by stripping it of its "*cytotoxicity*" via eliminating all the necessary organelles but the bacterial motor or suppressing the harmful biological agents by biochemically thus creating a "*ghost*" that can be controlled by introducing the necessary ions as Uenoyama and Miyata (2005) did with *Mycoplasma mobile.* Although direct steering of these ghosts was not reported, this study, along with the work presented by Dreyfus et al. (2005), stand out as a unique possible solution for the problem of biocompatibility if the cell is collected from the patient directly. As a matter of fact, red blood cells functionalized with antibodies to achieve target binding are already being used as biohybrid and passive cargo carriers (Shi et al., 2014; Alapan et al., 2018). This approach may also present the answer to the following difficulty.

There has been one more biocompatibility-related question that is not fully addressed yet: how to get the robot back out of the living organism once the task is completed. The answer, which could also be the response to how not to trigger the immune system, is to use biodegradable materials and agents to manufacture the micro-swimmers. Hong et al. (2010) demonstrated the use of biodegradable polymers, such as poly(DL-lactic-*co*-glycolic) (PLGA), in the manufacturing of a capsule to be used as cargo hull. The capsule is formed inside the T-junction of a fluidic microchannel system (see Fig. 10), and used to keep the potent chemicals safe and concentrated as it is being towed to the target area by *Serratia marcescens* species. A very recent and intriguing example of biodegradability and biocompatibility with steerable micro-swimmers is given by Yan et al. (2017). The authors exploited *Spirulina* microalgae (*Spirulina platensis*) cells by coating them with magnetite

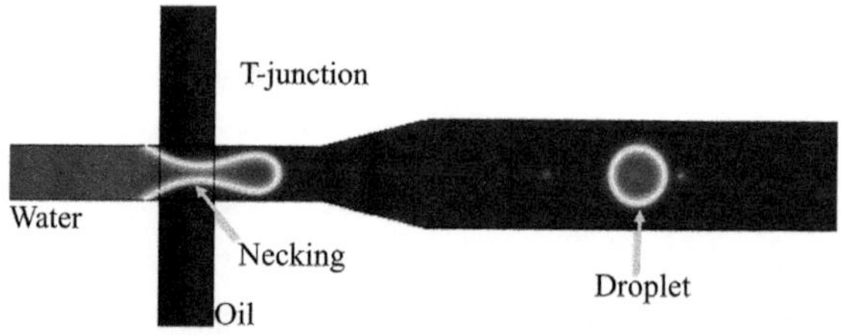

Fig. 10 The generation of liquid droplets in T-junction of a microchannel system that will be coated with the biodegradable polymer owing to the optimized-flow conditions and the surface energy at the liquid-liquid interface (Hong et al., 2010). The representative depiction presented here is obtained from the COMSOL Multiphysics FEM simulation.

(Fe_3O_4), which are known to be biocompatible after this process as they are rendered fixed, yielding helical micro-swimmers to be actuated via external magnetic fields in swarms. One of the key significances of this work is that the studied micro-swimmers are reported to degrade completely within a few days to a week given sufficient saline content in the buffer medium (Yan et al., 2017), although some species of the algae cells are relatively large to enter small cavities deep into the tissue. Biodegradation takes time which is needed for the duration of the desired flight time to complete the task after which the micro-swimmer could be left aside to slowly degrade into its constituent molecules and disperse in the system to be used in other natural cycles or to be get rid of by natural means. In the meantime, again, the proper precautions should be taken against the cytotoxicity of the natural micro-swimmers that are employed as the propulsive workforce. Otherwise, researchers should simply find a way to extract the components of the micro-swimmer from the living tissue as it may be the only choice left to eliminate the unforeseen complications in time. The other concern is the use of nanoparticles. Iron is the metal human beings have in abundance in the circulatory system. That makes iron nanoparticles a suitable agent as opposed to other magnetic materials. The implications of accumulation of any nanoparticle in the tissue, however, remains to be a long-term discussion with far and wide implications for material science and engineering in a more comprehensive manner as nanoparticles are being introduced in various daily-use consumables nowadays. The use of nanoparticles for actuation brings us to the next discussion on how to address the power demand of these systems.

3.3 Actuation and Energy Supply

As long as the proposed micro-swimmer is supposed to perform autonomously, that is, without any cables attached to it to gather information or to supply power, there must be built-in means to either consume stored energy or harvest energy from the environment. This subsection deals with the solutions proposed to overcome the issue of power supply in the literature. Having finished the introduction to biocompatibility before this section will make some of the hypothetical solutions to the energy crisis in the micro-robotic world more conceivable to the reader. Furthermore, most of the actuation methods being used to supply the required energy directly are already briefly introduced. Other exotic methods, such as contraction of implanted cardiac muscle cells, provide the required mechanical energy only

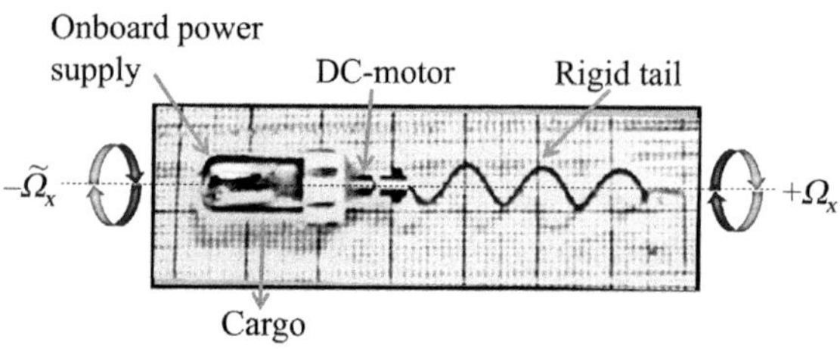

Fig. 11 Bacteria-mimicking micro-swimmer mock-up (Tabak and Yesilyurt, 2014b) with onboard power supply pack and electronics. The DC-motor constitutes a revolute joint, simulating the bacterial motor (Berg, 2003) in between the head and rotating tail. Due to the conservation of angular momentum, they rotate in opposite directions with different angular velocities, that is, Ω_x (rad/s) vs $\tilde{\Omega}_x$ (rad/s).

after a series of chemical reactions. So, in this subsection, we will briefly revisit what is already laid out and add the missing pieces in order to complete the picture.

Some systems can be fully autonomous with onboard control and energy supply (see Fig. 11). Furthermore, it is obvious that ad hoc decision making is a crucial factor for robotic applications. Certain bacteria can be employed based on their chemotaxis and magnetotaxis abilities toward a predefined target such as cancer tissue. Even further, bacteria can be modified to mute all toxic aptitudes so the bacterial motor could be employed as to propel the cybernetic system as already discussed (Uenoyama and Miyata, 2005). Unfortunately, it is not easy to differentiate actuation method and energy supply method in separate contexts at the first glance. Thus here they are briefly presented together. The reader will quickly realize that not all methods are easy to implement; however, it is important to explore and acknowledge all possibilities.

We have already seen that living cells can be integrated with the microswimmers transforming them into biohybrid and cybernetic systems with the onboard conversion from chemical energy or proton flux to mechanical energy. For instance, in one of the examples by Williams et al. (2014), the shape of the artificial elastic tail needed to be optimized to accommodate the cardiomyocytes while their contraction will only cause planar bending similar to the flagellum of spermatozoa cells. The other example of the natural means of energy use, although even more complicated and very tough to mimic, is the proton flux that makes the bacterial motor work. The size and the physics of the bacterial motor (Berg, 2003) led a series of studies, trying to exploit and, recently, imitate it as will be discussed later.

Some of the bacteria-steering studies have been already reviewed (Nogawa et al., 2009; Park et al., 2010; Uenoyama and Miyata, 2005), and additional examples can be given (Sakar et al., 2011; Alapan et al., 2018). In the former example, authors made use of static electric fields, that is, DC fields, to steer the single-celled organisms that swim otherwise autonomously. However, as a popular choice owing to their innate versatility, the steering is mostly achieved by magnetic and EM fields.

There is a wide selection of EM coil systems, rotating permanent magnets, or MRI systems to generate the required rotating or oscillating magnetic fields to actuate and manipulate the magnetic body or the magnetic tail of a micro-swimmer (Kummer et al., 2010; Dreyfus et al., 2005; Khalil et al., 2017a; Martel et al., 2009; Yu et al., 2010; Fountain et al., 2010; Martel, 2017). The origins of the magnetic property of the micro-swimmers in the literature are of either single magnets (Khalil et al., 2017a), deposited magnetic layers (Ghosh and Fischer, 2009; Zhang et al., 2009), or magnetic nanoparticles (Dreyfus et al., 2005; Khalil et al., 2017b; Yan et al., 2017). Magnetic manipulation depends on the alignment between the magnetization vector of the magnetic particles and field lines of the external magnetic field. Changing the orientation of magnetic vector on a particular location prompts out of alignment, which results in nonzero magnetic torque components acting on the magnetic body accordingly. Helmholtz coils are widely used to generate rotating magnetic fields, without gradients, to that effect (Beiranvand, 2013). However, the uniformity of the field is limited to a relatively smaller volume (Kirschvink, 1992). Furthermore, one has to use gradient coils or Maxwell coils to apply magnetic forces: magnetic gradient presents a potential which forces the particles to undergo linear translation in order to minimize the potential energy in that field (Choi et al., 2009). Moreover, much different geometric design can be implemented to generate different EM field configurations based on the application (Kummer et al., 2010; Kirschvink, 1992; Jeon et al., 2010). Fig. 12 illustrates the application of the EM field by coil pairs in order to actuate a magnetic micro-swimmer.

Some of the other examples are rather on the exotic side. One attempt is to generate torque on a sphere of 1 mm in diameter via directional control of surface acoustic waves (SAW) induced by interdigital transducers patterned on PZT wafer. Directional and modulated propagation of sound waves on the surface of the wafer is proved to exert a certain torque to induce rotation. The sphere is coupled to the surface by dry friction without any actual bearing structure (Tjeung et al., 2011); however, it can be improved further by

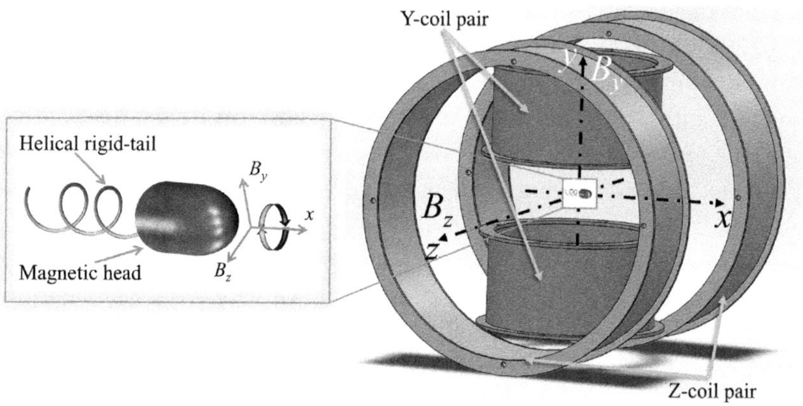

Fig. 12 Bacteria-like magnetic swimmer. The magnetic head and the tail are glued together, thus the joint in between is rigid. Hence, the entire swimmer rotates in the same direction with the same angular velocity. It is noted that the tail could be elastic or rigid depending on the manufacturing method (Honda et al., 1996; Khalil et al., 2016, 2017b). The electromagnetic coil pairs are designed to generate a homogeneous magnetic field along the respective axis. The actuation principle is the same as an electric motor: the magnetic field shall rotate due to in-phase sinusoidal currents running through the coils, thus the micro-swimmer is compelled to follow that rotation owing to the magnetic head.

inert gasses and "*hermetic sealing*," that is, prohibiting mass transfer to prevent any leakage. Again, lattice direction of the wafer has an important role in this application combined with the preferred actuation direction of piezoelectric effect.

Another group of study utilizes the angular momentum harnessed from light, either by means of generating force couples via incident photons deflected by the specially tailored surfaces making a nonperpendicular angle with the axis of rotation (Galajda and Ormos, 2001), or directly coupling with the orbital angular momentum of the circularly polarized light beams, also known as vortex beams, to rotate arbitrary complex geometries (Higurashi et al., 1997). If implemented to generate forward thrust in a sustainable manner, this method may be useful inside tissues and cavities where photons can penetrate, such as the globe of the eye.

One other option is to build actual molecular motors (Browne and Feringa, 2006), although efforts did not yield exact duplication of the bacterial motor at this point. The challenge with molecular motors is that direction of motion cannot be controlled at the moment. The motion can be linear or rotational driven by chemical reactions or Brownian noise, and go up to incredible speeds, for example, 100 MHz in some cases. The reader can find detailed information in the given reference, but, it is noted that this approach is inherently stochastic rather than depending on formulae pertinent to pure Newtonian mechanics, and dimensions can be much smaller than the limit of continuity assumption.

One of the sought after power supply techniques in MEMS is "*energy scavenging*" from the environment which usually contains stray mechanical, EM or electrostatic, and thermal energy in abundance (Cook-Chennault et al., 2008; Khaligh et al., 2010; Nielsen et al., 2003). Mechanical energy scavenging depends on sensing and capitalizing on mechanical vibrations via piezoelectric materials. Piezoelectric cantilevers of different lengths, thus with different natural frequencies, could be used in an array to detect a frequency band. However, the downside is that a power-regulation circuit is required to achieve continuous supply with the conditioned current. EM and electrostatic energy scavenging depends on a similar approach: The oscillation in the field could be used to generate current on coils by inductance or creating an alternating potential difference on capacitor plates owing to electrostatic forces. On top of these, the field vibrations could be detected by capacitor plates or magnetic particles attached to piezoelectric cantilever beams hanging freely thus generating electrical energy by piezoelectric effect invoked by the resultant structural vibrations. It is obvious that piezo materials can be used for both actuation, for example, consuming energy to generate wave propagation (Kósa et al., 2007) as presented in Fig. 13, and for transforming stray energy to electrical power to sustain that wave propagation.

Another scavenging method proposed to power MEMS devices is heat scavenging. This method is based on "*thermo-photovoltaic*" effect, which is very similar to field emission electron gun technology. Heat energy can be used to create photon emissions from a silicon-based source, and those photons could be collected by a photovoltaic cell on the collector side (Nielsen et al., 2003) which in turn may be used to harvest heat energy in the organism as the primary or an auxiliary means of the power generation.

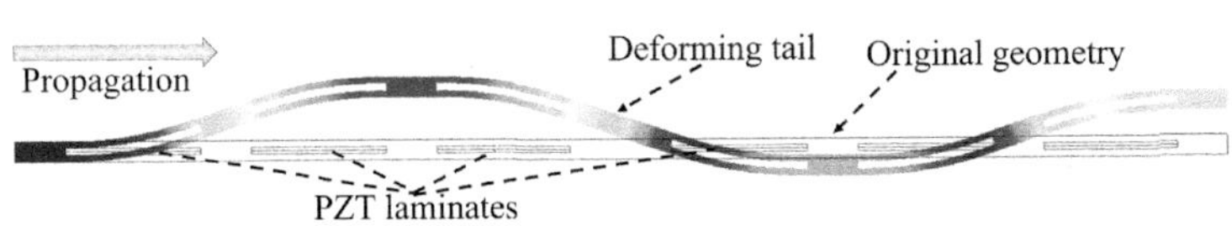

Fig. 13 COMSOL Multiphysics FEM simulation of piezoelectric-effect and linear structural deformation. Each shear-mode PZT laminate is driven by a in-phase electric field thus the convoluted local deformation results in planar wave propagation in a certain direction.

In addition, photovoltaic cells could be used separately at locations where light is available (Cook-Chennault et al., 2008; Bermejo et al., 2005), which could be separately classified as a light-scavenging method. All the scavenging approaches are based on "*regenerative sources*," much like the renewable energy sources of macroscale world. Other methods can be classified as nonregenerative sources, namely batteries and power packs.

One out of the ordinary nonregenerative power source idea is to use "*nuclear batteries*." Indeed, the idea of a nuclear battery is an old one: Braun et al. (1973) presented the theory for such a power supply in 1973. The proposed method was to collect the emissions from the nuclear source, that is, tritium (^{3}H), of very small quantities, for example, in micrograms, which can supply continuous energy in the range of microwatts and pico-Amperes, that is, 10^{-12} A, for a considerable time interval. Lal and Blanchard (2004) later published a brief discussion on how the proposed nuclear isotopes, in very small doses, will emit particles well below the acceptable radiation limits, and further elaborated on how the isotopes can be safely stored as the beta particles emitted by nickel-63 (^{63}Ni) or tritium (^{3}H) can only penetrate at most a depth of 25 μm in solids or liquids. Authors also presented a comparison between conventional Li-ion batteries, methanol fuel cells, tritium nuclear batteries, and polonium-20 nuclear batteries. Finally, they reported energy harvest of 3 nano-Watts, that is, 10^{-9} W, with 0.1 millicuries of ^{63}Ni, a supposedly much safer nuclear isotope. Although the notion of nuclear materials inside living organisms sounds somewhat eerie, with limited amounts and hermetic sealing with thick-enough shell, customized micro-swimmers designed for special applications might benefit them as power sources in the future. On the other hand, chemical batteries, such as Li-ion cells, are not really considered for micro-swimmers; however, there exist design examples intended for other MEMS devices (Cook-Chennault et al., 2008).

Energy supply remains an issue for the applications within the living organisms. Untethered artificial micro-swimmers cannot perform for relatively long flight times as the robot will either go deep into tissue where EM fields of high intensity should not penetrate for long time intervals for safety reasons. Also, during the therapeutic process, the patient is expected to remain as still as possible eliminating the mechanical energy scavenging possibility unless researchers find a way to generate continuous energy from the pulses in the bloodstream. Scavenging body heat sounds more suitable; however, it may not yield continuous power intensity as the temperature will change depending on the location of the organism the micro-swimmer is performing in. Carrying piezoelectric ceramics to generate power, or generate propulsion for that matter, could be problematic because of the immune response issues of the ceramic materials. The entire swimmer could be coated bioactive or biodegradable materials to minimize the risk although it is highly possible that the mechanical properties of the structure would be altered. It may be possible to use acoustic waves to visualize the robot and to control its final position; however, sound waves could present a disadvantage of impedance mismatch due to the composite arrangement of various living tissues.

A similar problem arises with MRI due to the interference of magnetic fields. Thus, techniques like ultrasound and MRI seem to be more suitable for visual servoing alone, if not improved with additional application-driven solutions. It is possible to use an onboard power source, such as nuclear materials as long as they are not in abundance and are properly sealed, however, might be the last resort for special cases and if the gait in order to leave the organism is planned for with absolute certainty. It is quite possible that two or more of these methods and materials will be used to actuate, visualize, and control the micro-swimmer while the power demand is simultaneously addressed. Nevertheless, there is a possibility that the end-effector will simply get lost in the tissue rendered nonresponsive because of any number of possible glitches. Thus, the design of the robot should be as optimized as can be, and one should be able to predict the state of the micro-swimmer as accurately as possible during an operation not to lose control completely. It would not be too much of an exaggeration if we say this system should be treated like a satellite sent into deep space where problems will occur and intervention is quite tough. This leads us to the last section; modeling the physics of the micro-swimmer robot for design optimization, gait planning, and real-time control purposes.

4 MODELING A MICRO-SWIMMER ROBOT AS AN END-EFFECTOR

4.1 Evolution of Equation of Motion

Assume that a design optimization study will be carried out. So, what if one wants to calculate the swimming velocity of the micro-swimmers without resorting to demanding computational models and conducting laborious physical experiments? In that case, one would require a fast model to obtain the velocity vector up to a certain error limit. In that effect, RFT is the method of choice within this chapter to represent the hydrodynamics of the propulsion system.

The equation of motion presented in this section is the constitutive relation, in a sense, for the micro-swimmer. It will be assumed that: (i) outer dimensions of the micro-swimmer are large enough that the continuity assumption holds; (ii) the actuation method is powerful enough to overcome the Brownian noise; (iii) viscous forces are large enough to reign over the inertia of the robot; (iv) force-free swimming will be valid at all times; (v) the force balance will be written in the reference of the frame of the swimmer; (vi) all external stimuli will be added in accordance with Newton's second law of motion; (vii) resultant velocity will never exceed the limit where Re number indicates transitional flow regime; and (viii) micro-swimmer will be fully submerged at all times.

4.2 Equation of Motion

The deterministic equation of motion for an untethered, nonholonomic, and fully submerged micro-swimmer in its own frame of reference is given by

$$\mathbf{0}=\mathbf{F}_p+\mathbf{F}_d+\mathbf{F}_c+\mathbf{F}_f+\mathbf{F}_m+\mathbf{F}_e+\mathbf{F}_g+\mathbf{F}_t \quad (17)$$

and

$$\mathbf{0}=\mathbf{T}_p+\mathbf{T}_d+\mathbf{T}_c+\mathbf{T}_f+\mathbf{T}_m+\mathbf{T}_e+\mathbf{T}_g+\mathbf{T}_t \quad (18)$$

To list the terms briefly; the subscripts "p," "d," "c," "f," "m," "e," "g," and "t" denote the effects of propulsion, viscous fluid drag, reaction force upon contact with surrounding boundaries, friction force on the point of contact, EM field, electric charges, gravitational attraction, and steady-periodic effects within the liquid, respectively. The very first term in both equations, $\mathbf{F}_p(t)$ (N) and $\mathbf{T}_p(t)$ (N m), denote the propulsive force and propulsive torque. Propulsive effects can be calculated by

$$\begin{bmatrix}\mathbf{F}_p(t)\\ \mathbf{T}_p(t)\end{bmatrix}=\mathbf{G}(t)\begin{bmatrix}\frac{\mathrm{d}\mathbf{P}(t)}{\mathrm{d}t}\\ \mathbf{0}\end{bmatrix} \quad (19)$$

$\mathbf{G}(t)$ is the 6-by-6 resistance matrix of the flagellum that may contain information about hydrodynamic interaction caused by nearby boundaries including the hydrodynamic impedance of the micro-swimmer (Tabak and Yesilyurt, 2014b); $\mathbf{P}(t)$ (m) is the position vector of an arbitrary point along the flagellum, which can be rigid or elastic, rotating or undulating.

Reader will find detailed derivations of $\mathbf{G}(t)$ and $\mathbf{P}(t)$ (m) in Purcell (1977), Lauga et al. (2006), Higdon (1979), Keller and Rubinow (1976), and Tabak and Yesilyurt (2014b).

The second terms, $\mathbf{F}_d(t)$ (N) and $\mathbf{T}_d(t)$ (N m), denote the drag force and drag torque due to the resultant rigid-body motion of the center of mass of the micro-swimmer. Hydrodynamic drag acting on the micro-swimmer can be estimated by

$$\begin{bmatrix} \mathbf{F}_d(t) \\ \mathbf{T}_d(t) \end{bmatrix} = \mathbf{K}(t) \begin{bmatrix} \mathbf{U}(t) \\ \mathbf{\Omega}(t) \end{bmatrix} \tag{20}$$

where $\mathbf{K}(t)$ is the 6-by-6 total resistance matrix of the micro-swimmer that also contains the same hydrodynamic interaction acting on both head and tail, and $\mathbf{\Omega}(t)$ (rad/s) is the rigid-body rotation of the center of mass of the micro-swimmer. Reader will find detailed derivations of $\mathbf{K}(t)$ in Purcell (1977), Lauga et al. (2006), Higdon (1979), Keller and Rubinow (1976), and Tabak and Yesilyurt (2014b).

The third terms on each row, $\mathbf{F}_c(t)$ (N) and $\mathbf{T}_c(t)$ (N m), denote the contact force and contact torque. Contact force vector is parallel to the surface normal of the boundary that the micro-swimmer collided with. Contact force vector is computed with respect to the center of mass of the micro-swimmer. The contact force can be modeled with the help of numerical estimation, that is, "*penalty method*" (Spong and Vidyasagar, 1989), or using the actual physical properties of the material. The contact effects could be given in the form of spring-damper model:

$$\begin{bmatrix} \mathbf{F}_c(t) \\ \mathbf{T}_c(t) \end{bmatrix} = \begin{bmatrix} kl(t)\mathbf{n}_{\text{wall}}(t) + b(t)(\mathbf{U}(t) + \mathbf{\Omega}(t) \times \mathbf{P}(t)) \cdot \mathbf{n}_{\text{wall}}(t) \\ \mathbf{p}(t) \times \mathbf{F}_c(t) \end{bmatrix} \tag{21}$$

where $\mathbf{n}_{\text{wall}}(t)$ is the surface normal of the foreign boundary at the point of contact, $\mathbf{p}(t)$ (m) is the location vector of any arbitrary point on the boundary of the micro-swimmer, k (N/m) is the stiffness coefficient, l (m) stands for the instantaneous penetration depth, and $b(t)$ (N s/m) accommodates the conditional damper coefficient, that is, the molecular viscous resistance associated with the material, with an option for adhesive surfaces and Van der Waal forces if applicable (Khalil et al., 2017a; Sadelli et al., 2017). The condition is that the molecular friction will not be in effect if the point of contact is not moving in the opposite direction of the surface normal, $\mathbf{n}_{\text{wall}}(t)$, unless the surface is somewhat sticky. If the surface is sticky, depending on the rate of peeling or detaching, $b(t)$ will always impose an additional force. Fig. 14

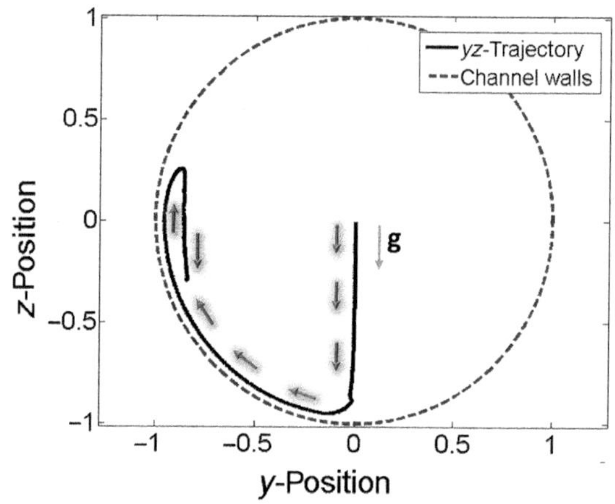

Fig. 14 Normalized *yz*-trajectory of a swimming micro-robot with forward velocity being along the *x*-axis. The micro-swimmer assumed to start swimming from total rest at the center of a channel of cylindrical cross section. Under the influence of gravitational pull, **g** (m/s^2), the robot converges to the channel wall after which it starts to follow it. At some point, gravity comes over the viscous traction from the wall and robot loses contact only to descend again. Results are obtained with the resistive force theory (RFT)-based model for magnetically actuated artificial micro-swimmer of dimensions that of the *Escherichia coli* minicell (Chattopadhyay and Wu, 2009).

represents an example of contact interaction between micro-swimmer and channel walls that it is confined by.

The fourth terms, $\mathbf{F}_f(t)$ (N) and $\mathbf{T}_f(t)$ (N m), are pertinent to the friction force. Friction is defined as the motion perpendicular to the surface normal at the point of contact. This motion can be abrasive or nonabrasive thus could be modeled in different ways. However, we can simply define it based on the contact force as

$$\begin{bmatrix} \mathbf{F}_f(t) \\ \mathbf{T}_f(t) \end{bmatrix} = \begin{bmatrix} \nu|\mathbf{F}_c(t)|\mathbf{t}_{\text{wall}}(t) \\ \mathbf{p}(t) \times \mathbf{F}_f(t) \end{bmatrix} \tag{22}$$

where ν is the effective friction coefficient at the point of contact and $\mathbf{t}_{\text{wall}}(t)$ signifies the tangent vector at the surface of the foreign boundary pointing the direction opposite to the motion. The following force terms, $\mathbf{F}_m(t)$ (N) and $\mathbf{T}_m(t)$ (N m), are the magnetic force and magnetic torque vectors, respectively, acting on the magnetic micro-swimmer which could be expressed as (Honda et al., 1996; Zhang et al., 2009; Khalil et al., 2016, 2017a,b)

$$\begin{bmatrix} \mathbf{F}_m(t) \\ \mathbf{T}_m(t) \end{bmatrix} = \begin{bmatrix} (\mathbf{M} \cdot \nabla)\mathbf{B}(t) \\ \mathbf{M} \times \mathbf{B}(t) \end{bmatrix} \tag{23}$$

if we assume that center of mass and volumetric center of the magnetic material coincide. The sixth term on both rows, $\mathbf{F}_e(t)$ (N) and $\mathbf{T}_e(t)$ (N m), denote the electrostatic effects, that is, if there is a charge potential, between the micro-swimmer and boundaries nearby. This effect can be simply represented as

$$\begin{bmatrix} \mathbf{F}_e(t) \\ \mathbf{T}_e(t) \end{bmatrix} = \begin{bmatrix} \dfrac{q^2}{4\pi\varepsilon h_c^2(t)} \mathbf{n}_{\text{wall}}(t) \\ \mathbf{p}(t) \times \mathbf{F}_e(t) \end{bmatrix} \tag{24}$$

where q (C) denotes the charge difference and $h_c(t)$ (m) is the proximity of the geometric center of the charged surfaces to the nearest boundary (Sadelli et al., 2017). Next effect to consider arises due to gravitational attraction and buoyancy which some bacteria species known to exploit to move simply up and down (Khalil et al., 2017a; Sadelli et al., 2017; Ning and Cannon, 1998):

$$\begin{bmatrix} \mathbf{F}_g(t) \\ \mathbf{T}_g(t) \end{bmatrix} = \begin{bmatrix} m_{\text{eff}}\mathbf{g}(t) \\ \mathbf{p}_{\text{cov}}(t) \times \mathbf{F}_g(t) \end{bmatrix} \tag{25}$$

where m_{eff} (kg) is the effective mass of the fully submerged micro-swimmer after buoyancy effect is included and $\mathbf{p}_{\text{cov}}(t)$ (m) stands for the position vector of volumetric center with respect to the center of mass of the micro-swimmer. In other words, it is assumed that the micro-swimmer is not a homogeneously distributed construct. It should be acknowledged that gravity, although the volumetric force acting on the mass of the micro-swimmer, cannot be lightly eliminated from the equations because gravitational acceleration is always present unless the experiment is conducted in an environment under microgravity conditions. Fig. 14 shows the effect of gravitational pull inside confined spaces for an artificial micro-swimmer of dimensions that of the *Escherichia coli* minicell (Chattopadhyay and Wu, 2009). Furthermore, the reader should keep in mind that the gravitational attraction is written in the frame of reference of the micro-swimmer that is experiencing complex rotations in the lab frame. Thus, the time-dependent nature of $\mathbf{g}(t)$ (m/s^2) is justified.

The final term in the equation of motion gives us the inertial effects within the flow field, which are present as long as the wave propagation exists, owing to the fact that the head most often follows a helical trajectory, that is, an oscillating motion, due to geometric aberrations or hydrodynamic asymmetries (Corkidi et al., 2008; Frymier et al., 1995). This type of rigid-body motion invokes the following combined effects of Basset history integral and added mass (Wang and Ardekani, 2012; Landau and Lifshitz, 2005), respectively, represented in the form of:

$$\begin{bmatrix} \mathbf{F}_t(t) \\ \mathbf{T}_t(t) \end{bmatrix} = \begin{bmatrix} 6R^2\sqrt{\pi\mu\rho} \int_{-\infty}^{t} \frac{\partial \mathbf{U}(t)}{\partial t} \frac{\mathrm{d}T}{\sqrt{t-T}} - \frac{2}{3}\pi R^3 \rho \frac{\partial \mathbf{U}(t)}{\partial t} \\ \mathbf{p}_{\mathrm{head}}(t) \times \mathbf{F}_t(t) \end{bmatrix} \tag{26}$$

where $\mathbf{P}_{\mathrm{head}}(t)$ (m) stands for the position vector of the center of volume of the head with respect to the center of mass of the micro-swimmer. It should be noted that this particular equation is given for a perfect sphere and is subject to geometry. History integral is important to be able to account for the phase information embedded in the induced forces due to the reciprocal or cyclic motion, which can also be referred as the hydrodynamic impedance. Added mass happens to be the effect of liquid volume replaced by the rigid-body translation of the micro-swimmer. These effects seem counterintuitive to the low Reynolds number flow assumptions; however, they will die out due to excessive shear the instant when the propulsive action is terminated. In other words, they are present as long as the micro-swimmer is in motion.

Although Eq. (26) seems to be violating the force-free swimming condition stated at the very beginning, it is important to emphasize that these two effects can be implicitly included in the resistance matrix $\mathbf{K}(t)$ in Eq. (20) by means of impedance analogy-based "*hydrodynamic interaction coefficients*" (Tabak and Yesilyurt, 2014a,b; Tabak, 2018) so that we will stay true to the initial assumptions. Especially, the integral term in Eq. (26) is the definition of fractional derivative (Ortigueira, 2011); therefore, the introduction of the so-called "*phase-angle correction*" is a necessity to account for the steady-periodic nature of the said forces and hydrodynamic interactions. Furthermore, the reader can pursue making use of Eq. (26) regardless; however, calculating the acceleration term, $\partial\mathbf{U}(t)/\partial t$, entails some finesse on the part of the numerical analysis.

After we add all the force and torque components, the instantaneous rigid-body velocity vector is found by

$$\begin{bmatrix} \mathbf{U}(t) \\ \mathbf{\Omega}(t) \end{bmatrix} = -\mathbf{K}^{-1}(t) \begin{bmatrix} \mathbf{F}_p + \mathbf{F}_c + \mathbf{F}_f + \mathbf{F}_m + \mathbf{F}_e + \mathbf{F}_g + \mathbf{F}_t \\ \mathbf{T}_p + \mathbf{T}_c + \mathbf{T}_f + \mathbf{T}_m + \mathbf{T}_e + \mathbf{T}_g + \mathbf{T}_t \end{bmatrix} \tag{27}$$

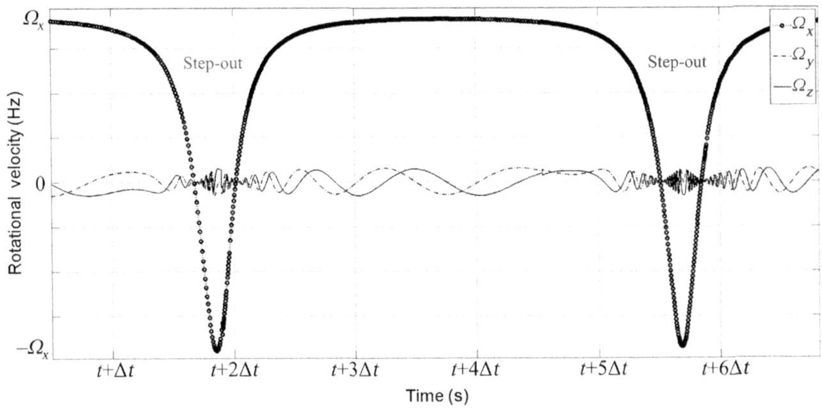

Fig. 15 Demonstration of step-out phenomenon of a rotating magnetic micro-swimmer going through complex rigid-body rotation: Depending on the viscous forces, the micro-swimmer might not be able to follow the rotation of the external magnetic field, Ω_x (rad/s). Once the lag reaches π radians, the micro-swimmer tends to rotate in the opposite direction momentarily. Once the magnetization vector of the head aligns with the external magnetic field, it follows it till step-out is experienced again.

This way one can predict whether the micro-swimmer will experience step-out phenomenon or not as demonstrated in Fig. 15. The $\mathbf{K}(t)$ matrix can be written for the entire swimmer body in the form of:

$$\mathbf{K}(t) = \int_0^{\ell_{\text{tail}}} \begin{bmatrix} \mathbf{R}(t)\mathbf{C}(t)\mathbf{R}^{\text{T}} & -\mathbf{R}(t)\mathbf{C}(t)\mathbf{R}(t)^{\text{T}}\mathbf{S}_{\text{tail}}(t) \\ \mathbf{S}_{\text{tail}}(t)\mathbf{R}(t)\mathbf{C}(t)\mathbf{R}(t)^{\text{T}} & -\mathbf{S}_{\text{tail}}(t)\mathbf{R}(t)\mathbf{C}(t)\mathbf{R}(t)^{\text{T}}\mathbf{S}_{\text{tail}}(t) \end{bmatrix} d\ell + \begin{bmatrix} \mathbf{D}_{\text{translation}}(t) & -\mathbf{D}_{\text{translation}}(t)\mathbf{S}_{\text{body}}(t) \\ \mathbf{S}_{\text{body}}(t)\mathbf{D}_{\text{translation}}(t) & \mathbf{D}_{\text{rotation}}(t) \end{bmatrix} \quad (28)$$

where ℓ_{tail} (m) is the cord length of the slender tail, $\mathbf{R}(t)$ is the 3×3 rotation matrix of an arbitrary location along the slender tail from local Frenet-Serret coordinates to the coordinate frame residing at the instantaneous center of mass of the micro-swimmer (Tabak and Yesilyurt, 2014b), that is, $\mathbf{R}(t) = [\mathbf{t}(t)\ \mathbf{n}(t)\ \mathbf{b}(t)]$ with $\mathbf{t}(t) = (\text{d}\mathbf{P}(t)/\text{d}x)/||\text{d}\mathbf{P}(t)/\text{d}x||$, $\mathbf{n}(t) = (\text{d}\mathbf{t}(t)/\text{d}x)/||\text{d}\mathbf{t}(t)/\text{d}x||$, and $\mathbf{b}(t)$ is found by the cross product of the first two as $\mathbf{b}(t) = \mathbf{t}(t)\times\mathbf{n}(t)$ for all three are expected to be orthonormal in order to constitute a local coordinate system. Furthermore, $\mathbf{C}(t)$ (N s/m^2) is the 3×3 matrix which holds the local resistive force coefficients, that is, c_t, c_n, and c_b, of the slender tail, $\mathbf{S}_{\text{tail}}(t)$ (m) is the 3×3 skew-symmetric matrix written with respect to the instantaneous center of mass and representing the local cross-products to calculate linear velocities from angular velocities and torques from forces. $\mathbf{S}_{\text{body}}(t)$ (m) is written for the location of the center of volume of the head of the micro-swimmer with respect to its center of mass. Furthermore, $\mathbf{D}_{\text{translation}}(t)$ (N s/m) and $\mathbf{D}_{\text{rotation}}(t)$ (N m s) are the 3×3 drag coefficient matrices of the head of the micro-swimmer (Berg, 1993; Tabak and Yesilyurt, 2014b). The matrix $\mathbf{C}(t)$ (N s/m^2) will encompass the wall effects (Brennen and Winet, 1977; Lauga et al., 2006; Khalil et al., 2017a) if a boundary present, so it can be time dependent even for a single perfect sphere following a straight path. Likewise, $\mathbf{D}_{\text{translation}}(t)$ (N s/m) and $\mathbf{D}_{\text{rotation}}(t)$ (N m s) will account for the wall effects if the swimmer is confined within a reservoir or a channel (Lauga et al., 2006; Khalil et al., 2017a; Tabak and Yesilyurt, 2014b; Higdon and Muldowney, 1995; Takano and Goto, 2003) thus they can also be time dependent. So, in the end, we end up with a full 6×6 matrix coupling the translational and rotational velocity vectors with each other and with force and torque vectors. The $\mathbf{K}(t)$ matrix is a collection of coefficients with three different units: The upper left part is in (N s/m), the upper right part is in (N s), the lower left part is in (N s), and finally the lower right part is in (N m s).

The integration of 6×6 matrix over the chord length of the tail is a necessity owing to geometric asymmetries. However, for the head to be towed, it is customary to use lumped coefficients obtained by experiments or theoretical analysis (Tabak and Yesilyurt, 2014b) unless the body itself is exhibiting some kind of deformation (Khalil et al., 2017c). Moreover, in order to abide by Newton's second law of motion, the location of the center of mass should be updated in the frame of reference of the micro-swimmer. If the micro-swimmer is a magnetically actuated artificial one, the total resistance matrix, $\mathbf{K}(t)$, will not be modified. However, if the micro-swimmer is a cybernetic and/or biohybrid with helical tail and controlled only by external steering, it means that the bacterial motor is already using one of the degrees of freedom of the tail. This can be called "*the tracking mode*" and the corresponding row and column of the integrant should be set to zero when calculating the velocities at the final step (Eq. 28) (Tabak and Yesilyurt, 2014b).

On top of these, if the tail is elastic, parameters of the wave geometry will also be expected to exhibit variations. Consequently, we need to model the deformation along the tail. The approach that will be presented here is going to be for an actively undulating flagellum actuated internally via piezo-laminates embedded in while simultaneously being subject to two-way coupling with the surrounding viscous liquid, for sake of completeness. However, in other cases, the elastic tail in question may be a just a passive and slender extremity (Khalil et al., 2016, 2017b). The preferred governing equation for structural deformation should take internal shear if the undulation amplitude is large, for example, comparable to the wavelength. Thus, Rayleigh-Timoshenko equations (Kaneko, 1975) are selected here:

$$EI\frac{\partial^4\varphi}{\partial s^4}+m\frac{\partial^2\varphi}{\partial t^2}+\frac{Jm}{k_TAG}\frac{\partial^4\varphi}{\partial t^4}=F_\perp+\left(J+\frac{mEI}{k_TAG}\right)\frac{\partial^4\varphi}{\partial t^2\partial x^2}+\frac{Jm}{k_TAG}\frac{\partial^2F_\perp}{\partial t^2}-\frac{mEI}{k_TAG}\frac{\partial^2F_\perp}{\partial x^2} \tag{29}$$

where φ (m) is the local planar wave deformation for an active elastic tail, E (Pa) is the Young's modulus, I (m^4) is the second moment of area, J (kgm^2) is the mass moment of inertia, m (kg) denotes the mass of infinitesimal slice around the mesh node, A (m^2) is the total cross-sectional area of the flagellum that is perpendicular to its long axis, G (Pa) is the shear modulus, k_T is the Timoshenko correction coefficients for internal shear, and with $F_\perp$ (N/m) being the unequally distributed load per unit length due to external viscous drag and internal piezoelectric effect combined as (Cook-Chennault et al., 2008; Jeon et al., 2005)

$$F_\perp=c_n\left(\frac{\partial\varphi}{\partial t}+u_{//}\right)+\frac{E_{PZT}Vd_{15}A_{PZT}}{t_{PZT}L_{PZT}} \tag{30}$$

with $u_{//}$ (m/s) being the velocity component parallel to the direction of deformation on that particular location with respect to the surrounding fluid owing to the rigid-body motion of the micro-swimmer. Furthermore, E_{PZT} (Pa) is the Young's modulus of the PZT-laminate, A_{PZT} (m^2) is the cross-sectional area of the PZT-laminate that is perpendicular to its long axis, L_{PZT} (m) is the length of the PZT-laminate, and d_{15} (C/N) is the piezoelectric coupling coefficient used to predict the mechanical strain while driven in the shear mode along the direction of wave propagation. Indeed, piezoelectric laminates can be designed to deform in different directions thus different coupling coefficients and loading conditions will be used instead (Preumont, 2006; Piefort, 2001) and it will be evident by the references given here that the stress term due to the piezoelectric effect presented by the second part of Eq. (30) is a very simplistic approximation for a far more complicated phenomenon. It should be also noted that the local resistive force coefficient associated with the surface normal, c_n (N s/m^2), attains a negative value by definition of the shear stress.

Accordingly, the solution to the equation of deformation will provide us $\mathbf{P}(t)$ (m) which will be used in the construction of hydrodynamic resistance matrix (Khalil et al., 2016, 2017b). The passive flagellum model is relatively simpler on the numerical effort without internal stress generation and velocity of the local deformation, however, can be expressed by the same governing equation. It is also possible to model the elastic tail based on the discrete-chain assumption (Gauger and Stark, 2006) but the details will not be given here. It is important to note that, the effect of bundling is not included in Eq. (29). Whatever the modeling approach might be, the apparent length and chord length will be different. In other words, if one solves Eq. (29) via finite difference formulation, it will be evident that the apparent length should be shorter than the actual length under the assumption of inflexibility. As a consequence, all the mesh nodes used in the finite difference formulation and in the calculation of $\mathbf{P}(t)$ (m), thus $\mathbf{K}(t)$, should have an updated position in the frame of reference of the micro-swimmer at each time step of the simulation. Indeed, it must also be noted that the first part on the right-hand side of Eq. (30) is a very simplified version of a rather complicated calculation because of the presence of large deformations (Khalil et al., 2017b).

The reader should bear in mind that other physical stimuli different than what has been discussed here may also be included in the model depending on the swimming conditions. For example, if the micro-swimmer is moving at the interface of two fluids than the capillary forces can be considered as to be a significant stimulus (Bai et al., 2017). Also, interfaces such as locally trapped bubbles may introduce Marangoni-flow effects owing to local species concentration gradients (Stricker, 2017). Finally, although it seems to be contradictory to the intuition, the Magnus force may be present as it can also be observed in micro-realm (Cipparrone et al., 2011). Likewise, a fretting-like behavior may be observed depending on the type of contact between the micro-swimmer and the environment, which requires a separate reaction force and torque calculation based on conservation of energy and momentum (Khalil et al., 2017a). It goes without saying that ascertaining the correct and accurate model for the scenario at hand dictates careful assessment of all the eventualities along the gait to follow.

The model laid out here can be deemed linear only because of the equation of motion; however, the expected behavior of the micro-swimmer is quite nonlinear. For instance, if we look closely at the resistance matrix for a magnetically actuated micro-swimmer equipped with a passive elastic tail, it should be obvious that the resistance matrix, with its 36 different elements, and the magnetic field, with its three components combined with nine gradients in total, will oscillate during a full period of actuation. The former should be so even if we use simple Euler-Bernoulli beam theory for deformation with no other external or internal physical stimuli other than gravity in presence. And, the latter is guaranteed to happen unless meticulous field-gradient calibration is done for a workspace of interest, preferably with much larger dimensions than that of the end-effector. Also keeping in mind the fact that for different dimensionless number groups various distinct changes in swimming and deformation behavior are already marked with varying actuation and design variables, one should almost always expect local and global extrema while studying swimming speeds within a design space of choice. It is further noted that the presented model is to predict the rigid-body velocity vector of the micro-swimmer under the influence of various forces strictly. If one to simulate the dynamics of the entire robotic system as a whole, the mathematical model should encompass the time-dependent behavior of the driving circuits, means of energy supply, chemical reactions, the performance of the sensors, physical environment, and control law, details of which can be found in the references provided throughout this chapter. However, next, we will talk about the latter for sake of completeness and to properly present a complete picture combined with all the aspects emphasized until this point.

4.3 Stability and Controllability in the Wake of Efficiency

It has been already established that efficiency of micro-swimmers is expected to be quite small due to viscous effects (Purcell, 1977, 1997; Ido et al., 2016). Although this represents a problem in terms of excessive energy loss through shear friction, it also means that inertia of the end-effector is not expected to be a source of instability. However, the geometry of the swimmer and direction of swimming will present instabilities. It has been shown that certain tail shapes, that is, the total number of waves along a flagellum, will result in wobbling trajectories of micro-swimmers (Tabak and Yesilyurt, 2014a) because of force unbalance along the flagellum. Being able to swim along a straight path when needed is an important asset for a micro-swimmer for several reasons: first of all, minimum energy loss due to unwanted lateral motion is a highly sought after feature, and second, the control law will

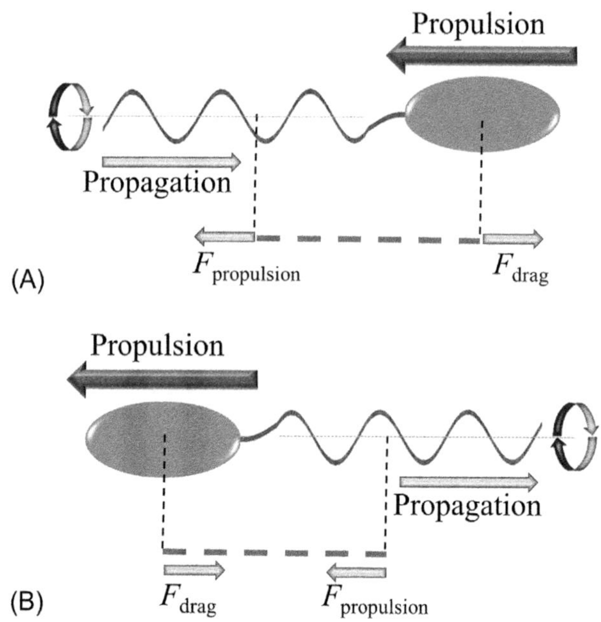

Fig. 16 Direction of helical-wave propagation and the resultant propulsion of the micro-swimmer. The force couple of drag and propulsion, as the head will be the main source of drag, will determine the stability of the swimming trajectory. (A) Stable pendulum-like swimming vs (B) inverted pendulum-like micro-swimmer. It is important to note that, the tail of the micro-swimmer depicted in (A) is a left-handed helix rotating in the clockwise direction while the tail of the micro-swimmer depicted in (B) is a left-handed helix rotating in the counterclockwise direction.

demand zero position error when actively manipulating the end-effector. Maneuvering will require backward and forward swimming which present different dynamics due to stability issues as presented in Fig. 16. The control law is an application-driven custom mathematical function executed to minimize the difference between the desired state of the system, such as position, velocity, etc., and the actual metrics registered by various sensors instantaneously.

One of the concerns here is that the step-out phenomenon could render the control law ineffective if it cannot be predicted properly. For instance, a magnetically actuated or maneuvered robot is expected to follow the computer controlled field lines, but this does not necessarily mean that the micro-swimmer will be able to align with the field at all times, as discussed earlier. This could seriously impede the gait control, meaning that the actuation frequency should be reduced when it is time to decide whether to go left or right at a narrow junction point. However, there may be a velocity constraint due to the presence of a background flow. Thus, obtaining the most faithful mathematical model and applying the best possible control law with an adequate sensory input is a high priority.

In addition, it is possible to cast observers of special design to eliminate high-order disturbances and to assume the micro-swimmer to be an over damped linear system. An overdamped system has considerable energy loss by means of heat due to nonconservative effects, for example, resistance in a circuit or soft-close damper in a door, so transient effects decay rapidly and uncontrollable oscillations are not experienced. This is the main reason why micro-swimmer stops immediately when the propulsive effect is no more. One profound control scheme is known as proportional-integral-derivative (PID) (Ogata, 1998) with which the instantaneous error, $\hat{e}$, between the extrinsic state of the system and its desired value is used to obtain the control signal, $Z(t)$, in the following way:

$$Z(t) = p_p\hat{e} + p_d\frac{\mathrm{d}\hat{e}}{\mathrm{d}t} + p_i\int\hat{e}\mathrm{d}t \tag{31}$$

where p_p, p_d, and p_i denote the proportional gain, the derivative gain, and the integral gain, respectively, although the derivative gain is not desired for overdamped first-order systems. The value in question may be velocity, and the core of the mathematical model being RFT, the instantaneous rigid-body velocity vector of the micro-swimmer can be predicted by a matrix inversion and product of that inverse with a vector as already presented in Eq. (27). This also means that, once the resistance matrix $\mathbf{K}(t)$ is obtained numerically, the nonlinear effects not contributing to $\mathbf{K}(t)$ can be compensated for by means of an appropriate observer. However, said observer does not necessarily have to be of model-based approach but of a genuinely designed numerical filter estimating and counteracting the higher-order effects, that is, noise, in a particular system (Huba, 2012). Here, the performance analysis, tuning requirements, computation time, and accuracy trade-offs will dictate the overall scheme. A hybrid approach may prove necessary to obtain the optimum control law in order to achieve desired robot control performance. A micro-swimmer with force and position controller is depicted in Fig. 17 with necessary sensors, for example, Hall effect sensors and ultrasonic sensors for the micro-swimmer (Abayazid et al., 2015) along with current and voltage sensors for the actuation system, and necessary observers, that is, mathematical models to predict states and intrinsic properties of the micro-robotic system interacting with the environment.

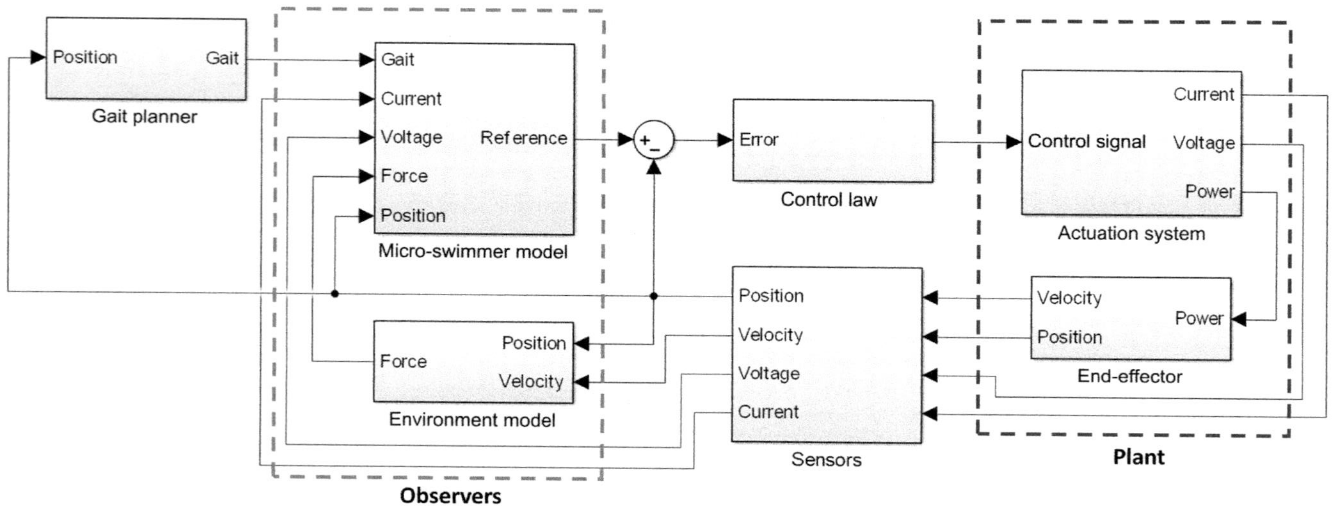

Fig. 17 The entire micro-robotic system with the actual plant, that is, micro-swimmer (end-effector) with power transfer (actuation system), sensors, observers, and with high- and low-level control systems. It is noted that the control signal could contain a set of multiplexed data equal to or less than the total states of the plant, in which case the system is called under-actuated. Observers are mathematical models of different orders to predict the necessary intrinsic states of the plant and filter nonlinear effects out when needed.

The ultimate task for these micro-devices is to perform inside the human body, such as in the bloodstream where there will be a flow in the background with a possible sinusoidal velocity component because of the heartbeat. Therefore, at this point, it is appropriate to mention a simulation-based control study addressing this very scenario: Sadelli et al. (2017) devised a control strategy to that effect, although the authors were only considering a magnetic robot of a spherical body for sake of simplicity. However, authors modeled magnetic forces, contact forces, gravitational attraction, viscous drag, and electrostatic forces together to represent the problem as accurate as possible. Blood flow was predicted by an observer and with the help of biophysical properties of the flow based on prior knowledge. Trajectory control was implemented with mean-value theorem based on the backstepping controller scheme. Authors also provided the state-space representation of such a system after some mathematical steps. The blood flow, indeed, is not easy to predict online for control studies, although fluid flow in confined spaces is expected to affect the resistance matrix (Happel and Brenner, 1965) of micro-swimmers. Although details of position and velocity control of such systems are beyond the scope of this text, at this juncture, the reader is urged to refer to "*system dynamics and control*" texts to be familiarized with the necessary terms, for example, such as observers, controllers, feedback and feed-forward loops, tuning, and state-space representations, if not experienced in the control theory and applications.

Finally, it is of utmost importance to strongly emphasize that the control input should be generated while observing the patient's safety at all times. The second highest concern is to see to it that the robotic system remains intact during operation as losing the micro-swimmer will probably not be a desirable scenario. One cannot just apply any arbitrary control signal without knowing the physical limits of the end-effector and the restrictions of the environment. To that effect, especially, if the end-effector will be actuated or steered by means of the external intervention of a computer- or human-controlled scheme, the meticulous design of the entire robotic system coupled with its surroundings will determine what can and what cannot be done without putting the patient and the end-effector in a risky situation.

5 EPILOG: FUTURE DIRECTIONS

The ultimate goal of this intriguing field of mechatronics is to help patients and medical doctors to improve the quality of medicine and quality of life by providing an alternative and reliable therapeutic tool of highly interdisciplinary but equally attractive nature. Such tools can proliferate easily in terms of the sheer number of total worldwide use given that they need very little material to construct. Consequently, their fully autonomous versions will definitely revolutionize the medicine in times of crises. Indeed, this type of advancement may happen only after the fundamental underlying problems are properly resolved. What has been presented throughout this chapter is a look at the culmination of these demanding problems followed by but a brief and limited summary of the state-of-the-art of the efforts to solve them. Today, it would be only too ambitious to claim that there is one way, one design, one method, or one material to address all these issues at once; however, the goal is to find the most suitable solution, whatever that may be, to avoid tragic and unpredicted incidents (Cohen, 1995).

To that end, the lookout for bioinspired and bio-mimicking micro-swimmers, in particular, seems to be quite promising along with numerous challenges. It is ill-advised to favor a material or a method without knowing details and risks of the therapeutic task at hand. Each approach presented here are mostly studied separately subject to in vitro situations or they are suffering the trade-offs between the visualization and actuation techniques. The reader should bear in mind that proof-of-concept studies are only to help us to figure out how to overcome numerous obstacles expected to occur in real applications and they are not free of disadvantages. The design of the overall system, as depicted in Fig. 17, is always subject to different constraints.

For instance, if the micro-swimmer will be used in tissue with the lack of blood vessels, then it may be possible to omit the problem of engineered look for the surfaces. Such an application might be sending nutrients to cartilage in different locations such as the kneecaps when they are reengineered for medical reasons (Schaefer et al., 2002), and consequently, the micro-swimmer will experience a viscoelastic medium to swim in. On the other hand, if the targeted area is in the neighborhood of the aorta, the chances are pretty high that the engineered surface will have a key role to success; however, the micro-swimmer will perform chaotically as the flow rate and blood pressure would imply transient and even turbulent flow conditions. In such a case, one may speculate, a risky strategy could be to switch the actuation system off, but not the visualization. Hence, letting the micro-robot soar with the flow till it is feasible to achieve addressable and controllable propulsion again may be feasible, however, only if there is no imminent risk of obstruction the flow.

As it should be apparent to the readers patient enough to follow the context thus far, biochemistry and materials science also became an integral part of the overall micro-swimmer research as to implement new functionalization approaches and to achieve nonimmunogenicity. It should always be strongly stressed out that the tendency is to go for biocompatibility and biodegradability as to build safer micro-swimmers in order to avoid triggering a chain reaction during any therapeutic operation. For instance, if the micro-swimmer is designed to accommodate elastic extremities, wear analysis will be an added task for long flight times as friction, internal or external, is expected to cause failure much easier in micro-realm. If the structure is not biocompatible and tends to break down before retraction, the design must be reevaluated. It is possible to use materials and cells gathered from the patient's own body for coatings to minimize the risk of allergic reactions and infections given that at each step the biological agents are handled meticulously in accordance with respective biological safety level regulations. Furthermore, the micro-swimmer may be tailored to be a soft robot or a rigid body concerning the risks, that is, rotating rigid helices could be harmful to the veins if they are too narrow.

In all cases, rigid and soft biocompatible and biodegradable materials should be available to the researchers with the ability to manufacture the desired aspect ratio and required resolution. Such a technique is already in use: two-photon polymerization method presents a wonderful opportunity to manufacture tiny structures with high aspect ratios and different end-product stiffness values (Servant et al., 2015; Nelson and Peyer, 2014). However, in such a design, printing the means of actuation based on manufactured intrinsic properties will be essential, for example, a magnetic micro-swimmer could be manufactured out of biological polymers, such as the gelatin used in food, pharmaceutical, and cosmetics industries, mixed with superparamagnetic particles, which could be one of the best options for building magnetically actuated artificial micro-swimmers. In a sense, by the introduction of high-end micro-prototyping systems, manufacturing complex geometries became much easier, however, along with the necessity of the vast knowledge in biochemistry for proper volume and surface functionalization efforts. In addition, different and elaborate material combinations will be needed for different actuation strategies some of which may be directly harvested from the patients (Chang, 2007) to present a patient-based micro-robotic solution for sake of medical and pharmaceutical safety. A very fruitful alternative would be deploying the micro-swimmers, especially cybernetic and biohybrid ones encapsulated in a droplet of some sort

(Park et al., 2010; Hong et al., 2010), or even in "*artificial cells*" with a more stable outer shell manufactured to the same end (Chang, 2007). Thus, the exposure time of the micro-swimmer to the immune system would be limited to reduce the risks. Of course, the means of motion control might not be active for such a pill will only move by naturally occurring local effects, until the individual micro-swimmers are released.

Furthermore, efficient swimming strategies hence energy efficiency are uncompromisable qualities. We are still struggling with various visualization and control problems as in some cases one tends to hinder the other and none performs perfectly in all cases all by itself. Commuting deep in the living tissue is a serious challenge as the human body is a perfect example of a composite structure with layers upon layers of biological tissue of different acoustic impedance, as pointed out several times, leading to obscured and low-resolution images, accompanied with abundance of ions susceptible to magnetic fields, leading to local heating due to magnetic friction should high strength and high-frequency EM fields be applied for a considerable duration. Functionalization of cybernetic micro-swimmers with suppressed toxicity seems more feasible at this juncture. We can just send them and, after the task is successfully completed, forget about them till all, somehow, dissolve into their much smaller components (Yan et al., 2017). Harvesting energy on demand from the molecules available in abundance in the tissue is very attractive although we are not fully capable of mimicking the internal structure of single-celled organisms. As a result, relying on ATP molecules being built and destroyed when they are called upon so utilizing the chemical energy would require integrating biological agents into the robotic systems as articulated by several studies reviewed here.

The not so straightforward alternative is to scavenge energy from stray heat, EM fields, and mechanical vibrations with fully artificial micro-swimmers, although it does not seem feasible to harness useful current without proper regulation. Any additional component will add to the overall volume which will reduce the maneuverability and hinder the chance of penetrating deeper into vulnerable regions of the tissue. Alarming as it may sound, should the other obstacles be properly resolved, low-energy-particle-emitting nuclear micro-batteries may pave the way to solve energy crisis for rare challenging applications with demand on high-energy density and long flight times. The problem of high-percentage energy dissipation due to shear losses in micro-realm forces us to focus on optimum gait planning and efficient control of motion.

The strategy for the control of the commissioned micro-swimmers, single or multiple, should be adaptable in order to account for different scenarios and situations, especially for swarming applications. This is utmost importance as almost all of the proof-of-concept studies lack 3D path planning and tracking. Furthermore, as the human body is a vast expense in comparison with the size of single-celled organisms, relying on a single robot will probably not be feasible for all applications. For instance, when the task is to carry potent chemicals to a specific location, a single robot may not be enough as to reach the temporal and spatial concentration threshold to impose any meaningful impact. On the other hand, if one is interested in opening a clogged capillary via chemical lysis or mechanical rubbing (Khalil et al., 2017a), assuming that the dimension is not a problem with the micro-swimmer being employed, multiple robots attacking the target together may not be desirable as to avoid further clogging issues. Control of a single micro-swimmer shall depend on the sensor and observer data on an individual end-effector whereas control of a swarm shall concentrate on the information about its periphery, that is, the total volume it disperses into, and collective center of mass with respect to the destination, both simultaneously fed into the control loop.

In addition, if only a number of the micro-swimmers are appointed for an individual task while the others are supposed to carry out a different but codependent task simultaneously, then a hybrid control strategy exploiting multiple actuation mechanisms could be needed. Hence, understanding the collective behavior (Polin et al., 2009) is also very important to ensure the reliability and efficiency of the overall strategy. Likewise, a single actuation system may be employed to steer micro-swimmers moving simultaneously in different directions should the design of the end-effector allow such applications (Khalil et al., 2017b). In addition, clinical MRI systems, as they already are ubiquitous, with proper software and hardware tweaks (Martel, 2017) can be extensively cast to visualize and control various types of end-effectors in a concurrent manner. At any rate, because of the interference between different coil pairs and discrepancies due to biological tissue of various types, meticulous calibration strategies (Petruska et al., 2017) and observer/predictor type mathematical formulae will be used to eliminate disturbances when the higher-order effects cannot be predicted within the acceptable error margins while making use of such EM fields.

REFERENCES

Abayazid, M., Moreira, P., Shahriari, N., Patil, S., Alterovitz, R., Misra, S., 2015. Ultrasound-guided three-dimensional needle steering in biological tissue with curved surfaces. Med. Eng. Phys. 37 (1), 145–150.

Abramson, S., Alexander, H., Best, S., Bokros, J.C., Brunski, J.B., Colas, A., Cooper, S.L., Curtis, J., Haubold, A., Hench, L.L., Hergenrother, R.W., Hoffman, A.S.,

Hubbell, J.A., Jansen, J.A., King, M.W., Kohn, J., Lamba, N.M.K., Langer, R., Migliaresi, C., More, R.B., Peppas, N.A., Ratner, B.D., Visser, S.A., von Recum, A., Weinberg, S., Yannas, I.V., 2004. Classes of materials used in medicine. In: Ratner, B.D., Hoffman, A., Schoen, F.J., Lemons, J.E. (Eds.), Biomaterials Science. second ed. Elsevier Academic Press, China (Chapter 2).

Alapan, Y., Yasa, O., Schauer, O., Giltinan, J., Tabak, A.F., Sourjik, V., Sitti, M., 2018. Soft erythrocyte-based bacterial microswimmers for cargo delivery. Sci. Robot. 3 (17), aa4423.

Bai, H., Cochet, N., Pauss, A., Lamy, E., 2017. DLVO, hydrophobic, capillary and hydrodynamic forces acting on bacteria at solid-air-water interfaces: their relative impact on bacteria deposition mechanisms in unsaturated porous media. Colloids Surf. B Biointerfaces 150, 41–49.

Beiranvand, R., 2013. Analyzing the uniformity of the generated magnetic field by a practical one-dimensional Helmholtz coil system. Rev. Sci. Instrum. 84 (7), 075109.

Berg, H.C., 1993. Random Walks in Biology, New expanded ed. Princeton University Press, NJ, USA (Chapter 1).

Berg, H.C., 2003. The rotary motor of bacterial flagella. Annu. Rev. Biochem. 72, 19–54.

Berg, H.C., Anderson, R.A., 1973. Bacteria swim by rotating their flagellar filaments. Nature 245, 380–382.

Bermejo, S., Ortega, P., Jimenez, J.J., Castañer, L., 2005. Response of c-Si PV arrays under monochromatic light for MEMS power supply. J. Micromech. Microeng. 15, 1446–1453.

Blakemore, R.P., 1982. Magnetotactic bacteria. Ann. Rev. Microbiol. 36, 217–238.

Bourot, J.-M., 1974. On the numerical computation of the optimum on the numerical computation of the optimum profile in Stokes flow. J. Fluid Mech. 65 (pt. 3), 513–515.

Braun, J., Fermvik, L., Stenbäck, A., 1973. Theory and performance of a tritium battery for the microwatt range. J. Phys. E Sci. Instrum. 6 (8), 727–731.

Brennen, C., Winet, H., 1977. Fluid mechanics of propulsion by cilia and flagella. Annu. Rev. Fluid Mech. 9, 339–398.

Brokaw, C.J., 1965. Non-sinusoidal bending waves of sperm flagella. J. Exp. Biol. 43, 155–169.

Brokaw, C.J., 1966. Effects of increased viscosity on the movements of some invertebrate spermatozoa. J. Exp. Biol 45, 113–139.

Browne, W.R., Feringa, B.L., 2006. Making molecular machines work. Nat. Nanotechnol. 1, 25–35.

Bruck, S., 1977. Interactions of synthetic and natural surfaces with blood in the physiological environment. J. Biomed. Mater. Res. A 11 (1), 1–21.

Chang, T.M.S., 2007. Artificial Cells: Biotechnology, Nanomedicine, Regenerative Medicine, Blood Substitutes, Bioencapsulation, and Cell/Stem Cell Therapy. World Scientific Publishing Co. Pte. Ltd., Singapore (Chapter 8).

Chattopadhyay, S., Wu, X.-L., 2009. The effect of long-range hydrodynamic interaction on the swimming of a single bacterium. Biophys. J. 96 (5), 2023–2028.

Chen, B., Jiang, S., Liu, Y., Yang, Y., Chen, S., 2010. Research on the kinematic properties of a sperm-like swimming micro robot. J. Bionic Eng. 7, S123–S129.

Choi, H., Choi, J., Jang, G., Park, J.-h., Park, S., 2009. Two-dimensional actuation of a microrobot with a stationary two-pair coil system. Smart Mater. Struct. 18, 055007.

Cipparrone, G., Hernandez, R.J., Pagliusi, P., Provenzano, C., 2011. Magnus force effect in optical manipulation. Phys. Rev. A 84, 015802.

Cohen, J., 1995. IL-12 deaths: explanation and a puzzle. Science 270 (5238), 908.

Cook-Chennault, K.A., Thambi, N., Sastry, A.M., 2008. Powering MEMS portable devices - a review of non-regenerative and regenerative power supply systems with special emphasis on piezoelectric energy harvesting systems. Smart Mater. Struct. 17 (4), 043001.

Corkidi, G., Taboada, B., Wood, C.D., Guerrero, A., Darszon, A., 2008. Tracking sperm in three-dimensions. Biochem. Biophys. Res. Commun. 373 (1), 125–129.

Diller, E., Zhuang, J., Lum, G.Z., Edwards, M.R., Sitti, M., 2014. Continuously-distributed magnetization profile for millimeter-scale elastomeric undulatory swimming. Appl. Phys. Lett. 104 (17), 174101.

Dreyfus, R., Baudry, J., Roper, M.L., Fermigier, M., Stone, H.A., Bibette, J., 2005. Microscopic artificial swimmers. Nature 437, 862–865.

Durrani, A., Hayward, J.A., Chapman, D., 1986. Biomembranes as models for polymer surfaces. II. The synthesis of reactive species for covalent coupling of phosphorylcholine to polymer surfaces. Biomaterials 7 (2), 121–125.

Einstein, A., 1956. Investigations on the Theory of the Brownian Movement. Dover Publications, Inc., NY, USA

Fauci, L.J., McDonald, A., 1995. Sperm motility in the presence of boundaries. Bull. Math. Biol. 57 (5), 679–699.

Felfoul, O., Martel, S., 2013. Assessment of navigation control strategy for magnetotactic bacteria in microchannel: toward targeting solid tumors. Biomed. Microdevices 15 (6), 1015–1024.

Feynman, R.P., 1992. There is plenty of room at the bottom. J. Microelectromech. Syst. 1 (1), 60–66.

Feynman, R.P., 1993. Infinitesimal machinery. J. Microelectromech. Syst. 2 (1), 4–14.

Flores, H., Lobaton, E., Méndez-Diez, S., Tlupova, S., Cortez, R., 2005. A study of bacterial flagellar bundling. Bull. Math. Biol. 67 (1), 137–168.

Fountain, T.W.R., Kailat, P.V., Abbott, J.J., 2010. In: Wireless control of magnetic helical microrobots using a rotating-permanent-magnet manipulator.Proceedings of the 2010 IEEE International Conference on Robotics and Automation, Anchorage Convention District, May 3–8, Anchorage, Alaska, USA, pp. 576–581.

Frymier, P.D., Ford, R.M., Berg, H.C., Cummings, P.T., 1995. Three-dimensional tracking of the motile bacteria near a solid planar surface. Proc. Nat. Acad. Sci. U. S. A. 92 (13), 6195–6199.

Galajda, P., Ormos, P., 2001. Complex micromachines produced and driven by light. Appl. Phys. Lett. 78 (2), 249–251.

Gauger, E.M., Stark, H., 2006. Numerical study of a microscopic artificial swimmer. Phys. Rev. E 74, 021907.

Ghosh, A., Fischer, P., 2009. Controlled propulsion of artificial magnetic nanostructured propellers. ACS Nano Lett. 9 (6), 2243–2245.

Gibbons, I.R., Gibbons, B.H., 1980. Transient flagellar waveforms during intermittent swimming in sea urchin sperm. I. Wave parameters. J. Muscle Res. Cell Motil. 1 (1), 31–59.

Gorbet, M.B., Sefton, M.V., 2004. Biomaterial-associated thrombosis: roles of coagulation factors, complement, platelets and leukocytes. Biomaterials 25 (26), 5681–5703.

Gray, J., Hancock, G.J., 1955. The propulsion of sea-urchin spermatozoa. J. Exp. Biol. 32 (4), 802–814.

Guyon, E., Hulin, J.P., Petit, L., Mitescu, C.D., 2015. Physical Hydrodynamics, second ed. Oxford University Press, London, UK (Chapter 4).

Hancock, G.J., 1953. The self-propulsion of microscopic organisms through liquids. Proc. R. Soc. Lond. A Math. Phys. Sci. 217 (1128), 96–121.

Happel, J., Brenner, H., 1965. Low Reynolds Number Hydrodynamics With Special Applications to Particulate Media. Prentice-Hall, Inc., NJ, USA (Chapter 7).

Higdon, J.J.L., 1979. A hydrodynamic analysis of flagellar propulsion. J. Fluid Mech. 90 (part 4), 685–711.

Higdon, J.J.L., Muldowney, G.P., 1995. Resistance functions for spherical particles, droplets and bubbles in cylindrical tubes. J. Fluid Mech. 298, 193–210.

Higurashi, E., Ohguchi, O., Tamamura, T., Ukita, H., Sawada, R., 1997. Optically induced rotation of dissymmetrically shaped fluorinated polyimide micro-objects in optical traps. J. Appl. Phys. 82 (6), 2773–2779.

Honda, T., Arai, K.I., Ishiyama, K., 1996. Micro swimming mechanisms propelled by external magnetic fields. IEEE Trans. Magn. 32 (5), 5085–5087.

Hong, S.-J., Koo, K.-I., Lee, S.-M., Park, H.-S., Yoo, H.-J., Kim, J., Park, S., Park, J., Cho, D.-I., 2010. In: Biodegradable polymer droplet for efficient drug delivery using flagellated bacteria.Proceedings of the 10th IEEE International Conference on Nanotechnology Joint Symposium with Nano Korea 2010, August 17–20, Seoul, Korea, pp. 1072–1075.

Hsu, T.-R., 2002. MEMS & Microsystems: Design and Manufacture, International ed. McGraw Hill, NY, USA (Chapter 7).

Huba, M., 2012. In: Modular disturbance observer based constrained PI-controller design. Proceedings of the 12th IEEE Workshop on Advanced Motion Control, March 25–27, 2012, Sarajevo, Bosnia-Herzegovina, pp. 1–6.

Ido, Y., Li, Y.-H., Tsutsumi, H., Sumiyoshi, H., Chen, C.-Y., 2016. Magnetic microchains and microswimmers in an oscillating magnetic field. Biomicrofluidics 10 (1), 011902.

Jeon, Y.B., Sood, R., Jeong, J.-H., Kim, S.-G., 2005. MEMS power generator with transverse mode thin film PZT. Sens. Actuators A 122 (1), 16–22.

Jeon, S., Jang, G., Choi, H., Park, S., 2010. Magnetic navigation system with gradient and uniform saddle coils for the wireless manipulation of micro-robots in human blood vessels. IEEE Trans. Magn. 46 (6), 1943–1946.

Johnson, R.E., Brokaw, C.J., 1979. Flagellar hydrodynamics. A comparison between resistive-force theory and slender-body theory. Biophys. J. 25 (1), 113–127.

Kane, T.R., Levinson, D.A., 1985. Dynamics: Theory and Applications. McGraw-Hill Book Company, NY, USA (Chapter 2).

Kaneko, T., 1975. On Timoshenko's correction for shear in vibrating beams. J. Phys. D Appl. Phys. 8 (16), 1927–1936.

Kasyap, T.V., Koch, D.L., Wu, M., 2014. Hydrodynamic tracer diffusion in suspensions of swimming bacteria. Phys. Fluids 26 (8), 081901.

Keller, J.B., Rubinow, S.I., 1976. Swimming of flagellated microorganisms. Biophys. J. 16 (2, pt. 1), 151–170.

Khaligh, A., Zeng, P., Zheng, C., 2010. Kinetic energy harvesting using piezoelectric and electromagnetic technologies - state of the art. IEEE Trans. Ind. Electron. 57 (3), 850–860.

Khalil, I.S.M., Tabak, A.F., Klingner, A., Sitti, M., 2016. Magnetic propulsion of robotic sperms at low-Reynolds number. Appl. Phys. Lett. 109 (3), 033701.

Khalil, I.S.M., Tabak, A.F., Sadek, K., Mahdy, D., Hamdi, N., Sitti, M., 2017a. Rubbing against blood clots using helical robots: modeling and *in vitro* experimental validation. IEEE Robot. Autom. Lett. 2 (2), 927–934.

Khalil, I.S.M., Tabak, A.F., Hamed, Y., Elwi, M., Tawakol, M., Sitti, M., 2017b. Swimming back and forth using planar flagellar propulsion at low Reynolds numbers. Adv. Sci. 5 (2), 1700461.

Khalil, I.S.M., Tabak, A.F., Hageman, T., Ewis, M., Picheli, M., Mitwally, M.E., El-Din, N.S., Abelmann, L., Sitti, M., 2017c. In: Near-surface effects on the controlled motion of magnetotactic bacteria.Proceedings of the 2017 IEEE International Conference on Robotics and Automation, May 29–June 3, Singapore, pp. 5976–5982.

Kim, B., Kim, D.-H., Jung, J., Park, J.-O., 2004. A biomimetic undulatory tadpole robot using ionic polymer-metal composite actuators. Smart Mater. Struct. 14 (6), 1579–1585.

Kirschvink, J.L., 1992. Uniform magnetic fields and double-wrapped coil systems. Bioelectromagnetics 13 (5), 401–411.

Kósa, G., Shoham, M., Zaaroor, M., 2007. Propulsion method for swimming microrobots. IEEE Trans. Robot. 23 (1), 137–150.

Kummer, M.P., Abbott, J.J., Kratochvil, B.E., Borer, R., Sengul, A., Nelson, B.J., 2010. OctoMag: an electromagnetic system for 5-dof wireless micromanipulation. IEEE Trans. Robot. 26 (6), 1006–1017.

Lal, A., Blanchard, J., 2004. Nuclear batteries: the daintiest dynamos. IEEE Spectrum 41 (9), 36–41.

Landau, L.D., Lifshitz, E.M., 2005. Fluid mechanics. In: Course of Theoretical Physics. second English ed., Revised vol. 6. Elsevier Butterworth Heinemann, China (Chapter 24).

Larm, O., Larsson, R., Olsson, P., 1983. A new non-thrombogenic surface prepared by selective covalent binding of heparin via a modified reducing terminal residue. Biomater. Med. Devices Artif. Organs 11 (2–3), 161–173.

Lauga, E., DiLuzio, W.R., Whitesides, G.M., Stone, H.A., 2006. Swimming in circles: motion of bacteria near solid boundaries. Biophys. J. 90 (2), 400–412.

Lighthill, J., 1976. Flagellar hydrodynamics: the John von Neumann lecture, 1975. SIAM Rev. 18 (2), 161–230.

Lighthill, J., 1996. Helical distributions of Stokeslets. J. Eng. Math 30 (1), 35–78.

Liu, B., Powers, T.R., Breuer, K.S., 2011. Force-free swimming of a model helical flagellum in viscoelastic fluids. Proc. Natl. Acad. Sci. U. S. A. 108 (49), 19516–19520.

Liu, X., Yuan, L., Li, D., Tang, Z., Wang, Y., Chen, G., Chen, H., Brash, J.L., 2014. Blood compatible materials: state of the art. J. Mater. Chem. B 2 (35), 5718–5738.

Lovalenti, P.M., Brady, J.F., 1993. The hydrodynamic force on a rigid particle undergoing arbitrary time-dependent motion at small Reynolds number. J. Fluid Mech. 256, 6561–6605.

Mahoney, A.W., Sarrazin, J.C., Bamberg, E., Abbott, J.J., 2011. Velocity control with gravity compensation for magnetic helical microswimmers. Adv. Robot. 25 (8), 1007–1028.

Maniyeri, R., Suh, Y.K., Kang, S., Kim, M.J., 2012. Numerical study on the propulsion of a bacterial flagellum in a viscous fluid using an immersed boundary method. Comput. Fluids 62, 13–24.

Martel, S., 2017. Beyond imaging: macro- and microscale medical robots actuated by clinical MRI scanners. Sci. Rob. 2(3)eaam8119.

Martel, S., Felfoul, O., Mathieu, J.-B., Chanu, A., Tamaz, S., Mohammadi, M., Mankiewicz, M., Tabatabaei, N., 2009. MRI-based medical nanorobotic platform for the control of magnetic nanoparticles and flagellated bacteria for target interventions in human capillaries. Int. J. Robot. Res. 28 (9), 1169–1182.

Merrill, E.W., Salzman, E.W., 1983. Polyethylene oxide as a biomaterial. Am. Soc. Artif. Intern. Organs 6, 60–64.

Nelson, B.J., Peyer, K.E., 2014. Micro- and nanorobots swimming in heterogeneous liquids. ACS Nano 8 (9), 8718–8724.

Nielsen, O.M., Arana, L.R., Baertsch, C.D., Jensen, K.F., Schmidt, M.A., 2003. In: A thermophotovoltaic micro-generator for portable power applications.The 12th International Conference on Solid-State Sensors, Actuators and Microsystems, June 8–12, Boston, MA, USA, vol. 1. pp. 714–717.

Ning, L., Cannon, M.C., 1998. Gas vesicle genes identified in *Bacillus* megaterium and functional expression in *Escherichia coli*. J. Bacteriol. 180 (9), 2450–2458.

Nogawa, K., Kojima, M., Nakajima, M., Kojima, S., Homma, M., Fukuda, T., 2009. Rotational speed control of Na^+-driven flagellar motor by dual pipettes. IEEE Trans. Nanobiosci. 8 (4), 341–348.

Ogata, K., 1998. System Dynamics, third ed. Prentice-Hall, Inc., NJ, USA (Chapter 8).

Ortigueira, M.D., 2011. Fractional Calculus for Scientists and Engineers. Springer, Heidelberg, Germany (Chapter 2).

Park, S.J., Bae, H., Kim, J., Lim, B., Park, J., Park, S., 2010. Motility enhancement of bacteria actuated microstructures using selective bacteria adhesion. Lab Chip 10 (13), 1706–1711.
Peppas, N.A., Huang, Y., Torres-Lugo, M., Ward, J.H., Zhang, J., 2000. Physicochemical foundations and structural design of hydrogels in medicine and biology. Annu. Rev. Biomed. Eng 2, 9–29.
Petruska, A.J., Edelmann, J., Nelson, B.J., 2017. Model-based calibration for magnetic manipulation. IEEE Trans. Magn. 53 (7), 1–6.
Phan-Thien, N., Tran-Cong, T., Ramia, M., 1987. A boundary-element analysis of flagellar propulsion. J. Fluid Mech. 184, 533–549.
Piefort, V., 2001. Finite Element Modelling of Piezoelectric Active Structures. (Ph.D. dissertation)Faculty of Applied Sciences, Université Libre De Bruxelles.
Polin, M., Tuval, I., Drescher, K., Gollub, J.P., Goldstein, R.E., 2009. *Chlamydomonas* swims with two "gears" in a eukaryotic version of run-and-tumble locomotion. Science 325 (5939), 487–490.
Preumont, A., 2006. Mechatronics: Dynamics of Electromechanical and Piezoelectric Systems. Springer, Dordrecht, Netherlands (Chapters 4 and 5).
Probstein, R.F., 2003. Physicochemical Hydrodynamics. Wiley-Interscience, NJ, USA (Chapter 2).
Purcell, E.M., 1977. Life at low Reynolds number. Am. J. Phys 45 (3), 3–11.
Purcell, E.M., 1997. The efficiency of propulsion by a rotating flagellum. Proc. Natl. Acad. Sci. U. S. A. 94, 11307–11311.
Remes, A., Williams, D.F., 1992. Immune response in biocompatibility. Biomaterials 13 (11), 731–743.
Roper, M., Dreyfus, R., Baudry, J., Fermigier, M., Bibette, J., Stone, H.A., 2006. On the dynamics of magnetically driven elastic filaments. J. Fluid Mech. 554, 167–190.
Rusu, M., Ursu, M., Rusu, D., 2006. Poly(vinyl chloride) and poly(e-caprolactone) blends for medical use. J. Thermoplast. Compos. Mater. 19, 173–190.
Sadelli, L., Fruchard, M., Ferreira, A., 2017. 2D observer-based control of a vascular microrobot. IEEE Trans. Autom. Control 62 (5), 2194–2206.
Saito, M., Kitamura, N., Terauchi, M., 2002. Ultrasonic manipulation of locomotive microorganisms and evaluation of their activity. J. Appl. Phys. 92 (12), 7581.
Sakar, M.S., Steager, E.B., Kim, D.H., Julius, A.A., Kim, M.-J., Kumar, V., Pappas, G.J., 2011. Modeling, control and experimental characterization of microrobots. Int. J. Robot. Res. 30 (6), 647–658.
Salvagnini, C., Michaux, C., Remiche, J., Wouters, J., Charlier, P., Marchand-Brynaert, J., 2005. Design, synthesis and evaluation of graftable thrombin inhibitors for the preparation of blood-compatible polymer materials. Organ. Biomol. Chem. 3 (23), 4209–4220.
Schaefer, D., Martin, I., Jundt, G., Seidel, J., Heberer, M., Grodzinsky, A., Bergin, I., Vunjak-Novakovic, G., Freed, L.E., 2002. Tissue-engineered composites for the repair of large osteochondral defects. Arthritis Rheumatol. 46 (9), 2524–2534.
Servant, A., Qiu, F., Mazza, M., Kostarelos, K., Nelson, B.J., 2015. Controlled in vivo swimming of a swarm of bacteria-like microrobotic flagella. Adv. Mater. 27 (19), 2981–2988.
Shi, J., Kundrat, L., Pishesha, N., Bilate, A., Theile, C., Maruyama, T., Dougan, S.K., Ploegh, H.L., Lodish, H.F., 2014. Engineered red blood cells as carriers for systematic delivery of a wide array of functional probes. Proc. Natl. Acad. Sci. U. S. A. 111 (28), 10131–10136.
Shum, H., Gaffney, E.A., Smith, D.J., 2010. Modelling bacterial behaviour close to a no-slip plane boundary: the influence of bacterial geometry. Proc. R. Soc. Lond. A 466 (8), 1725–1748.
Spagnolie, S.E., Lauga, E., 2010. The optimal elastic flagellum. Phys. Fluids 22 (3), 031901.
Spong, M.W., Vidyasagar, M., 1989. Robot Dynamics and Control. John Wiley & Sons, NY, USA (Chapter 9).

Stricker, L., 2017. Numerical simulation of artificial microswimmers driven by Marangoni flow. J. Comput. Phys. 347, 467–489.

Tabak, A.F., 2018. Hydrodynamic impedance of bacteria and bacteria-inspired micro-swimmers: a new strategy to predict power consumption of swimming micro-robots for real-time applications. Adv. Theory Simul. 1 (4), 1700013.

Tabak, A.F., Yesilyurt, S., 2014a. Computationally-validated surrogate models for optimal geometric design of bio-inspired swimming robots: helical swimmers. Comp. Fluids 99, 190–198.

Tabak, A.F., Yesilyurt, S., 2014b. Improved kinematic models for two-link helical micro/nanoswimmers. IEEE Trans. Robot. 30 (1), 14–25 (special issue on nanorobotics).

Tabak, A.F., Temel, F.Z., Yesilyurt, S., 2011. In: Comparison on experimental and numerical results for helical swimmers inside channels.Proceedings of the 2011 IEEE/RSJ International Conference on Intelligent Robots and Systems, September 25–30, San Francisco, California, pp. 463–468.

Takano, Y., Goto, T., 2003. Numerical analysis of small deformation of flexible helical flagellum of swimming bacteria. JSME Int. J. Ser. C Mech. Syst. Mach. Elem. Manuf. 46 (4), 1234–1240 (special issue on bioengineering, released on June 2004).

Taylor, G., 1951. Analysis of the swimming of microscopic organisms. Proc. R. Soc. Lond. A 209, 447–461.

Teran, J., Fauci, L., Shelley, M., 2010. Viscoelastic fluid response can increase the speed and efficiency of a free swimmer. Phys. Rev. Lett. 104, 038101.

Tjeung, R.T., Hughes, M.S., Yeo, L.Y., Friend, J.R., 2011. Surface acoustic wave micromotor with arbitrary axis rotational capability. Appl. Phys. Lett. 99 (21), 214101.

Uenoyama, A., Miyata, M., 2005. Gliding ghosts of *Mycoplasma mobile*. Proc. Natl. Acad. Sci. U. S. A. 102 (36), 12754–12758.

Wang, S., Ardekani, A.M., 2012. Unsteady swimming of small organisms. J. Fluid Mech. 702, 286–297.

Wiggins, C.H., Goldstein, R.E., 1998. Flexive and propulsive dynamics of elastica at low Reynolds number. Phys. Rev. Lett. 80 (17), 3879–3882.

Williams, B.J., Anand, S.V., Rajagopalan, J., Saif, M.T.A., 2014. A self-propelled biohybrid swimmer at low Reynolds number. Nat. Commun. 5, 3081.

Woolley, D.M., Vernon, G.G., 2001. Study of helical and planar waves on sea urchin sperm flagella, with a theory of how they are generated. J. Exp. Biol. 204 (pt. 7), 1333–1345.

Yan, X., Zhou, Q., Vincent, M., Deng, Y., Yu, J., Xu, J., Xu, T., Tang, T., Bian, L., Wang, Y.-X.J., Kostarelos, K., Zhang, L., 2017. Multifunctional biohybrid magnetite microrobots for imaging-guided therapy. Sci. Robot. 2(12). eaaq1155.

Yu, C., Kim, J., Choi, H., Choi, J., Jeong, S., Cha, K., Park, J.-O., Park, S., 2010. Novel electromagnetic actuation system for three-dimensional locomotion and drilling of intravascular microrobot. Sens. Actuators A: Phys. 161 (1–2), 297–304.

Zhang, L., Abbott, J.J., Dong, L., Kratochvil, B.E., Bell, D., Nelson, B.J., 2009. Artificial bacterial flagella: fabrication and magnetic control. Appl. Phys. Lett. 94 (6), 064107.

Zhang, L., Ye, M., Giataganas, P., Hughes, M., Bradu, A., Podoleanu, A., Yang, G.-Z., 2017. From macro to micro: autonomous multiscale image fusion for robotic surgery. IEEE Robot. Autom. Mag. 24 (2), 63–72.

CHAPTER THIRTEEN

Artificial Hearts

Graham Brooker
Australian Centre for Field Robotics, University of Sydney, Sydney, NSW, Australia

Contents

1 INTRODUCTION

The heart can be considered to be a pair of pumps folded together to form a single unit as shown in Fig. 1. The right half of the heart pumps blood only to the lungs in which deoxygenated blood enters the right atrium through the superior and inferior vena cavae, and then out from the right ventricle into the pulmonary arteries at low pressure (25 mmHg). The left half of the heart pumps blood to the rest of the body with oxygenated blood entering the left atrium from the lungs through the pulmonary veins, and then out from the left ventricle via the aorta at high pressure (120 mmHg).

A series of valves ensure the flow remains in the correct direction with the left and right atria receiving the incoming blood and pumping it into the ventricles through the tricuspid and bicuspid (mitral) valves, respectively. Following that, the two ventricles produce enough pressure to push the

Handbook of Biomechatronics
https://doi.org/10.1016/B978-0-12-812539-7.00013-1

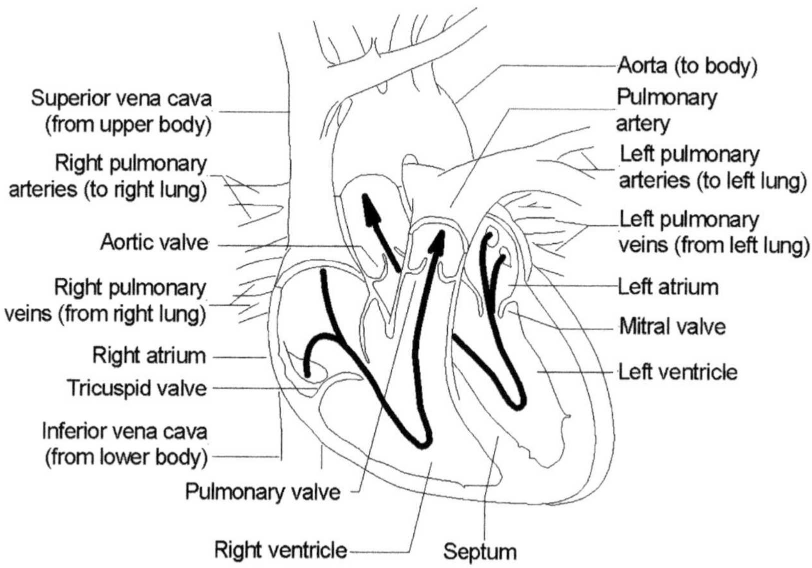

Fig. 1 Structure of the heart.

blood out through the semilunar valves and through the pulmonary and systemic circulations. There are generally no valves where the vena cavae join the right atrium or where the pulmonary veins enter the left atrium, because pressures in the atria are small and valves are not needed.

Cardiac output is normally about 5 L/min, but can triple during strenuous exercise. This is the product of the stroke volume (the volume of a single output) and the heart rate. These are nominally 70 cm^3 per stroke at 70 beats per minute (bpm). As the heart rate increases, the total output increases proportionally, until about 200 bpm after which the heart chambers do not have time to fill properly, limiting the maximum flow rate to about 15 L/min.

The blood pressure (BP) in the aorta alternates between a high pressure (systole) of about 120 mmHg, and low pressure (diastole), about 80 mmHg. This lower limit is determined by the elasticity of the ventricular walls. Pressure will obviously be a function of the flow rate and the resistance to flow, with the body changing both the cardiac output and the resistance to maintain the required level, as described schematically in Fig. 2. The baroreceptor reflex makes short-term adjustments to the BP with a time constant of <1 min and often much shorter than that. An example of this is the increase in heart rate and stroke volume that occurs when you stand up suddenly. Pressure is measured by sensors in the arch of the aorta and the carotid sinus

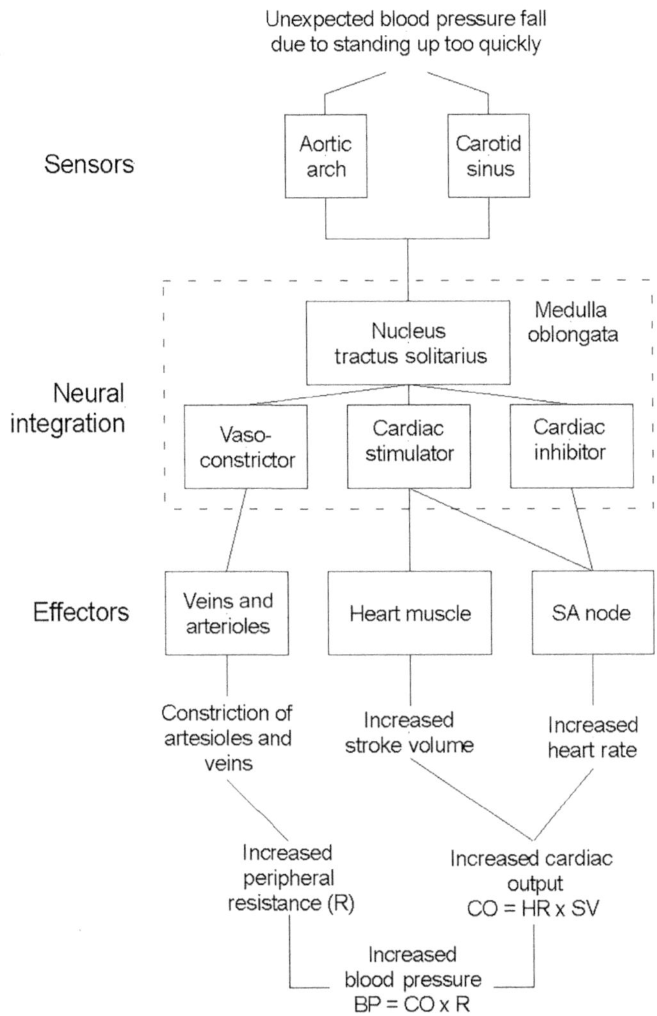

Fig. 2 Short-term pressure regulation. *(Based on Orme, F., 2002. Human Physiology—Lecture Notes. http://members.aol.com/Bio50/index.html (Retrieved September 2008).)*

from which nerves convey the information to the nucleus tractus solitarius of the medulla oblongata, in the brain. Feedback is provided via the vagus nerve that slows the heart, and the accelerator nerve that speeds it up (Orme, 2002).

Long-term regulation is determined mostly by the kidneys as they regulate salt and water content in the body, and these control BP. Sodium retention is controlled by the Na pump as regulated by the hormone aldosterone. This is in turn regulated by the hormones rennin and angiotensin. If sodium is retained the blood osmotic pressure rises and this causes water to be retained, also due to osmotic pressure. Water reabsorption is through water channels in the kidney tubules, and these are controlled by the anti-diuretic hormone (ADH). If ADH is present in high concentrations, water absorption will be high and BP will rise.

2 HISTORICAL BACKGROUND

Aristotle considered the heart to be "the source of all movement, since the heart links the soul with the organs of life." The Greeks called the pulse "sphygmos," and therefore sphygmology deals with the theory of the pulse. At about the same time in ancient China, Wang Shu-he wrote 10 books about the pulse. The Greek physician, Galen, knew that blood vessels carried blood and identified venous (dark red) and arterial (brighter and thinner) blood, each with distinct and separate functions. He believed that growth and energy were derived from venous blood created in the liver, while arterial blood gave vitality by containing pneuma (air) and originated in the heart. Blood flowed from both of these organs to all parts of the body where it was consumed, and there was no return of blood to the heart or liver (Boylan, 2007).

In 1242, the Arab scholar Ibn Nafis became the first person to accurately describe the circulatory process (Al-Ghazal, 2002). Three hundred years later, in 1552, Michael Servetus described the same, and Realdo Colombo proved the concept, but it remained largely unknown in Europe. It was another 100 years before William Harvey performed a sequence of experiments and announced, in 1628, the discovery of the human circulatory system as his own. Harvey published an influential book, the *Exercitatio Anatomica de Motu Cordis et Sanguinis in Animalibus*, about it and the rest of the medical world started to take note (Shackleford, 2003).

From a circulatory support perspective, history only begins in 1934 when Michael DeBakey developed the concept of a peristaltic (or roller) pump to facilitate blood transfusions. A few years later William Sewell Jr. and William Glenn, Yale University medical students, built a section of a heart pump using a toy erector set which they used in experimental bypass surgery on dogs. The initial design using a roller pump driven by the Erector set motor was unsuccessful as the motor was underpowered for the blood pumping function. Instead Sewell built the device shown in Fig. 3 to drive eccentric

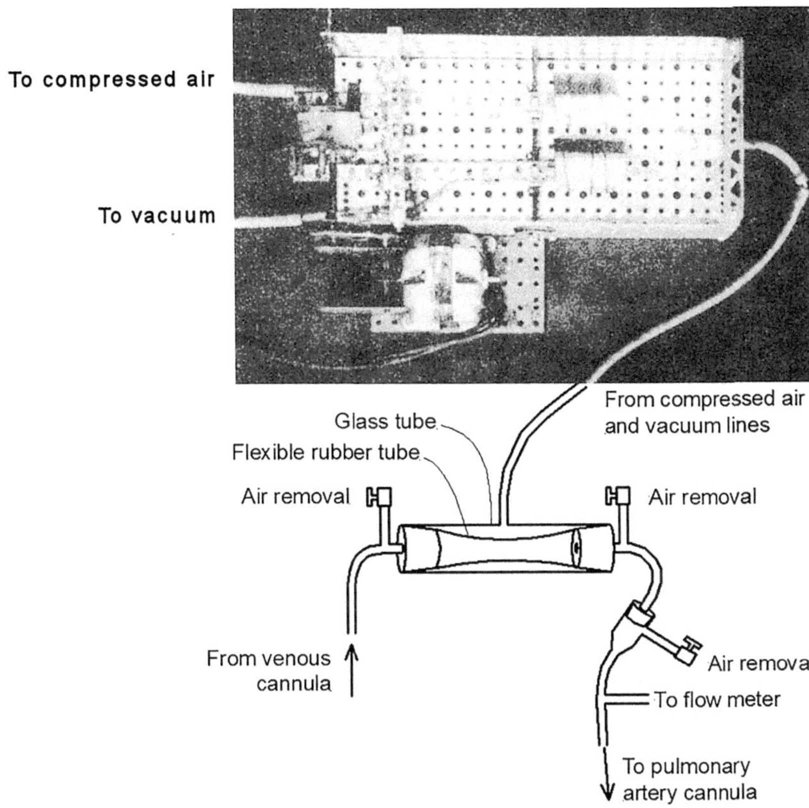

Fig. 3 The first artificial heart pump built by William Sewell and William Glenn. *(Based on Glenn, W., 1993. Seawell's Pump. Guthrie J. 63 (1).)*

cams that occluded and released small rubber tubes leading to the compressed air and vacuum lines that actually drove the pump (Glenn, 1993).

In 1949, the pump was used to bypass the right heart of dogs in two experiments lasting for 61 and 82 min, respectively, with the right ventricle wide open. After restoration of normal circulation, removal of the pump and closure of the chest, the dogs made uneventful recoveries.

Four years later in 1953, the first heart surgery was undertaken on a human subject by John Gibbon using a similar pump for cardiopulmonary bypass (a primitive heart-lung machine). This ushered in the era of open heart surgery, but also started researchers investigating the possibility of providing augmentation or even replacement of the natural heart (Joyce et al., 2012).

Search for an optimal internal circulatory-support device began in 1964, with the National Institutes of Health (NIH) artificial-heart program. That year, DeBakey inserted the first left ventricular assist device (LVAD) between the left ventricle and the descending aorta in a human patient using a pump almost identical in design to Sewell's (Glenn, 1993). Following this partial success DeBakey went on to develop paracorporeal LVADs for temporary support.

Another early method of providing circulatory assistance was the auxiliary artificial ventricle developed by Kantrovitz and his colleagues in 1963. This consisted of a valveless bulb between the ascending and descending aorta that provided compressed air-driven counterpulsation to augment blood flow and reduce left ventricular workload. In a first, the patient fitted with the device was able to return home where he survived for about 3 months (Joyce et al., 2012).

Many different variations of the pulsatile LVAD were developed in the 1980s and 1990s and are examined from a mechanical perspective in more detail later in this section. Probably the most successful of these is the pneumatically powered Thoratec HeartMate, which gained approval as a destination device. This occurred as a result of the successful outcome of the REMATCH (Randomized evaluation of mechanical assistance for the treatment of congestive heart failure) trial, shown in Fig. 4, which compared

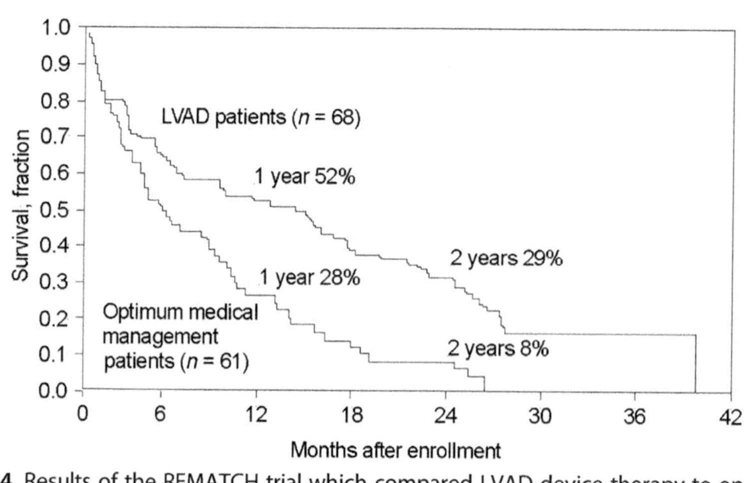

Fig. 4 Results of the REMATCH trial which compared LVAD device therapy to optimal medical management. *(Based on Kormos, R., Miller, L. (Eds.), 2012. Mechanical Circulatory Support. Elsevier, USA.)*

device therapy to optimal medical management and found the former to be superior in the longer term.

The large size and limited lifetime of pulsatile LVADS drove research into alternative pumping technologies that would be more suited to long-term implant. Rotary devices, both axial, and centrifugal were considered ideal as they were smaller, valve free and because the internal volume did not change they did not need a compliance chamber or external vent tube. In addition, they were more efficient and lower cost.

Axial-flow pumps consist of a simple spiral impellor that rotates at around 10,000 rpm to pump the blood in an axial path (the inlet and outlet cannulae aligned). The configuration of the axial pump, as shown in Fig. 5, occurs where no radius change exists between the streamlines moving from the inlet to the outlet, and centrifugal action plays no part. These second-generation pumps were first used in 1998 and included the Berlin Heart Incor, the MicroMed DeBakey, and HeartMate II devices among others.

With centrifugal pumps, fluid enters the central portion, called the eye, flows radially outward and is discharged around the entire circumference of the pump into a casing. During flow through the rotating impeller, the fluid receives energy from the vanes resulting in an increase in both pressure and absolute velocity. Since a large portion of the energy leaving the impeller is kinetic, it is necessary to reduce the absolute velocity and transform most of this energy to pressure head. This is accomplished in the volute casing surrounding the impeller or in flow through diffuser vanes. Centrifugal pumps operate at lower speeds than axial pumps, typically 2000–3000 rpm to generate blood flows of around 5 L/min against a 120 mmHg head. Because they operate at lower speeds, bearing life is typically better than it is for axial pumps.

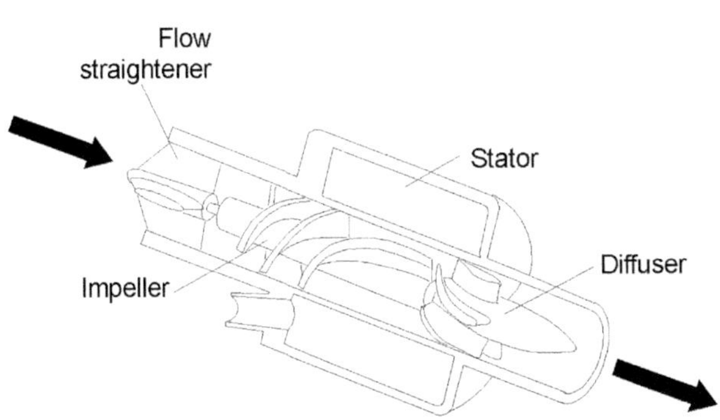

Fig. 5 Generic axial blood pump type.

If a higher capacity is required without an increase in the diameter, the pump dimensions in the direction parallel to the shaft, must be increased (Daugherty and Franzini, 1977). This becomes a mixed-flow pump, and is shown in Fig. 6 along with an example of a centrifugal pump.

The third-generation mixed flow and centrifugal pumps eliminate the shaft connecting motor to rotor by coupling the two magnetically and providing magnetic or hydrodynamic levitation (Yamane, 2016). This eliminates the shaft connection with its associated bearings and sealing issues and so results in a more compact and reliable device. The first clinical application was the successful implantation of a VentraCor VentrAssist device in 2005. Others in this generation include the WorldHeart Levacor and Terumo DuraHeart devices.

The fourth generation of VADs, either axial or mixed flow, is sufficiently small and light to be implanted above the diaphragm. These have masses ranging from <100 to about 150g, and include the HeartWare HVAD, the DeBakey HeartAssist 5, and the Jarvik-2000.

To provide an indication of progress made in the development of LVADS over the last 15 years, photographs of a range of these spanning the first to fourth generation are shown to scale in Fig. 7.

In cases where biventricular failure has occurred, a total artificial heart (TAH) is required. The requirement for these devices is limited compared with those for LVADs, but from a biomechatronic perspective their development has provided significant input into the technology overall.

The first attempt at the implantation of a TAH was by a Soviet researcher, V. Demikhov who performed a number of experiments on animals in 1937. The implanted pumps were driven via a shaft that passed

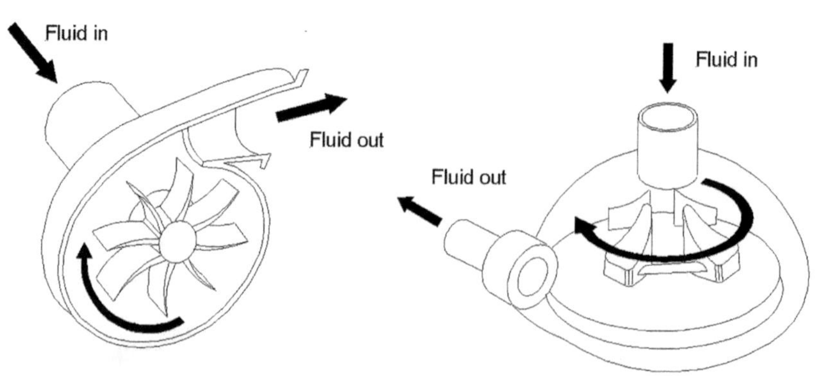

Fig. 6 Generic centrifugal and mixed-flow blood pump types.

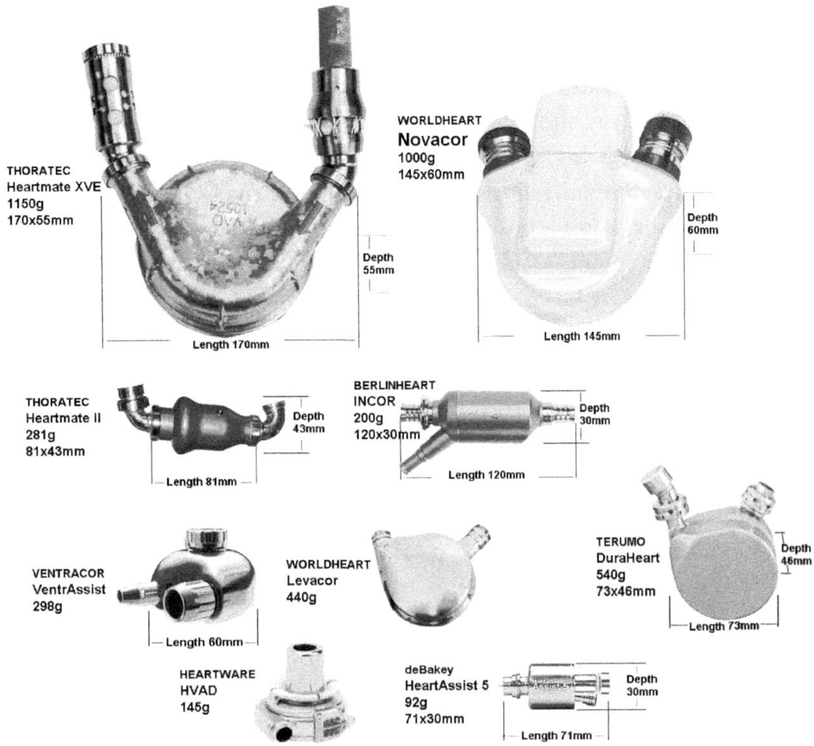

Fig. 7 Ventricular assist devices grouped by generation (Brooker, 2012).

through the chest wall. He continued his research until 1958 but did not publish. The individual who is now recognized as the main pioneer in the development of artificial hearts is Willem Kolff. After emigrating from Holland, he joined the Cleveland Clinic as a research assistant, and within 7 years he and Tetsuzo Akutsu were testing primitive artificial hearts in animals to identify problems that might be encountered if such devices were to be later implanted in a human patient. Their first success was a hydraulically driven TAH implanted into a dog which survived for 90 min. In collaboration with Thompson Woolridge they also developed a solenoid-driven heart. Apart from its power supply, this was the first totally implantable TAH. Unfortunately, it used substantial amounts of power requiring five solenoids which transferred their force through an oil bath into the elastic pumping chamber.

In 1963, the first patented artificial heart was developed by Paul Winchell with help from Henry Heimlich (of Heimlich maneuver fame).

Winchell subsequently assigned the patent to the University of Utah where Robert Jarvik ultimately used it as the model for the Jarvik-7. This generation of artificial hearts were all powered pneumatically and required large external compressors and control systems.

The Liotta-Cooley heart was the first temporary artificial heart implanted in a human being. It was developed by Domingo Liotta and implanted by surgeon Denton Cooley of the Texas heart Institute on April 4, 1969. The recipient, Haskell Karp, lived for 64 h with the artificial heart until a human heart became available for transplant (Cooley, 2003). The heart was a pneumatic double-ventricle pump with Wada-Cutter hingeless valves to control the direction of blood flow. The interior of the pump was lined with a material that promoted the formation of a smooth cellular surface (pseudoneointima-forming surface). The flexible inflow and outflow tracts were made of Dacron fabric, and the pump chambers were made of a combination of Dacron fabric and Silastic. The pumps were connected to the external compressor with Silastic tubing covered by Dacron. The console was about the size of a large tumble dryer in which two pneumatic power units generated the pumping and vacuum actions needed to move blood through the artificial heart. An adjacent control panel could be used to adjust pumping rate and pumping pressure (Cooley et al., 1969).

The first permanent implantation of a TAH occurred in 1982 by William DeVries. It was a Jarvic-7 made by Kolff Medical (later Symbion) and was implanted into Barney Clark by DeVries and Lyle Joyce. Owing to his poor general health Clark had not been a candidate for a heart transplant and after receiving the TAH he was never able to leave the hospital. The system was open to infection, so Clark, and subsequent Jarvik-7 recipients, got sick. Patients had to be kept on blood thinners to prevent clots and strokes. Although Clark was reported to be in a stable condition 48 h after the implant, his subsequent postoperative condition was not good and he died after 112 days (Hajar, 2005).

The Jarvik-7, shown in Fig. 8, is made of polyurethane, polyester, plastic, and aluminum and was designed for permanent implantation. It consists of two identical pneumatically driven pumps that replaced the left and right ventricles of the heart.

The identical pump sections consist of a chamber divided by a flexible multilayered diaphragm as shown in Fig. 9. Direction of blood flow is controlled by tilting-disk valves, held in place by a polycarbonate ring structure. External arterial and venous connections are made using "quick connect" cuffs. One of the innovations of the device was the inner coating of rough

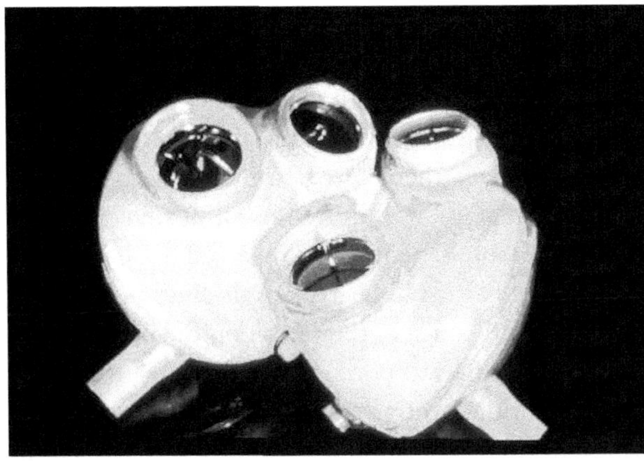

Fig. 8 Photograph of the Jarvik-7 artificial heart. *(Courtesy NIH archives.)*

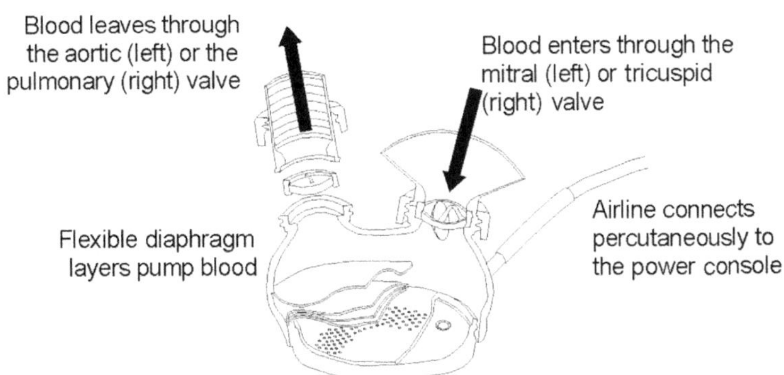

Fig. 9 Schematic diagram of one pump of the Jarvik-7. *(Based on Hajar, R., 2005. The artificial heart, Heart View 8(2):70–76.)*

material, developed by David Gernes which helped the blood to clot and coat the inside of the device, enabling a more natural blood flow. The total weight of the device is about 800 g and it required 520 cm^3 of space within the thorax (Jaron, 1990).

A 2-m long external air-line fed compressed air to the chamber where changes in the pressure flexed the diaphragm cyclically to drive blood flow. These airlines were attached to the large compressor and control console, which regulated the pump stroke and dictated the pumping rate.

The Jarvik-7 heart was implanted many times with the longevity record held by William Schroeder, who was hooked to a Jarvik-7 in 1985. He lived for 18 months with an apparently good quality of life though he did suffer strokes, sudden hemorrhages and infections during his final days. A 1985 study of 15 patients with active infections from device implantation showed mortality rates as high as 70%. Most infections developed from the site of the percutaneous tubes passing into the body, though some arose from blood clots that developed from the internal surfaces of the pump (Lemelson-MIT, 2002). As materials and knowhow improved, the survival rate improved, but after about 90 people had received the Jarvik device, the permanent implantation of artificial hearts was discontinued for use in patients with heart failure, because most of the recipients did not live more than half a year and their quality of life was poor. However, for some time after the ban it was still used as a bridge to transplant device.

Other bridge devices started to take the place of the Jarvik-7 with the CardioWest (now Syncardia) temporary total artificial heart (TAH-t) approved for use in 2004. It is the first implantable artificial heart to be approved by the US Food and Drug Administration (FDA). Compared with the Jarvik-7, it is very light, weighing only 160 g, but in other respects it is very similar to its predecessor.

In September of 2006, the FDA approved the first totally implanted permanent artificial heart for patients with advanced bi-ventricular failure. According to the FDA, the "AbioCor Implantable Replacement Heart is intended for people who are not eligible for a heart transplant and who are unlikely to live more than a month without intervention." This system consists of a 900-g electrohydraulic mechanical heart implanted in the chest cavity along with a controller and an internal battery, which are implanted in the patient's abdomen, and a transcutaneous power transfer coil that recharges the internal battery.

3 HEART PUMPS AND MOTORS

A pulsatile pump maintains a regular flow rate by driving a fixed amount of blood with each stroke (typically 70 mL) and adjusting the stroke rate to suit the total flow requirements. All of the early VADs and TAHs were pulsatile in nature with the pump chamber consisting of a multilayered polymer sack or diaphragm connected to an inlet and an outlet valve. Compression and expansion of the sac/diaphragm was usually achieved using compressed air, by a pusher plate usually driven by an electric motor actuated ball/roller screw, by a single or multiple solenoids, or by a linear actuator.

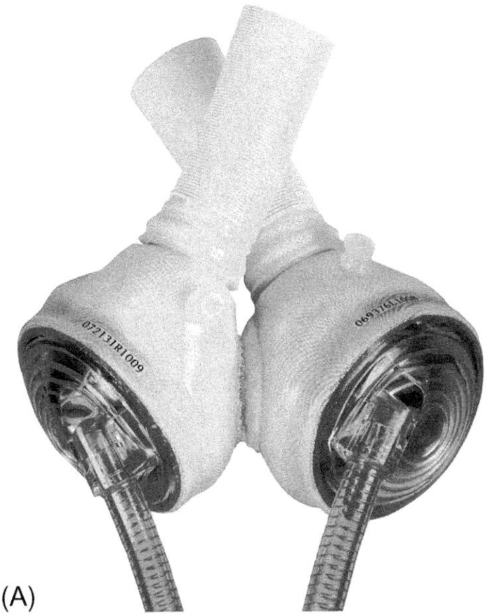
(A)

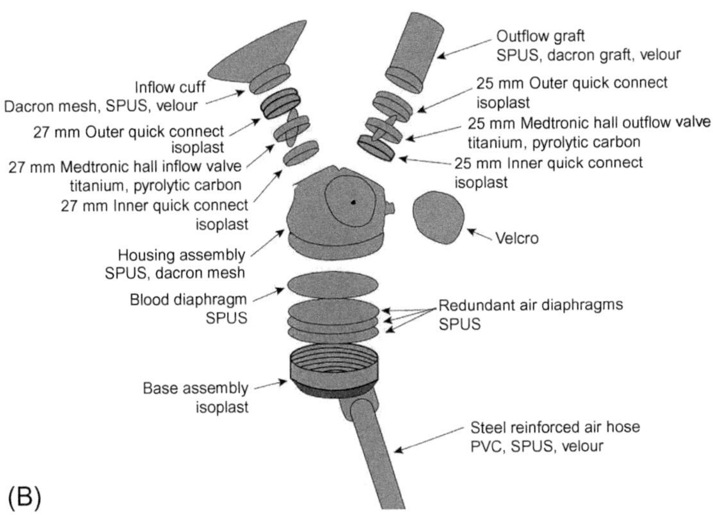

(B)

Fig. 10 CardioWest (now Syncardia) temporary total artificial heart (ScienceDaily, 2006; Slepian et al., 2006). (A) Photograph of the artificial heart. (B) Exploded view of the components of the artificial heart. *(Courtesy of Syncardia.com, with permission.)*

3.1 Pulsatile Pumps

The Syncardia TAH-t shown in Fig. 10 is an example of the genre of pneumatically powered pulsatile pumps. The plastic housing is lined with polyurethane and has a four-layer pneumatically driven diaphragm. Four Medtronic-Hall mechanical valves ensure blood flows correctly through the device.

The pneumatic control system for the heart, shown graphically in Fig. 11, is based on two principles; partial fill and full ejection. The pressures are set to fully eject all of the blood from each ventricle with each beat which is achieved by setting the ejection pressure of the right ventricle to 30 mmHg higher than the pressure in the pulmonary artery, and that of the left ventricle 60 mmHg higher than systemic pressure. At the end of systole, the diaphragms are therefore fully distended at which time the driver opens a valve to release the air pressure in the drivelines. Air is pushed out of the drivelines as blood enters the ventricles and its volume is measured as indicative of the amount of blood entering. The ventricles are adjusted to fill to between 50 and 60 mL to allow for some overhead during the fill phase. This produces between 7 and 8 L/min while maintaining the correct Starling law pressure differential. At the maximum stroke volume of 70 mL and a rate of 130 bpm, the artificial heart can pump over 9 L/min (Slepian et al., 2006; Joyce et al., 2012).

Pusher plate devices are more common than pneumatics with a number of different manufacturers producing similar units. These include the Thoratec HeartMate and the LionHeart VAD among many. A good example of a planetary roller-screw-driven LVAD similar to that used in the Arrow LionHeart device is discussed in Takatani et al. (2001). This device uses a brushless DC motor to drive the roller screw to produce a stroke length of 12 mm and volume of 55 cm^3. The maximum pump output is 8 L/min at an electrical power of 8 W and a 24% electrical-to-hydraulic efficiency. The pump is housed within a titanium alloy shell 90 mm in diameter and 56 mm thick with a total volume of 285 cm^3 weighing 552 g. As shown in Fig. 12, the LVAD consists of a miniature 14-pole Y-wound brushless DC motor from Kollmorgan Inc. and a planetary roller screw from SKF. Motor rotation is converted into rectilinear motion using the roller screw attached to a pusher plate which compresses the diaphragm. It is then reversed after the completion of each ejection cycle to allow passive filling of the blood chamber. Hall-effect sensors monitor the position of the pusher plate so that the stroke volume and beat rate can be controlled. The diaphragm is made from polyurethane manufactured by Polymer Technology Inc. using a dip-coating method. The housing is manufactured from a titanium alloy.

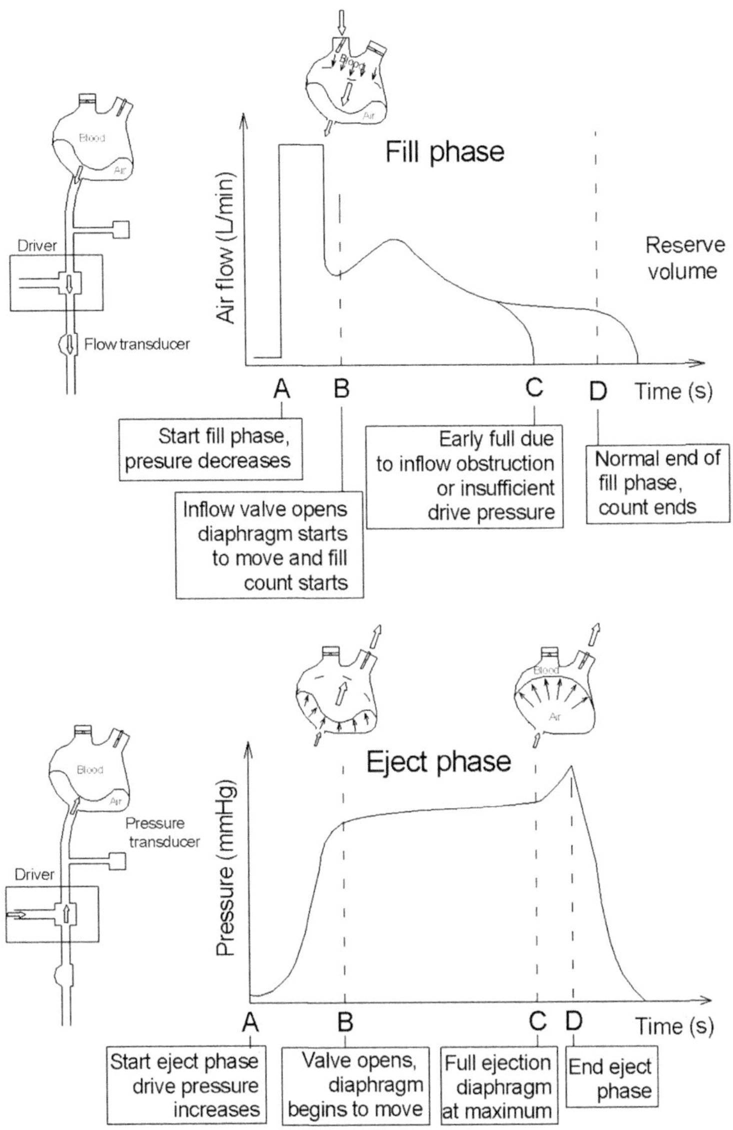

Fig. 11 Details of the CardioWest TAH fill and eject phases. *(Based on Slepian, M., Smith, R., Copeland, J., 2006. The syncardia cardiowest total artificial heart. In: Baughman, Baumgartner. (Eds.), Treatment of Advanced Heart Disease. Taylor and Francis, pp. 473–490 (Chapter 26); Joyce, D., Joyce, L., Loebe, M. (Eds.), 2012. Mechanical Circulatory Support. McGraw Hill Medical, New York.)*

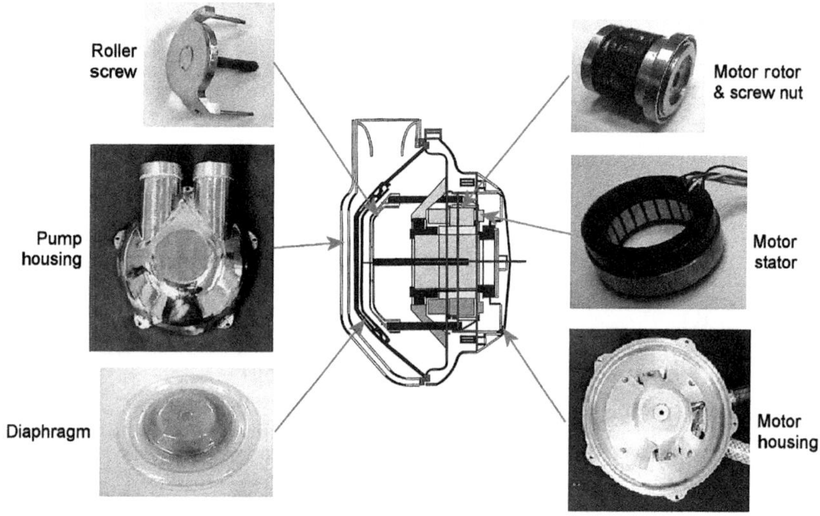

Fig. 12 Schematic diagram of the roller-screw LVAD components. *(Reproduced with permission Takatani, D., Ouchi, K. Nakamura, M., Sakamoto, T., 2001. Ultracompact, totally implantable, permanent, pulsatile VAD system. J. Congestive Heart Fail. Circ. Support 1 (4): 407–412.)*

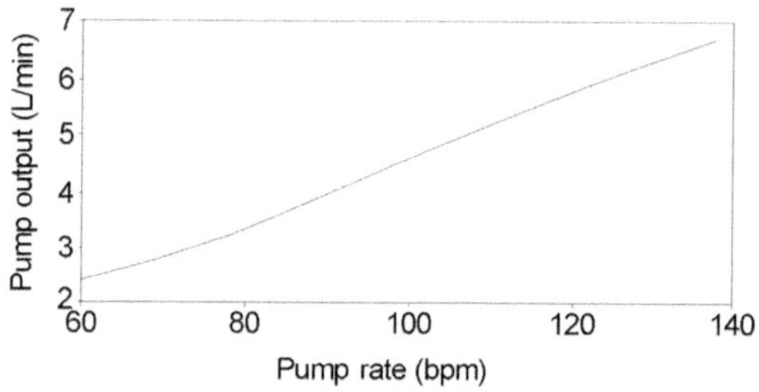

Fig. 13 Roller-screw pump output power as a function of pump rate. *(Based on Takatani, D., Ouchi, K. Nakamura, M., Sakamoto, T., 2001. Ultracompact, totally implantable, permanent, pulsatile VAD system. J. Congestive Heart Fail. Circ. Support 1(4): 407–412.)*

Pump output is a reasonably linear function of pump rate as shown in Fig. 13 with the highest pump output of about 7 L/min obtained at 140 bpm. The power requirements shown in Fig. 14 also bear a reasonably linear relationship to the output flow rate.

Pump efficiencies increased with flow rate up to 140 bpm with the best efficiency of 23% obtained using polyurethane valves. The use of conventional Bjork-Shiley and St Jude valves resulted in slightly lower efficiencies (about 22% and 20%, respectively).

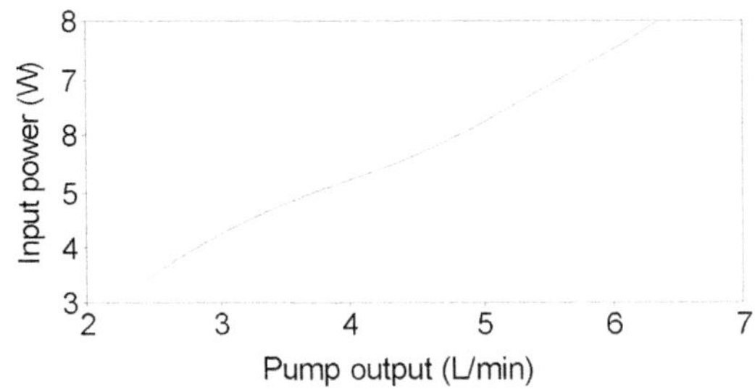

Fig. 14 Roller-screw pump input power as a function of pump rate. *(Based on Takatani, D., Ouchi, K. Nakamura, M., Sakamoto, T., 2001. Ultracompact, totally implantable, permanent, pulsatile VAD system. J. Congestive Heart Fail. Circ. Support 1(4): 407–412.)*

A simpler mechanism is described in Fukui et al. (2004). In this design, a moving magnet linear actuator uses conventional voice-coil technology powered by an AC signal to drive pusher-plate directly as shown in Fig. 15.

The housing is made from epoxy using a rapid-prototyping machine and the diaphragm is made from segmented polyurethane from PTG medical Co. and formed using a dipping method. All of the blood-contact area within the housing is coated with the same material. The diaphragm, pusher plate, linear guide, and actuator are integrated into a single unit. A Hall-effect device is set into the actuator housing to detect the displacement of the mover. A pair of Bjork-Shiley tilting-disc valves are mounted on the inlet and outlet ports of the pump housing. The complete pump has a diameter of 101 mm and a thickness of 49 mm making its volume 320 cm^3.

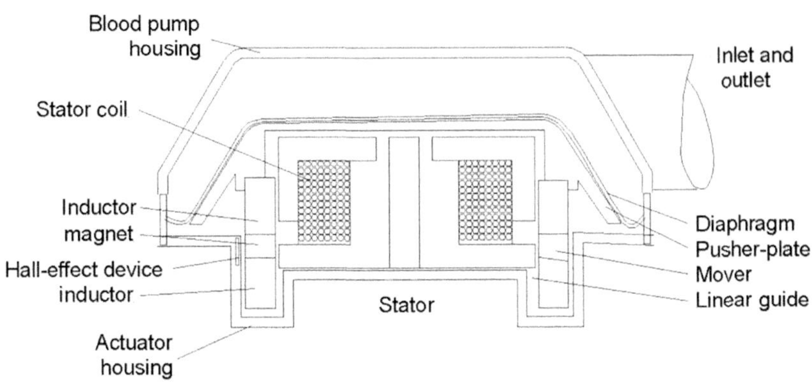

Fig. 15 Structure of a linear actuator-based LVAD. *(Based on Fukui, Y., Funakubo, A., Fukunaga, K., 2004. Development of the assisted artificial heart with linear motor actuator. In: SICE 2004 Annual Conference.)*

The mass is 770 g which is significantly heavier than the roller-screw-based device. The controller comprises a microprocessor that reads the Hall-effect signals and drives a MOSFET-controlled H-bridge that powers the actuator coil from a 9-V DC source. The control unit has two modes, a fixed rate mode where a constant frequency signal is provided to the actuator, and a full fill/full eject (FFFE) mode that uses the Frank-Starling control mechanism to govern the pump speed.

The measured pump performance is shown in Fig. 16. For a head of 100 mmHg the maximum output was 6.1 L/min at 155 bpm with a power consumption of 8 W. The maximum efficiency was 16.3% at a pump rate of 135 bpm. In the FPPP mode the pump output was 7.9 and 5.1 L/min for loads of 60 and 120 mmHg, respectively with the pump rate decreasing by 54 bpm over that range. The stroke volume remained between 38 and 43 cm^3.

An alternative to the pneumatic drive method is the one used by the AbioMed TAH which is symmetrical dual-cavity hydraulically driven blood pump replacing both the right and left hearts. Each pump is capable of delivering more than 8 L/min. Blood is pumped from the superior and inferior vena cava to the lungs through the pulmonary artery by the right pump and from the pulmonary veins to the rest of the body via the aorta by the left pump.

The heart pump consists of the following components as shown in Fig. 17:

- **Hydraulic pump**—An efficient electric motor spins the impeller inside the centrifugal pump at 10,000 rpm to create the required hydraulic pressure in a silicone hydraulic fluid.

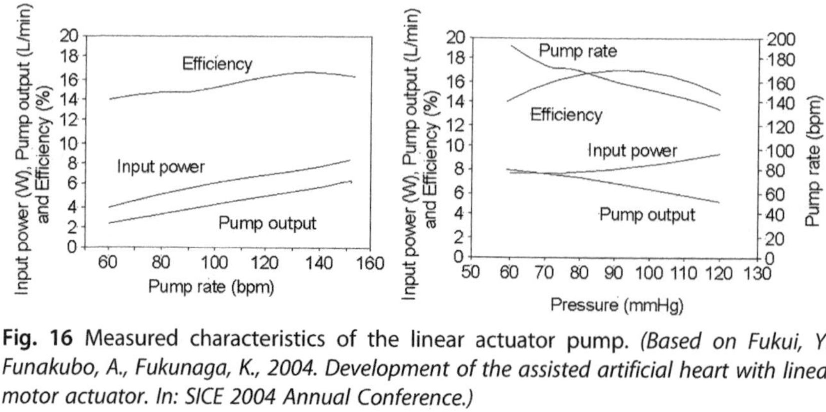

Fig. 16 Measured characteristics of the linear actuator pump. *(Based on Fukui, Y., Funakubo, A., Fukunaga, K., 2004. Development of the assisted artificial heart with linear motor actuator. In: SICE 2004 Annual Conference.)*

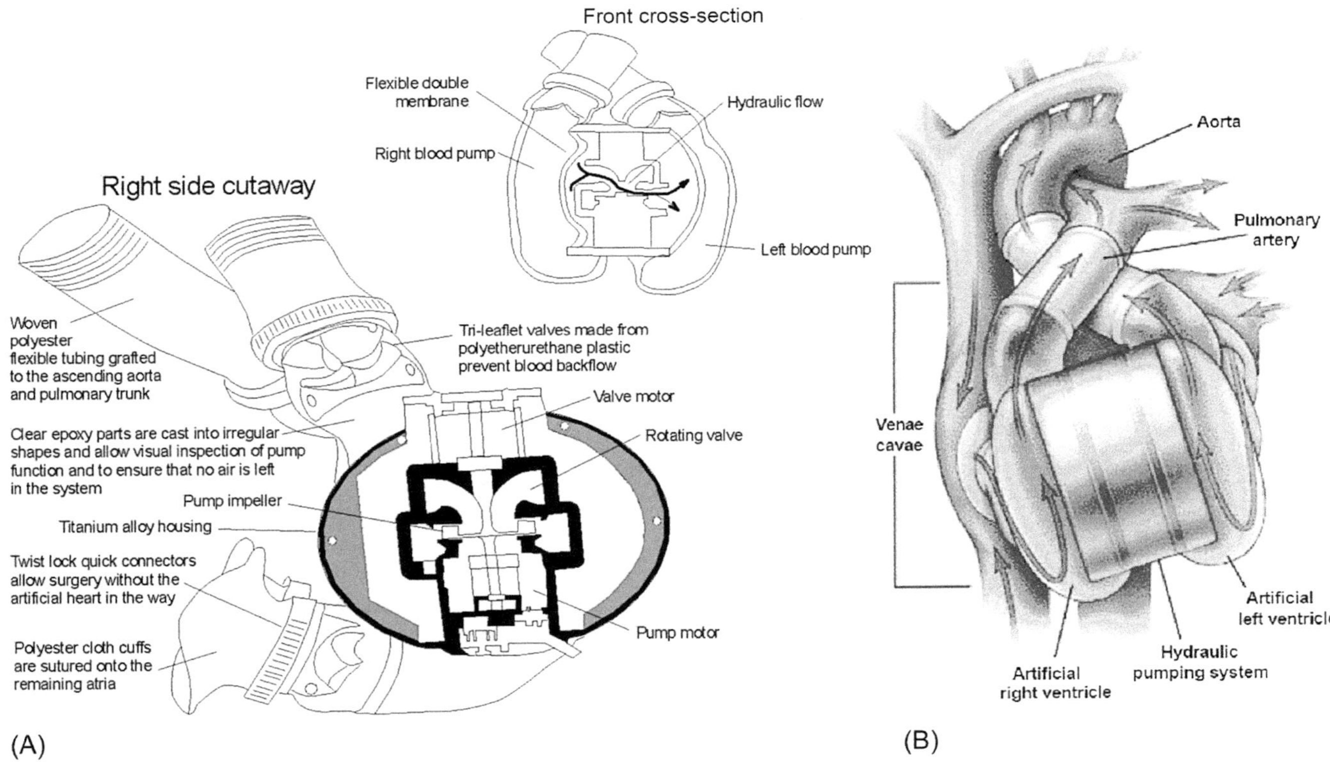

Fig. 17 Cutaway view of the internal workings and installation of the AbioCor heart (Sherief, 2007).

- **Porting valve**—A separate motor rotates the valve which opens and closes to let the hydraulic fluid flow from one side of the artificial heart to the other in turn.
- **Artificial ventricles**—When the fluid moves to the right, it compresses the flexible membrane on the inner surface of the right artificial ventricle (pump sac) and blood is pumped to the lungs via a nonreturn valve. When the fluid moves to the left, a similar sac is compressed and blood is pumped to the rest of the body via a separate nonreturn valve.

3.2 Dynamic (Continuous Flow) Pumps

Most second-generation VADs were based on axial pump technology. This was made possible by advances in pump design that minimized damage to blood constituents, as well as surface treatments on the surfaces in contact with the blood to minimize clotting. The primary advantages of these designs was the higher pumping efficiency coupled with significant reductions in size and mass. A wide variety of different design configurations have been developed by different parties to try to minimize damage to red blood cells (hemolysis) and to maximize efficiency.

A good example of the genre is the Thoratec HeartMate II with a volume of 124 mL and a length of about 70 mm as shown in Fig. 18. This size advantage gives it the potential to help a larger range of patients with the end-stage

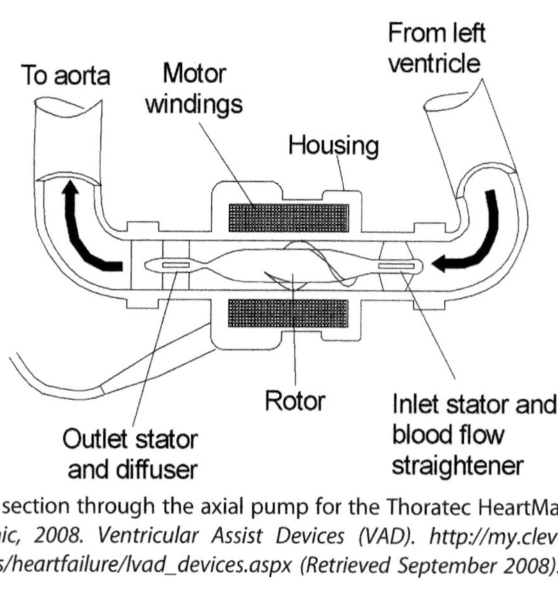

Fig. 18 Cross section through the axial pump for the Thoratec HeartMate II. *(Based on Cleveland_Clinic, 2008. Ventricular Assist Devices (VAD). http://my.clevelandclinic.org/heart/disorders/heartfailure/lvad_devices.aspx (Retrieved September 2008).)*

heart failure whose bodies are not big enough for other devices. The pump is driven by a cylindrical magnet within the rotor excited by a rotating magnetic field generated by the stator coils surrounding the core. Blood flows from the inlet conduit past three neutral aerofoil-shaped guide vanes that straighten the blood flow before it encounters the rotor. Three curved blades on the rotor impart a radial velocity to the blood before it passes into the outlet stator vanes. These are twisted and convert the radial velocity to an axial one. The exit orifice narrows to convert flow velocity to pressure. The inlet and outlet conduits are made from woven Dacron and require pre-clotting while the pump rotor and cowling are made from smooth titanium and the intraventricular conduits are textured with titanium microspheres.

The performance of continuous flow pumps such as this one is determined primarily by the speed of the rotor and the pressure difference across the pump. As shown in Fig. 19, flow rate is inversely proportional to the pressure differential across the pump. These characteristics are obtained by measuring the pressure differential and the flow rate as outflow resistance is gradually increased until pump shutoff.

Unlike rates of flow in pulsatile devices that are easily evaluated, the pressure-flow characteristics of dynamic pumps require a different interpretation. During the cardiac cycle the pump differential pressure equals aortic pressure minus left ventricular pressure plus a combined pressure loss across the inlet and outlet conduits. Because these pumps are nonocclusive, they must operate at a sufficiently high speed to avoid pressure differentials that fall below normal expected aortic pressures, as these would result in the reverse flow (Griffith et al., 2000).

Maximum flow occurs during ventricular systole when the inlet-to-outlet pressure differential is the smallest and minimum flow occurs during

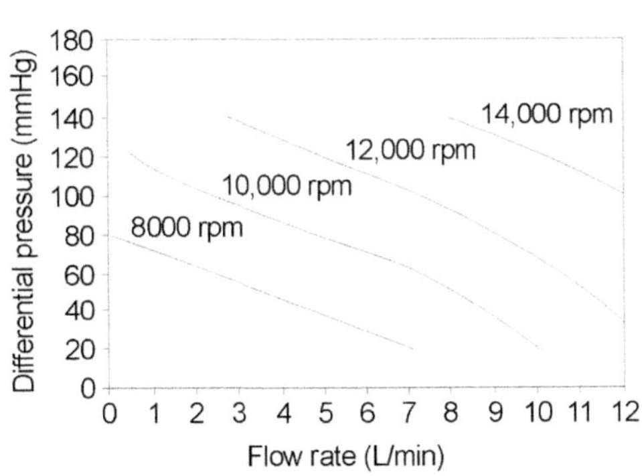

Fig. 19 Differential pressure as a function of flow rate characteristics of the HeartMate II.

left ventricular diastolic filling when the inlet pressure is lower and the pressure differential is a maximum. Minimal pulsatility occurs in patients with very poor left ventricles or if the pump speed is too high and the ventricle is driven to collapse.

For various reasons mostly related to problems with hemolysis and bearing support the next-generation pumps were mostly centrifugal or mixed flow types as these pumps could run at much lower speeds. An example of a typical centrifugal pump is the Mohawk Technology MiTi*Heart* shown in Fig. 20.

The cylindrical pump consists of four components: the pump housing, the stator, rotor with integrated vanes, and an end cap. It is 80 mm long with a diameter of 50 mm and a total mass of 640 g. The pump is designed with a nominal operating point of 5 L/min and 100 mmHg occurring at 4000 rpm as shown in Fig. 21.

The VentraCor VentrAssist centrifugal pump shown in Fig. 22, combines the motor and impellor into an integrated unit to reduce size and

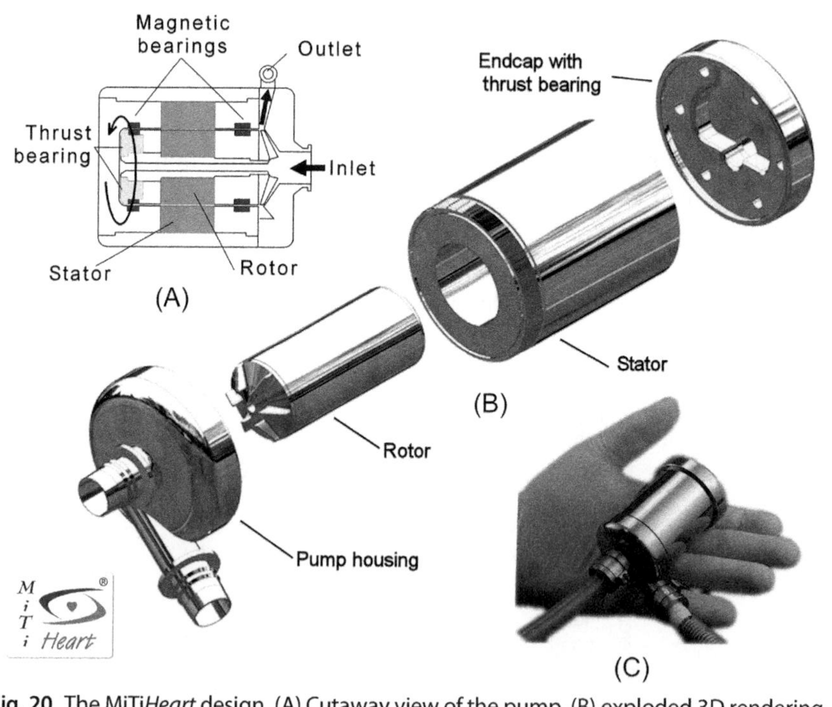

Fig. 20 The MiTi*Heart* design. (A) Cutaway view of the pump, (B) exploded 3D rendering of the pump components, and (C) a photograph of the pump. *(Reproduced with permission, copyright MiTi*Heart *corporation.)*

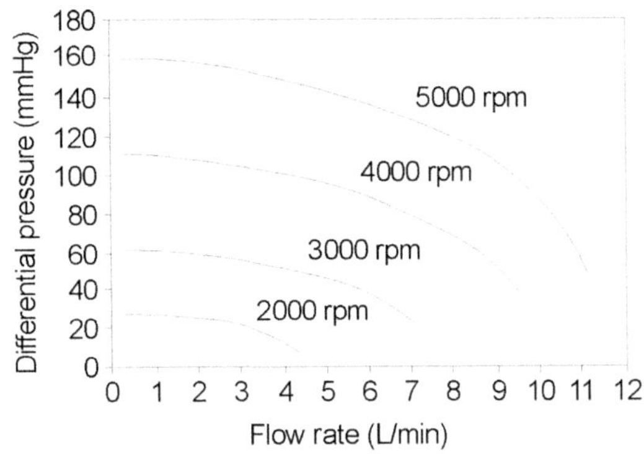

Fig. 21 Measured performance of the latest MiTi*Heart* LVAD design. *(Based on MiTiHeart Corporation.)*

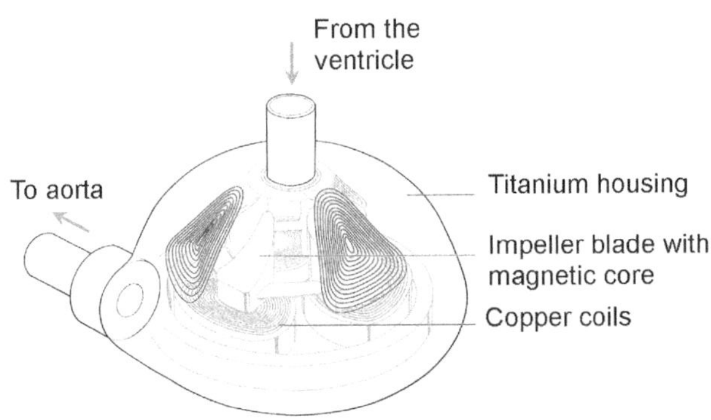

Fig. 22 Schematic diagram showing the VentrAssist pump. *(Based on Gosline, A., 2004. Simpler pump boosts failing hearts. New Sci. 28 July.)*

weight substantially compared to the MiTi*Heart*. It weighed only 298 g and was <60 mm in diameter, making it suitable for both children and adults. The device had only one moving part, a hydrodynamically suspended impeller made from a titanium alloy and covered with a diamond-like coating. Because of the large impeller size, it rotated relatively slowly and the impeller had no shaft seals or bearings ensuring clean flow lines with no stagnant zones.

Rotating magnetic fields generated by the six copper coils in the base and walls of the unit interact with the permanent magnets mounted within the rotor and cause it to spin rapidly. Hydrodynamic forces, which result from the small clearances between the outside surfaces of the impeller and the pump walls, support it. These small clearances range from ~50 to 230 μm (Chung et al., 2004). Blood enters the center of the pump and is spun up

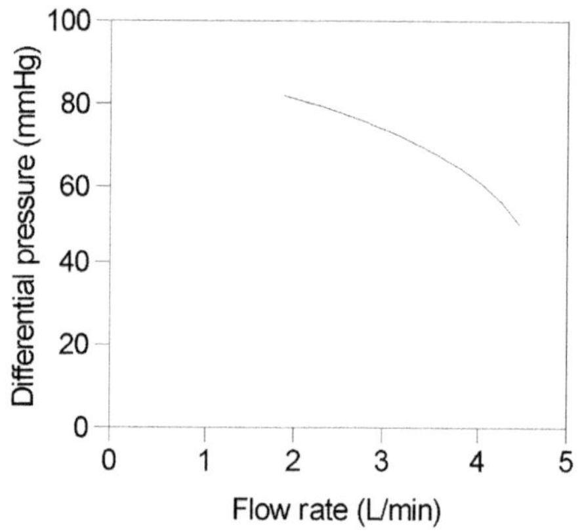

Fig. 23 Measured performance of the VentraCor VentrAssist at 2000 rpm.

by the rotors and forced outwards by centrifugal force where it exits through a pipe on the outside edge.

Some performance data from Stanfield and Selzman (2013a, b) shows the differential pressure-flow rate curve for the VentrAssist at 2000 rpm in Fig. 23.

Motors to drive continuous flow heart pumps need to use brushless commutation to avoid releasing wear products into the mechanism. These can use the conventional Hall-effect switch-based commutation in which the rotor angle is sensed from its magnetic field, and field effect transistors excite the required coils to maintain the correct rotation. Alternatively, a synchronous motor configuration can be used in which an external variable frequency drive excites the stator coil to generate a rotating magnetic field that causes the magnetic rotor to rotate at the correct speed.

The primary differences between centrifugal and axial-flow pumps are in the design of their rotary elements. A centrifugal pump operates as a "thrower" meaning that the blood is captured and thrown tangentially off the blade tips. In contrast, axial pumps operate as "pushers" or as an auger, screwing through the fluid (Joyce et al., 2012).

Both pump types generate a pressure differential by creating a vortex and then converting using either flow straighteners in axial pumps or a volute in centrifugal pumps. An intermediate configuration, known as a mixed flow pump, combines the pusher and thrower mechanisms in a continuum. This continuum is quantified by the specific speed, n_s, of the pump type that is a function of the rotation rate, the flow rate and the pressure head.

The specific speed determines the general shape of a centrifugal pump impeller. As the specific speed increases, the ratio of the impeller outlet

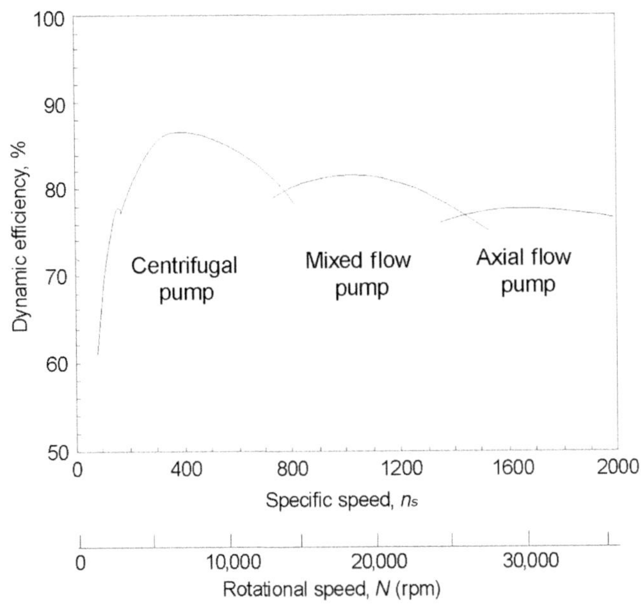

Fig. 24 Relationship between the pump configuration and specific speed. *(Based on Yamane, T., 2016. Mechanism of Artificial Heart. Springer, Japan.)*

diameter to the inlet of the eye diameter decreases. This ratio becomes unity for an axial-flow pump. The relationship between specific speed and fluid dynamic efficiency is shown in Fig. 24.

Specific speed, n_s, can be determined using.

$$n_s = \frac{N\sqrt{Q}}{h^{3/4}} \tag{1}$$

where N (rpm) is the rotation rate, Q (m^3/min) is the flow rate, and h (m) is the pressure head delivered by the pump.

Radial flow impellers develop head through centrifugal force and are low-flow high-head designs, while pumps with higher specific speeds develop head partly by centrifugal force and partly by axial force. In the limit, an axial pump develops head using axial forces only. It can be shown that the ratio of the radial to tangential velocity is equal to the square of the specific speed, therefore high n_s pump impellers have inlet diameters that approach or equal the outlet diameter, and relatively large open flow passages (axial-flow pumps). Low n_s pump impellers have outlet diameters that are much larger than the inlet diameters and relatively narrow flow passages (centrifugal pumps). It can be seen from Fig. 24, that there are clear optima, as regards pump efficiency, for the different types.

The Euler turbomachine equation states that the torque τ (Nm) required to drive a pump is equal to the change in angular momentum of the fluid from the inlet to the outlet. This is common for all turbo pumps and is described by.

$$\tau = \rho Q(r_2 V_{t2} - r_1 V_{t1}) \tag{2}$$

where r_1 and r_2 (m) are, respectively, the inlet and output radii and V_{t1} and V_{t2} are the fluid tangential (circumferential) velocities (m/s) at the inlet and the outlet flow boundaries, respectively.

It is clear that only the difference between $r_2 V_2$ and $r_1 V_1$ at the outlet and inlet sections is important in determining the torque applied to the rotor. No restriction is made regarding the geometry in regard to fluid entering at the same or different radii, therefore this equation can be used to describe centrifugal, mixed-flow, or axial-flow pumps.

The power in watts required to produce this torque is $P_w = \omega\tau$ where ω (rad/s) is the rotation rate of the pump.

$$\begin{aligned} P_w = \omega\tau &= \rho Q(\omega r_2 V_{t2} - \omega r_1 V_{t1}) \\ &= \rho Q(U_2 V_{t2} - U_1 V_{t1}) \end{aligned} \tag{3}$$

where $U_1 = \omega r_1$ and $U_2 = \omega r_2$ are the tangential speeds of the turbine blades at the inlet and outlet of the pump.

It is useful to describe the relationships between the flow vectors and the pump geometry using the velocity diagram shown in Fig. 25.

It is clear from the figure that $V_{t1} = 0$ under most circumstances as there is no tangential component in the fluid velocity at the pump inlet unless swirl is introduced by inlet vanes. Therefore Eq. (3) simplifies to.

$$P_w = \rho Q U_2 V_{t2} \tag{4}$$

The power at the output of the pump is equal to the product of the flow rate Q and the increase in pressure ΔP. This is equal to the product of the power required to drive the pump and the pump efficiency assuming that there are no other losses.

$$\eta P_w = \Delta P Q \tag{5}$$

Rewriting Eq. (3) in terms of the pressure change.

$$\eta P_w = \Delta P Q = \eta \rho Q U_2 V_{t2}$$

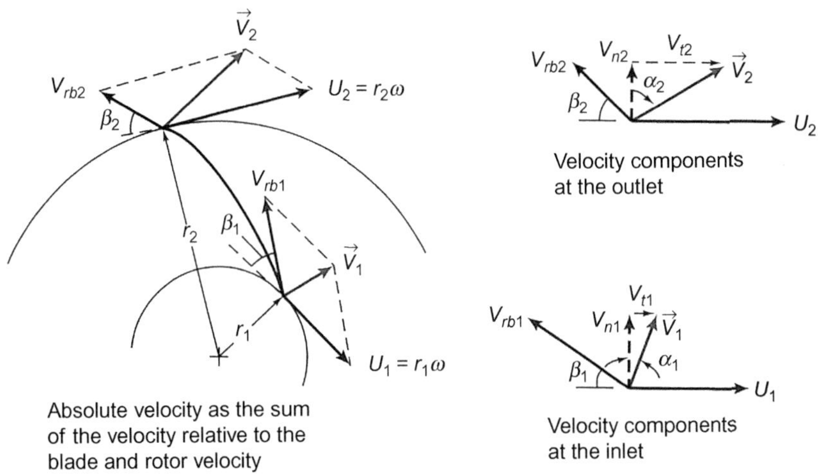

Fig. 25 Geometry used to develop the velocity diagram for a centrifugal pump where r_1 and r_2 are the inlet and outlet radii of the pump, and the various velocity vectors are as defined in the text. *(Based on Fox, R., McDonald, A., 1998. Introduction to Fluid Mechanics. John Wiley & Sons, Inc., New York.)*

and simplifying.

$$\begin{aligned}\Delta P &= \frac{\eta\rho Q U_2 V_{t2}}{Q} \\ &= \eta\rho U_2 V_{t2}\end{aligned} \tag{6}$$

This means that the differential pressure, ΔP is proportional to the fluid tangential velocity V_{t2} at a constant rotational speed U_2.

The relationship between V_{t2} and U_2 is a function of the normal fluid velocity V_{n2} and the outflow angle β_2 relative to the blade.

$$V_{t2} = U_2 - \frac{V_{n2}}{\tan\beta_2}$$

The normal velocity V_{n2} is equal to.

$$V_{n2} = \frac{Q}{A_2}$$

where A_2 (m^2) is pump cross section over which outflow takes place and is proportional to r_2.

Substituting

$$V_{t2} = U_2 - \frac{Q}{A_2 \tan\beta_2} \tag{7}$$

It can be seen that if the outflow angle β_2 is close to 90 degrees, the tangential outlet velocity of the fluid is approximately equal to the impeller tangential velocity making.

$$\Delta P \approx \eta \rho U_2^2 \tag{8}$$

According to Yamane (2016), in the case of axial-flow pumps, their characteristics can be determined using the Kutta-Joukowski theorem which describes the lift of an aerofoil as.

$$L = \frac{1}{2}\rho c_m W^2 C_L = \rho W \Gamma \tag{9}$$

for

$$\Gamma = V_{t2}\frac{2\pi r_m}{b} \tag{10}$$

where W (m/s) is the mean relative velocity, C_L is the lift coefficient, c_m (m) is the blade chord length, r_m (m) is the typical radius of turbine blades, and b is the number of blades.

The fluid tangential velocity at the outlet V_{t2} can be written as.

$$V_{t2} = \frac{1}{2}\sigma_m C_L W \tag{11}$$

where

$$\sigma_m = \frac{b c_m}{2\pi r_m} \text{ and } W^2 = U_2^2 + \left(\frac{Q}{A_2}\right)^2$$

where A_2 (m^2) is the flow cross-sectional area and the lift coefficient C_L is a function of the aerofoil incidence angle, i (rad).

$$C_L = \frac{dC_L}{d\alpha}\alpha \approx 5.73\alpha \text{ where } \alpha = i - \tan^{-1}\frac{Q}{A_2 U_2}$$

This suggests that as Q increases, the relative angle α decreases with a resulting decrease in the tangential fluid velocity V_{t2} and the pump pressure ΔP until they reach zero (Yamane, 2016).

Again, according to Yamane (2016), for a pressure rise of at least 100 mmHg (13.3 kPa) and a flow rate of 5 L/min, a rotational velocity of

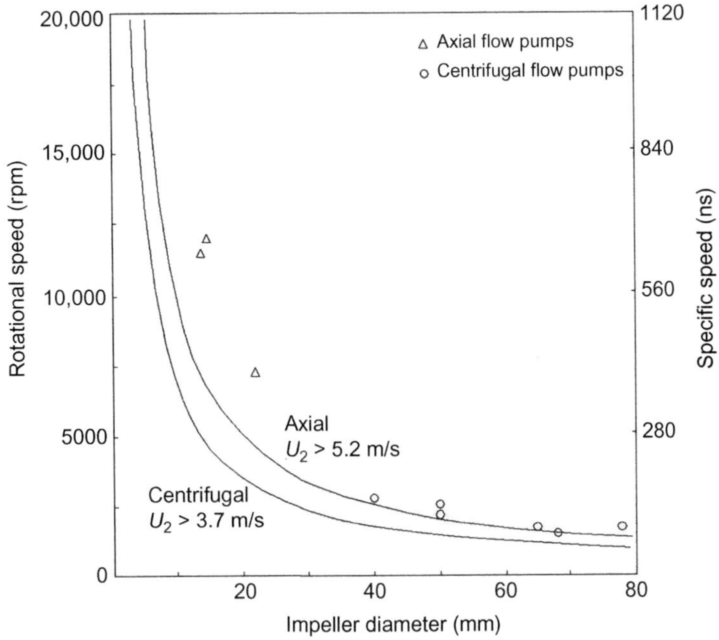

Fig. 26 Some design parameters for suitable centrifugal and axial-flow pumps for a flow rate $Q = 5$ L/min and a differential pressure $\Delta P = 100$ mmHg. *(Based on Yamane, T., 2016. Mechanism of Artificial Heart. Springer, Japan.)*

at least 3.7 m/s is required for a centrifugal pump, and 5.2 m/s for an axial-flow pump. These limits are shown in Fig. 26 along with examples of some pump specifications.

The most important difference between centrifugal and axial-flow pumps is that the former have a flat head curve where they operate over a wide range of flows for a very small change in differential pressure across the pump. The curve for axial-flow pumps is more linear and much steeper. The result is that flow rate changes significantly for typical changes in pulsatile pressure with centrifugal pumps but less so for their axial counterparts. This relationship is shown for representative axial and centrifugal pumps in Fig. 27.

The example shows that for one cardiac cycle in which the differential pump pressure swings from 40 to 80 mmHg centrifugal pumps have a large swing in flow (from 0 to 10 L/min), acting like pulsatile pumps with high-peak systolic flows and often negligible diastolic flows. In contrast, axial pumps with their steep head curves produce a low pulsatility swing of only 3–7 L/min for the same pressure swing. These differences affect control feedback strategies used by the different pump types.

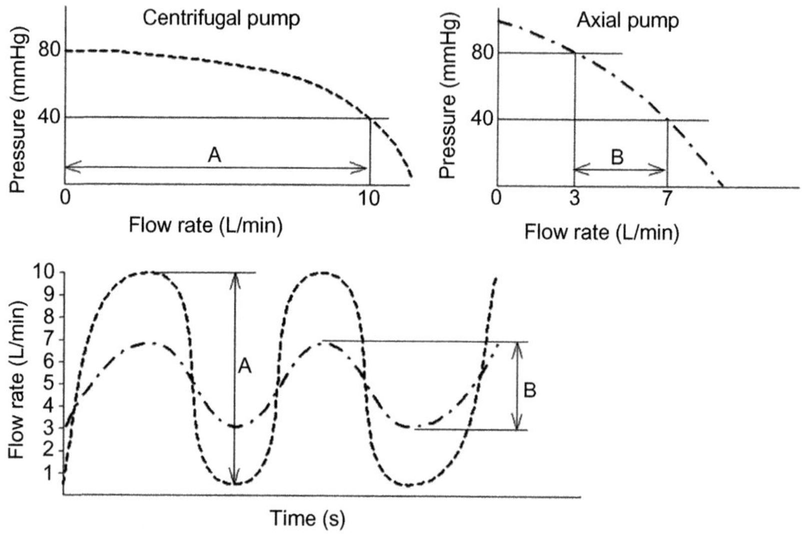

Fig. 27 Changes in flow rate as a function of time for a ventricular pressure range between 40 and 80 mmHg. *(Based on Moazami, N., Fukamachi, K., Kobayashi, M., Hoercher, S.N.K., Massiello, A., Lee, S., Horvath, D., Starling, R., 2013. Axial and centrifugal continuous-flow rotary pumps: a translation from pump mechanics to clinical practise. J. Heart Lung Transplant. 31(1).)*

4 BEARINGS

There are a number of methods that can be used to support the rotor, as listed below (Moazami et al., 2013):

- Mechanical/pivotal: The rotor is suspended on one or more mechanical bearings with spherical surfaces rotating in sockets.
- Hydrodynamic: The rotor is separated from the housing by a thin film of blood based on the relative motion of the two surfaces.
- Electromagnetic: Electromagnets support the rotor based on electronic control.
- Permanent magnet: Repelling magnets in the rotor and stator suspend the rotor at low speeds, but additional support is required when the rotor spins up.

4.1 Pivot Bearings

Most second-generation VADs are axial flow with their rotors supported on pairs of ceramic pivot bearings as shown in Fig. 28. These include the

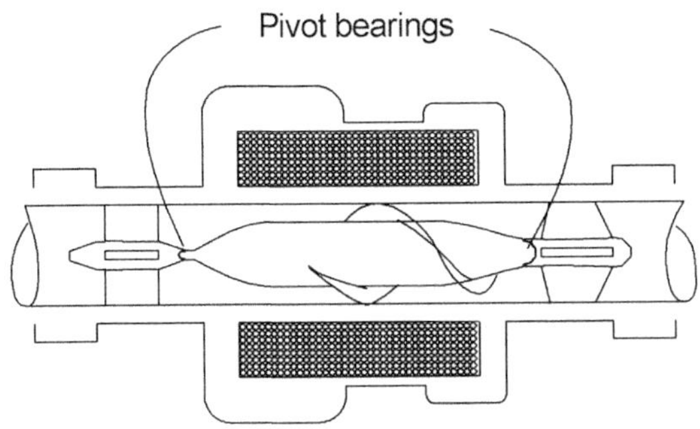

Fig. 28 Generic axial-flow VAP pump showing the positions of the mechanical pivot bearings.

Jarvik-2000 which originally used ceramic pin bearings but later changed to cone bearings, with the latter found to be superior on most counts (Stanfield and Selzman, 2013a, b). The HeartMate II uses cup-socket ruby bearings. Centrifugal pumps of this generation were generally supported on a single ceramic pivot.

From a practical perspective, mechanical bearings have tiny precise components which can position the rotor assembly in all six directions and are stable at any speed and operating condition. However high load concentrations during normal use, and potential shock impacts can result in a limited life. Additionally, though these bearings are designed to minimize damage to red blood cells (hemolysis) in the contact area, it does occur with increasing contact area tending to increase the hemolysis index.

4.2 Hydrodynamic Bearings

Most third-generation VADs used centrifugal pumps with noncontact bearings (either hydrodynamic or electromagnetic). Hydrodynamic bearings use the pressure distribution of the fluid squeezed into a wedge shaped or stepped channel to support the rotor. Asymmetrical forces as illustrated in Fig. 29 will force blood convergence and as a result will increase the hydrodynamic pressure on one side of the bearing to counteract any asymmetries in the rotor position. Analysis of these bearings relies of the Reynolds lubrication equation and is beyond the scope of this chapter.

From a practical perspective these are simple and reliable, and once the motor has started spinning, it slides on a thin film of blood. However, the load-bearing film of blood suffers from high shear stress and potentially more hemolysis can occur than with other bearing types. In addition, as the motor

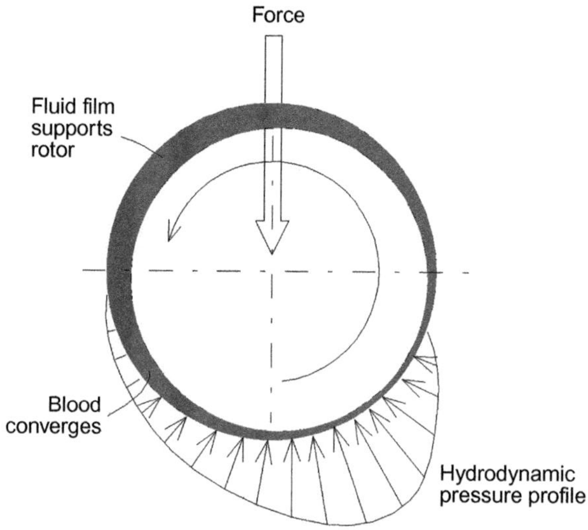

Fig. 29 Operational principle of a hydrodynamic radial bearing. *(Based on Moazami, N., Fukamachi, K., Kobayashi, M., Hoercher, S.N.K., Massiello, A., Lee, S., Horvath, D., Starling, R., 2013. Axial and centrifugal continuous-flow rotary pumps: a translation from pump mechanics to clinical practise. J. Heart Lung Transplant. 31(1).)*

starts up or shuts down, the solid-bearing surfaces come into contact and must be designed to handle these situations with minimal wear (Moazami et al., 2013).

4.3 Electromagnetic Bearings

Magnetic suspension has advantages from the viewpoints of hydraulic efficiency, wear life, and blood damage. Power lost due to bearing friction is extremely low, especially when compared to hydrodynamic rotor support systems. Most active magnetic-bearing supported pumps use a magnetic-bearing system with five active axes (one axial, two radial, and two tilt) to provide complete control of the pump rotor during operation. An example of a magnetically levitated rotor is shown in Fig. 30.

From a practical perspective magnetic bearings provide low shear associated with the relatively large gaps between the rotating surfaces. Because they are also mostly used in centrifugal and mixed flow pumps, they operate at lower speeds (4000 rpm compared with 10,000 rpm in axial pumps) which together ensure low hemolysis. Equally important is the use of noncontact bearings which eliminate the lifetime reducing wear and tribocompatibility issues that are present in rolling element or point contact-bearing systems

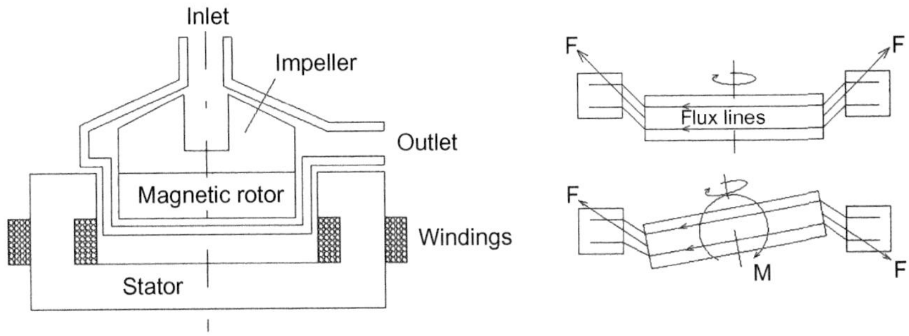

Fig. 30 Operational principles of electromagnetic bearings. *(Based on Joyce, D., Joyce, L., Loebe, M. (Eds.), 2012. Mechanical Circulatory Support. McGraw Hill Medical, New York.)*

(Jahanmir et al., 2008). However, the tradeoff is increased complexity with additional electromagnets, position sensors, and electronics required. In addition, to accommodate the high risk of pump damage should the electronics fail, most electromagnetic systems are backed up with hydrodynamic bearings to ensure a soft landing.

5 CONTROL AND POWER TRANSMISSION

5.1 Control

Most LVAD systems receive blood through an inflow cannula that is inserted into the apex of the left ventricle through a "cored" hole. The device then pumps the blood back into circulation through an outflow cannula grafted into the aorta. Thus the LVAD operates in parallel with the left ventricle. Since the pressure in the left ventricle varies as the heart beats, the LVAD is subjected to cyclical variations in loading that must be accommodated.

System models for simulation often replace the mechanical components with their electrical equivalents to form a circuit in which current corresponds to blood flow and voltage corresponds to BP. An example of such a model for the left ventricle is shown in Fig. 31.

The electrical model of the LVAD is placed in parallel with the two series resistor-diode sections.

Pulsatile systems regulate flow and/or arterial pressure synchronously with the natural beating of the heart, using the passive filling of the LVAD pumping chamber to regulate stroke volume and stroke rate. In centrifugal and axial-flow pumps only the rotation speed can be controlled to maintain the required pressure and flow. In these cases flow is mostly governed by supply from the venous system and innate feedback provided by resistance

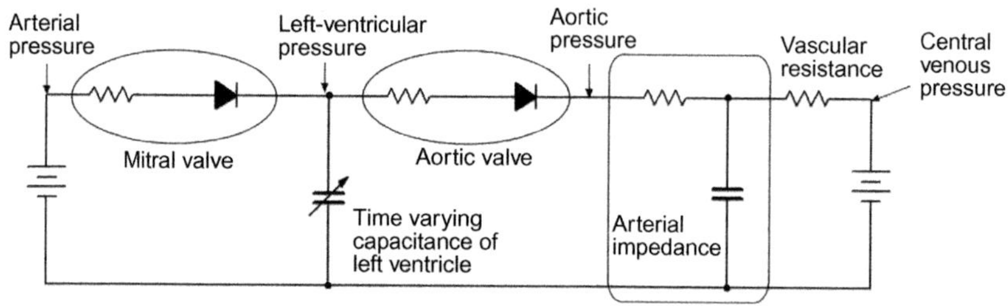

Fig. 31 Circuit model for the left ventricle and its load. *(Based on Paden, B., Ghosh, J., Antaki, J., 2000. Control system architecture for mechanical cardiac assist devices. In: Proceeding 2000 American Control Conference.)*

to blood flow. This open-loop control is adequate over a small operating range, but as soon as the patient becomes rehabilitated and starts to resume a normal lifestyle, LVADs must be capable of responding to demand.

Flow rate and differential pressure are key variables needed in the control of dynamic blood pumps. However, use of flow and/or pressure probes can decrease reliability and increase system power consumption and expense so are seldom used. For a given fluid viscosity, the flow state is determined by any two of the four pump variables: flow, pressure differential, speed, and motor input power. Thus, if viscosity is known or if its influence is sufficiently small, flow rate and pressure difference can be estimated from the motor speed and motor input power (Tanaka et al., 2001).

Centrifugal pumps provide a linear current (or power) to flow relationship across a wide range of flow rates. In addition, the characteristic changes in current due to pulsatility allow the pump to provide an excellent sensorless index of both flow and heart rate. In contrast, axial pumps only provide a linear relationship between changes in flow and current over a narrow range of operating conditions and so monitoring current is only indicative of flow and pulsatility at best. Axial pumps are therefore generally operated in a constant speed mode with only suction-detection control algorithm based on motor current feedback. To overcome this limitation some axial pumps now rely a flow probe incorporated into the outlet cannula (Moazami et al., 2013).

As discussed by Paden et al. (2000) and Antaki et al. (2003), any cardiac augmentation system must function within the following constraints:

- Cardiac output should be above a minimum value. This is nominally 5L/min, but will vary between 3 and 6L/min depending on the size of the patient.

- Left atrial pressure should be maintained below 10–15 mmHg to avoid pulmonary edema, and above 0 mmHg to avoid suction.
- Systolic arterial pressure should be maintained within specific limits to ensure an adequate oxygen supply while avoiding risks associated with hypertension.
- The system should be maximally efficient in terms of blood flow and pump power.

In general, it is not possible to minimize all of these simultaneously, so control systems are designed that optimize performance based on cost functions associated with deviation from the constraint. These cost functions are normally asymmetrical because of hard minima below which the patient cannot function. An intelligent controller based on multiobjective optimization of these parameters as well as information about the patient's activity level can be used to control the pump motor speed as shown in Fig. 32.

Other control architectures have been developed. For example, a closed-loop controller for a DeBakey VAD uses venous return based on flow pulsatility as well as the available return derived from the patient's own heart rate (desired flow) along with power use and minimal flow as inputs. These are analyzed on a beat-to-beat basis within a 10 s moving window before the motor speed is adjusted (Vollkron et al., 2006).

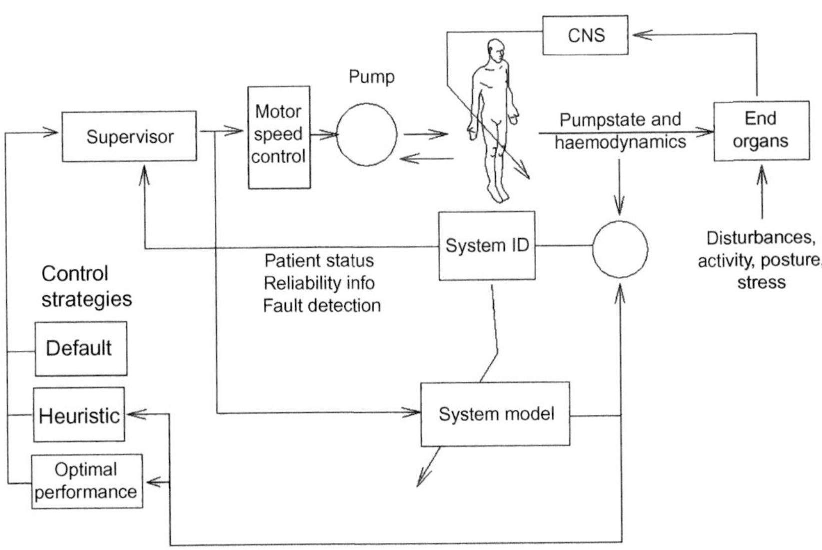

Fig. 32 Control architecture for an axial pump-based LVAD. *(Based on Antaki, J.F., Boston, J.R., Simaan, M.A., 2003. Control of heart assist devices. In: Proceeding 42nd IEEE Conference on Decision and Control.)*

5.2 Power Transmission

Existing methods of power transmission to a heart pump use either pneumatics or electrical energy. The former require external air pumps and controllers which vary in size from a large tumble dryer to a briefcase, as can be seen in Fig. 33.

The mechanisms required to regulate the air pressure for effective operation are discussed earlier.

Most LVADS still use a transcutaneous leads that transfer external power, usually as a variable frequency drive signal from an external controller to the pump attached to the heart, as shown in the Fig. 34 schematic for the DeBakey VAD.

A transcutaneous energy transfer (TET) system to power an artificial heart consists of components both external and within the body of the patient as shown in Fig. 35. The external components include the following:

- DC Power Source—a battery pack with sufficient capability to provide 10 W to power a VAD and 20 W to power a TAH for between 6 and 8 h.

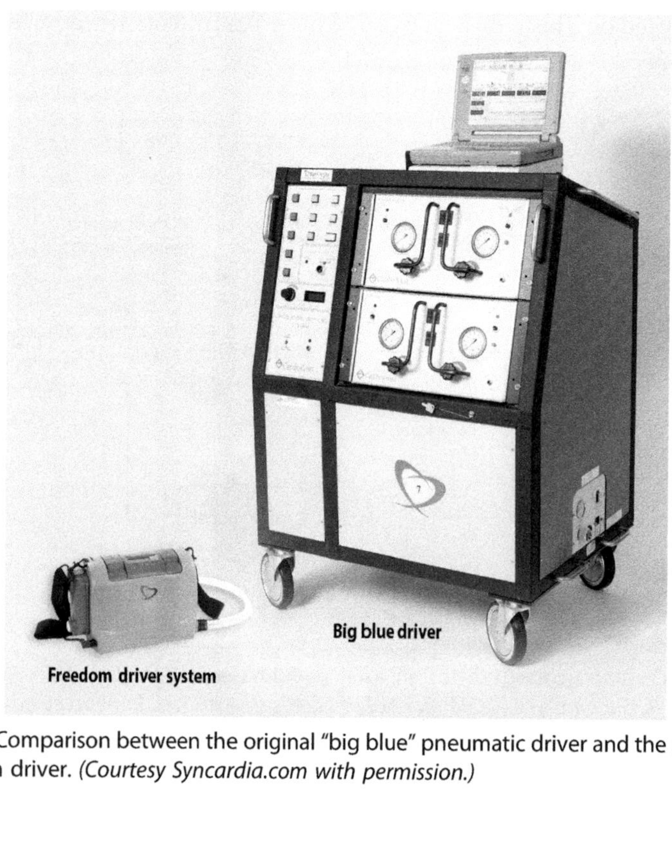

Fig. 33 Comparison between the original "big blue" pneumatic driver and the portable Freedom driver. *(Courtesy Syncardia.com with permission.)*

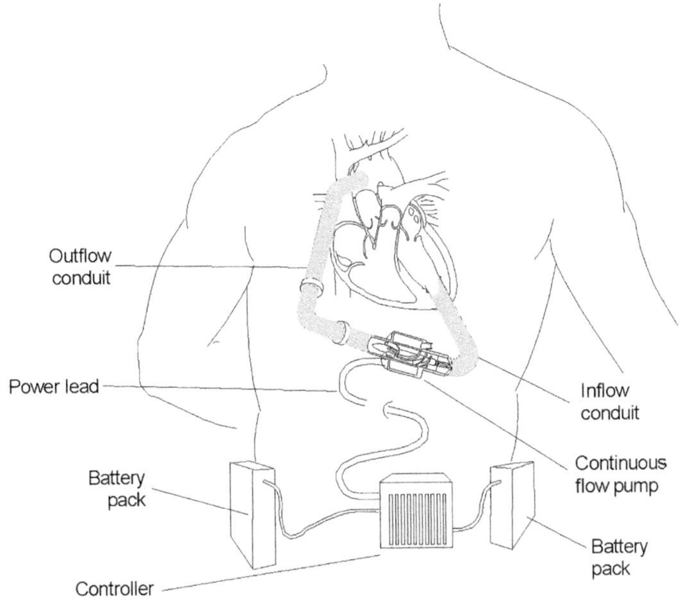

Fig. 34 Transcutaneous power leads provide power to the DeBakey axial heart pump.

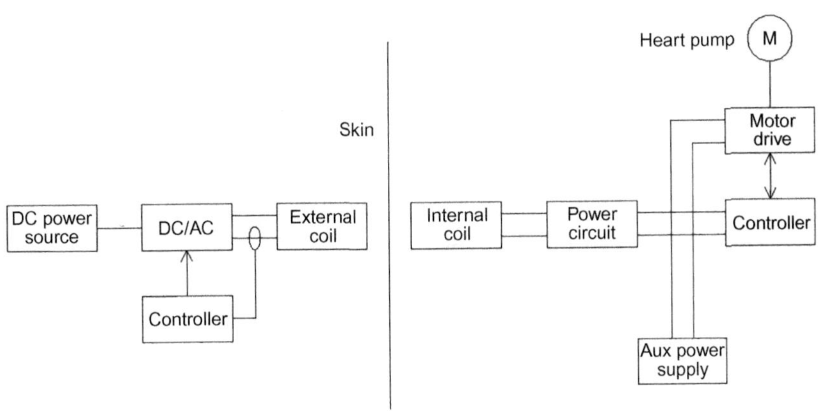

Fig. 35 Block diagram showing the components of a transcutaneous energy transfer system.

It uses rechargeable cells with the highest power density possible to minimize the mass that the patient needs to carry.

- DC/AC converter—this consists of an inverter or power oscillator that typically generates an adjustable square-wave signal at a frequency of between 100 kHz and 1 MHz.

- Controller—the energy transfer efficiency depends on the alignment of the internal and external coils and their mutual resonant frequency. The controller monitors the current to the coil and adjusts the DC/AC converter parameters to maintain optimum conditions for transfer. It also detects faults, and closes down the circuit if the coils are completely out of alignment.
- External coil network—this is the primary coil, between 5 and 10 cm in diameter, that is excited by the AC signal and generates a changing magnetic field that couples through the skin into the subcutaneous secondary coil. It is generally part of a tuned LC resonant circuit.

The internal components include the following:

- Internal coil network—this coil is hermetically sealed and inserted just below the skin. It is typically a little smaller than the primary to make alignment as simple as possible. It is also part of the tuned LC circuit so that it captures the highest possible proportion of the radiated magnetic energy.
- Power circuit—it consists of a rectifier and regulator to produce the required DC voltages to power the control electronics and the pump.
- Auxiliary power supply—it is an internal battery that allows the device to operate for a reasonable period (typically 30 min) without external power. It also includes the electronics to recharge the internal battery when external power is available.
- Controller—most heart pumps consist of brushless DC motors that must be controlled to suit the patient's requirements. The controller monitors the rotation rate and the current requirements and adjusts the supply voltage to suit.
- Drive circuit—depending on the motor types, the drive circuit is usually a half- or full-bridge pulse-width modulated MOSFET circuit that chops the DC supply to drive the heart pump.

The AbioCor is an example of a heart pump that uses TET technology. It is shown in Fig. 36. It consists of various components (Bonsor, 2008) including the pump, the energy transfer system, internal and external batteries, and a controller.

The original AbioCor and LionHeart TET systems highlighted some problems with energy transfer and storage. Their coils had to be very well aligned and in close proximity with the result that power transmission could be disrupted by small positional changes in the coils and even weight gain. In addition, internal batteries had limited lifetimes and could only provide run times of as little as 30 min.

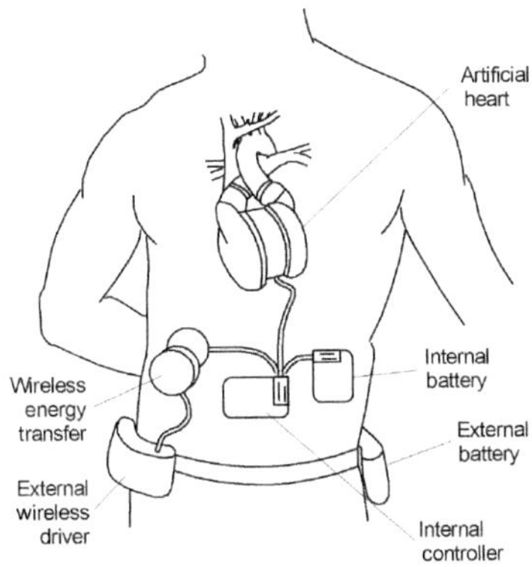

Fig. 36 Schematic showing wireless transcutaneous energy and control transfer across the skin.

New nonradiative resonant magnetic coupling has allowed high-efficiency alignment tolerant wireless energy transfer over greater distances than was previously available. These devices use near-field strong coupling modes that arise between two high-efficiency resonances tuned to slightly different frequencies. This technology also makes use of high permeability materials in the resonators that concentrate the transmitted magnetic flux. The latest systems can be separated by distances comparable to their diameters (Kyo, 2014).

Robust new battery technologies specifically tailored for implantable electronics provide higher power densities and therefore longer run times without external power as well as lifetimes of up to 3 years before replacement is required.

6 OTHER CONSIDERATIONS

Modern pump design relies heavily on computational fluid dynamic (CFD) analysis to ensure that fluid shear is minimized and stagnation points are avoided, as excessive shear results in increased hemolysis and stagnation results in clotting. Additional techniques to identify problems include visual observations using scaled up transparent pump models that maintain the Reynolds similarity law to examine flow. These models use particle tracking velocimetry techniques with very fine spatial resolutions to track particle movements close to the pump walls.

A final consideration is the selection of biocompatible materials for pump lining and pump housing. Some of these have been addressed earlier.

In summary, for pulsatile pumps polyurethane is often used as it has good biocompatibility and the required structural capabilities, whereas most implantable rotary VADs use titanium polished to submicron surface roughness, again because of its strength and biocompatibility. Coating materials can include heparin-containing polymers or diamond-like carbon coatings (Yamane, 2016).

7 FUTURE DIRECTIONS

The ultimate aim of LVAD and TAH technology is to provide a device that is "invisible" to the patient by being mechanically reliable and by adapting to their unique physiology. This would require completely implantable devices on both the left and right sides of the heart with appropriate feedback to automatically regulate pump functions.

To achieve this for TAHs, a shift in modality from pulsatile to continuous flow is required. However, to date the only single-piece CFTAHs in development are the Texas Heart Institute's biVACOR and Cleveland Clinic's CFTAH device, both of which are still in the preclinical studies.

In the long term, implants will require less invasive procedures as has occurred with heart surgery, which can now be performed through catheters inserted in arteries near the groin. Already, the first minimally invasive LVADs such as the CircuLite Synergy are implanted under the skin. These are currently about the size of an AA battery and are connected to from the heart to one of the big arteries via a mini-thoracotomy without cutting the breast bone. At present they can only provide partial support of between 2.5 and 3.5 L/min, but there is scope for improvement (Joyce et al., 2012).

The next step would be a device that is completely inserted through the blood vessels without requiring surgery. However, such dramatic size reduction requires significant increases in impeller speed. This, in turn, requires completely wear-free operation offered only by magnetic levitation along with improved CFD-based design techniques and material coatings to eliminate hemolysis and clotting.

Improvements in electric motor and battery technology are also required, but these are already being supported to a large extent by the burgeoning micro-drone market. Improvements in TET, augmented perhaps by internal energy scavenging mechanisms will also occur. Advances in this technology are also being driven by commercial factors including wireless charging of mobile phones and battery-free power for Internet of Things devices.

Finally, the introduction of stem cell treatments to aid with cell repair or regeneration used in conjunction with biomaterial scaffolds and in conjunction with LVADs may hasten recovery of patients suffering from advanced heart failure.

REFERENCES

Al-Ghazal, S., 2002. Ibn Al-Nafis and the Discovery of Pulmonary Circulation. http://www.islamonline.net/english/Science/2002/08/article06.shtml. (Accessed September 2008).

Antaki, J.F., Boston, J.R., Simaan, M.A., 2003. In: Control of heart assist devices.Proc. 42nd IEEE Conference on Decision and Control.

Bonsor, K., 2008. HowStuffWorks: How Artificial Hearts Work. http://health.howstuffworks.com/artificial-heart.htm/printable. (Accessed September 2008).

Boylan, M., 2007. Galen: on blood, the pulse and the arteries. J. Hist. Biol. 40, 207–230.

Brooker, G., 2012. Introduction to Biomechatronics. SciTech Publishing, Raleigh, NC.

Chung, M., Zhang, N., Tansley, G., Quin, Y., 2004. Experimental determination of dynamic characteristics of the ventrassist implantable rotary blood pump. Artif. Organs 28 (12), 1089–1094.

Cooley, D., 2003. The total artificial heart. Nat. Med. 9, 108–111.

Cooley, D., Liotta, D., Hallman, G., Bloodwell, R., Leachnam, R., Milam, J., 1969. Orthotopic cardiac prosthesis for two-staged cardiac replacement. Am. J. Cardiol. 24 (5), 723–730.

Daugherty, R., Franzini, J., 1977. Fluid Mechanics with Engineering Applications. McGraw Hill Kogakusha, Tokyo.

Fukui, Y., Funakubo, A., Fukunaga, K., 2004. In: Development of the assisted artificial heart with linear motor actuator.SICE 2004 Annual Conference.

Glenn, W., 1993. Seawell's pump. Guthrie J. 63(1).

Griffith, P., Kormos, R., Borovetz, H., Litwak, K., Antaki, J., Poirier, V., Butler, K., 2000. In: HeartMate II left ventricular assist system: from concept to first clinical use.Fifth International Conference on Circulatory Support Devices for Severe Cardiac Failure, New York.

Hajar, R., 2005. The artificial heart. Heart Views 8 (2), 70–76.

Jahanmir, S., Hunsberger, A., Henshmat, H., Tomaszewski, M., Walton, J., Weiss, W., Lukic, B., Pae, W., Zapanta, C., Khalapyan, T., 2008. Performance characterization of a rotary centrifugal left ventricular assist device with magnetic suspension. Artif. Organs 32 (5), 366–375.

Jaron, D., 1990. In: Current status of the artificial heart and cardiovascular assist devices.Proc. 1990 IEEE on Colloquium in South America.

Joyce, D., Joyce, L., Loebe, M. (Eds.), 2012. Mechanical Circulatory Support. McGraw Hill Medical, New York.

Kyo, S. (Ed.), 2014. Ventricular Assist Devices in Advanced-Stage Heart Failure. Tokyo, Springer.

Lemelson-MIT, 2002. Inventor of the Week: Robert Jarvik. http://web.mit.edu/invent/iow/jarvik.html. (Accessed September 2008).

Moazami, N., Fukamachi, K., Kobayashi, M., Hoercher, S.N.K., Massiello, A., Lee, S., Horvath, D., Starling, R., 2013. Axial and centrifugal continuous-flow rotary pumps: a translation from pump mechanics to clinical practise. J. Heart Lung Transplant. 31 (1), 1–11.

Orme, F., 2002. Human Physiology—Lecture Notes. http://members.aol.com/Bio50/index.html. (Accessed September 2008).

Paden, B., Ghosh, J., Antaki, J., 2000. Control system architecture for mechanical cardiac assist devices.Proc. 2000 American Control Conference.

ScienceDaily, 2006. VCU Medical Team Implants Total Artificial Heart. http://www.sciencedaily.com/releases/2006/04/060405022627.htm. Accessed September 2008.

Shackleford, J., 2003. William Harvey and the Mechanics of the Heart (Oxford Portraits in Science). Oxford University Press, Oxford.

Sherief, H., 2007. Biomedical Engineering: Artificial Heart & Heart Assist Devices. http://mybiomedical.blogspot.com/2007/08/artificial-heart-heart-assist-devices.html. (Accessed September 2008).

Slepian, M., Smith, R., Copeland, J., 2006. The syncardia cardiowest total artificial heart. In: Baughman, Baumgartner, (Eds.), Treatment of Advanced Heart Disease. Taylor and Francis, pp. 473–490 (Chapter 26).

Stanfield, J., Selzman, C., 2013a. In vitro hydrodynamic analysis of pin and cone bearing designs of the Jarvik 2000 adult ventricular assist device. Artif. Organs 37 (9), 825–833.

Stanfield, J., Selzman, C., 2013b. In vitro pulsatility analysis of axial-flow and centrifugal-flow left ventricular assist devices. J. Biomech. Eng. 135 (3), 345051–345056.

Takatani, D., Ouchi, K., Nakamura, M., Sakamoto, T., 2001. Ultracompact, totally implantable, permanent, pulsatile VAD system. J. Congestive Heart Fail. Circ. Support 1 (4), 407–412.

Tanaka, A., Yoshizawa, M., Yamada, T., Abe, K., Takeda, H., Yambe, T., Nitta, S., 2001. In vivo evaluation of pressure head and flow rate estimation in a continuous-flow artificial heart.Proc. 23rd Annual International Conference of the IEEE Engineering in Medicine and Biology Society.

Vollkron, M., Schima, H., Huber, L., Benkowski, B., Morello, G., Wieselthaler, G., 2006. In: Control of implantable axial blood pumps based on physiological demand.American Control Conference.

Yamane, T., 2016. Mechanism of Artificial Heart. Springer, Japan.

FURTHER READING

Cleveland_Clinic, 2008. Ventricular Assist devices (VAD). http://my.clevelandclinic.org/heart/disorders/heartfailure/lvad_devices.aspx. (Accessed September 2008).

Fox, R., McDonald, A., 1998. Introduction to Fluid Mechanics. John Wiley & Sons, Inc., New York

Gosline, A., 2004. Simpler pump boosts failing hearts. New Sci. 28 July.

Kormos, R., Miller, L. (Eds.), 2012. Mechanical Circulatory Support. Elsevier, USA.

CHAPTER FOURTEEN

Pacemakers

Graham Brooker
Australian Centre for Field Robotics, University of Sydney, Sydney, NSW, Australia

Contents

1 INTRODUCTION

The heart is made up of atrial and ventricular smooth muscle that generate contractions as well as conductive and excitation fibers that aid with the essential synchronization process. The excitation and conduction system shown in Fig. 1 consists of the sinoatrial (SA) node, intermodal tracks, Bachmann's bundle, and the atrioventricular (A-V) node as well as the bundle of His, bundle branches, and Purkinje fibers. Several cell types in the

Handbook of Biomechatronics
https://doi.org/10.1016/B978-0-12-812539-7.00014-3

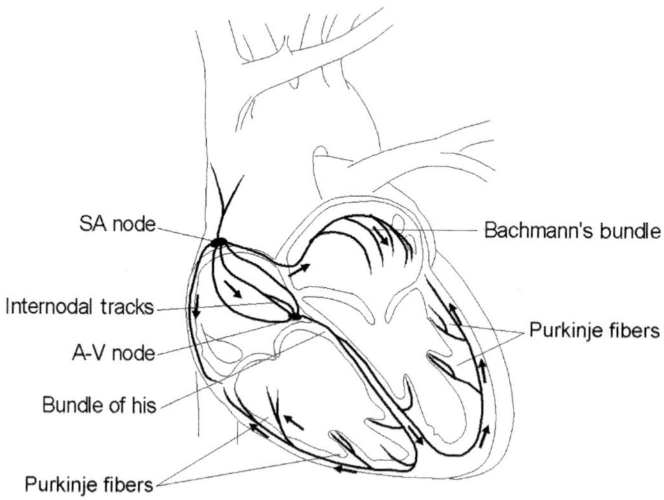

Fig. 1 Drawing of the heart showing details of the excitation and conduction system.

heart can generate rhythmic and repeated excitation pulses if required. These include the A-V mode with an inherent rate of about 50 beats per minute (bpm), and the Purkinje fibers (about 40 bpm), both at a rate lower than the SA node (60–80 bpm). The SA node therefore controls the heart rate, all being well. The intermodal tracks and Bachmann's bundle transmit the SA node excitation throughout the atria and trigger a coordinated contraction of the atrial walls. The excitation impulse also reaches the A-V node that provides only the electrical connection between the atria and the ventricles. The reason for this is that A-V node must introduce a delay to allow the atrial contractions to finish before the ventricular contractions begin. After the delay, electrical impulses are conveyed rapidly along the bundle of His and via the bundle branches to the Purkinje fibers in the left and right ventricles. The Purkinje fibers radiate throughout the ventricular muscle and ensure that ventricular contraction is coordinated in each ventricle and synchronized with the other one (Guyton and Hall, 1996; Haddad et al., 2006).

Cardiac arrhythmias are due to heart problems that upset its coordination, with a resultant reduction in the haemodynamic (pumping) performance of the organ. There are myriad causes for this but are commonly caused due to damage to the normal conduction pathways that result in a different portion of the heart taking over. For example, in the cases of A-B block, ventricular pacing is required because the disease is in the A-V node or His-Purkinje system. However, with sick sinus syndrome the electrodes must stimulate the right atrium (Guyton and Hall, 1996).

This pacing function is performed using a "pacemaker" which consists of an electrical pulse generator, a power source, and a set of leads and electrodes. Each pulse needs to provide sufficient energy to depolarise the heart cells in the vicinity of the electrodes. This is determined by the pulse area (the product of its amplitude and duration).

2 HISTORICAL BACKGROUND

To be properly appreciated, the evolution of pacemakers must be placed in a historical context, starting with publications speculating on the electrical nature of the cardiovascular system that began to appear in the middle of the 18th century, as physicians began to experiment with the effects of static electrical discharges on the heart. In 1775 a Danish physicist, Nickolev Abildgaard, was able to electrocute a chicken and then use a similar discharge to reanimate the bird, apparently by defibrillating the heart (Aquilina, 2006). Toward the turn of the century Luigi Galvani published his now often repeated experiments showing the effects of electrical stimulation on frog leg muscles and frog hearts (Galvani, 1791). He attributed the phenomenon to *animal electricity,* believing that the frog muscle was similar to that of the electric eel. The publication of these experiments resulted in correspondence between him and Alessandro Volta. In his attempts to prove that Galvani was wrong and that the phenomenon was the reaction to a small electric charge, Volta performed a research that resulted in the invention of the battery (voltaic pile). Much of this early work using both electrostatics and voltaic piles, generated by the interaction of dissimilar metals, is documented by Wilkinson (1804). A few years later Alexander von Humboldt found a dead bird in his garden and placed a blade of zinc in the beak and a shaft of silver into the rectum. When a circuit was completed, the electric shock caused the bird to flap its wings and attempt to walk. He also tried the experiment on himself, with unpleasant consequences (Westaby and Bosher, 1997). Toward the end of the 19th century a 46-year-old patient arrived in the clinic of Hugo Von Ziemssen. Her name was Catarina Serafin, and she had had a chest tumor excised, along with part of the thoracic wall which exposed her heart. This allowed Von Ziemssen to confirm that electrical stimulation applied directly to the heart could change its rate (Lilderitz, 1998). The results, using a cuff that recorded arterial pressure changes, are shown in Fig. 2.

At about the same time, an English Doctor, John MacWilliam collated all the available data and laid down the basic concepts of cardiac pacing. In 1889

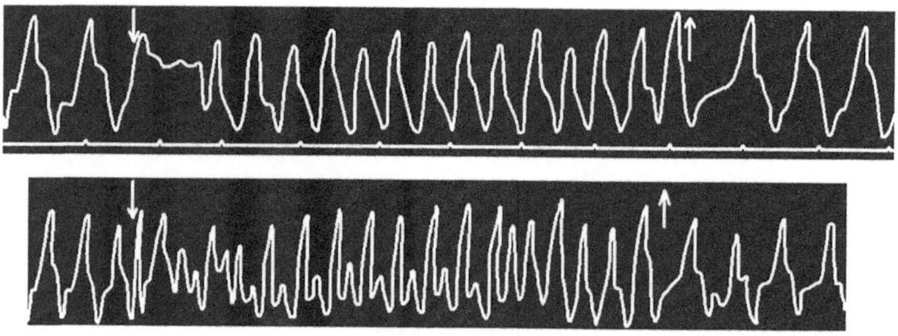

Fig. 2 Arterial pressure recordings showing that Caterina Serafin's heart rate is altered by electrical stimulation. *(Modified from Lilderitz, B. (Ed.), 1998. History of Disorders of Cardiac Rhythm, Futura Publishing Company.)*

he described the application of electricity across the chest to "excite rhythmic contraction … to stimulate by direct means the action of the heart which has been suddenly enfeebled or arrested" (Nelson, 1993).

2.1 The First Pacemakers

Credit for the invention of the external pacemaker in 1928 is shared by an Australian anaesthesiologist, Mark Lidwill, and an American physiologist, Albert Hyman. Lidwell's pacemaker, which had been developed in conjunction with Edgar Booth, a physicist from Sydney University, consisted of an external electrode and a long needle insulated along its length except at the tip, which was inserted into the patient's ventricle. The voltage could be varied from 1.5 to 120 V (which seems rather excessive) and the pulse rate varied from 80 to 120 ppm. The apparatus was used to revive a stillborn infant at Crown Street Women's Hospital in Sydney whose heart continued "to beat on its own accord, at the end of 10 minutes" of stimulation (Nelson, 1993). Hyman had been experimenting with the intracardiac injection of stimulants, like epinephrine (adrenaline), to restart hearts in fibrillation, when he soon became aware that it was actually the insertion of the needle that set up an action current to restart the heart. He developed a portable generator in 1932 that he described as an "artificial pacemaker." It consisted of a spring driven DC dynamo which was chopped by a rotating interrupter disc to produce pulses that were introduced into the right atrium through a bipolar needle electrode.

2.2 Table-Top Pacemakers

Nearly 20 years later in 1949 William Bigelow and John Callaghan from Toronto, Canada had started with the experimental use of hypothermia to reduce metabolism to slow and even stop the heart to permit cardiac surgery. Rewarming could not always restore cardiac contractions, so the surgeons started experimenting with SA node stimulation. By the 1950s, pacemakers were in common use in hospitals. They used vacuum tubes and were mains powered so were too large for portable use. The device of choice at this time was the Grass stimulator. It used a thyratron to rectify the AC which could then be controlled to produce pulses of various amplitudes, pulse widths (monophasic, rectangular, 2–20 ms duration), and rates.

Paul Zoll, a Boston cardiologist had read the work of Callaghan and realized that the pacemaker would be suitable for the treatment of heart block. Unfortunately, in 1951, the technology was hardly up to developing portable hardware and he had to settle on using mains-powered tabletop devices. His device delivered a 2-ms pulse with amplitude between 50 and 150 V to metal electrodes strapped to the patient's chest. Unfortunately they were uncomfortable and the electrodes irritated the skin (Zoll et al., 1956; Nelson, 1993). At about the same time C. Walton Lillehei was pioneering open heart surgery at the University of Minnesota where he soon discovered that about 10% of his open heart surgery patients experienced problems when the sutures interrupted the proper transmission of electrical impulses through the heart. He also used a mains powered, vacuum-tube pacemaker to restore normal heart rhythms for these patients. Although this pacemaker was effective, patient recovery was subject to power failures and occasional electrocution. In fact, one of Lillehei's pediatric patients died in a city-wide blackout. In addition to power source problems, these pacemakers were also bulky devices that rendered patients immobile because they had to stay tethered to a wall socket while they recuperated (Nelson, 1993).

3 MODERN PACEMAKERS

3.1 The First Portable Pacemaker

To solve the problem of reliance on mains power, Lillehei knew he needed a battery powered pacemaker and he approached Earl Bakken for help. Bakken and his brother-in-law Palmer Hermundslie had cofounded Medtronic in 1949 in a garage in northeast Minneapolis. The company had had a rough start operating as a repair service for hospital electrical equipment and the regional distributor for other manufacturers' medical appliances. Within 6 weeks Bakken had modified a metronome circuit

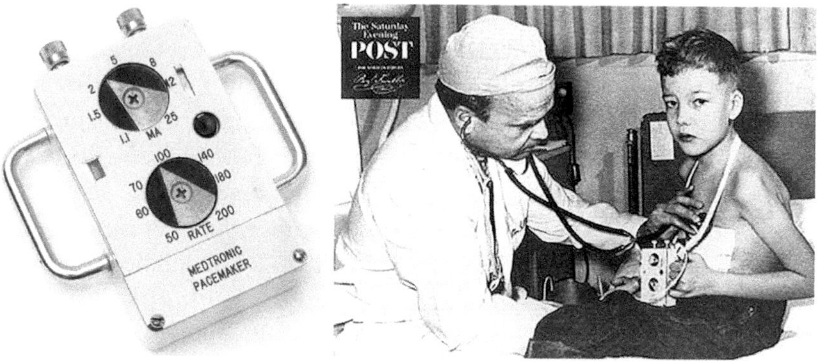

Fig. 3 Medtronic wearable pacemaker (Aquilina, 2006). *(Courtesy Medtronic with permission.)*

design he had found in Popular Electronics and created a box powered by mercury batteries to deliver a 9-V DC pulse. A day after the device was first tested, Lillehei used it to restore a child's heartbeat to a normal rhythm.

The first production run of 10 units shown in Fig. 3 were more refined versions of the original prototype, and they went into clinical use soon after. The dials had been recessed to stop children fiddling and a neon light flashed with each stimulus. The pacemaker could be hung around the neck of the patient, making it truly portable. It became known as the 5800 (because it was made in 1958), and the product literature stated: "So small and light that it may be attached to and worn by the patient, the Medtronic Cardiac Pacemaker stimulates ventricular function in cases of A-V dissociation that are induced during the surgical repair of septal defects, or that occur spontaneously as in Stokes-Adams syndrome. The Pacemaker is designed for internal applications with at least one wire attached directly to the myocardium for temporary stimulation or with a bipolar patch for prolonged stimulation. Created with imagination and originality, the transistorised circuit completely removes the hazards and nuisance associated with AC powered instruments. Its self-contained miniature power source will operate the instrument for approximately 1000 hours." (Aquilina, 2006).

The chosen pacemaker output was a 2-ms square wave, variable in amplitude from 1 to 20 mA into a 1000-Ω load. The blocking oscillator repetition rate was variable from 60 to 180 pulses per minute.

Lillehei referred many of his colleagues to Bakken with the result that Medtronic quickly expanded, and their pacemaker came to be used throughout the world. Medical historians regard Bakken's pacemaker as one of the first successful applications of transistor technology to medical devices, helping to launch the new field of medical electronics. Prior to 1957, and the recent invention of the transistor, there had never been a partly or completely implantable electrical device.

While the wearable pacemaker enabled heart surgeons to save more lives on the operating table, it was also becoming clear that heart attacks were frequently caused by problems with the heart's natural electrical system and that artificial pacemakers could be used for long-term treatment as well as during surgical recovery periods. The wearable pacemaker was uncomfortable and inconvenient for patients' long-term use. In addition transcutaneous leads made it difficult to prevent infections from taking hold. To solve these problems, various physicians and engineers across the country started trying to develop an implantable pacemaker.

3.2 The First Implantable Pacemakers

On October 8, 1958, the first pacemaker implantation was performed in Sweden. The system had been developed by the surgeon Ake Senning and the physician inventor Rune Elmqvist and was implanted in a 43-year-old engineer called Arne Larsson. This first experience with a fully implantable pacemaker system was reported at the Second International Conference on Medical electronics in 1959 and published as an abstract in 1960 (The_Bakken, 2007).

The pulse generator delivered impulses at an amplitude of 2 V and a pulse width of 1.5 ms. The pulse rate was fixed at a constant rate of 70–80 bpm. The energy utilized was low since Elmqvist managed to obtain a few of the first silicon transistors imported to Sweden. These had a higher efficiency than the older germanium transistors, and with them Elmqvist was able to design a stable and efficient blocking oscillator. The circuit of this device and the first prototype are shown in Fig. 4 (Elmqvist et al., 1963).

Several types of primary battery cells could have been used. Ruben-Mallory cells with zinc as an anode and mercuric-oxide as depolarizer were a possible choice. These had been invented during World War II for army field telephones, and although the cell potential remained constant they had a short lifetime and released hydrogen during operation. The effect of this gas in a cell encapsulated in plastic was not known, so nickel-cadmium rechargeable cells were chosen instead. Two cells of 60 mAh each were sealed, encapsulated, and connected in series as shown in the figure.

Recharging was accomplished inductively via a coil antenna with a diameter of about 50 mm that was connected to the cells through a silicon diode. This was inductively coupled transcutaneously to an external flexible

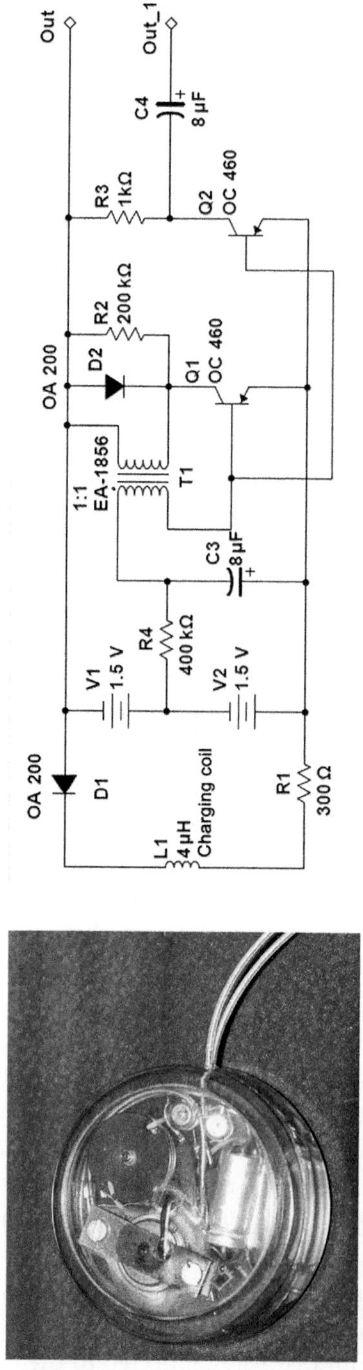

Fig. 4 Photograph of a reproduction of Elmqvist's pacemaker and the circuit diagram.

coil about 250 mm in diameter taped to the patient's abdomen. Recharging was accomplished weekly for 12 h using a 150-kHz radio frequency vacuum tube oscillator that fed the external coil.

The unit was entirely handmade and consisted of the batteries, the electronic circuit, and the charging coil encapsulated in a new epoxy resin (Araldite) produced by Ciba-Geigy, which had been found to have excellent biocompatibility properties. The approximate diameter and thickness, 55 and 16 mm, respectively, were determined by the dimensions of a Kiwi shoe polish tin in which the device was cast. A photograph of a reproduction of the device and the circuit diagram are shown in Fig. 4.

The patient suffered from Stokes-Adams attacks that required resuscitation many times daily and his situation was considered hopeless. The implantation was a desperate measure and very risky. The first prototype failed within 8 h and the second started to fail within a week, but everyone persevered and in the end the patient outlived them all, having survived 25 pacemaker changes.

In the United States, Doctors William Chardack and Andrew Gage were collaborating with electrical engineer Wilson Greatbatch to design an implantable pulse generator. His serendipitous installation of the incorrect resistor in a circuit that he was building caused it to start oscillating regularly and he realized that he had constructed the perfect pulser for a pacemaker. The circuit diagram for this device is shown in Fig. 5. After 2 years of

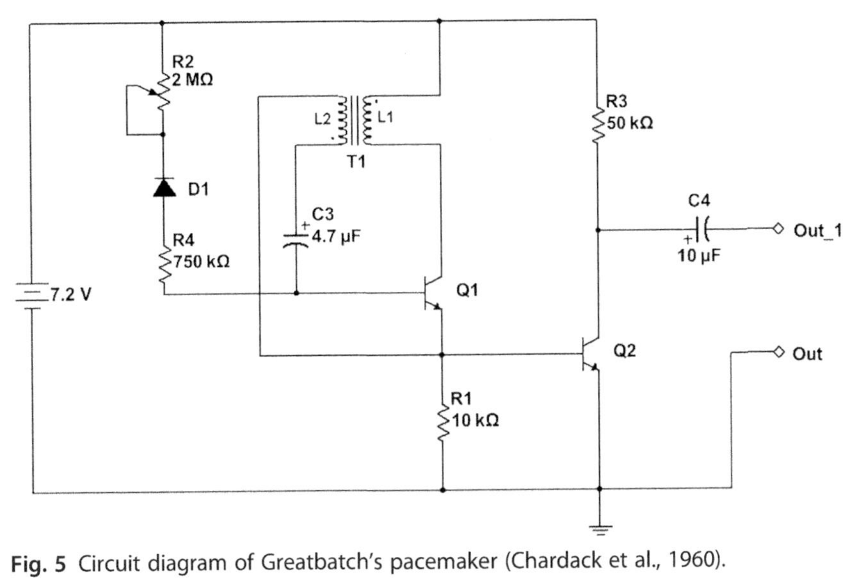

Fig. 5 Circuit diagram of Greatbatch's pacemaker (Chardack et al., 1960).

refinements and myriad animal trials, it was successfully implanted in a human patient in June 1960. Medtronic then met with the developers and obtained exclusive rights to produce and market it (Chardack et al., 1960).

Examination of Elmqvist and Greatbatch's circuits show that their pulsing operation is generated in the same way. They are similar to a two-transistor astable multivibrator in which one of the transistors is replaced by the pulse transformer which provides positive feedback. At turn on, R1 and R2 bias Q1 into conduction allowing the current to flow through winding L1 of transformer T1. This induces a voltage across the secondary winding, L2, which charges C1 to bias Q1 even harder until it saturates. The rate of current increase after that is determined by the time constant of L1 and R3. After time t_1 (the pulse time) the current through L1 saturates and the induced voltage in L2 reduces to zero allowing the capacitor C1 to discharge through R1 and switch Q1 off after time t_2 (the relaxation time). Q2 is switched by the voltage across Q1 and in turn generates a pulse to the heart.

At about this time, Medtronic engineer Norman Roth in conjunction with Dr. Samuel Hunter from St. Joseph's hospital in St. Paul devised an electrode that could be positioned strategically and sutured securely within the heart's chambers. This innovation resulted in the pacemaker requiring far less current for stimulation. Medtronic combined this electrode design with their newly acquired implantable pulse generator and began production immediately to meet the large demand for implantable pacemakers.

Because of the poor component reliability and hence numerous failures of the electronics of the early pacemakers, some researchers followed a different approach. This involved implanting the electrodes into the epicardium and connecting these to a subcutaneous induction coil. The less reliable pulsing electronics remained outside and drove an external coil taped to the patient's skin which induced the required voltage into the internal coil (Aquilina, 2006).

A major problem that continued to plague embedded pacemakers for years was electrode-lead fracture. These leads flex about 30×10^6 times per year and hence often failed from fatigue. In 1959 a new cable was developed by Elema Schonander while working at Ericsson. It consisted of four thin bands of stainless steel wound around a core of polyester braid and insulated with soft polyethylene as shown in Fig. 6. It was designed to survive over 184 million flex cycles to last at least 6 years.

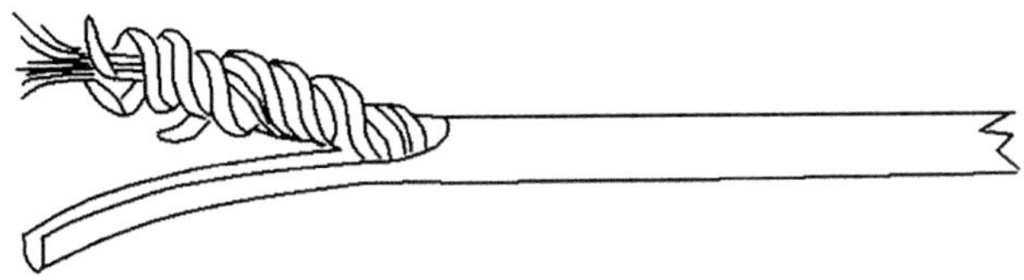

Fig. 6 Construction of the braided stainless steel Elema lead.

4 PACEMAKER EVOLUTION

In 1959, Dr. Seymour Furman a cardiac surgeon first inserted a pacemaker lead through the basilic vein and secured the electrode to the endocardium, and by the early 1960s, this procedure was being applied to internal pacemakers. Combining the advancement of implantable pacemakers with the transvenous access technique not only reduced skin burn and infections, but also avoided major surgery. This technique remains the standard for the majority of patients who require pacemakers, though some pediatric patients with complex congenital cardiac abnormalities still require thoracotomies.

4.1 Demand Pacing

Early pacemakers delivered a fixed pulse at a preset frequency to the ventricle irrespective of any spontaneous cardiac activity. These asynchronous (fixed rate) pacemakers can compete with the natural beat frequency of the heart and can induce arrhythmias or even ventricular fibrillation. The addition of a sensing amplifier ensures that the pacemaker only operates when there is no natural heartbeat and in doing so avoids the problems associated with fixed rate pacing, while also reducing battery drain. Berkovitz introduced the demand concept in 1964 which is the basis of all modern pacemakers. This is illustrated schematically in Fig. 7, though typically, the same set of electrodes are used to monitor cardiac activity and to stimulate it.

The sensing amplifier consists of the preamp, a filter to remove 50/60 Hz mains noise, and a comparator to normalize the signal. This rate discriminator generates unipolar pulses with a fixed amplitude and duration while discriminating against signals that are faster than the expected maximum heart rate. In modern devices, a microcontroller examines these synchronization pulses and either passes them through to the pacing pulse generator or in their absence, generates pacing pulses itself.

By the early 1970s implanted pacemakers were becoming more reliable and were implanted with both rechargeable and high-energy density

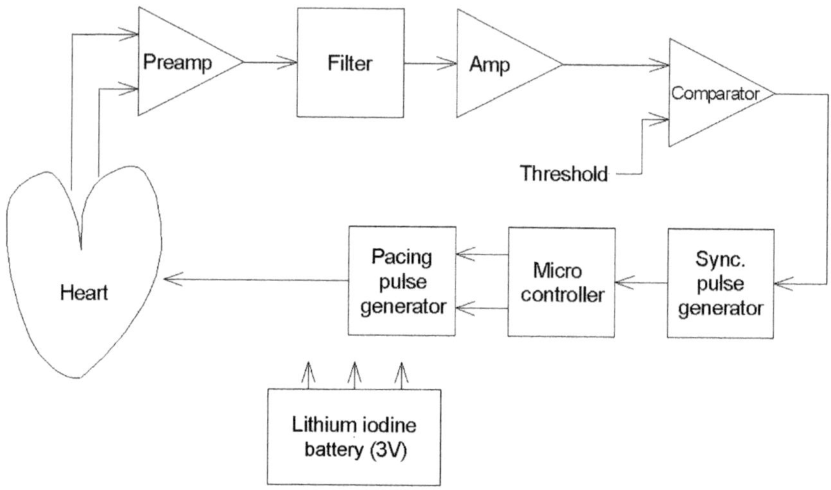

Fig. 7 Schematic diagram showing the principle components required for demand pacing.

primary cells as well as nuclear power sources. The nuclear option was developed to allow young people to receive a single implant that would last a lifetime (some are still operating after nearly 30 years). However, the advent of good lithium batteries, with lifetimes in excess of a decade and the fact that as technology progresses, it is often expedient to upgrade, resulted in the demise of the nuclear option. A range of pacemaker types is shown in Fig. 8.

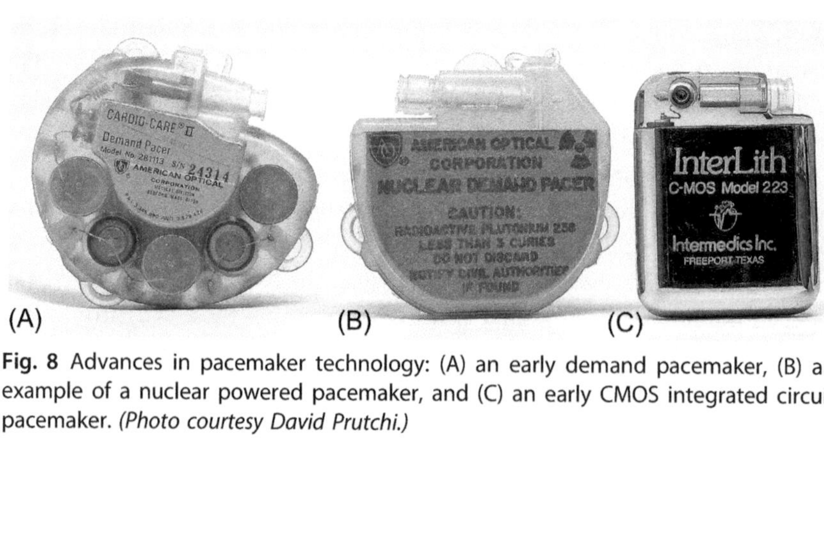

Fig. 8 Advances in pacemaker technology: (A) an early demand pacemaker, (B) an example of a nuclear powered pacemaker, and (C) an early CMOS integrated circuit pacemaker. *(Photo courtesy David Prutchi.)*

4.2 Housings and Size

Progress continued with titanium casings replacing epoxy and silicone, and by the end of the 1970s pacemakers had become programmable through radiofrequency (RF) links. Rate response pacing was introduced using piezo transducers or accelerometers to detect motion and breathing. This presupposes that physical activity or increased breathing rate relate to increasing oxygen requirements and a commensurate increase in the required heart rate, as seen in normal people. Multiple electrodes were introduced to improve synchronization of the various portions of the heart, with the result that contraction efficiency improved. As the sophistication of devices improved exponentially their size and mass continued to decrease due improved battery technology and smaller electronics. Indicative of this is the evolution of Medtronic devices shown in Fig. 9. The latest devices are only about 30 mm in diameter and 5 mm thick and include a data logging capability, the accumulated results of which can be downloaded to a cardiologist by phone or over the web.

With single-component leadless systems, the entire pacemaker (battery, electronics, stimulating electrodes, and sensors) is compressed into a tiny capsule that is delivered into the heart. Advantages of eliminating the leads include greater energy efficiency, system simplicity, and ease of implantation. However, their primary limitations include short battery life, issues with retrieval once the battery has run down, and an uncertain thrombus and infection risk (Madhavan et al., 2017).

5 MODERN PACEMAKER TECHNOLOGY

A modern pacemaker consists of a number of distinct modules as shown in Fig. 10. These include the sensing amps for demand pacing, motion sensing for rate control of the primary, and backup pulse generators as well the usual batteries and electrodes. A wireless telemetry interface allows the pacemakers parameters to be adjusted in situ.

Modern pacemakers use complementary metal oxide semiconductor (CMOS) technology comprising a few Kbytes of read-only memory (ROM) containing the program and some random-access memory (RAM) to store diagnostics and history. All of the components are housed within a hermetically sealed titanium case with a connector block to attach the leads. The outer casing is generally laser etched with manufacturing and

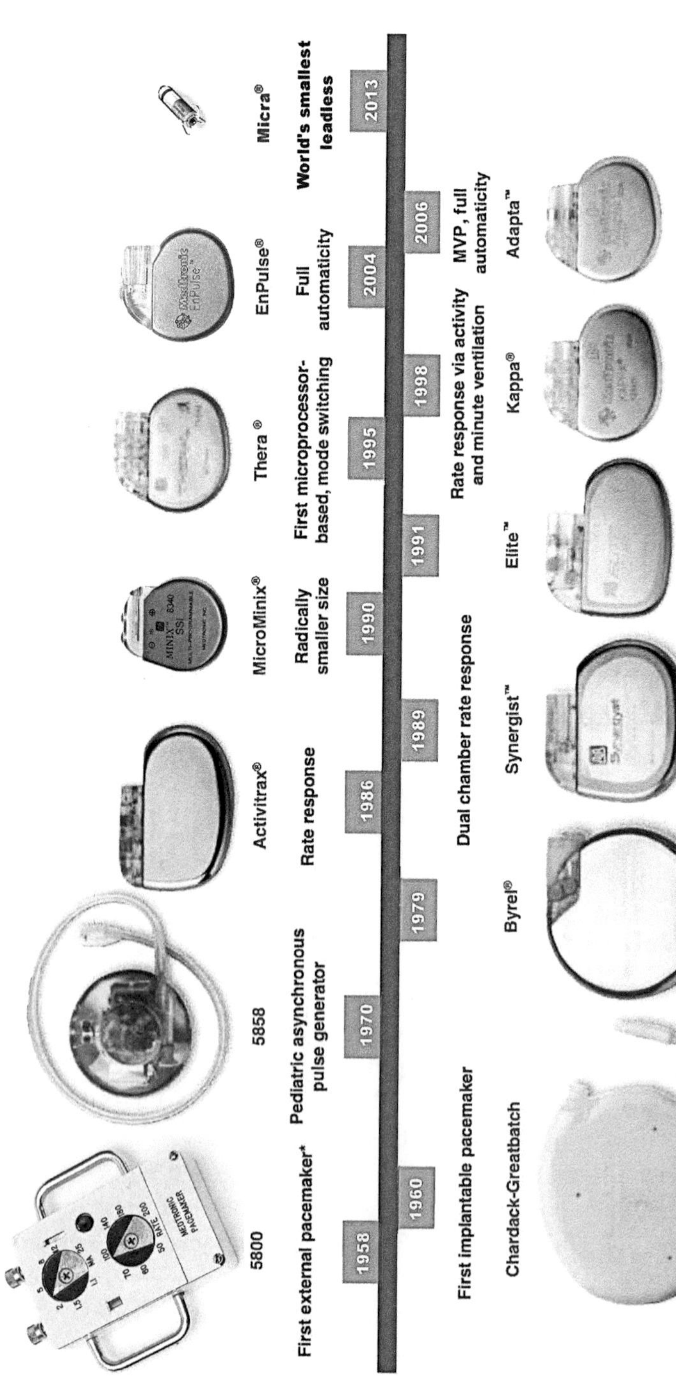

Fig. 9 Evolution of Medtronic pacemakers (Medtronic, 2008). *(Courtesy Medtronic with permission.)*

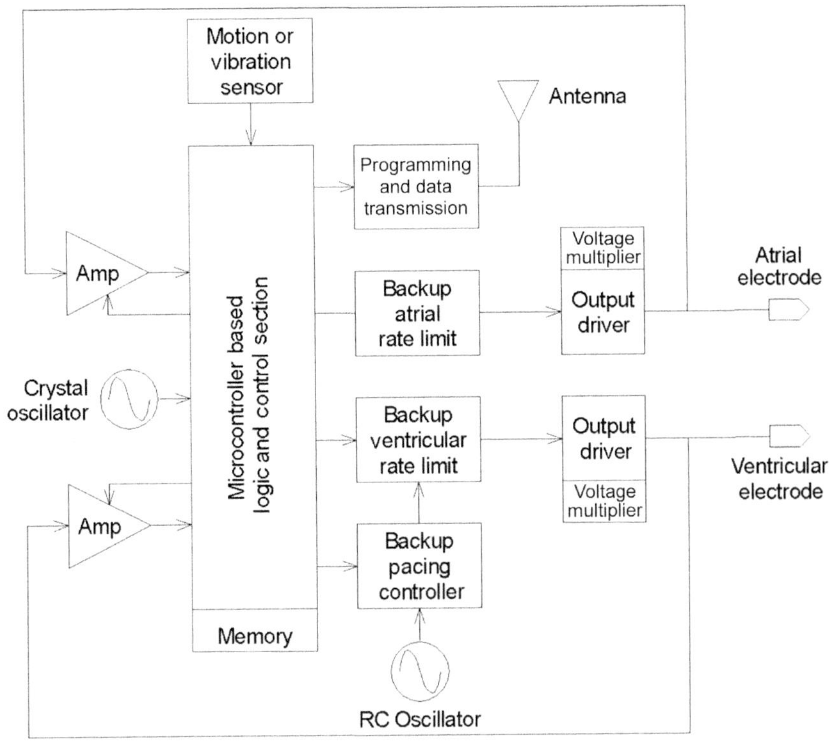

Fig. 10 Block diagram of a modern pacemaker.

model details. Some models also include radio opaque symbols of a characteristic shape so that the device can easily be identified from an X-ray image.

5.1 Pulse Generators

The pulse generator contains a power source, output electronics, and sensing and timing electronics, with rate-responsive pulse generators including movement or breathing sensors and the relevant processing electronics.

5.2 Sensing Electronics

Sensing refers to the identification of heart-generated signals. The specific signal of interest is the intrinsic cardiac depolarization from the chamber, or chambers, in which the leads are placed (the R- and P-waves). It is critical that the electronics can discriminate between these events and those caused by electrical interference such as diastolic potentials, skeletal muscle contraction, and, of course, pacing stimuli.

Sensing and pacing can be accomplished using one of two different configurations: bipolar and unipolar. In bipolar, the anode and the cathode are close together, with the anode at the tip of the lead and the cathode as a ring about 2 cm back from the tip. In unipolar, the anode and the cathode can be 5–10 cm apart with the anode being the lead tip, and the cathode, the housing of pulse generator itself (usually in the pectoral region). In terms of pacing performance, they produce similar results, but in terms of sensing, the unipolar configuration is more likely to sense noncardiac signals, particularly myoelectric signals. Most of the new-generation pacemakers can be configured to sense and pace in either configuration.

The electrogram measured by the electrodes is passed through a band-pass filter, with characteristics shown schematically in Fig. 11, as spectral power is concentrated between 10 and 30 Hz. The main frequency component within the R-wave is about 25 Hz and so the sensing filter is centered at this frequency. T-waves have slower rise times and contain frequency components of <10 Hz. Far-field signals are generally low frequency, while the signals from skeletal muscles fall in the range between 10 and 200 Hz (Haddad et al., 2006).

The sensing electronics measure the amplitude of the R-wave, and the P-wave in the case of dual-chamber pacing. R-wave amplitudes are typically 3–20 mV and P-wave amplitudes slightly smaller at 0.5–7 mV. These signals are amplified, and then compared to an adaptive threshold that provides a trigger to the pacer electronics. Input protection in the form of a Zener diode-based clamp and an EMI filter ensures that high-voltage signals or

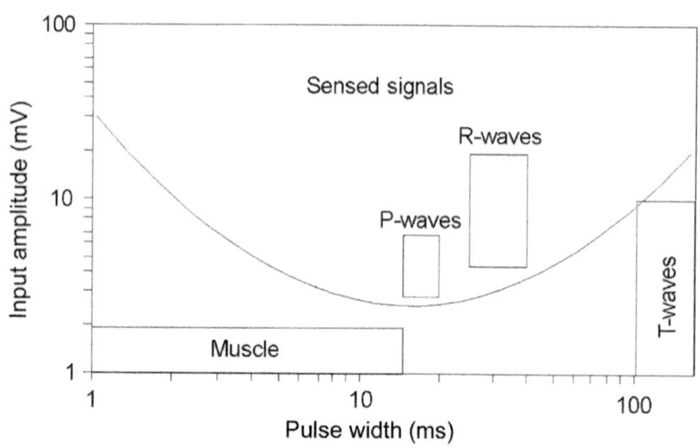

Fig. 11 Band-pass filter to discriminate between wanted and unwanted signals. *(Based on Bronzino, J. (Ed.), 2006. Medical Devices and Systems, CRC Press.)*

rapid voltage transients do not damage the sensitive electronics (Haddad et al., 2006).

As an additional precaution the slew rates of the input signals are also measured. These should be between 0.1 and 4 V/s.

The last consideration is a blanking period. This includes a true blanking period where the sensing circuitry is turned off, and a refractory period, when it can observe the inputs, but does not react. These times are programmable, and vary between 10 and 400 ms, and are particularly helpful in dual-chamber pacemakers where it is possible that the pacing output on the atrial side could interfere with the ventricular pacing signal.

With regard to external sources of electromagnetic interference (EMI), users of pacemakers should avoid strong sources of electromagnetic radiation like arc welders and radar antennas. In general, signal levels in public places, like anti-theft devices, airport security, microwave ovens, and ham-radio equipment, are too low to penetrate the pacemaker's metal case. However, there are a number of medical appliances that can cause interference. These include defibrillators, electrocautery devices, magnetic resonance imaging (MRI), transcutaneous electrical nerve stimulation (TENS) units, and radiation therapy devices.

5.3 Output Electronics

The pacing process uses a lot of power and therefore the electronics must be carefully designed to provide an appropriate signal level larger than the stimulation threshold, while still minimizing current consumption. Modern pacemakers apply a constant voltage pulse, and the current varies in relation to the terminal impedance. Output energy is controlled by two programmable parameters: pulse amplitude and pulse duration. Typically the pulse amplitude can vary between 0.8 and 5 V and the duration between 0.05 and 1.5 ms. The output pulse is generated by the discharge of a capacitor that has been charged from the 2.8-V battery via a voltage multiplier circuit, as shown in Fig. 12.

Voltage multipliers are used as they are much more efficient than DC/DC converters, and result in almost all of the battery energy being provided to the output pulse.

5.4 Timing Circuitry

In modern pacemakers, a crystal oscillator of the kind found in electronic watches drives CMOS divider circuits to produce the specific timing signals

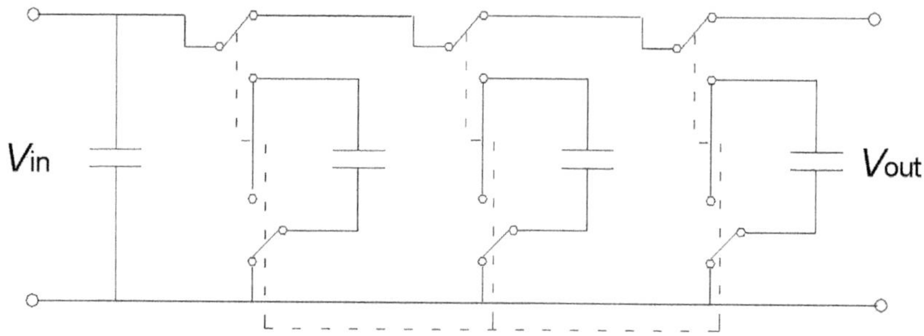

Fig. 12 Voltage multiplier circuit used in a pacemaker.

for refractory and blanking periods; the pulse duration and the interval between atrial and ventricular events. Runaway protection circuitry is included in the case of component failure. This limits the maximum pulse rate to between 180 and 200 ppm.

5.5 Telemetry

Pacemakers are capable of both transmitting and receiving information via a wireless telemetry link. This two-way communication has a baud rate of about 300 bps. The pacemaker can provide real-time information on pulse amplitude and duration, as well as lead impedance and battery condition, lead current, charge, and energy. The programmer provides encoded instructions to alter any of the programmable parameters and to retrieve diagnostic data. These signals are encoded to minimize the likelihood of extraneous interference signals reprogramming the device. There are concerns that inadequate password protection and encryption are incorporated in the link which is allowing pacemakers to be "hacked."

5.6 Power Source

Over the years, a range of different battery technologies has been tried in pacemakers, these include mercury-zinc, rechargeable silver-modified mercuric oxide-zinc, nickel-cadmium, and even radioactive plutonium batteries. Some were associated with corrosion and early failure while others produced hydrogen and required venting. Early batteries lasted weeks or months, which is insignificant compared to modern lithium-iodide devices with reported lives of up to 15 years.

Battery longevity is, of course, determined to a large extent by the activity expected of the pacemaker electronics; this includes pulse amplitude and duration, single or dual chamber pacing, and power use by the sensing circuitry, among others. Improvements in lead design have also had a major effect on battery life. In the 1960s effective stimulation required a pulse generator output of 675 μJ, while modern leads only require only 3–6 μJ, a reduction of more than 100 times (Bronzino, 2006).

In modern batteries, the anode and the cathode are separated by a layer of semisolid lithium iodide. This gradually thickens as power is consumed, increasing the internal resistance of the cell. The output voltage drops in a linear fashion from 2.8 V to about 2.4 V as the cell capacity decreases from 100% to 10%. It then drops exponentially as the cathode is consumed. This is shown in Fig. 13.

Pacemakers that do not include telemetry are designed to change some specified characteristic to indicate that the battery voltage is low (typically between 2 and 2.4 V depending on the manufacturer). This is called an elective replacement indicator (ERI) and it could include changes to the pacing mode or pulse duration. If the battery voltage is allowed to drop as low as 1.8 V, the behavior of the pacemaker electronics becomes erratic, and the batteries or the pacemaker should be changed immediately.

5.7 Leads

Each lead comprises five main components as shown in Fig. 14. In the monopolar configuration they include a single electrode, stabilization tines, the conductor, insulator, and terminal pin. The uninsulated electrode at the tip is in direct contact with the myocardium. In the bipolar case, they contain an uninsulated tip and a ring about 20 mm back from the tip. The smaller

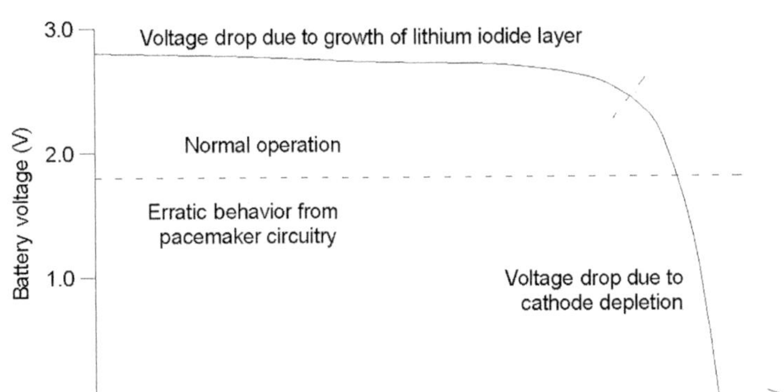

Fig. 13 Battery output voltage as a percentage of charge remaining. *(Based on Bronzino, J. (Ed.), 2006. Medical Devices and Systems, CRC Press.)*

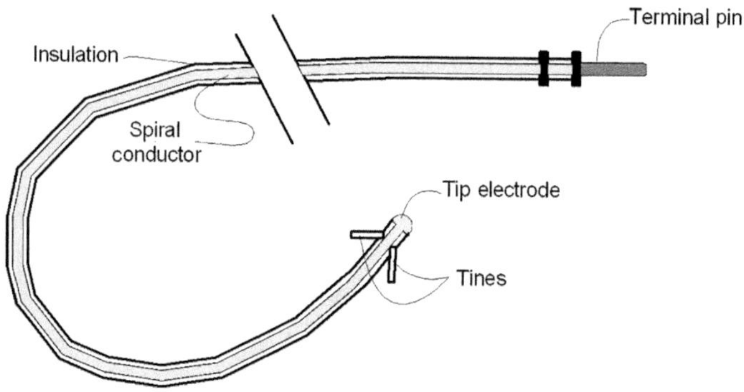

Fig. 14 Schematic diagram of a pacemaker lead.

the tip radius, the larger is the current density and the lower is the stimulation threshold, so in most modern pacemakers a tip area of only 6–8 mm^2 is used.

Small electrodes result in inferior sensing capability, so modern designs include porous-tips containing hundreds of pores between 20 and 100 μm in diameter to allow the ingrowth of tissue with the result that the sensing area increases while the pacing area remains small. Commonly used electrode materials include platinum-iridium, platinum coated with platinized titanium, or vitreous carbon coating a titanium or graphite core.

Many modern electrodes include about 1 mg of corticosteroid contained within a silicone core. This leaks out into myocardium over several years and reduces inflammation as well as retard the growth of a fibrous sac that forms around the electrode and reduces its effectiveness (Bronzino, 2006).

Fixing the lead in place in the myocardium includes active corkscrew mechanisms, barbs, or hooks. Passive fixation using tines that get tangled in the net-like lining of the heart are also used, but these electrodes can be difficult to remove.

Modern cables contain spiral multifilar nickel alloy conductors wound onto a fiber core and then sheathed with a biocompatible insulating plastic such as silicone or polyurethane.

5.8 Programmers

To be effective, a pacemaker must be able to adapt to the user's needs during the many years that it is implanted. In the 1960s, programming rates were adjusted using a Keith needle manipulated through the skin into a knob on the device. Later, magnetic adjustment became possible by embedding reed switches into the pacemaker. By the late 1970s wireless telemetry became available. This allowed bidirectional communication and because of encryption, reduced the possibility of unintended programming.

5.9 New Features in Modern Pacemakers

In pacemakers one of the main challenges is to reduce unnecessary therapies delivered to the patient when the heartrate dynamics look similar to those which could be lethal. For example, a typical pacemaker cannot discriminate between supraventricular tachycardia (SVT) that may occur as a result of atrial fibrillation and normal VT that does not, if it only looks at heart rate. Research is underway to examine the R-wave morphology in more detail to help with this discrimination. An additional problem is the one related to memory use. Pacemakers have small amounts of RAM to log the underlying waveform information during triggered episodes and it is important that false positives caused by electromyological (EMG) signals do not waste this. Haddad et al. (2006) suggests that morphological aspects of the electrogram identified using a wavelet transform (WT) and stored could contribute significantly to effective postevent analysis.

The WT facilitates the analysis of an electrogram over multiple scales and has been able to discriminate between the features related to different electrophysiological phenomena. It also provides a better method of noise removal (denoising) than a conventional smoothing filter. Offline ECG analysis like Holter analysis already uses a multirate filter bank digital implementation of the discrete wavelet transform (DWT). However, in its current form this is far too energy expensive for application within a pacemaker. To overcome this, very low-power analog circuitry has been developed to implement the WT in real time. To achieve this, an analog domain signal-processing circuit which approximates the output of the DWT for use in low-power wearable sensors has been developed. Analog filters which generate DWT-like information to emulate Butterworth and Daubechies maximally flat mother wavelet responses have been developed (Casson, 2015).

Alternative approaches using machine learning algorithms to identify ventricular arrhythmias have been successfully tested, but such power hungry digital algorithms are still not amenable to inclusion in implanted pacemakers.

6 FUTURE DIRECTIONS

Future directions of cardiac pacing can be divided into three separate technologies (Madhavan et al., 2017):

- Improving cardiac resynchronization and cardiac efficiency.
- Reducing hardware to eliminate lead failure, injury to heart valves, and infection.
- Elimination of batteries and other hardware.

Cardiac resynchronization therapy using multisite pacing allows the simultaneous pacing of multiple sites at the same time. By improving the electrical (and hence mechanical) coordination, pump efficiency is improved with a consequent improvement in stroke volume and ejection fraction. An adjunct to this process is to recruit the His-Purkinje system to mimic normal cardiac activation and the restoration of a narrow QRS complex.

For the last half century the basic construction of pacemakers has remained unchanged, with an implanted extravascular pulse generator connected to a lead that makes contact with the endocardium. Though effective, complications are generally caused by the lead which is prone to fracture. Additionally, because pulse generator pockets are extravascular, they are prone to infection, with the lead then serving as a conduit for infection to the vascular system. Finally, because the lead must cross the tricuspid valve, that can result in some backflow and mechanical damage in the longer term. The obvious solution is a single component leadless pacemaker, and a number of these are now entering the market, as discussed earlier in this chapter. However, they suffer from a retrieval problem once the battery has run down. An alternative, under development is a multicomponent leadless system where a tiny "seed" is placed within the left ventricle to act as an energy transducer. A second, more accessible and larger component nearby beams ultrasonic or electromagnetic energy to the seed which is then converted to a pacing pulse.

Cardiac and pulmonary motion provide a reliable source of energy for scavenging during the life of a patient. For example, piezoelectric nanowires attached to a flexible substrate are able to generate voltages as large as 1–2 V and currents up to 100 nA, sufficient to power the microelectronics of a pacemaker. Such motion harvesting pacemakers have already been built and tested in animal models and should become available for human trials in the near future.

REFERENCES

Aquilina, O., 2006. A brief history of cardiac pacing. Images Paediatr. Cardiol. 27, 17–81.

Bronzino, J. (Ed.), 2006. Medical Devices and Systems. CRC Press, Boca Raton, London, New York.

Casson, A., 2015. An analog circuit approximation of the discrete wavelet transform for ultra low power signal processing in wearable sensor nodes. Sensors (Basel) 31914–31929.

Chardack, W., Gage, A., Greatbach, W., 1960. A transistorised, self-contained, implantable pacemaker for the long-term correction of complete heart block. Surgery 1960 (48), 643–654.

Elmqvist, R., Landegren, J., Pettersson, S., Senning, A., William-Olsson, G., 1963. Artificial pacemaker for treatment of adams-stokes syndrome and slow heart rate. Am. Heart J. 1963 (65), 731–748.

Galvani, L., 1791. De Viribus Electricitatis in Motu Musculari. Bononiae.

Guyton, A., Hall, J., 1996. Textbook of Medical Physiology, ninth ed. W.B. Saunders Company, Philadelphia, PA.

Haddad, S., Houben, R., Serdijn, W., 2006. The evolution of pacemakers—an electronics perspective. IEEE Eng. Med. Biol. Mag. 38–48 May/June 2006.

Lilderitz, B. (Ed.), 1998. History of Disorders of Cardiac Rhythm. Futura Publishing Company, Armonk, NY.

Madhavan, M., Mulpuru, S., McLeod, C., Cha, Y., Friedman, P., 2017. Advances and future directions in cardiac pacemakers—Part 2. J. Am. Coll. Cardiol. 69 (2), 211–235.

Medtronic, 2008. Pacemakers. http://www.medtronic.com/our-therapies/pacemakers/index.htm. Accessed September 2008.

Nelson, R., 1993. A brief history of cardiac pacing. Tex. Heart Inst. J. 20 (1), 12–18.

The_Bakken, 2007. The Bakken Library and Museum—First Implanted Pacemaker. http://www.thebakken.org/artifacts/Elmqvist.htm. (Accessed September 2008).

Westaby, S., Bosher, C., 1997. Landmarks in Cardiac Surgery. Informa Health Care, London.

Wilkinson, C., 1804. Elements of Galvanism, in Theory and Practise. John Murray, Oxford.

Zoll, P., Linenthal, A., Gibson, P., 1956. Termination of ventricular fibrillation in man by externally applied countershock. NEJM 254, 727–729.

Index

Note: Page numbers followed by *f* indicate figures, and *t* indicate tables.

C

D

G

H

I

M

N

O

P

Q

R

T

U

V

W

9780128125397